Y0-DLU-703

WITHDRAWN
Stafford Library
Columbia College
1001 Rogers Street
Columbia, MO 65216

PRINCIPLES OF ENERGY

PRINCIPLES OF ENERGY

Editor
Richard M. Renneboog, MsC

SALEM PRESS
A Division of EBSCO Information Services, Inc.
Ipswich, Massachusetts

GREY HOUSE PUBLISHING

Cover photo: Light bulb surrounded by energy innovation graphics. By Blue Planet Studio. (Via iStock)

Copyright © 2021, by Salem Press, A Division of EBSCO Information Services, Inc., and Grey House Publishing, Inc.

Principles of Energy, published by Grey House Publishing, Inc., Amenia, NY, under exclusive license from EBSCO Information Services, Inc.

All rights reserved. No part of this work may be used or reproduced in any manner whatsoever or transmitted in any form or by any means, electronic or mechanical, including photocopy, recording, or any information storage and retrieval system, without written permission from the copyright owner. For permissions requests, contact proprietarypublishing@ebsco.com.

For information contact Grey House Publishing/Salem Press, 4919 Route 22, PO Box 56, Amenia, NY 12501.

∞ The paper used in these volumes conforms to the American National Standard for Permanence of Paper for Printed Library Materials, Z39.48 1992 (R2009).

Publisher's Cataloging-In-Publication Data
(Prepared by The Donohue Group, Inc.)

Names: Renneboog, Richard, editor.
Title: Principles of energy / editor, Richard M. Renneboog.
Description: Ipswich, Massachusetts : Salem Press, a division of EBSCO Information Services, Inc. ; Amenia, NY : Grey House Publishing, [2021] | Series: Principles of | Includes bibliographical references and index.
Identifiers: ISBN 9781642657647
Subjects: LCSH: Force and energy. | Power (Mechanics) | Power resources. | LCGFT: Reference works.
Classification: LCC QC73 .P75 2021 | DDC 531.6—dc23

FIRST PRINTING
PRINTED IN THE UNITED STATES OF AMERICA

Contents

Publisher's Note vii
Introduction ix
Contributors xv

Aerobic Cellular Respiration 1
Air Conditioning 4
Alternative Energy 7
Ampère, André-Marie 12
Animal Power 14
Atomic Energy Commission 18
Batteries 21
Biodiesel 26
Biomass Energy 29
Boilers 33
Breeder Reactors 37
Building Envelope 38
Capacity (Electricity) 41
Cellulosic Ethanol 43
Chernobyl 45
China Syndrome (Nuclear Meltdown) 49
Clausius, Rudolf 51
Climate and Weather 53
Climate Neutrality 58
Coal and Energy Production 60
Cogeneration and Electricity Generation 69
Cold Fusion 72
Communication to Garner Support for
    Energy Development 74
Computers and Energy Use 79
Corn Ethanol 81
Dams 85
Daylighting 89
Demand-Side Management (DSM) 91
Diesel, Rudolf 93
Einstein, Albert 97
Electric Grids 100
Electric Potential 103
Electricity Energy Transmission, Secondary . . . 106
Embodied energy 108
Energy and Power 111
Energy Conservation 113
Energy Independence and Security
    Act of 2007 116
Energy Intensity 118

Energy Payback 120
Energy Policy 122
Energy Poverty 130
Energy Storage Techniques 132
Enthalpy 135
Entropy 137
Exergy 138
External Combustion Engine 140
Faraday, Michael 145
Flex-Fuel Vehicles (FFV) 147
Flywheels 149
Fossil Fuels 151
Franklin, Benjamin 156
Frequency 159
Fuel Cells and Energy Efficiency 161
Fundamentals of Energy 164
Gas Energy Transmission 171
Gasoline and Other Petroleum Fuels 172
Geothermal and Hydrothermal Energy 175
Geothermal Energy 182
Gibbs, Josiah Willard 185
Green Buildings 187
Green Energy Certification 190
Greenhouse Gases and Human Industry 193
Heat Transfer 201
Helmholtz, Hermann von 204
Hertz, Heinrich 206
High-Intensity Light-Emitting Diodes
    (LEDs) and Energy Conservation 208
Hybrid Vehicles and Energy Security 211
Industrial Revolution and Machine Power 215
Internal Combustion Engine 221
Isotopes, Radioactive 224
Isotopes, Stable 226
Joule, James 229
Kinetic Energy 233
Liquid Fluid Energy Transmission 235
Mechanical Energy Transmission 239
Natural Energy Flows 243
Natural Gas 244
Nuclear Power Plants 250
Ocean Current Energy 255
Ocean Thermal Energy Conversion 256
Ocean Wave Energy 258

Oil and Petroleum	261
Oil Shale and Tar Sands	267
Otto, Nikolaus	269
Photovoltaics (PVs)	273
Potential Energy	277
Propane	280
Rankine, William	283
Renewable Energy	284
Renewable Energy Resources	289
Rotational Power	294
Sadi Carnot	299
Solar Concentrator	300
Solar Energy	303
Solar Thermal Systems	309
Steam and Steam Turbines	312
Steam Engines	313
Stirling, Robert	315
Sugar Beet Ethanol	317
Sugarcane Ethanol	319
Sun Day	322
Tesla, Nikola	323
Thermochemistry	326
Thermodynamics and Energy	328
Thompson, Benjamin (Count Rumford)	330
Thomson, Joseph John	332
Thomson, William (Lord Kelvin)	334
Tidal Power Generation	336
Units of Measurement	341
U.S. Department of Energy (DOE)	344
Volta, Alessandro	349
Waste Heat Recovery	353
Watt, James	355
Watt	357
Wave Properties	359
Wavelength	360
Westinghouse, George	363
Wheat Ethanol	366
Willis Carrier and Air Conditioning	368
Wind Energy	370
Work and Energy	375
Work-Energy Theorem	378
Bibliography	381
Glossary	405
Organizations	429
Subject Index	431

Publisher's Note

Grey House Publishing is pleased to add *Principles of Energy* to its Salem Press collection of Principles of Science titles. Covering more than twenty topics, including *Marine Science, Astronomy, Computer Science, Physical Science, Biology, Botany, Mathematics,* and *Information Technology*, this series introduces students and researchers to the fundamentals of these scientific topics using easy-to-understand language for a solid background and a deeper understanding and appreciation. Major categories in *Principles of Energy* include principles, historical discoveries, alternative energy, sustainable energy, biographies of energy scientists, and the future of energy.

The 122 entries contain the following helpful sections:
- **Fields of Study** to illustrate the connections between the topic and the various branches of science related to the study of energy;
- **Abstract** provides a brief, concrete summary of the topic and how the entry is organized;
- **Key Concepts** introduce the terminology used in the entry;
- **Images and charts** illustrate key concepts such as fuel cells, alternative energy sources, energy intensity, energy grids, historical inventions, conservation energy ratings, and pipelines, plus photos of individuals;
- **Further Reading** lists additional sources related to the study of energy.

This work begins with a comprehensive Editor's Introduction to the topic of energy written by volume editor Richard Renneboog. This discussion starts with the Sun—the biggest source of energy known to man—and goes on to explain the four main types of energy identified throughout the centuries: chemical energy, mechanical energy, nuclear energy, and gravitational energy. Entries, arranged in A to Z order, range from Aerobic Cellular Respiration to Work-Energy Theorum and everything in between.

Following the entries, *Principles of Energy* ends with helpful appendices: Bibliography; Glossary; Organizations; and Subject Index.

Salem Press and Grey House Publishing extend their appreciation to all involved in the development and production of this work. A list of contributors follow the Editor's Introduction.

Principles of Energy, as well as all Salem Press reference books, is available in print and as an e-book. Please visit www.salempress.com for more information.

INTRODUCTION

The story of energy in the modern world is a saga that spans literally billions of years. But the question of what energy is is not easily, or simply, answered. Of all concepts in physics, the concept of energy is perhaps the most elusive and difficult of them all. Everyone has an innate sense of what broadly constitutes energy. A fully charged battery has a lot of energy, a very active child has a lot of energy, an atomic bomb releases a lot of energy, as does a firecracker, water rushing down a dry riverbed in a flash flood, or a big, slow-moving river. The dictionary shows two official definitions: the strength and vitality required for sustained physical or mental activity; and, power derived from the utilization of physical or chemical resources, especially to provide light and heat or to work machines.

To begin to understand what energy is, it is necessary to examine the biggest source of energy known to man—the Sun. That brilliant orb, and the great amount of heat and light it emits, has been the subject of much speculation in past centuries.

In the Ancient Egyptian pantheon, the Sun was worshiped as Ra, the Sun God who gives heat and light to the world. In the Ancient Greek pantheon, the Sun was seen as the chariot of Helios, the Sun God who drove the chariot of the Sun across the sky each day to drive away the darkness of night, and the Roman pantheon believed the same, but with Latin names. The Ancient Greek philosophers, however, posited the more pragmatic view that the Sun itself was something more than a god or a glowing chariot, although they had no real idea of what. Their thoughts focused on the familiar sources of intense heat and light that they knew and understood—wood, oil, and Greek fire, the most intense of the three.

In the Americas, the many indigenous peoples also saw the Sun as the most important entity of worship. Asian civilizations—China, India, Assyria, Persia, and others—held the Sun to be the most essential feature of the physical world and accorded it a corresponding reverence. In the Middle Ages, enlightened scientific approaches to the physical world were ignored or neglected. Educated alchemists, clerics, and natural philosophers of the Middle Ages believed the Sun and the stars were holes in the sky through which light from Heaven found its way to Earth.

Following the European Renaissance, scientific investigation regained its importance and emerged in the mid-eighteenth century as the Industrial Revolution. The most abundant and important source of heat and light in the world at that time was that of a white-hot coal fire, and scientists speculated that the Sun was a huge mass of burning coal in space. Given what was known about coal at the time, calculations were made about the mass of that big pile of burning coal, as well as when it would burn out.

More than a century after the Industrial Revolution, experiments demonstrated the existence of the subatomic particles that make up the atom. Development of atomic theory led to the mathematics of quantum mechanics and quantum electrodynamics by such notables as Niels Bohr, Ernest Rutherford, J.J. Thomson, Albert Einstein, Paul Dirac, and many others. With those two pieces of the puzzle in place, the concepts of nuclear fission and nuclear fusion soon followed, bringing us to our current understanding of how the Sun is the source of energy for Earth. Like all stars, the Sun is a massive collection of simple hydrogen atoms, so massive, in fact, that the high internal temperatures and the magnitude of the gravity of that mass squeezes hydrogen atoms together so powerfully that they undergo nuclear fusion. The fusion process releases a huge amount of energy, primarily as heat and the full spectrum of light, and the cumulative effect of thousands of tons of hydrogen fusing continuously into helium every minute is the heat and light—the energy—that comes from the Sun. It is an awesome, complex phenomenon, without which we could not exist.

The Sun radiates that energy into space in a spherically symmetric and incredibly immense outpouring of electromagnetic waves and subatomic particles. As Earth passes through this constant flood of energy as it orbits the Sun, it intercepts no more than 0.01 percent—just one ten-thousandth—of that energy. But it is enough.

Earth itself has existed for some 4.5 billion years, slowly becoming a hot, planet-size ball of rock and gases composed of the 92 naturally occurring elements that were themselves formed in the hearts of stars that had existed long before. Heat energy radiating into space from the infant Earth eventually allowed it to cool sufficiently, after about a billion years, for an outer crust to consolidate and liquid water to develop. After about another billion years, bacteria and algae became the only life-forms on the planet for another billion years. The algae used sunlight to manufacture the basic foodstuff of the planet, a simple sugar called glucose. With neither predators nor scavengers to recycle the miniscule corpses, the dead organic matter accumulated in immense quantities for millions of years, locking away all its captured solar energy—with very important ramifications billions of years later in the present day.

All manner of plant life eventually covered the available dry land areas, living and dying for hundreds of millions of years before animal life came to inhabit the Earth. Like the bacterial detritus, plant detritus would securely trap the solar energy it had acquired—again, with very important ramifications hundreds of millions of years later in the present day.

All the while, Earth has continued to radiate away heat from its formative years, the cooling crust continually contracting and placing pressure on the interior of the planet, helping to keep its temperature high enough to maintain a dynamic, molten state. More important than this, however, are some of those 92 natural elements noted earlier which, through complex processes, release large amounts of energy. In the atmosphere surrounding the exterior of the planet, certain gases capture infrared—that is, heat—energy and radiate a portion back into the atmosphere. This greenhouse effect serves to raise and maintain the average atmospheric temperature. Also, the energy released in radioactive decay processes in the molten rock of the interior is captured by other atoms serving to maintain the elevated temperature of the molten interior. Again, some of those multibillion-year-old elements have very significant ramifications in the present day.

Based on that 4.5-billion-year history of the planet, the stage is now set for the modern-day saga of energy: massive deposits of crude oil from long-dead bacteria and algae, immense coal formations from ancient swamps and forests, moving water in liquid and gas forms, radioactive elements, wind, natural gas from decaying organic matter, plants, animals, chemical and biochemical processes, and behind them all, the Sun. What is most important about these energy stores, however, is how they are used and controlled.

In order to use energy, it must be released so that it can perform work. For thousands of years, humans collected dry wood and other natural combustibles, burning them in open fires to cook food and to provide heat and light. The creation of ovens and other structures that made good use of the fire's heat by absorbing and holding it in place, allowed baking and other food preparations. Wood that has been burned with insufficient air is converted to charcoal, which burns at a more intense heat to smelt some metals from ores and to shape metal objects. Wood was therefore the primary heat source for centuries.

In some places coal (rocks that burn) was found where a coal seam had been revealed at the surface by erosion. In other places, crude oil and tar seeped to the surface and could be burned for heat and light. The scarcity of those resources, however, reserved them for the wealthiest and most powerful members of society. Among the common folk of a tribe, the dark smoke and smells of these burning products gave them an unholy reputation, and they were regarded with suspicion. There may have been something prophetic in such fears, because at some point those materials became valuable as weapons. In China, finely ground coal and charcoal became the carbon base for the explosive black powder propellant for cannons, various types of bombs and fireworks, and eventually guns. In Greece and other Middle East empires, crude oil became the principal ingredient of Greek fire, a fire weapon, something like modern-day napalm, that could be spewed like a flame-thrower.

With agriculture came olive oil and other plant-based oils, tallow and other animal fats that were used for light and heat as well as for food. This remained the status quo for energy essentially until the mid-1700s and the invention of the steam engine. Almost overnight, heat energy was converted into mechanical work, and the Industrial Revolution was born. With the discovery of thermodynamics, scientists discovered that energy cannot be destroyed, only changed from one form to another.

Widespread deforestation for shipbuilding occurred, and the superior energy qualities of coal assured it as the fuel of choice for steam engines. Metal refineries and transportation and manufacturing facilities all required a constant supply of coal, driving the development of a growing coal mining industry. Machinery with moving parts, of course, requires lubrication, but the animal and vegetable oils that were used satisfactorily for low-speed applications were no match for high-speed operations, quickly becoming thick, gummy residues that fouled up the machinery. Mineral oils refined from coal and petroleum replaced them, and the invention of the internal combustion engine soon made petroleum deposits the black gold of the world. Fuels, lubricants, and a vast assortment of other products became available, driving the development of a burgeoning chemical industry that itself required a great deal of energy. From that also came fertilizers and pesticides that enhanced agricultural output immensely, enabling a positive feedback loop for population growth and the concomitant increase in demand for consumer goods and the consumption of resources. The nations and corporations that were best able to control the energy resources, and reap the greatest monetary gains, worked to secure their hold on existing resources and less powerful nations, as well as develop their own resources.

The development of ways to generate and utilize electricity added an entirely new facet to the saga of energy in the present day. The big question was how to generate enough electricity to meet the demand. Once again, the value of coal and oil as energy sources was in high demand. Steam engines converted the heat energy of those fuels into mechanical work used to drive the generators that converted mechanical work into electrical energy. Still the demand for electricity could not be met in many local regions, and the growing awareness of pollution from burning those fuels drove the need to find other means of turning the generators. The mechanical power of water falling under the force of gravity provided that means, first naturally at Niagara Falls and then artificially at dams constructed for that purpose in many different locations.

In the late-nineteenth and early-twentieth centuries, physicists assembled a working model of atomic structure. This opened the door to understanding the nature of electricity and magnetism, and led to the development of nuclear physics. It is unfortunate that the immense amount of energy available within a small amount of matter through the process of nuclear fission was first realized during WWII. Be that as it may, the age of nuclear energy had arrived. In an atomic bomb, nuclear fission is an uncontrolled process. When that process is moderated, however, the energy released can be captured and controlled as heat energy, which can then be transferred from the moderating fluid to the working fluid used to drive electrical generating systems—the basic operating system of all nuclear power plants. The fuel for those plants is billions-year-old radioactive minerals found in Earth's crust. As with all major energy sources, the nations that control that energy source not only reap the economic benefit, but also the possibility of nuclear weapon proliferation.

The advantage of nuclear power generation, despite its inherent dangers, is that it is not a direct source of atmospheric pollutants such as carbon dioxide. With growing concern over climate change and global warming, other non-polluting means of generating electricity continue to be developed. Wind power is used to turn the generators atop windmills in many lands. In some places, the energy of moving water in river currents, ocean currents, rising and falling tides, and in the action of waves is harnessed. Today, a concerted effort is being made to build the first nuclear power plant utilizing the more powerful nuclear fusion (rather than nuclear fission), the same fusion that produces the power of the Sun. The Sun is also being used to generate electricity directly through the photoelectric arrays. Those factors that have accumulated on and in the planet from billions of years ago in capturing the energy of sunlight are now in play.

There is another factor of the energy saga not yet discussed here—the power of animals. It is obvious that animals and plants get energy from the food they consume, but how exactly does the energy happen? People have been figuring out how living beings get the energy they need for thousands of years, always looking for better foods and food supplies. We understand that if we do not get food and water, we die; if we do not get air, we die; if we sustain serious injury or sickness, we die. But what is it that allows us to live? During the Renaissance Period, serious scientific investigation of animal physiology began, at first to bring realistic detail into works of art. When the nature of atomic structure and the involvement of electrons in chemical and biochemical reactions was understood, so was bioenergy—renewable energy produced by living organisms.

Bioenergy begins with the Sun, and the conversion of sunlight into energy stored in the simple sugar called glucose that is formed during photosynthesis. Glucose is found in many forms, ranging from the simple sweet-tasting molecule itself to highly complex carbohydrates and carbohydrate-based molecules. Within aerobic living systems (air-breathing animals) glucose is absorbed when food is digested and enters into a biochemical process called respiration. In this multistep process, the glucose molecule is disassembled step by step until it is once again just the carbon dioxide and water molecules from which it was made during photosynthesis. Anaerobic living organisms—simple organisms like the bacteria that once occupied Earth and live without oxygen—also utilize a variation of the respiration process. While the steps in the anaerobic process differ from those of the aerobic process, the end result is still the same—the energy of sunlight captured in the formation of glucose is released and used to promote the life of these one-celled anaerobes.

Respiration in aerobic life-forms provides the electrochemical bioenergy that allows nerves to signal and muscles to function. Muscle function allows animals to convert that energy into mechanical work that is applied in human activities. Horses, oxen, mules, burros, elephants and other animals were, and in many parts of the world still are, traction engines to pull plows and power machinery. Dogs were also once used as cart dogs for transportation and as spit dogs to drive the rotation of roasting spits in residential and commercial fireplaces.

The other part of the energy saga is how energy is monitored and used responsibly by societies. This is in a constant state of flux as governments seek to balance the use of resources with the protection of resources. Needs and demands change, availability and appropriateness of resources change, popular and scientific attitudes toward various types of energy and energy policies change. Both national and international policies need to agree in content while maintaining a level of respect for the policy requirements of individual nations. It is a demanding requirement that often takes many years of research and negotiation to achieve.

This volume presents a broad knowledge of the many aspects of energy, including the historical figures who played a significant role in the understanding and development of energy. Historical context is important in revealing the incredibly rapid growth of energy-related technology. In close examination of the history of energy since the Industrial Revolution, it becomes an interesting debate as to whether the development of energy has driven the development of technology, or whether it is technology that has driven the development of energy—a debate that could consume a great deal of energy in its own right.

Physics is another aspect to the concept of energy—the energy an object possesses as a function of its mass and velocity. All physical objects have mass, and because of this they have potential energy, energy the object does not actually have but would if it were to move through space. Potential energy is always relative to some fixed point in a gravitational field. For example, if one were to hold a rock motionless in one hand a couple of feet above the ground, and let it fall, its velocity would change from zero to however fast it was going when it hit the ground. Its potential energy is calculated based on its height above the ground, its mass, and the force of whatever gravitational field it happens to be in. But as soon as the rock is released, it begins to fall. It is in motion, with velocity, and if objects are in motion they are possessed of kinetic energy.

From the four main types of energy: chemical energy (energy stored in the bonds of atoms and molecules); mechanical energy (energy stored in objects by tension); nuclear energy (energy stored in the nucleus of an atom); and gravitational energy (energy stored in an object's height) to other types, like bioenergy, thermal energy, radiant energy, electrical energy, kinetic energy, and many more, it's all part of the energy saga.

—*Richard M. Renneboog, MsC, Editor*

Contributors

Ronju Ahammad
Charles Darwin University, Australia

Emily Alward
College of Southern Nevada

Aileen Anderson
University of California, Irvine

Kaushik Ranjan Bandyopadhyay
Indian Institute of Management

Grace A. Banks
Independent Scholar

John H. Barnhill
Independent Scholar

Charles Barnhart
Western Washington University

Harlan H. Bengtson
Southern Illinois University

Alvin K. Benson
Professor Emeritus, Brigham Young University

Richard D. Besel
Grand Valley State University

Lisa Blagg
Independent Scholar

Sarah Boslaugh
City University of New York

Matthew Branch
Independent Scholar

Richard Brent
Professor Emeritus, Australian National University
University of Newcastle

Casey Lance Brown
P-Rex Think Tank, MIT

Marcella Bush
Independent Scholar

Kiril Caral
Shell Philippines, Exploration

Roger V. Carlson
Jet Propulsion Laboratory

Rochelle Caviness
Independent Scholar

Bernard Cohen
Professor Emeritus, University of Pittsburgh

Danielle J. Daly
Independent Scholar

C.S. Eigenmann
Independent Scholar

Reza Fazeli
ANU College of Engineering & Computer Science, Australia

Rodrigo Marcussi Fiatikoski
Londrina State University, Brazil

Fausto Freire
University of Coimbra, Portugal

Angel G. Fuentes
Chandler-Gilbert Community College

Peter Garforth
Garforth International Energy Management

Sophie Gerber
TRAction Fintech, Australia

Soraya Ghayourmanesh
Queensborough Community College

Curt Gilstrap
University of Southern Indiana

Kyle Gracey
Carnegie Mellon University

Joanta Green
Writer, Renewable Energy Systems in Southeast Asia

Leon Gurevitch
Victoria University of Wellington, Australia

Gavin Harper
University of Birmingham

Cary Hendrickson
The Azek Company

Jordan P. Howell
Rowen University

Andrew Hund
Independent Scholar

Dinseh Kapur
Independent Scholar

Bill Kte'pi
Science Writer, New Hampshire

Alok Kumar
University of Miami Herbert Business School

Ralph L. Langenheim, Jr.
Professor Emeritus, University of Illinois

Kirk S. Lawrence
St. Joseph's College

Manja Leyk
Independent Scholar

Guido Mattei
Independent Consultant on Energy, Environment, and Climate Change

Emily McGlynn
PhD Candidate, Climate, Energy, and Environmental Policy
University of California Davis

Russell McKenna
University of Aberdeen, Scotland

Saurabh H. Mehta
Cornell University

Nicole Menard
Independent Scholar

Francesca Musiani
French National Center for Scientific Research

Alice Myers
Independent Scholar

Hazel Nash
Independent Scholar

Josef Nguyen
University of Texas, Dallas

Emmanuel U. Nzewi
A & M University

Swati Ogale
Science Writer, Green Technology

Sean G. O'Keefe
Syracuse University

Tiago Oliveira
Universidade Nova de Lisboa, Portugal

Judith Otto
Framingham State University

Soji Oyeranmi
Olabisi Onabanjo University, Nigeria

Ana Penha
Researcher, Brazil

Julia C. Pfaff
Wuppertal Institute for Climate, Environment, and Energy, Germany

Elizabeth Rholetter Purdy
Writer and Editor

William O. Rasmussen
University of Arizona

Richard M. Renneboog, MSc
Professor Emeritus, The University of Western Ontario

Jörn Richert
University of St. Gallen, Switzerland

Michael K. Rulison
Oglethorpe University

Harold H. Schobert
Chief Scientist, Schobert International

Richard Sheposh
Writer and Editor, Olyphant, Pennsylvania

Laura J. Stroup
St. Michael's College

Marcella Bush Trevino
Florida SouthWestern State College

Madeline Tyson
Rocky Mountain Institute

Leonardo Fonseca Valadares
Institute of Chemistry, University of Brasilia, Brazil

Gilbert Valentine
La Sierra University

Kenrick Vezina
Science Writer, Boston, Massachusetts

Stacy Vynne
Washington State Department of Ecology

Robert J. Wakulat
Green Energy Lawyer

Andrew J. Waskey
Dalton State College

Thomas W. Weber
Founder and Chief Engineer, Webertech, Inc.

Edwin G. Wiggins
Webb Institute, College of Engineering

Damon Eric Woodson
Inventor, Georgia Power

Qingjiang Yao
Lamar University

Aerobic Cellular Respiration

FIELDS OF STUDY
Biochemistry; Bioenergetics; Thermodynamics

ABSTRACT
Aerobic cellular respiration is one of two processes the cells of living organisms use to break down food and convert it into energy. Aerobic respiration occurs in the presence of oxygen. The four-step process takes the molecular sugars stored in food and transforms them into chemicals that cells use as energy. This energy is what allows cells to perform their basic functions. Due to the presence of oxygen, aerobic cellular respiration produces a larger amount of energy. Anaerobic respiration, another type of respiration, occurs without oxygen. This process also creates energy used to power cells but is less efficient and produces less energy.

KEY CONCEPTS
aerobic: processes that require the presence of oxygen

anaerobic: processes that either do not require or that do not occur in the presence of oxygen

eukaryote: a cell that has a central nucleus structure containing the cell's deoxyribonucleic acid (DNA) and maintains its life through aerobic respiration

mitochondria: organelles within eukaryotic cells, where energy is extracted from the glucose obtained from food

prokaryote: a cell that does not have a central nucleus structure containing the cell's DNA and maintains its life through anaerobic respiration

respiration: the process of extracting energy from glucose by converting it back to the carbon dioxide and water from which it was photosynthesized

WHERE ENERGY COMES FROM
The energy of sunlight is captured by photosynthetic processes that produce one molecule of glucose from six molecules of atmospheric carbon dioxide and six molecules of water taken in from the environment. This process is the foundation of all bioenergy on this planet. Once captured, this energy must be recovered as food by living animal systems in order to obtain the energy that sustains life.

Aerobic cellular respiration is one of two processes the cells of living organisms use to break down food and convert it into energy. Aerobic respiration occurs in the presence of oxygen. The four-step process takes the molecular sugars stored in food and transforms them into chemicals that cells use as energy. This energy is what allows cells to perform their basic functions. Due to the presence of oxygen, aerobic cellular respiration produces a larger amount of energy. Anaerobic respiration, another type of respiration, occurs without oxygen. This process also creates energy used to power cells but is less efficient and produces less energy.

BACKGROUND
Cells are the fundamental building blocks that make up all living things. While hundreds of types of cells serve various functions within an organism, they are generally divided into two basic types. Prokaryotes are typically single-celled organisms that lack a central nucleus. The cell's genetic material, or DNA, floats in a gel-like substance called cytoplasm found within the cell. Bacteria are prime examples of prokaryotic cells.

Eukaryotic cells are found in more complex life forms such as humans, animals, plants, and fungi. These cells have a central nucleus that houses the cell's genetic material. They also have various membrane-enclosed structures within the cytoplasm called organelles. Organelles perform different functions within the cell. For example, ribosomes use the cell's genetic instructions to reproduce proteins. Mitochondria are organelles that act as the cell's power plant, taking the nutrients from food and converting it into energy the cell can use.

Aerobic Cellular Respiration

Aerobic cellular respiration is the primary way that eukaryotic cells produce energy. These cells run on a form of chemical energy provided by a chemical compound known as adenosine triphosphate (ATP). Because ATP molecules are larger, delivering them directly to the body's cells would be a difficult and an inefficient task. The smaller sugar molecule glucose acts as the "fuel tank" that carries the stored energy to the cell. Cellular respiration is the process of converting the glucose into the energy-transferring ATP. In

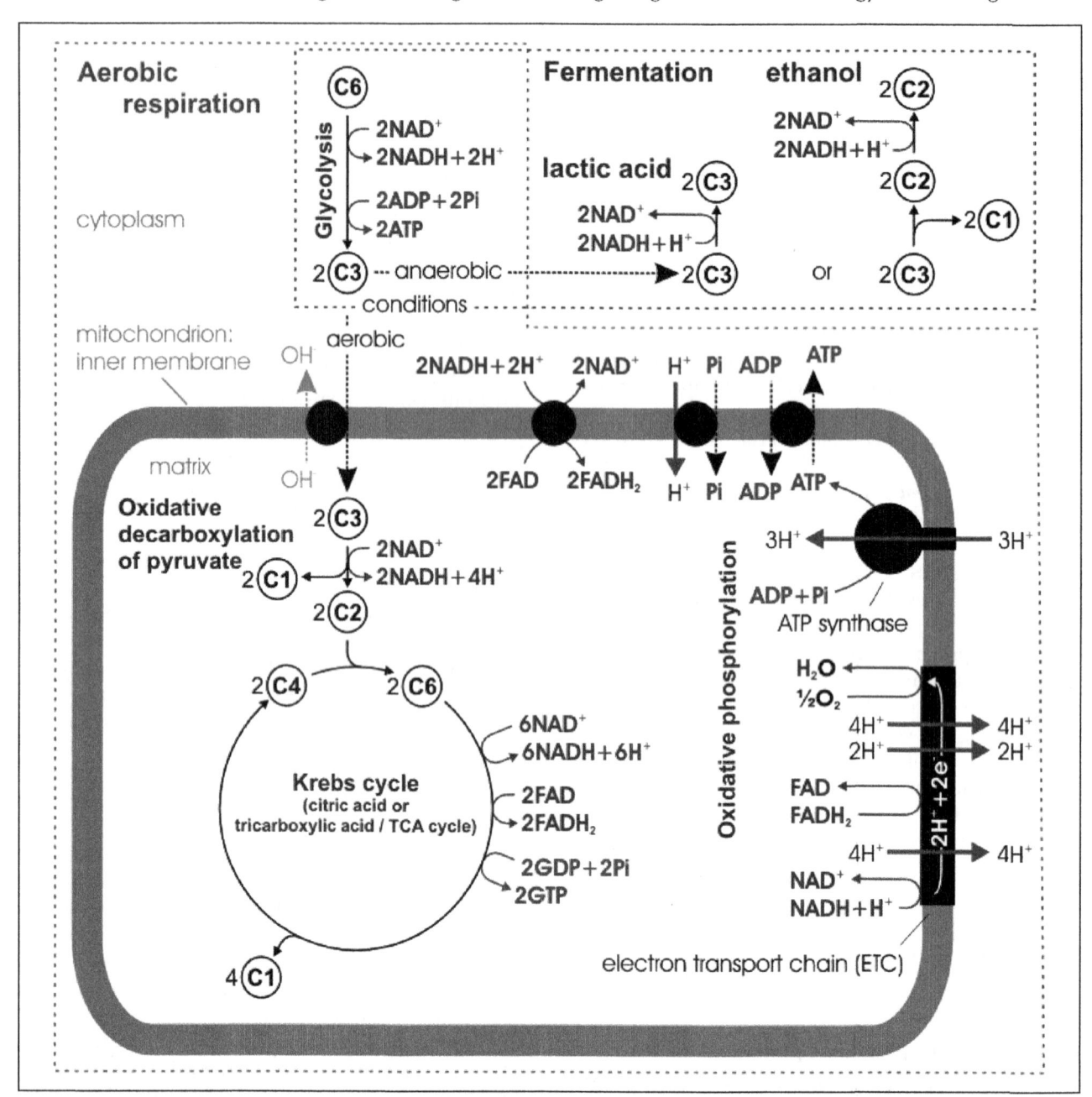

Stoichiometry of aerobic respiration and most known fermentation types in eukaryotic cell. Numbers in circles indicate counts of carbon atoms in molecules, C6 is glucose $C_6H_{12}O_6$, C1 carbon dioxide CO_2. Mitochondrial outer membrane is omitted. (Darekk2 via Wikimedia Commons)

aerobic cellular respiration, oxygen is used as a catalyst to transfer the energy stored in glucose into ATP.

The molecular formula of glucose is written as $C_6H_{12}O_6$, meaning that each molecule is made up of six carbon atoms, twelve hydrogen atoms, and six oxygen atoms. In the first step of cellular respiration, each glucose molecule is broken down into two units of pyruvic acid, a compound with the molecular formula $C_3H_6O_3$, as its pyruvate anion. This process is called glycolysis—a term meaning "sugar splitting"—and takes place in the cell's cytoplasm. To create the energy needed during glycolysis, each glucose molecule also produces two molecules of ATP and two molecules of nicotinamide adenine dinucleotide (NADH). NADH is another energy-transferring molecule and a chemical compound used as a catalyzing agent. The glycolysis process occurs in both aerobic and anaerobic cellular respiration.

For aerobic cellular respiration to proceed, the remaining steps of the process require the presence of oxygen. Through a process called oxidation, oxygen strips away electrons from atoms or molecules by way of a chemical reaction. After glycolysis, the pyruvate is transferred into the cell's mitochondria, specifically, an interior section of the structure known as the mitochondrial matrix. There, the pyruvate reacts with an enzyme called coenzyme A to form a two-carbon molecule called acetyl-CoA. During this part of the process, one carbon atom is stripped away from the pyruvate and combines with oxygen to form carbon dioxide (CO_2), which is released as a waste product to eventually be breathed out. In addition, more NADH molecules are also produced.

The next step of the process is called the citric acid cycle or the Krebs cycle, named after Hans Krebs, the biologist who discovered it in the 1930s. The acetyl-CoA reacts with a molecule called oxaloacetate (OAA), which produces citric acid. The citric acid then goes through a series of reactions to produce energy and carbon dioxide. The end of each cycle produces additional OAA to begin the cycle over again. Because the process to this point has produced two molecules of acetyl-CoA for every glucose molecule, the cycle must repeat itself twice for each glucose molecule. At the end of the cycle, each glucose molecule has been completely broken down. It has produced six carbon dioxide atoms—which are expelled as waste—four ATP molecules, ten NADH, and two molecules of flavin adenine dinucleotide ($FADH_2$), another type of energy-transferring molecule.

The last step of aerobic cellular respiration is called the electron transport chain. This occurs in the christae, a folded area in the inner membrane of the mitochondria. During this step, energy from NADH and $FADH_2$ is transferred into ATP. The electron transport chain is a series of proteins that transfers high-energy electrons along the inner membrane of the mitochondria. This process creates hydrogen ions—hydrogen atoms that have lost their electrons. As the hydrogen ions build up, a protein known as ATP synthase channels them through the membrane, capturing their kinetic energy and transforming it into chemical energy found in ATP. At the end of the aerobic cellular respiration process, each glucose molecule can produce from thirty-six to thirty-eight molecules of ATP. The leftover electrons that have passed through the chain combine with the oxygen to form water (H_2O).

Cells can also produce energy without oxygen, but this process is far less efficient. Anaerobic cellular respiration shares the process of glycolysis with its aerobic counterpart. However, without oxygen, the process splits off into fermentation in which the glucose is broken down by microorganisms such as bacteria or yeast. Fermentation is the same process used to make beer, wine, and cheese. In humans, a lack of cellular oxygen can break down glucose into lactic acid. During strenuous exercise, in which the body uses oxygen very quickly, a buildup of lactic acid can cause a burning feeling in the body's muscles.

—*Richard Sheposh*

Further Reading

Bailey, Regina. "All About Cellular Respiration." *ThoughtCo.*, 6 May 2019, www.thoughtco.com/cellular-respiration-process-373396. Accessed 3 Mar. 2020.

Campbell, A. Malcolm, and Christopher J. Paradise. *Cellular Respiration*. Momentum Press, 2016.

"Cellular Respiration." *IUPUI Department of Biology*, February 2004, www.biology.iupui.edu/biocourses/N100/2k4ch7respirationnotes.html. Accessed 3 Mar. 2020.

Kimball, John W. "Cellular Respiration." *Kimball's Biology Pages*, 1 Aug. 2019, www.biology-pages.info/C/CellularRespiration.html. Accessed 2 Mar. 2020.

Markgraf, Bert. "Cellular Respiration in Humans." *Sciencing*, 23 Apr. 2019, sciencing.com/cellular-respiration-humans-5438875.html. Accessed 3 Mar. 2020.

Nave, R. "Cellular Respiration." *HyperPhysics*, 2016, hyperphysics.phy-astr.gsu.edu/hbase/Biology/celres.html. Accessed 2 Mar. 2020.

"Powering the Cell: Cellular Respiration" *CK-12 Foundation*, www.boyertownasd.org/cms/lib/PA01916192/Centricity/Domain/743/B.%20Chapter%204-Lesson%202-Powering%20the%20Cell-Cellular%20Respiration.pdf. Accessed 3 Mar. 2020.

"What Is a Cell?" *US National Library of Medicine*, 3 Mar. 2020, ghr.nlm.nih.gov/primer/basics/cell. Accessed 3 Mar. 2020.

Air Conditioning

FIELDS OF STUDY
Electromechanical Engineering; HVAC Trades; Thermodynamics

ABSTRACT
Air conditioning (AC) is the process of conditioning several air properties simultaneously, including temperature, humidity, velocity, and quality (amount of fresh air), to ensure that the space matches desired conditions. Modern air conditioning is used to control indoor environments, providing thermal comfort to occupants and suitable environmental conditions for process applications. Air conditioners often consume large amounts of available electrical energy.

KEY CONCEPTS
dehumidification: the artificial removal of water vapor from the air inside of a building in order to maintain human comfort level and prevent damage due to condensation and mold growth

energy efficiency ratio: a rating according to the amount of heat energy removed from the air per watt-hour of electricity consumed by the particular system

humidification: the artificial addition of water vapor to the air inside of a building in order to maintain human comfort levels and avoid physical damage due to dehydration

humidity: the amount of water vapor in ambient air, relative to the maximum amount of water vapor possible in air at specific temperatures and pressures (100 percent humidity); for example, 25 percent humidity means the amount of water vapor present in air is 25 percent of the maximum amount possible

water vapor: water that exists in the air of the atmosphere as a gas

USE OF AIR CONDITIONING
Air conditioning (AC) is the process of treating, or conditioning, air. Through air conditioning, several air properties are controlled simultaneously, including temperature, humidity, velocity, and quality (amount of fresh air), to ensure that the space matches desired conditions. Requirements for air conditioning are usually divided into those needed for comfort applications, or the satisfaction of the human and animal occupants of the environment, and those needed for process applications, or the satisfaction of prescriptive conditions necessary to accommodate the technologies involved with a specific process.

A BRIEF HISTORY OF AIR CONDITIONING
From the very first rudimentary techniques of ancient times to the highly complex systems of today, human beings have always been searching for ways to ensure comfort in built environments. In ancient Rome, for example, water was conveyed from aqueducts through the walls of selected houses to achieve a cooling effect. The same effect was produced in medieval

In ancient times, the Egyptians hung wet reeds in windows and thresholds to chill the incoming breeze. (Wikimedia Commons)

Persia with a different technique, using a combination of water cisterns and wind towers; during hot seasons, evaporating water from the cisterns provided a cooling effect to the air stream blown into the buildings by the wind towers. In several Chinese dynasties, the use of specific apparatuses aimed conditioning air has been documented; these ranged from manually powered rotary fans to water-powered fan wheels and water jets from fountains (between the second and thirteenth centuries CE).

The pioneer of modern air-conditioning was the American engineer and inventor Willis Carrier. In 1902, Carrier designed the first large-scale electrical air-conditioning unit in the world, the "apparatus for treating air," to address a manufacturing problem in a lithography company. With Carrier's machine, the temperature and humidity in the plant could be controlled, improving the manufacturing process. Carrier was granted a patent for his invention in 1906. Installation of air-conditioners started in the following years, mainly targeting industrial applications, due to the size and cost of the first machines.

Household electric "box" fan with a propeller-style blade. (Wikimedia Commons)

Over the years, smaller and safer equipment brought the air-conditioning industry into the residential and transport sectors, focusing on thermal comfort. In the 1920s central air-conditioning systems were running in office and recreational buildings, and in the 1930s the first packaged and room air conditioners were installed. The first automotive air conditioner was introduced by Packard in 1939, although its operation was not very user-friendly. Nowadays, with the advent of electronic miniaturization and complex control algorithms, the satisfaction of people and process requirements can be achieved at a very detailed level and, most important, with increased energy efficiency.

General Requirements for Indoor Air Conditioning	
Temperature (Celsius)	20°–25°
Relative humidity (%)	35%–65%
Air Velocity[a] (ms-1)	0.1–0.25 m.sec^{-1}
Fresh Air Supply Rate[b] (m3h-1 per pax)	25–35

a. In the occupied zone.
b. Ventilation requirements depend on the space to be conditioned and the level of activity, and they are often regulated by national or federal codes.

AIR-CONDITIONING PROCESSES

Because atmospheric air is a mixture of dry air and water vapor, AC systems must be capable of handling heat- and water-transfer processes. (An air-conditioning process is described by changes in the thermodynamic properties of atmospheric air between the initial and final states of the conditioning process. Heat and moisture supply and removal rates are calculated on the basis of mass and energy balances.) Basic processes include sensible heating, wherein heat is added to atmospheric air, with a corresponding temperature increase; sensible cooling, wherein heat is removed from the air (resulting in a drop in ambient temperature); humidification, the transfer of water vapor to the air, usually by steam injection or evaporation from a water spray; and dehumidification, the removal of water vapor from the air using a cooling coil or a desiccant unit. A combination of these processes is often required to achieve the desired temperature and humidity lev-

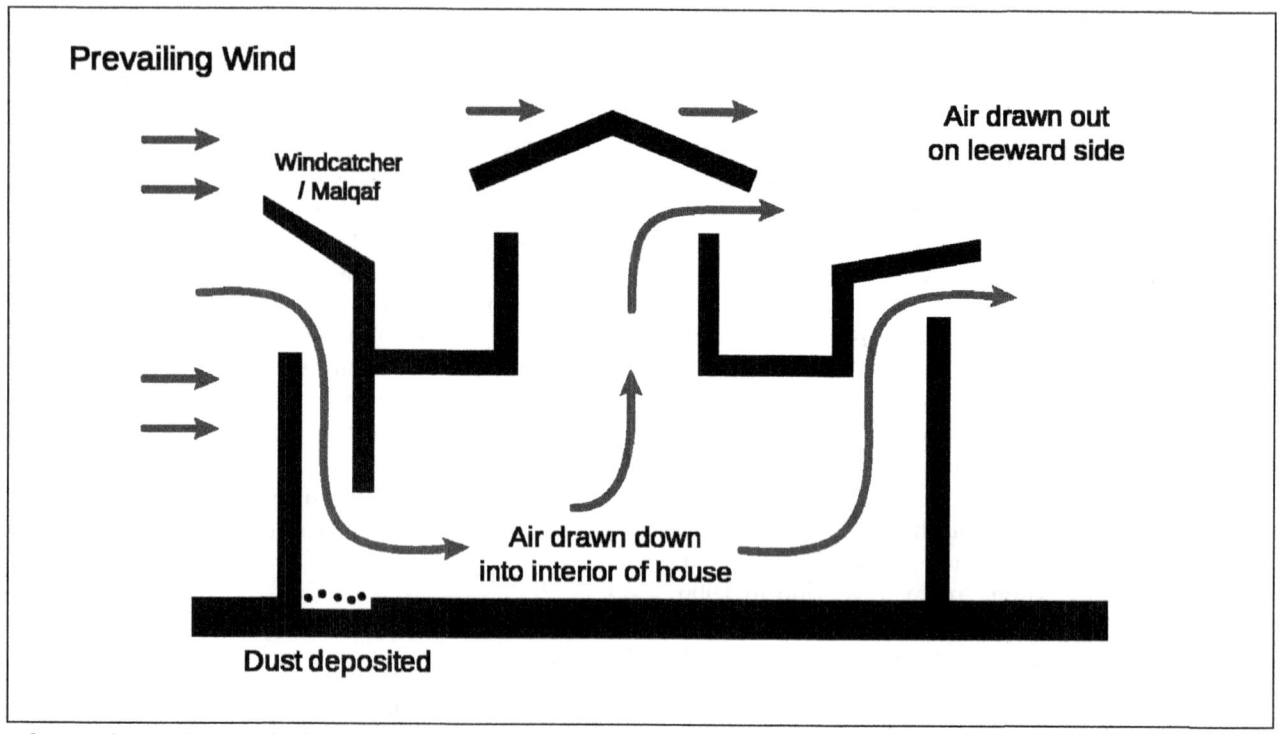

A diagram showing the principle of a malqaf or windcatcher for natural ventilation in traditional Arabic architecture. (Fred the Oyster via Wikimedia Commons)

els—namely, heating and humidification in the winter and cooling and dehumidification in the summer.

Moreover, AC systems must also comply with ventilation requirements (providing minimum fresh air and air change rates), usually achieved by a mix of fresh (outside) and recirculated airstreams, in varying proportions, supplying the conditioned space. (Recirculated air is air returned from a conditioned space and treated by the AC system according to specific set points, before being supplied again to the space.) The removal of contaminants from ambient air, or air filtration, is another important feature of AC units, contributing to a comfortable and healthy environment, or good indoor air quality.

ENERGY PERFORMANCE OF AC SYSTEMS

The energy performance of AC systems is expressed by the ratio between the useful effect (cooling or heating effect) provided by the system and the energy expended to obtain this effect. By convention, it is called the coefficient of performance (COP) or energy efficiency ratio (EER) for cooling. These are dimensionless coefficients, calculated for rated conditions at full load. A higher COP or EER means that a given amount of heating or cooling is achieved with less work input—that is, the system is more energy efficient if these ratings are higher. Common AC units have COP and EER values ranging from 2.5 to 4. In the United States, the performance of AC systems is usually given by the amount of heat (in British thermal units, or Btu) added or removed from a space per watt-hour of electricity consumed by the system.

Nowadays, variable speed compressors and fans (inverter technology) allow AC units to adjust in real time to varying heating and cooling loads, maximizing the energy performance at part-load operation. Weighted indexes have been proposed that take into account variations in ambient temperature and part-load operation within a season. Examples of coefficients for seasonal averaged conditions include the European seasonal energy efficiency ratio (ESEER), and the seasonal energy efficiency ratio (SEER) in the United States.

—*Tiago Oliveira*

Further Reading
Cook, N. *Refrigeration and Air-Conditioning Technology.* Macmillan, 1995.
Ghenai, Chaouki, and Tareq Salameh, eds. *Sustainable Air Conditioning Systems.* IntechOpen, 2018.
Hundy, G.F., A.R. Trott, and T.C. Welch. *Refrigeration and Air Conditioning.* Elsevier, 2008.
Jon, Chua Kian, Md Raisul Islam, Ng Kim Choon, and Muhammad Wakil Shahzad. *Advances in Air Conditioning Technologies: Improving Energy Efficiency.* Springer, 2021.
Khazaii, Javad. *Energy-Efficient HVAC Design: An Essential Guide for Sustainable Building.* Springer, 2014.
Lun, Y.H. Venuss, and S.L. Dennis Tung. *Heat Pumps for Sustainable Heating and Cooling.* Springer, 2020.
Owen, M.S., ed. *ASHRAE Handbook: Fundamentals.* American Society of Heating, Refrigerating and Air-Conditioning Engineers, 2009.

ALTERNATIVE ENERGY

FIELDS OF STUDY
Chemical engineering; Electrical Engineering; Geology; Physics

ABSTRACT
Alternative energy focuses on types of energy that are distinct from those generally employed (namely petroleum). Its boundaries are not precisely defined, and a variety of definitions are used worldwide. Some include nuclear energy, natural gas, propane, and oil shale, while others exclude them. There is, however, a general acceptance that renewable energies are included under the umbrella of alternative energy.

KEY CONCEPTS
alternative: indicates some feature or process that is intended to displace and be used instead of the historical standard feature or process
biogas: gases, typically hydrogen and methane, that are produced as fuels through the degradation and decomposition of organic matter
geothermal: the literal meaning of the word is "earth heat," or heat from below the ground surface; can refer to heat exchanged from the constant temperature within the soil at depths below all frost, or to heat from volcanic sources
hydrocarbon: compounds in which the component molecules are composed only of hydrogen atoms and carbon atoms, hence hydrocarbon
petroleum: also known as crude oil, bitumen and other terms, typically obtained from deep drilled wells or extraction from oil sands

ALTERNATIVE AND RENEWABLE ENERGY VS. FOSSIL FUEL ENERGY
Although the boundaries of alternative energy are not commonly defined, alternative energy is deemed to be the set of energies put in place as a substitute for conventional fossil fuel-based energy. It is generally accepted that the renewable energies are the main alternative ones.

Alternative energy focuses on types of energy that are distinct from those generally employed (namely oil and coal). Its boundaries are not precisely defined, and a variety of definitions are used worldwide. Some include nuclear energy, natural gas, propane, and oil shale, while others exclude them. There is, however, a general acceptance that renewable energies are included under the umbrella of alternative energy.

It is worth noting that, for many centuries, renewable energies were the only energy options available and so these types of energy were the conventional ones. Nowadays, solar, wind, and geothermal sources of energy, as well as other forms, are deemed to be alternatives to the predominant fossil fuels. In fact, wind and water were used for a long time to power sailing ships and to turn windmills and waterwheels for mechanical needs. Also, until the 1880s, wood was the most significant energy source in advanced civilizations.

The Industrial Revolution and consequent increased energy requirements led to the development and use of the first hydrocarbon-based fuel: coal. When the use of this and other fossil fuels became possible on a large scale, few of the renewable-energy techniques could compete.

There was a renewed interest in energy alternatives in response to oil embargoes and a spike in oil prices in the 1970s, but interest in alternative energy sources was not effectively sustained. The basic reason for the failure to develop renewable energy sources during those years was either a concern over the environment or their high costs, which became difficult to justify once oil prices began to decline.

However, in the first decades of the twenty-first century huge improvements in the performance and affordability of solar cells, wind turbines, and biofuels have paved the way for mass commercialization of these alternative technologies.

By 2014 around 11 percent of global final energy consumption derived from renewables, according to estimates by the U.S. Energy Information Administration (EIA). The International Energy Agency (IEA) reported that renewable sources accounted for 22 percent of global power generation in 2013 and was expected to exceed 26 percent in 2020.

BENEFITS OF ALTERNATIVE ENERGIES

- *Energy security:* Enhancing diversity in energy supply and therefore strengthening energy security.
- *Gas emissions:* Considering their entire life cycle, even if most forms of alternative energy are not carbon-free they make a major contribution to the reduction of greenhouse gas emissions and other pollutants on a local and global scale.
- *Local development:* Creates job opportunities and spins off economic benefits for local communities in infrastructure manufacturing, installation, and maintenance.
- *System flexibility:* Increases flexibility of power systems.
- *Energy source in rural areas:* Provides energy in underdeveloped rural regions where electricity grids are absent.
- *Environmental effects:* In general, compared to fossil fuels, alternative energies have significantly fewer environmental effects, but they do exist: visual impact of wind farms, effects in the ecosystem of dams, deforestation in the case of biomass, and noise, among others.
- *Global business:* Renewable energy is becoming a global business, and in this sense mobilizes important financial resources and creates millions of new jobs worldwide.

BARRIERS TO ALTERNATIVE ENERGY

- *Technology costs:* The initial capital cost is probably the most significant barrier to the use of alternative energies on a large scale.
- *Storage and distribution:* Natural fluctuations of energy sources (such as consistency of available sunlight due to cloud cover, and variability of wind) create the need for storage. Also, the geographic availability may require long-distance transportation to be used.
- *Technicalities:* The conversion of energy from alternative sources to practical applications needs further development. Also, the vulnerability of infrastructures to elements (such as the potential for offshore wind farms to be damaged in storms) can be critical.
- *Social barriers:* The NIMBY (not in my backyard) phenomenon, in which populations claim to support a development in general but oppose it when it threatens to disrupt their direct neighborhood, is also a problem to be dealt with by alternative energies. Some examples of this are the competition between biofuel and food and the displacement of population caused by the creation of water reservoirs.

WIND ENERGY

Wind energy is one alternative source used to produce electricity, and the IEA estimated in 2016 that it accounted for about 2.5 percent of global demand for electricity, making it the second-largest source of renewable electricity. Wind power is considered a relatively mature technology. Its costs have been declining significantly and the efficiency of wind power technologies has steadily increased, making it more feasible for large-scale energy generation. Still, wind power is in most cases dependent on public subsidies.

In a wind turbine, the wind flows over airfoil-shaped blades, causing lift, which causes them to turn. The blades are connected to a drive shaft that turns an electric generator in order to produce electricity.

Wind power does have an environmental impact. The visual effects of wind turbines (which many deem unsightly) are often deemed the most important impact. Moreover, the impact on wildlife—particularly migratory birds and bats—has been debated, and studies with contradictory results have been published. Mark Jacobson ranked wind power as the winner among alternative energy sources in terms of its overall benefits with regard to the environment and human health. Jacobson also ranked wind-battery electric vehicles as the best vehicle solution.

The potential of wind power is largest at remote sites, but the costs of the grid connection often hamper the development of wind sites. Because of the limited availability of land as well as the fact that winds are stronger and more stable at sea, several countries are looking with increasing interest at offshore wind power stations.

SOLAR ENERGY

Solar energy produces both electricity and heat. Like wind power, the solar power industry saw rapid growth in the late-twentieth and early-twenty-first centuries as technology improved and prices fell drastically. According to the IEA global photovoltaic (PV) solar capacity in 2013 was 128 gigawatts, and by 2016 many estimates reached 200 gigawatts after several years of major industry expansion, especially in China and India.

The most familiar form of solar energy production is solar panels. Also known as photovoltaics (PV), these are arrays of cells with certain semiconductor material that converts solar radiation into direct current electricity. Solar photovoltaics are particularly flexible because they can be installed in different places: on the roofs or walls of houses and office buildings, or even sewn onto clothing to feed portable electronic devices with power. They can be combined into farms as well.

Also, sunlight can be harnessed through solar-thermal technologies, which collect sunlight to generate heat. These systems include solar water heaters that have long been used to provide hot water.

A third type of technology corresponds to the concentrated solar power in which sunlight is focused (concentrated) by mirrors or reflective lenses to heat a fluid in a collector at high temperature in order to produce electricity.

Solar technologies have a moderate environmental impact, but in no case is the impact of any renewable zero. It is important to note that solar devices need an external surface, which may result in user conflicts. Also, dismantled systems require proper waste management. Furthermore, in the thermal solar power plants, the water that is required in the steam system and in cooling towers can be a limited resource in certain locales.

The solar cell market has seen a huge increase over the last few years, but there are some associated disadvantages. The most important one is that the access on a short-term basis is uncertain—it is not easy to predict the efficacy and output of solar systems on a daily basis.

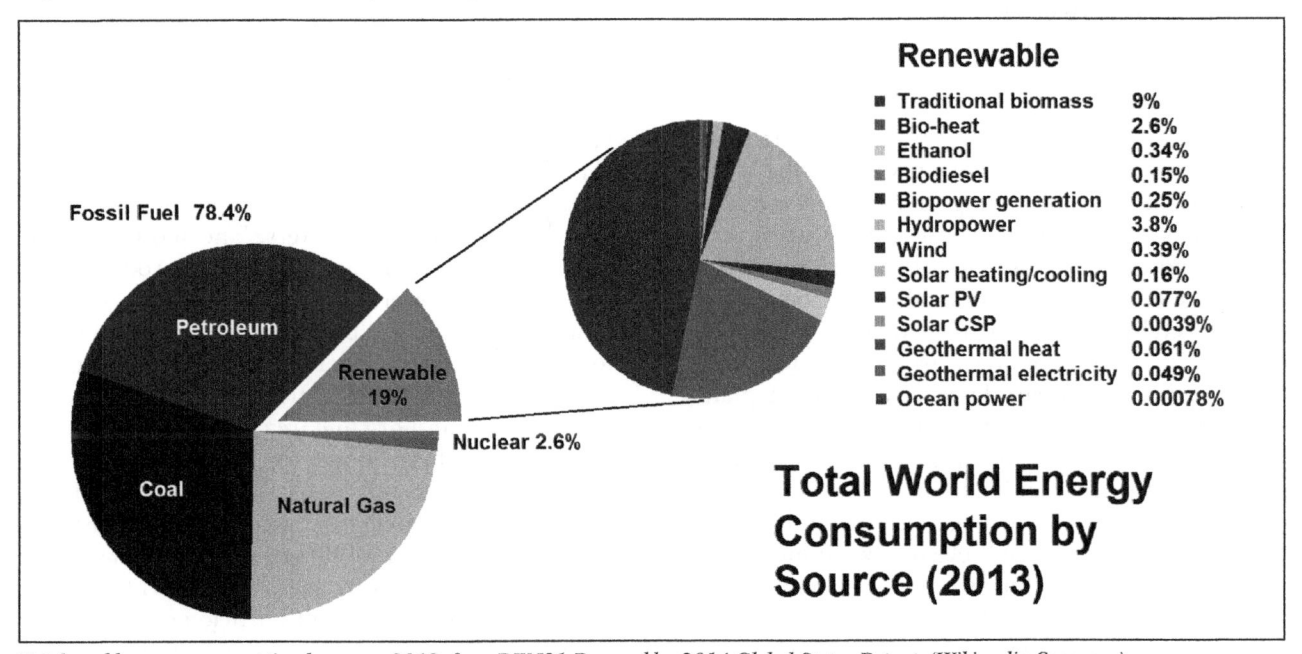

Total World Energy Consumption by Source (2013)

Fossil Fuel 78.4%
- Petroleum
- Coal
- Natural Gas

Renewable 19%

Nuclear 2.6%

Renewable
Traditional biomass	9%
Bio-heat	2.6%
Ethanol	0.34%
Biodiesel	0.15%
Biopower generation	0.25%
Hydropower	3.8%
Wind	0.39%
Solar heating/cooling	0.16%
Solar PV	0.077%
Solar CSP	0.0039%
Geothermal heat	0.061%
Geothermal electricity	0.049%
Ocean power	0.00078%

Total world energy consumption by source 2013, from REN21 Renewables 2014 Global Status Report. (Wikimedia Commons)

GEOTHERMAL ENERGY

Geothermal energy is heat energy from Earth's interior. This heat has two main sources: the original heat from when Earth was created and the energy released as a result of ongoing radioactive decomposition in Earth's crust. Geothermal sources of energy are used to produce both electricity and heat in geothermal power systems. Geothermal energy had a world capacity of over 12 gigawatts in 2014, with power generation estimated at 77 terawatt hours (TWh) according to the IEA. Geothermal sources with flow temperatures higher than 120°C can be exploited to produce electric power; other temperatures are used for heating (and even cooling). Technologies for the utilization of geothermal energy are established and mature, and their costs are competitive with those of conventional fossil fuels. Projects for the utilization of geothermal energy for electricity production involve a high level of economic risk, however. Research and experimental drilling must be carried out to know if the use of that geothermal resource is viable. The exploitation of low-temperature thermal energy, as opposed to high-temperature geothermal energy, is much less complex and is feasible all over the world.

Thermal pollution from the return of cooling water and the emissions of chemicals presented in hydrothermal liquid or steam are examples of the potential environmental impacts of these projects.

HYDROPOWER

Hydropower produces electricity. In 2016 its world capacity was estimated at 936 gigawatts by the World Energy Council, representing the largest share of all renewables. Hydropower is considered a mature technology and currently supplies around 15 percent of the total electricity production. Most hydroelectricity is produced by water falling from dams. It is also produced by water flowing down rivers and water stored in reservoirs. In standard hydroelectric power plants water drops gravitationally, driving a turbine and generator. It is simple to adjust production and can provide energy in seconds and therefore is often used for peaking power. The driving force behind the construction of dams is not only electricity production but also mitigation of flooding, support to irrigation, and water supply protection. Hydropower is a renewable energy form but is not always deemed to be a "green," or environmentally friendly, category of energy. Massive new projects are controversial for a variety of reasons. Changes in the flow of water can severely affect ecosystems, especially migrating fish and other aquatic organisms, and water reservoirs can occupy large areas of land, displacing humans as well as animals and plants. Another big challenge for hydropower is its dependence on rainfall. Moreover, the economic risk in hydropower projects can be large, since they are capital intensive.

OCEAN ENERGY

The oceans have been used to produce electricity, though ocean energy technologies are considered immature. Much research and development is needed before the force of waves and tidal power become competitive with fossil energy sources. Of the two sources of ocean energy, waves and tides, wave power may have the technology closest to a commercial viability. Although no substantial commercial wave energy plants exist, there are a few small ones. Tidal power may see a breakthrough in the next decade, based on the successful operation of tidal power projects in France, Canada, Russia and other nations.

BIOMASS

Biomass as a source of energy is used to produce both electricity and heat. In 2016 the IEA reported that it accounted for about 10 percent of the total world energy supply, with bioenergy electricity production reaching 370 TWh, or 1.5 percent of global electricity generation in 2012. Bioenergy is used extensively all over the world. Biomass can be directly used as fuel or can be decomposed by microorganisms to produce biogas fuel in biogas digesters. The most common area of application for bioenergy is the production of heat. Electric power and liquid biofuel, such as biodiesel, can also be produced. Commercial bioenergy resources are derived mainly from forestry, farming, and waste (incineration plants where materials are burned and organic treatment plants or sanitary landfills where biogas is produced). Biomass includes wood, wood chips, bark, or refined products, such as briquettes, pellets, charcoal, and sawdust.

The combustion of solid fuels such as biomass generates bottom ash and fly ash. Fireplaces are also an important source in terms of particle emissions. It is important to note, however, that the combustion and gasification of biomass generally have lower emissions of certain pollutants (such as nitrogen oxides and sulfur than the combustion of fossil fuels. Large biomass plants must have a gas treatment stage in order to control harmful emissions to the atmosphere.

BIOFUEL

The use of vegetable oils in internal combustion engines dates back to the beginning of the 20th century, when a compression ignition engine, first developed by Rudolf Diesel, worked on peanut oil at the 1900s World Exhibition in Paris. Experiments conducted by Diesel proved that diesel engines could run on vegetable oils, with results almost as effective as with mineral oil.

Diesel indeed stated that "they make it certain that motor-power can still be produced from the heat of the sun, which is always available for agricultural purposes, even when all our natural stores of solid and liquid fuels are exhausted." However, vegetable oils were used in diesel engines for only a few years, until manufacturers optimized the engine for low-grade fractions of petroleum in the 1920s. Oil shortages in the 1930s, in the 1940s during World War II (when vegetable oils were used as emergency fuels), and in the 1970s during the oil crises of 1973 and 1979 revived interest in the use of vegetable oil as fuel.

Biofuels in the present day originate from biomass such as plants or organic waste and can be used directly as fuels or blended into diesel fuel or gasoline. Consequently, they are an important alternative source of transport fuel. Ethanol, for example, was produced at 25.68 gallons (97.2 billion liters) per year in 2015.

There are two main forms of biofuel today: ethanol and biodiesel, also known as fatty acid methyl esters (FAME). Ethanol is made from basic fermentation plants, such as corn and sugarcane (the most important ethanol crops), as well as sugar beets and wheat. Biodiesel is produced through a process called transesterification, which uses various raw material inputs, including corn, cottonseed, peanuts, sunflower seeds, and soybeans.

The impact of biofuel depends not only on the technology but also on the use of a raw material. Current biofuel technologies are already a viable renewable energy source but are not always as "green" as they originally appeared. Bioenergy made headlines for negative reasons in 2008, particularly for its association with rising food prices. In fact, biofuel demand was one of many factors contributing to the rise in food prices. Also, rising demand for biofuel and increasing world food consumption are considered to be leading to an expansion of world cropland at the expense of tropical rain forest and natural grasslands, which has a significant impact on biodiversity and consequences regarding climate change issues.

In addition, some studies mention that, in certain cases, turning some plants into fuel can use more energy than the resulting ethanol or biodiesel generates when all the energy inputs—including production of pesticides and fertilizer, running farm machinery and irrigating, and grinding and transporting the crop—are considered. On the other hand, biofuel derived from the correct feedstocks, grown on the right land with an efficient use of fertilizers and other resources, and produced in proper working conditions can be considered a sustainable biofuel.

Nonfood-based biofuels produced from inedible cellulosic material—leafy materials, stems, stalks, or even algae—are referred to as second-generation biofuels. They are currently under extensive research and development.

HYDROGEN

Hydrogen is used to produce transport fuel and power. There is no monitoring of the world hydrogen production and consumption; however, 45 million metric tons (500 million cubic meters) are estimated to be produced annually.

Even if some experts consider that hydrogen is a renewable resource that comes from renewable sources such as water, others categorize hydrogen as a carrier (like electricity), because it must be produced from other substances. Although hydrogen is the most abundant element in the universe, it does not exist on Earth as a readily available gas like nitrogen or oxygen. Hydrogen atoms must therefore be separated from other elements. Hydrogen can be produced from a variety of resources, including water, fossil fuels, and biomass. At present, a high percentage comes from fossil fuels.

The two most common methods of producing hydrogen are steam reforming and electrolysis. Steam reforming is less expensive. Normally it is used in industry to separate hydrogen from methane (in natural gas). In this process, greenhouse gas emissions result. Electrolysis is a process that separates hydrogen from oxygen (water splitting). It results in no emissions; however, currently it is very expensive.

Hydrogen can be used to fuel internal combustion engines. Its most promising application, however, is in fuel cells. Fuel cells create an electric current when they cross a charged membrane, stripping the hydrogen of electrons. Small fuel cells can power electric vehicles.

Hydrogen fuel cells are considered an immature technology but are already in use. Iceland is a leading example in type of application. Niche applications for the use of hydrogen are abundant. These include laptop computers and cell phones. However, the costs remain high.

CONCLUSION
On a global scale, more than 60 percent of oil is consumed in transportation. In this sense it is worth noting that alternative fuel is generally considered to include all alternatives to gasoline or diesel. It encompasses biodiesel, ethanol, and hydrogen, as described above, but also other resources, such as methanol, natural gas, propane, coal-derived liquid fuels, and electricity, as recognized by the U.S. Department of Energy.

—*Ana Penha and Fausto Freire*

Further Reading
Gallagher, Ed. "Making Biofuels Sustainable." *Our Planet: The Magazine of the United Nations Environment Programme*, Dec. 2008. http://www.unep.org/pdf/Ourplanet/2008/dec/en/OP-2008-12-en-FULLVERSION.pdf.

Inderwildi, Oliver, and David King. "Quo Vadis Biofuels." *Energy and Environmental Science*, vol. 2, 2009.

Jacobson, Mark. "Review of Solutions to Global Warming, Air Pollution, and Energy Security." *Energy and Environmental Science*, vol. 2, 2009.

Kammen, Daniel. "The Rise of Renewable Energy." *Scientific American*, Sept. 2006.

Pimentel, David, and Tad Patzek. "Ethanol Production Using Corn, Switchgrass, and Wood; Biodiesel Production Using Soybean and Sunflower." *Natural Resources Research*, vol. 14, no. 1, Mar. 2005.

"Renewable & Alternative Fuels." *US Energy Information Administration*. EIA, 2016. Web. 13 May. 2016.

Renewable Energy Policy Network for the 21st Century (REN21). *Renewables 2010 Global Status Report*. Paris: REN21 Secretariat, 2010. http://www.ren21.net/REN21Activities/Publications/GlobalStatusReport/tabid/5434/Default.aspx.

"Renewables." *International Energy Agency*. OECD/IEA, 2016. Web. 13 May. 2016.

Shales, Stuart. "Biofuels: The Way Forward?" *Science for Environmental Policy*, no. 1, Feb. 2008. http://ec.europa.eu/environment/integration/research/newsalert/pdf/1si.pdf.

Smith, Zachary, and Katrina Taylor. *Renewable and Alternative Energy Resources: A Reference Handbook*. ABC-CLIO, 2008.

United Nations Development Programme. "The World Energy Assessment: Overview, 2004 Update." http://www.undp.org/energy/weaover2004.htm.

U.S. Department of Energy. "Alternative and Advanced Fuels." http://www.afdc.energy.gov/afdc/fuels/indes.html.

U.S. Department of the Interior. *Ocean Energy*. Minerals Management, 2007.

AMPÈRE, ANDRÉ-MARIE

FIELDS OF STUDY
Electrical Engineering; Electronics; Physics

ABSTRACT
French physicist and chemist hailed as the founder of electromagnetism, André-Marie Ampère also created the vocabulary of electricity and pioneered the modern reflection on the atomic constitution of matter.

KEY CONCEPTS
alternating current: electrical current produced by magnetic oscillation, which reverses direction periodically rather than flowing continuously in one direction; for example, normal household current in North America oscillates at a frequency of 60 Hertz (60 cycles per second) and therefore reverses direction 120 times per second

ampere: the standard unit of electrical current, defined as the movement of 1 coulomb of electric charge through a potential difference of 1 volt in 1 second

direct current: electrical current that flows in one direction only, as from a relatively negative potential to a relatively positive potential

electrical current: normally, the flow of electrons through a conductor under the influence of a potential difference between two points in the conductor

electromagnetism: magnetism, or a magnetic field, produced by the functioning of an electrical current or field

A GENIUS IN THE MAKING

Born in Lyon, France, on January 20, 1775, André-Marie Ampère developed a passion for mathematics at an early age. His father, Jean-Jacques, a merchant and a judge, and a keen follower of philosopher and social theorist Jean-Jacques Rousseau, schooled his son at home. At the age of 13, young Ampère showed the first signs of his precociousness by composing an essay on conical sections. A friend of his father, struck by his talents, instructed him in differential and integral calculus, while young Ampère systematically studied and memorized large portions of Denis Diderot and Jean Le Rond d'Alembert's *Grande Encyclopédie*.

The French Revolution brought death and grief into Ampère's life: In 1793, his father was executed after Lyon was taken by troops of the National Convention and proclaimed a "liberated town." The rest of Ampère's life would be deeply affected by this event, by the death of his first wife, and by a second, unsuccessful marriage. The tragedies in his early personal life led him to dedicate himself completely to science. He entered the French Academy of Sciences in 1814 and obtained a full professorship in physics at the Collège de France in 1824. Worn down by work, he died in Marseille on June 10, 1836.

Ampère's legacy to science is important for many disciplines. In the field of chemistry, he was among the pioneers of modern reflection on the atomic constitution of matter. While Joseph Louis Gay-Lussac had already noted, in 1809, that gaseous substances combine in simple values of their volume ratios, Ampère deducted from this, in 1814, that the number of molecules in the same volume of gas is constant regardless of the gas. The law of Avogadro-Ampère takes its name from the French physicist's work and similar work conducted in the same years by the Italian scholar Amedeo Avogadro (1811).

The hypothesis of the existence of fluorine was, likewise, due to Ampère's intuition. Struck by the analogies between hydrochloric acid and hydrofluoric

André-Marie Ampère. (Wikimedia Commons)

acid, he deduced—according to an autobiographical note of 1810—the existence of a chemical element called *phlore*. However, the official announcement of the discovery was left to British scientist Humphry Davy, three years later.

AMPÈRE AND ELECTRICITY

The glory of naming the international unit of measurement of electric current after him, however, is linked to Ampère's main contribution to the science of physics, electromagnetism (or "electrodynamics," as he called the discipline). In 1820, Ampère attended a reenactment of the Danish scholar Hans Christian Ørsted's famous experiment of 1819, during which a magnetic needle was deflected according to variations in the current passing through a wire placed in its vicinity. François Arago repeated this experiment before the Academy some time later. Ampère concentrated on this phenomenon and shortly thereafter discovered the source of the magnetic actions in a stream, studied the interactions between the magnets, and was able to show that two closed currents act on each other.

Ampère's work was also a precursor of the electronic theory of matter, given his speculations on the existence of the current of particles. James Clerk Maxwell would later call him the Newton of electricity. Noting that electric current creates effects similar to those of a magnet, Ampère then laid the groundwork for a new discipline, electromagnetism, and provided its first mathematical formulas. He showed that two currents can act on each other, thereby laying the foundation for electrodynamics. All these results were published in his 1827 book *Mémoire sur la théorie mathématique des phénomènes électrodynamiques uniquement déduite de l'experience* (*Memoir on the Mathematical Theory of Electrodynamic Phenomena, Uniquely Deduced from Experience*).

Based on his theories, during the course of his career Ampère developed several devices, such as the galvanometer, an instrument for detecting and measuring electronic current; the electric telegraph, an instrument for the long-distance transmission of messages using electrical signals; and the electromagnet, a magnet whose magnetic field is produced by the flow of electric current. He is hailed as the creator of the vocabulary of electricity; he invented the terms *current* and *tension/voltage*. Ampère is thus considered today as one of the greatest scientists of the nineteenth century, the father of an entire branch of physics.

—*Francesca Musiani*

Further Reading

Ampère, André-Marie, and Henriette Cheuvreux. *The Story of His Love: Being the Journal and Correspondence of André-Marie Ampère, with His Family Circle During the First Republic, 1793-1804*. R. Bentley and Son, 1873.

Erlichson, H. "André-Marie Ampère, the 'Newton of Electricity' and How the Simplicity Criterion Resulted in the Disuse of His Formula." *Physis*, vol. 37, no. 1, 2000.

Gardiner, K.R. "André-Marie Ampère and His English Acquaintances." *British Journal for the History of Science*, vol. 2, 1964/65.

Hofmann, James R. *André-Marie Ampère*. Blackwell, 1996.

Steinle, Friedrich. *Exploratory Experiments: Ampère, Faraday, and the Origins of Electrodynamics*. Alex Levine, tr. U of Pittsburgh P, 2016.

Animal Power

FIELDS OF STUDY
Agriculture; Animal Husbandry; Anthropology; Zoology

ABSTRACT
By using animals' muscle power for traction and transport, humans expanded the efficiency of these processes immensely. Animal power, essential for heavy hauling or rapid travel until the mid-1800s, remains important to much of the world's agriculture.

KEY CONCEPTS

agricultural revolution: the period in which nomadic hunter-gatherer societies established agriculture in place, leading to the establishment of permanent settlements where the products of agriculture would be readily available

domestication: conversion of a wild species to one that is able to coexist with humans in a mutually beneficial relationship

draft animals: domesticated animals used primarily as "traction engines" to pull loads

pack animals: domesticated animals used primarily as "trucks" to carry loads

BACKGROUND

The dog was the first animal domesticated, tamed and bred from wolf ancestors. Archaeological sites showing this development date back approximately 11,000 years in both northern Europe and North America. The first dogs may have helped Stone Age hunters chase and exhaust game. They also may have pulled snow sleds and hauled loads via travois, as they did for American Indians in later centuries. If dogs were so employed in this era—and it has not been proved that they were, although remains of domesticated dogs dating to 30,000 years ago have been found in Europe—these would be the first intentional uses of animal power as an energy resource. Other important tasks using their senses and group instincts to help humans—tracking, scavenging, and guarding—probably meant that dogs were seldom kept primarily for their muscle power. In Victorian times, however, the muscle power of dogs was put to use in many ways, not the least of which was as "spit dogs" trained to walk in a treadmill that turned the roasting spit of an adjacent fireplace in both private homes and commercial establishments. In the modern day, dogs are used to perform many different functions, although converting their muscle energy into the mechanical work of machines is not one of them.

The next successfully domesticated animals were sheep, goats, pigs, and cattle. This process is shown in remains and artifacts from Jericho that document the origins of agriculture. Centuries after grain was first cultivated, people began to keep livestock. At first the animals were probably loosely controlled and were seen as "walking meat larders" and occasionally providers of milk, fiber, and leather. Once the process was under way, around 8000 to 7000 BCE, people must have experimented with riding and other ways to use the animals in their farming.

Of these anciently domesticated species, only cattle proved to have the combination of strength and malleability to do useful work under human direction. Shifting from a plow pushed by a man or woman to one pulled by an ox multiplied the traction enormously and enabled much more food to be produced with the same investment of human time. This was a major step in the ongoing "agricultural revolution" that created a growing population, town life, and a material surplus to support specialized trades. Using cattle in the fields also called forth other innovations. Harnesses and/or yokes had to be created to control the animals, implements had to become larger, and castration of young male animals had to be practiced to produce oxen that were both strong and docile. Later (sometime around 3000 BCE), the wheel was invented. Hitching such animals to wheeled carts and wagons, humans could travel farther and more easily and could move bulkier goods. Draft animals (animals used for hauling) thus served not only as a direct resource in agriculture and transportation but also as a source of synergy, expanding their owners' geographic and trade horizons and inspiring further inventions.

While most of the evidence for this sequence of events exists in the ancient Near East, some of the same steps took place independently, perhaps several times, elsewhere in Eurasia. For example, the working cattle native to Asia—water buffaloes and yaks—were bred from wild species different from European domestic cattle's ancestors.

HORSES AND RELATED SPECIES

Horses and their kin, the most versatile of hauling and riding animals, were domesticated later. Wild horses roamed much of the world during the last ice age but had become extinct in the Americas by 10,000 BCE and rare in Western Europe and the Mediterranean region about the same time. How much this disappearance was due to climate change and how much to humans' overhunting is uncertain.

Many prehistorians believe that horses were first tamed and trained for riding north of the Black Sea, where they survived in large numbers. Hence, they were reintroduced into Europe and western Asia, between 3000 and 2000 BCE, by successive invasions of mounted tribesmen from the central Asian steppes. However, as with much of prehistory, the evidence is unclear. Horses never disappeared completely in Europe, and they may have been domesticated in several places from local stock.

During the rule of Tiglath-Pileser III (c. 745–727 BCE), camels started to appear in large numbers. The camel was a valuable source of animal power between the 8th to 6th centuries b.c.e., living and working longer than donkeys, mules and oxen. (Jjron via Wikimedia Commons)

What is remarkable about the first 50 years of the 20th century is actually the lack of change in the world energy regime. Depicted is a member of the German Wehrmacht, then the world's most modern army, which invaded the Soviet Union with its infantry on foot and its logistics supported by 600,000 horses. (German Federal Archives via Wikimedia Commons)

The donkey or ass, which is native to North Africa, was brought into use in the same millennium. Donkeys loaded with packs or wearing saddlecloths are shown in Egyptian friezes from 2500 BCE; they also appear in early Sumerian and Assyrian records. Horses and asses were already being interbred at this time to produce vigorous offspring, notably the mule, with the traits of both species. From then until the end of the nineteenth century, the equids were the most widely used animals in the world for transporting people and goods. They also became immensely important in agricultural processes.

Because of their gait, asses cannot be ridden at high speeds, but their adaptability to harsh conditions makes them good pack animals and beasts-of-all-trades on small farms. Mules, hybrids from mare mothers and donkey sires, combine the horse's strength with the ass's stamina. Mules have been known to carry 450 kilograms each, going as far as 80 kilometers between water stops.

Horses' special qualities include speed, herd hierarchical instincts that dispose them to follow human leadership, and relative intelligence. Horses have been bred to strengthen various traits: the Arabian and the modern thoroughbred for speed; the medieval war horse and the modern shire and Clydesdale for strength and stamina; the Shetland pony for multiple tasks in damp and ferocious weather. As an example of the speed gained from using horses, over short distances (up to 5 kilometers) a horse can travel in the range of 48 kilometers per hour. A horse carrying rider and saddle might make a trip of 480 kilometers in sixty hours, and the time can be shortened by frequent changes of mount. A person in top condition—for example, a soldier accustomed to long marches—typically can walk around 65 kilometers a day.

Horse-related technology also developed continuously over time, adding to the effectiveness of horse and rider, and horse and vehicle. Bit and reins, stirrups (not adopted in Europe until the early Middle Ages), horseshoes, saddle and carriage designs, and modern veterinary medicine, all brought new capacities to the horse power on which humankind relied.

Not only was the horse an essential energy resource, but its presence also repeatedly changed history and society. The rise of cavalry as a mobile force in warfare and the change to a horse-based economy and culture by North American Plains Indians when horses were reintroduced by Spanish invaders are only two of the many transformations wrought by horse power.

OTHER ANIMALS AS ENERGY SOURCES

Humans have attempted to put many kinds of large animals to work, but only a few other species have proved useful. Of these, the most important have been those already adapted to extreme climates and terrains.

The camel is called the "ship of the desert" because of its ability to travel for long distances between water holes. Some desert nomad tribes organize their way of life around the use of camels. Normally employed as pack and riding animals, camels were also occasionally used in war in the ancient world, partly to frighten the enemy's men and horses. Llamas, members of the camelid family native to South America, serve as pack animals in the Andes mountain region.

Reindeer, adapted to living in an Arctic environment, can find forage on the barren tundra and survive temperatures of -50°C. Laplanders who live in far northern areas have used them in the roles filled by cattle and horses in warmer climates, including riding, pulling carts, and carrying loads as pack animals.

Elephants, native to the Indian subcontinent and to Africa, have been trained in both regions to lift and carry extremely heavy objects, although their use as riding beasts has been largely confined to ceremonial occasions and entertainment.

ANIMAL POWER TODAY

From prehistory through the nineteenth century, much of the work of civilization depended upon animal power. With the coming of steam power and the internal combustion engine, animals gradually became less essential for transport and traction, at least in the developed world. Yet as late as the 1930s, horses or mule teams, rather than tractors, were used by many American farmers.

In Asia and Africa, most of the farmland is still worked with draft animals. For a small farmer of limited means or in an isolated area, animal power has several advantages. Unlike machines, animals do not need complex networks to supply their fuel or parts for repair. They produce their own replacements, and their malfunctions sometimes heal themselves without special knowledge or tools on the owner's part. Their byproducts can be recycled into agricultural use.

For these reasons, and because they bring less devastation to land and air, some members of "back to the Earth" movements in the United States choose animal power. Heavy horses are also used in selected logging operations to avoid the clear-cutting and other environmental damage that machines bring.

—*Emily Alward*

Further Reading

Ableman, Michael. *From the Good Earth: A Celebration of Growing Food Around the World.* H.N. Abrams, 1993.

Chamberlin, J. Edward. *Horse: How the Horse Has Shaped Civilizations.* BlueBridge, 2006.

Chenevix Trench, Charles. *A History of Horsemanship.* Doubleday, 1970.

Clutton-Brock, Juliet. *Domesticated Animals from Early Times.* U of Texas P, 1981.

———. *Horse Power: A History of the Horse and the Donkey in Human Societies.* Harvard UP, 1992.

———. *A Natural History of Domesticated Mammals.* 2nd ed. Cambridge UP, Natural History Museum, 1999.

Greene, Ann Norton. *Horses at Work: Harnessing Power in Industrial America.* Harvard UP, 2008.

Kalof, Linda, and Brigitte Resl, eds. *A Cultural History of Animals.* 6 vols. Berg, 2007.

Pynn, Larry. "Logging with Horse Power." *Canadian Geographic*, vol. 111, no. 4, Aug.-Sept. 1991, p. 30.

Schmidt, Michael J., and Richard Ross. "Working Elephants: They Earn Their Keep in Asia by Providing an Ecologically Benign Way to Harvest Forests." *Scientific American*, vol. 274, no. 1, Jan. 1996, p. 82.

Tiwari, G.N., and M.K. Ghosal. "Draught Animal Power." *Renewable Energy Resources: Basic Principles and Applications*. Alpha Science International, 2005.

Watts, Martin. *Working Oxen*. Shire, 1999.

ATOMIC ENERGY COMMISSION

FIELDS OF STUDY
Modern History; Policy Studies

ABSTRACT
The Atomic Energy Commission was a civilian agency of the United States government responsible for administration and regulation of all aspects of the production and use of atomic and nuclear power from 1946 to 1974.

KEY CONCEPTS
- **atomic energy:** energy derived from the use of radioactive materials undergoing nuclear fission
- **high-level radioactive waste:** radioactive source materials such as spent nuclear fuel and other radioactive sources emitting "hard" radiation such as X-rays
- **low-level radioactive waste:** materials that have been exposed to radioactive materials and so have acquired some radioactivity, as well as other radioactive sources that do not emit "hard" radiation such as X-rays
- **radiological hazards:** the range of possible dangers to humans and the environment that may arise through exposure to nuclear radiation

BACKGROUND
In July, 1945, an interim committee formed by President Harry S. Truman drafted legislation to establish a peacetime organization similar to the Manhattan Project. This proposed legislation, the May-Johnson bill, proposed a nine-member part-time board of commissioners that included a significant military contingent and continued government control over atomic research and development. The bill was opposed by most U.S. atomic scientists because it established military control over research and would thereby stifle the free exchange of ideas. In late 1945, as support for the May-Johnson bill collapsed, Senator Brien McMahon introduced substitute legislation with reduced security requirements and diminished military involvement. This bill was signed into law by President Truman on August 1, 1946. The McMahon Act, officially the Atomic Energy Act (AEA) of 1946, transferred control over atomic research and development from the Army to the Atomic Energy Commission (AEC), which consisted of a five-member full-time civilian board assisted by general advisory and military liaison committees.

IMPACT ON RESOURCE USE
While the main mission of the AEC was to ensure national defense and security, the Atomic Energy Act also called for the development of atomic energy for improving the public welfare, increasing the standard of living, strengthening free enterprise, and promoting world peace. The commission was also authorized to establish health and safety regulations for possessing and using fissionable materials and their byproducts.

In 1953, President Dwight Eisenhower's famous "atoms for peace" speech to the United Nations called for the development of peaceful applications of atomic energy, and in particular for nuclear reactors that would produce power. This goal required eliminating the AEC's monopoly on nuclear research; Congress passed the Atomic Energy Act of 1954, which continued the AEC's role as sole regulator of nuclear activities, allowed licensing of privately owned facilities for production of fissionable materials, and imposed several safety and health requirements. To transfer technology from government to private industry, the AEC established the Power Demonstration Reactor Program, under which industries designed, constructed, owned, and operated power reactors with financial and other assistance from the AEC.

As the nuclear power industry grew during the 1960s, the Atomic Energy Commission came under increasing criticism for an inherent conflict of interest in its roles as promoter of nuclear power and regulator of environmental and reactor safety. At the end of the decade, the growing environmental movement charged that AEC regulations, which addressed only potential radiological hazards to public health and safety, were not consistent with the National Environmental Policy Act (NEPA) of 1970. In 1971, courts ruled that the commission was required to assess environmental hazards beyond radiation effects, such as thermal pollution. More stringent licensing requirements increased the costs associated with new reactor construction. The commission was simultaneously faced with the growing problem of disposal of high-level radioactive waste.

Under the Energy Reorganization Act of 1974, the AEC was abolished. The Nuclear Regulatory Commission (NRC) was created to handle commercial aspects of nuclear energy, while responsibility for research and development and the production of fissionable materials was transferred to the Energy Research and Development Administration.

—*Michael K. Rulison*

Further Reading

Ford, Daniel F. *The Cult of the Atom: The Secret Papers of the Atomic Energy Commission.* Simon & Schuster, 1984.

Hewlett, Richard G., Oscar E. Anderson, and Francis Duncan. *A History of the United States Atomic Energy Commission.* U of California P, 1990.

U.S. Department of Energy. "About the Department of Energy: Origins and Evolution of the Department of Energy." http://www.energy.gov/about/origins.htm.

U.S. Department of Energy. "Office of Science: The Atomic Energy Commissions (AEC), 1947." http://www.ch.doe.gov/html/site_info/atomic_energy.htm.

B

Batteries

FIELDS OF STUDY
Electrical Engineering; Electrochemistry; Environmental Engineering; Resource Management

ABSTRACT
A battery refers to an array of electrochemical cells. Cells in turn are comprised of a pair of half cells. Batteries pose both challenges and opportunities when they reach the end of their life. On the one hand, batteries contain hazardous materials that are harmful to the environment and human health; on the other hand, the increasing demand for batteries means that more metals and finite resources are captured for use in their production and manufacture. As a result, environmental protection and resource security is driving changes to the ways in which batteries are managed at the point of disposal.

KEY CONCEPTS

cell: a construct of two different half-cells that produces an electrical potential, or voltage, as the sum of the relative potentials of the two half-cells

half-cell: a metal in contact with a solution of a metal salt, having a specific voltage potential relative to a defined standard half-cell

landfilling: disposing of discarded batteries and other items in landfill operations

primary battery: a battery that cannot be recharged by driving an electrical current through it in the opposite sense of its natural voltage

pyrolysis: reduction of batteries and other items by decomposing them to their component materials through the application of heat

reconditioning: repair and replacement of the components of a poorly functioning battery in order to return it to a fully functional state

secondary battery: a battery that can be recharged by providing an electrical current from an external source in the reverse sense of the battery current

BATTERIES AS AN ENERGY SOURCE

A battery refers to an array of electrochemical cells. Cells in turn are comprised of a pair of half cells. Each half-cell must have an electrode and an electrolyte. Half-cells can share a common electrolyte. Cells are connected in various configurations to form batteries, which can produce more power than individual cells. Cells are an electrochemical energy source; that is, they produce electrical energy from a chemical reaction. This chemical reaction involves the electrodes and the electrolyte—or in the case of fuel cells, a special kind of battery, an external fuel source.

With global moves to decarbonize energy supplies, battery technologies are likely to play an increasing role by providing portable power, even without discrepancies in supply and demand of utility power, enabling new technologies. A battery refers to an array of electrochemical cells. Cells in turn are comprised of a pair of half-cells. Each half-cell must have an electrode and an electrolyte. Half-cells can share a common electrolyte. Cells are connected in various configurations to form batteries, which can produce more power than individual cells. Cells are an electrochemical energy source; that is, they produce electrical energy from a chemical reaction. This chemical reaction involves the electrodes and the electrolyte—or in the case of fuel cells, a special kind of battery, an external fuel source.

Batteries can be classified into two kinds: primary batteries, or nonrechargeable batteries; and secondary batteries, or rechargeable batteries. In primary batteries, a nonreversible chemical reaction occurs, which converts the chemical reactants into electricity; however, this reaction cannot easily be reversed. Conversely, secondary batteries use a reversible chemical reaction. This allows the battery to be charged and discharged, allowing the battery to be used as a temporary store for electricity. Usually, batteries are supplied in their discharged state and must be charged before use. Primary batteries result in an environmental burden, so there are moves in portable electronics to encourage the use of rechargeable batteries.

What makes secondary batteries particularly relevant in the present energy context is their ability to store energy produced by renewable sources. Fossil fuels and nuclear power are both finite energy sources. For this reason, in the medium to long term there is a need to move toward renewable forms of energy that will offer sustainable energy supplies in perpetuity. However, there are some challenges with renewable energy technologies that produce power from natural energy flows: Solar power, wind power, some implementations of hydropower technology, and wave power are all intermittent energy sources. To some degree, this problem of intermittency can be addressed by using battery technologies—storing energy in chemical form when it is abundant for later use. This is particularly relevant for rural communities without grid infrastructure. Furthermore, battery technologies may also be used as a substitute for hydrocarbon fuels in mobile applications, particularly in vehicles and transportation.

BATTERIES FOR LOW-CARBON VEHICLES

One of the particular areas where batteries can play a significant role in helping address energy challenges is in storing energy for mobility applications. At present, many vehicle technologies rely on a combination of the internal combustion engine and hydrocarbon fuels to provide motive power for movement. This presents challenges, because in burning hydrocarbon fuels, carbon emissions are produced as a byproduct; furthermore, hydrocarbon fuels are in finite supply. Battery technologies can help to address this energy challenge. By storing electricity produced from clean, renewable sources, batteries can act as a vector for using electricity in mobile applications. Batteries enable the production of vehicles that have no emissions at

Old batteries destined for collection and recycling. (Jaoquin Corbalin via iStock)

the point of use; however, there may be emissions produced at the point of electricity generation.

In a conventional vehicle, a battery is used to start the engine, provide lighting, and power ancillary convenience features. This requires a much smaller battery than the type of battery required for an all-electric vehicle; these are also known as cyclical batteries, because the alternator continuously helps to charge conventional vehicle batteries. An electric vehicle (EV) battery is what is known as a deep-cycle battery. It is designed to go from being fully charged to fully discharged. There are a range of battery chemistries that can be used in EV applications. Early electric vehicles used lead-acid batteries, a simple battery technology. While cheap to produce, the low power density of these batteries meant that early EV vehicle performance was poor by the standard of modern electric vehicles. Nevertheless, lead-acid battery powered vehicles found favor in some applications. Later electric vehicles employed nickel-metal hydride batteries, a battery technology also used in home rechargeable batteries.

There have also been some technologies developed more recently that offer the promise of greater power density and improved EV performance. The Zebra battery uses molten sodium chloroaluminate as the electrolyte. This requires the battery to be constantly heated for use, which presents some challenges for vehicle design and operation. One technology that holds great promise for future electric vehicles is lithium ion and lithium polymer battery technology. These types of batteries have been deployed in consumer electronics because of their high power density, which in mobile applications allows for small, light devices.

These characteristics (the batteries' energy density, power density, and charge/discharge efficiency) also lend themselves to electric vehicle applications. Batteries remain the most expensive component of battery electric vehicles, so significant effort is being invested into reducing the cost of producing EV batteries.

DEAD BATTERIES
Batteries pose both challenges and opportunities when they reach the end of their life. On the one hand, batteries contain hazardous materials that are harmful to the environment and human health; on the other hand, the increasing demand for batteries means that more metals and finite resources are captured for use in their production and manufacture. As a result, environmental protection and resource security is driving changes to the ways in which batteries are managed at the point of disposal. Waste batteries can be harmful to human health and the environment if disposed of improperly. The recovery of waste batteries is likely to increase in the future as finite resources diminish and battery use grows with technological advances.

There are many variations in batteries, depending upon their purpose. The composition and construction of batteries can be equally varied, however they usually comprise heavy metals such as lead, mercury, cadmium, silver, nickel, cobalt, and manganese. While some of these metals are toxic, they are also valuable economic resources, and as the number of batteries used grows, so does the amount of metals required. Recycling batteries diverts metals and plastics from landfill and captures these materials as a resource to be used again. There are a number of different techniques available for battery recycling, including thermal-metallurgical reprocessing, hydrometallurgical recycling, pyrolysis, and mechanical separation. The technology option often depends upon the type of battery. One of the central elements to achieving sustainable economic growth is efficient resource substitution. Harnessing metals in waste batteries can substantially reduce the mining of virgin material, environmental damage caused by mining, and energy used in these processes. According to the European Commission, recycled cadmium and nickel requires respectively 46 and 75 percent less energy in extraction than mining virgin metals.

The benefits of reusing or recycling waste batteries are not only embedded in the value of the composite metals, but also in the need to control and protect the environment and human health from the threat of harm posed by the substances contained in batteries. Lead, mercury, and cadmium are toxic substances that, if disposed of irresponsibly, can pollute environmental media, destroy ecosystems, damage the wider environment, and harm human health. For example, exposure to lead can cause fatigue, impaired learning, and seizures. Similarly, health effects of mercury include depression, nausea, and autism, and in wildlife mercury toxicity can result in physical deformities, abnormal behaviors, and endocrinological disorders that lead to reduced reproduction.

The surge in demand for batteries has been stimulated by the rapid uptake of electrical and electronic equipment such as mobile phones, laptops, battery-powered toys, hearing aids, and household gadgets such as toothbrushes and bread knives. HighBeam Business estimates that on a global level every year, more than 20 billion batteries are sold. This trend is set to continue as new technological innovations and mobile devices emerge onto the market. For instance, C. Trudell calculates that the volume of automotive batteries will increase to make up 15 percent of sales by 2025 as the transition from combustion engines to hybrid or electric drive train passenger vehicles increases.

MANAGING BATTERY WASTE

The management of waste batteries is regulated through law in many countries, although not all. Australia landfills over 8,000 tons (7,257 tonnes) of waste batteries per year, per Clean Up Australia, Ltd., but has not put in place government legislation that obliges manufacturers and retailers to share responsibility for the management of waste batteries through a financial contribution toward their separate collection and environmentally sound disposal. Much of the legislation that exists involves reducing the amount, or prohibiting altogether, the use of mercury in certain types of batteries and the setting up of separate collection facilities for waste batteries.

In the United States, the Mercury-Containing and Rechargeable Battery Management Act was passed by the 104th Congress in 1996. This phased out mercury from certain types of batteries (Title II, SEC 202) and provided for efficient and cost-effective collection and recycling of used nickel cadmium (Ni-Cd) batteries and other sealed lead-acid batteries. Under SEC 2, Congress found that it was in the public interest to phase out the use of mercury in batteries and educate the public concerning the separate collection, recycling, and proper disposal of batteries. It also established a uniform national labeling requirement for regulated batteries and emphasized the importance of rechargeable consumer products and packaging in steering consumer choices and responsible consumer behavior. Following the 1996 act, many states have introduced laws prohibiting incineration and landfilling of mercury-containing and lead-acid batteries and/or regulations that require battery recycling and encourage manufacturers to make their products in such a way that allows easy removal of batteries for separate management at the end of the product's life. This improves the potential for reuse and recycling.

The role of manufacturers and retailers in: (1) reducing the toxic constituents of batteries; (2) reducing their environmental impact; and (3) increasing their ability to be reused, recycled, or disposed of in an environmentally sound manner has been seized upon by the European Commission. The European Union (EU) adopted Directive 2006/66/EC on batteries and accumulators and waste batteries and waste accumulators. Its main purpose is to reduce the harmful effects posed by batteries and protect, preserve, and improve the quality of the environment through a requirement of producers to be financially responsible for the batteries they put on the market throughout their life cycle. The Batteries Directive: (1) requires member states to take action to encourage the separate collection and management of waste batteries and accumulators (Article 7); (2) prohibits the disposal by incineration or landfill of industrial and automotive batteries (Article 14); and (3) prioritizes the reuse or recovery of batteries and their component parts, material, and substances. The directive requires each member state to set up collection schemes for the collection of portable waste batteries. These collection schemes should enable end users to discard their waste batteries conveniently, in proximity to where they live, at no cost (Article 8). Targets for minimum collection have been established for September 26, 2012, at 25 percent; and by September 26, 2016, at 45 percent of the total annual sales of batteries (Article 10). A national interim target set by the United Kingdom of 10 percent by 2010 was narrowly missed, although data collection systems were weak.

In India, the Ministry of Environment and Forests introduced the Batteries Management and Handling Rules in 2001. These rules establish a responsibility on manufacturers, importers, assemblers, reconditioners, and dealers to ensure that the used batteries are returned (Rule 4 and 7) by setting up collection systems either individually or jointly in places for separate collection of used batteries from consumers or dealers. While these rules appear similar to the EU Batteries Directive, what sets them apart is Rule 10,

Batteries properly disposed of. (Chepko via iStock)

which places a duty on consumers to ensure that used batteries are not disposed of in any manner other than depositing with the dealer, manufacturer, importer, assembler, registered recycler, reconditioner, or at the designated collection centers.

The consumption of rechargeable batteries, also referred to as secondary batteries, is gaining momentum. The National Electrical Manufacturers Association has estimated that U.S. demand for rechargeables is growing twice as fast as demand for nonrechargeable batteries. Rechargeable batteries offer the opportunity to: (1) reduce the number of waste batteries disposed of by households (which currently stands at an average of 21 batteries per household per annum), (2) capture metals as a valuable secondary raw material, and (3) reduce the environmental impacts associated with waste battery disposal and mining of virgin resources.

The role that batteries play in many aspects of postmodern Western society should not be underestimated, nor should their place in the development of a sustainable future. Nowhere can this better be highlighted than in the automotive sector. The industry and legislators are placing reliance on the development of powerful, yet lightweight batteries for the successful market penetration of the electric vehicle. Reprocessing and recycling of these batteries will need to form part of a manufacturer's life-cycle assessment. The volume of batteries recycled, whether nonrechargeable or rechargeable, will increase across those countries with regulatory controls in place. As the need to secure natural resources is accepted, recycling behaviors become part of the social norm and consumer awareness of battery collection systems improves. It is likely that laws on batteries and waste batteries will continue to develop with more stringent collection and recycling targets and the phasing out of other hazardous metals from batteries.

—*Hazel Nash and Gavin Harper*

Further Reading

Berlow, Lawrence. "Battery." *How Products Are Made*. 2012. http://www.madehow.com/Volume-1/Battery.html.

Buchmann, Isidor. *Batteries in a Portable World: A Handbook on Rechargeable Batteries for Non-Engineers*. 3rd ed. Cadex Electronics Inc.

Clean Up Australia, Ltd. "Battery Recycling Fact Sheet Clean Up." May 2010. http://www.cleanup.org.au/PDF/au/cua_battery_recycling_factsheet.pdf.

Desmond, Kevin. *Innovators in Battery Technology. Profiles of 95 Influential Electrochemists*. McFarland & Company, 2016.

Gerrsen-Gondelach, Sarah J. "Performance of Batteries for Electric Vehicles on Short and Longer Term." *Journal of Power Sources*, vol. 212, no. 15, 2012.

Grünewald, Philipp. "The Role of Large-Scale Storage in a GB Low Carbon Energy Future: Issues and Policy Challenges." *Energy Policy*, vol. 39, no. 9, 2011.

High Beam Business. "Primary Batteries, Dry and Wet: Industry Report." 2012. http://business.highbeam.com/industry-reports/equipment/primary-batteries-dry-wet.

Hyman, Mark, M.D. "The Impact of Mercury on Human Health and the Environment." *Alternative Therapies*, vol. 10, no. 6, 2004.

International Panel for Sustainable Resource Management and UNEP. "Metal Stocks in Society: Scientific Synthesis." 2010. http://www.unep.fr/shared/publications/pdf/DTIx1264xPA-Metal%20stocks%20in%20society.pdf.

Jha, A.R. *Next-Generation Batteries and Fuel Cells for Commercial, Military and Space Applications*. CRC Press, 2016.

Joerissen, Ludwig, et al. "Possible Use of Vanadium Redox Flow Batteries for Energy Storage in Small Grids and Stand-Alone Photovoltaic Systems." *Journal of Power Sources*, vol. 127, no. 1-2, 2004.

Reddy, Thomas. *Linden's Handbook of Batteries*. 4th ed. McGraw-Hill Professional, 2010.

Rydh, Carl Johan, and Magnus Karlström. "Life Cycle Inventory of Recycling Portable Nickel-Cadmium Batteries." *Resources, Conservation and Recycling*, vol. 34, 2002.

Thaler, Alexander, and Daniel Watzenig, eds. *Automotive Battery Technology*. Springer, 2014.

Trudell, Craig. "Tesla Surges as Morgan Stanley Says Electric Cars Will Gain Market Share." Bloomberg.com, 31 Mar. 2011, http://www.bloomberg.com/news/2011-03-31/tesla-surges-as-morgan-stanley-says-electric-cars-will-gain-1-.html.

Zhao, Guangjin. *Reuse and Recycling of Lithium-Ion Power Batteries*. Wiley, 2017.

Biodiesel

FIELDS OF STUDY
Agriculture; Aquaculture; Chemical Engineering

ABSTRACT
There are a number of feedstock products that can be used to make biodiesel, such as canola, sunflower seeds, palm oil, hemp, jatropha, castor, and algae. The highest potential yield (i.e., 47,500 liters per hectare, or 5,000 gal. per acre) of the feedstock is algae. Biodiesel is the only alternative fuel or fuel additive that is registered under Section 211(b) of the Clean Air Act of 1990 and it significantly reduces air pollutants such as carbon monoxide, hydrocarbons, and other particulate matter. Biodiesel fuels have increased in popularity in the United States since the passage of the Energy Policy Act of 1992 and the increased cost of diesel fuel.

KEY CONCEPTS
feedstock: a material used as an input to a production process

immiscible: the condition in which two different liquids do not dissolve each other and so do not mix, as for example oil and water

particulates: solid microparticles that are produced during combustion of fuels, such as carbon soot, metal oxides and fly ash

transesterification: a chemical reaction process in which the portion of an ester molecule derived from an alcohol is replaced by that of a different alcohol to produce a different ester molecule. In biodiesel production, the tri-alcohol 1,2,3-propanetriol (glycerol) is replaced in the tri-ester (or triglyceride) molecule by the monoalcohol methanol, resulting in the formation of three monoester molecules

viscous, viscosity: the "thickness" of a liquid, expressed in its resistance to flowing and the extent to which it adheres to a solid surface as it flows

BIODIESEL SOURCES AND REGULATIONS
Biodiesel is a renewable and biodegradable fuel that can be domestically produced and manufactured from vegetable oils, animal fat/tallow, and waste res-

taurant grease. There are a number of feedstock products that can be used to make biodiesel, such as canola, sunflower seeds, palm oil, hemp, jatropha, castor, and algae. The highest potential yield (i.e., 5,000 gal per acre) of the feedstock is algae. Biodiesel is the only alternative fuel or fuel additive that is registered under Section 211(b) of the Clean Air Act of 1990 and it significantly reduces air pollutants such as carbon monoxide, hydrocarbons, and other particulate matter. Biodiesel fuels have increased in popularity in the United States since the passage of the Energy Policy Act of 1992 and the increased cost of diesel fuel.

Biodiesel is a combination of components derived from sources typically including restaurant oils, vegetable oil, and/or animal fat, with a varied blend of petroleum diesel. Biodiesel in technical terms is a fuel comprised of mono-alkyl esters of a long-chain of fatty acids that meets the American Society of Testing and Materials (ASTM) D6751 standards. Biodiesel that meets the 11 standard specifications of the ASTM D64751 can be run in an unmodified diesel engine, and if registered with the Environmental Protection Agency (EPA), can be legally sold and distributed in the United States. The international equivalent to ASTM D64751 is EN 14214. Biodiesel should not be confused with straight vegetable oil (SVO) systems, because SVOs require significant modifications to a diesel automobile that include a heat tank and a tank for the regular diesel fuel.

PRODUCTION

The major ingredients for producing biodiesel include oil, methanol or ethanol, and either sodium hydroxide (lye) or potassium hydroxide. Biodiesel is produced by the chemical reaction of a fat with a simple alcohol (methanol or ethanol) in the process of transesterification. Transesterification is a fairly basic organic chemistry reaction where more complex fat and/or oil molecules known as triglycerides break down into simpler fat and/or oil molecules known as monoglycerides and replace one form of alcohol (glycerine) with another (methanol). The rate of the chemical reaction is sped up by the use of a catalyst. In addition to the production of biodiesel, the transesterification reaction also yields a small amount of glycerol. For example, 45.5 kg (100 lb) of oil plus 4.5 kg (10 lb) of methanol or ethanol will yield approximately 45.5 kg (100 lb) of biodiesel and 4.5 kg (10 lb) of glycerol. As a result of increased global biodiesel production, there is a surplus of glycerol with little or no market value. The color of biodiesel ranges from golden to dark brown with most feedstocks. However, hemp-based biodiesel results in a green-colored fuel.

According to the report "Global Biodiesel Market (2009-2014)" by Markets and Markets in 2009, global biodiesel production was 19.4 bullion liters (5.1 billion gal), worth $8.6 billion, which was a 17.9% increase over 2008 production levels. The biodiesel market is expected to grow to $12.6 billion by 2014. Globally, Germany is the largest producer of biodiesel, with 2.27 million tonnes (2.8 million tons). One limitation of biodiesel is the availability of feedstocks to meet the production demands. Moving away from food feedstocks could reduce the cost of biodiesel globally. Specifically, the development of algae and other existing nonfood feedstocks holds the best promise for meeting the increased production demand of biodiesel and reducing the cost. Presently, automobiles account for 70 percent of biodiesel demand. The International Air Transport Association

Biodiesel sample. (Shizhao via Wikimedia Commons)

(IATA) has estimated that second-generation biodiesel will account for 6 percent of aviation fuel by 2020.

NONFOOD FEEDSTOCKS FOR BIODIESEL/ BIOHEAT POLICY AND PRODUCTION

There have been a number of U.S. federal and state incentive programs seeking to promote biodiesel and alternative fuels, such as the Energy Policy Act of 1992 (Public Law 102-486), which includes provisions for purchasing alternative fuel vehicles at the state and federal level. Another incentive program is the Cellulosic Biofuel Producer Tax Credit. A number of incentive programs expired in December of 2011, causing concern among producers and investors. Some states such as Vermont, Massachusetts, and Maine, as well as several major cities, such as Baltimore, Boston, and New York have implemented biodiesel/bioheat programs. Most of these programs are mandating a certain level (generally a few percent) of biodiesel/bioheat. Even though the science of nonfood feedstock biodiesel is sound, the production levels have been below expectations. For example, the U.S. Department of Energy estimated that cellulosic output would produce 3.8 billion liters (1 billion gallons) of biodiesel by 2018, which is far below the congressional mandate of 26.6 billion liters (7 billion gallons). The reasons for the low production levels range from limited financial and governmental policy support to challenges of the production and usability of cellulosic biodiesel.

ADVANTAGES AND DRAWBACKS OF BIODIESEL

There are a number of advantages to using biodiesel. Biodiesel is more viscous than diesel and it is a better lubricant for reducing fuel injection wear. Biodiesel has almost no sulfur emissions, has a higher boiling point, and low vapor pressure. Biodiesel and water are immiscible (i.e., do not mix). Biodiesel vaporizes at a lower temperature (130°C, 266°F) than diesel (64°C, 147°F) and gasoline (minus 45°C, minus 52°F). There are also a number of different blends of biodiesel, listed as a B followed by the percentage of biodiesel it contains. For example, a B100 blend means that it is 100 percent biodiesel and has no diesel fuel content. A B2 blend would have 98 percent diesel fuel and 2 percent biodiesel, and a B20 blend would have 20 percent biodiesel. These blends enable biodiesel to be used in various diesel-operated automobiles and equipment, such as diesel-based generators, most marine vessels, and even diesel-operated all-terrain vehicles (ATV). It can also be used for heating oil with a B5 blend.

There are several drawbacks to using biodiesel, which include cold temperature issues (crystals and gelling), water in the fuel, and the breaking down of rubber gaskets and hoses in vehicles built before 1992. Cold temperatures present a couple of problems for biodiesel. In cold temperatures, biodiesel starts to form crystals, and as the temperature decreases, biodiesel begins to gel. Biodiesel gels at different temperatures depending upon the oil used for fuel as well as the mix of esters. For example, soybean-based biodiesel gels at about 0°C (32°F), canola at about minus 10°C (14 degrees Fahrenheit), and animal fat biodiesel gels at around 16°C (61°F). For comparison, 100 percent nonwinterized diesel fuel starts to cloud at minus 21°C (20°F), with a gel point of minus 32°C (0°F). There are additives that can be used in biodiesel to lower fuel crystallizing and slow down gelling. Another process involves insulating the biodiesel tank, as well as wrapping it in a heating coil. The manufacturing process can result in enough water mixing with biodiesel to lower combustion, clog the fuel system, and produce corrosion.

—*Andrew Hund*

Further Reading

Blewitt, Laura, and Mario Parker. "Think Crude's Cheap? Biodiesel's Going for Free in Some Places." *Bloomberg*. Bloomberg, 2 Feb. 2016. Web. 13 May 2016.

Huang, Haibo, Stephen Long, and Vijay Singh. "Techno-economic Analysis of Biodiesel and Ethanol Co-production from Lipid-producing Sugarcane." *Biofuels, Bioproducts & Biorefining*, vol. 10, 2016, pp. 299-315. Print.

Lamba, Bhawna Yadav, et al. *Preparation and Characterization of Biodiesel from Vegetable Oils*. Lambert Academic, 2011.

Pahl, Greg. *Biodiesel: Growing a New Energy Economy*. 2nd ed. Chelsea Green, 2008.

Purcella, Guy. *Do It Yourself Guide to Biodiesel: Your Alternative*. Ulysses, 2008.

"US Biodiesel and Renewable Diesel Imports Increase 61% in 2015." *US Energy Information Administration*. Department of Energy, 11 Apr. 2016. Web. 13 May 2016.

U.S. Department of Energy. *Biodiesel Fuel Handling and Use Guidelines for Users, Blenders, Distributors.* GPO, 2010.

Biomass Energy

FIELDS OF STUDY
Agriculture; Biotechnology; Forestry

ABSTRACT
Biomass energy is derived from materials of recent biological sources and is considered a renewable energy resource. Biomass resources can be classified into three categories based on the origins of the feedstock: natural vegetation, residues and wastes, and purposefully grown energy crops. Biomass can be used to generate heat directly through combustion, and can be harnessed to produce electricity, providing an alternative to oil and other kinds of energy fuels.

KEY CONCEPTS
acetogens: bacteria that acts to break down cellulosic material, producing acetate (or acetic acid) as a product of the digestion process

biofuels: fuels produced from organic matter using either biological processes such as fermentation or chemical processes such as transesterification; principal biofuels are ethanol from fermentation and biodiesel from transesterification

feedstock: materials employed as the principal component of a production process

methanogens: bacteria that consume acetate or acetic acid, producing methane and carbon dioxide as products of the digestion process

pyrolysis: decomposition of an organic material to its simpler components by the application of heat in an oxygen-free environment

transesterification: a chemical reaction process in which the portion of an ester molecule derived from an alcohol is replaced by that of a different alcohol to produce a different ester molecule. In biodiesel production, the tri-alcohol 1,2,3-propanetriol (glycerol) is replaced in the tri-ester (or triglyceride) molecule by the monoalcohol methanol, resulting in the formation of three monoester molecules

BIOMASS AND ENERGY PRODUCTION
Biomass energy describes a diverse category of alternative energy resources derived from recent biological sources, which can be used to produce various gaseous and liquid biofuels or to produce heat directly through several different processes. Biomass energy is derived from materials of recent biological sources and is considered a renewable energy resource. Unlike fossil fuels, which are energy resources that result from ancient biomasses that have undergone geologic transformation into such forms as petroleum or coal, biomass energy specifically refers to biological feedstock for energy, the material that will be used to generate energy, produced recently by organisms. Broadly, biomass resources can be classified into three categories based on the origins of the feedstock: natural vegetation, residues and wastes, and purposefully grown energy crops. Energy derived

Denis Papin, whose invention of the steam digester in 1679 inspired many to come with the idea of machinery that was not driven by human or animal muscular power. (Wikimedia Commons)

from feedstock harvest from natural vegetation, such as through the burning of lumber from forests, represents one of the oldest sources of energy used by humans. Residues are byproducts of agriculture and food production, forestry operations, and other manufacturing and production practices that process organic materials, while waste used for biomass comes from animal farm waste, sewage, and municipal solid waste from agricultural, residential, commercial, and industrial sectors. Particular grasses, such as switch grass or fescue, and aquatic crops, such as kelp and algae, can be purposefully grown to derive biomass energy. Biomass can be used to generate heat directly through combustion, and can be harnessed to produce electricity, providing an alternative to oil and other kinds of energy fuels.

THERMAL CONVERSION

Thermal conversion of biomass resources into viable fuel involves the application of increased heat in four processes: direct combustion, pyrolysis, gasification, and liquefaction. Industrial conversion of biomass to viable fuel may use a combination of these processes. Direct combustion of biomass materials such as wood can be used to produce heat, which is either used for heating purposes or for the generation of electrical energy.

Electrical power generation from direct combustion, which is harnessed indirectly through turbines, is a generally inefficient means of producing electricity, since there is an often significant amount of waste heat (heat energy that is not harnessed for use in some capacity). Another method of converting biomass

Bagasse is the remaining waste after sugar canes have been crushed to extract their juice. (Ji-Elle via Wikimedia Commons)

feedstock into a viable fuel is through the process of pyrolysis, the process of inducing decomposition in organic materials by applying high heat without oxygen. Depending on the specific origins of the biomass feedstock and the conditions of the pyrolysis, such as temperature or pressure, the resulting viable fuel can be in the form of gas, liquid, solid, or a combination of the three. An analogue to crude oil called "bio-oil," or pyrolytic oil, can be produced through pyrolysis. Pyrolytic gas that is produced through pyrolysis is often fed back into the pyrolysis system as fuel in order to reduce the need for additional fuel to perform the conversion process. Solid residue called "char" may result from pyrolysis of organic materials and is usable as fuel.

Through gasification, the biomass feedstock converts into syngas by undergoing exposure to high temperatures and specific quantities of hydrogen and oxygen. Before gasification, the biomass might be pretreated by undergoing pyrolysis. The syngas that results from gasification can be used as fuel. The last major form of thermal conversion that can be applied to biomass feedstock is the process of liquefaction. Liquefaction produces liquid fuels such as ethanol from biomass resources.

BIOLOGICAL CONVERSION

Biological conversion consists of the feedstock undergoing processes of anaerobic digestion, digestion without the presence of oxygen, by biological agents such as yeast and various bacteria. A product of anaerobic digestion by bacteria and other biological agents is biogas, a usable gaseous fuel that is rich in methane. A digester is a tank where biomass feedstock is added, fed to biological agents that consume the resource and subsequently produce biogas. Biomass feedstock must be degraded by three kinds of bacteria before becoming usable fuel. First, the feedstock is broken down by fermentative bacteria and acetogens, organisms that produce acetate. Acetate and other simpler intermediary organic compounds that are produced by these two groups of organisms are converted into carbon dioxide and methane that become biogas by methanogens. The combination of digested biomass feedstock and anaerobically digesting organisms housed in the digester form a slurry or broth. Liquid biofuels that are mixtures of various alcohols, such as ethanol, can be derived from biomass feedstock through anaerobic digestion as well through additional distillation processes. As is the case with thermal conversion, the resulting viable fuel from biological conversion depends on the specific components of the biomass feedstock as well as the conditions of biological conversion, such as the biological agents used, pH levels, and general temperature. High moisture feedstock such as waste, sewage, and crop residues are especially well suited to produce fuel through anaerobic digestion. Biological conversion of biomass feedstock is currently performed in countries such as China, India, South Korea, Brazil, and Thailand.

CHEMICAL CONVERSION

Biodiesel is a specific diesel fuel derived from vegetable oil or animal fat feedstock through chemical conversion. Oils containing triglyceride esters, such as vegetable seed oils, olive oil and sunflower oil, can be combined with a simple alcohol such as methanol or ethanol for conversion to simpler monoesters through a process called "transesterification." The reaction produces biodiesel and glycerol, a viscous liquid used in pharmaceutical and soap manufacturing. The biodiesel produced through transesterification can be used directly or can be mixed with regular diesel before use. In addition to its use as a heating fuel, biodiesel can be used to fuel vehicles such as automobiles, trains, and airplanes. While plant oils can be used directly in diesel engines, doing so will result in shortened operating life of the engine. Biodiesel is a biodegradable fuel alternative to petroleum and produces fewer carbon emissions. There are currently many biodiesel refineries worldwide in such countries as Brazil, Argentina, Cambodia, Malaysia, Indonesia, Canada and the United States.

The United States and Brazil produce approximately 85 percent of the world's ethanol and biodiesel. The top consumers of biofuels include Germany, Brazil, France, Italy, and the United States. The United States in particular has increased its consumption of biodiesel as federal and state legislation has provided incentives for biodiesel use. Oregon and Minnesota, for example, are implementing legislation that will require the incorporation of biofuel into gasoline sold in the state; gasoline will contain 10 percent bioethanol, and diesel fuel will contain 20 percent biodiesel. In many northeastern states, buildings and homes have increasingly reduced the burning of

Steven's Croft, opened in 2008—the biggest biomass plant in Scotland with an output of 44mw. (Chris Newman via Wikimedia Commons)

oil for heating purposes by consuming biofuels instead. In most states, biodiesel consumption has increased because of its usage in government and civic transportation and vehicle needs, such as waste management, postal service, and military vehicles, as well as buses for public transit and school systems.

ENVIRONMENTAL EFFECTS
There are both environmental benefits and concerns regarding the use of biomass as a source of energy. The principal benefit of biofuels is due to the fact that the carbon they contain is all derived from atmospheric carbon dioxide through photosynthesis. They are therefore incapable of releasing more carbon dioxide when burned than was taken in from the atmosphere in photosynthesis. As such, biofuels are "net zero carbon" fuels and cannot contribute to increasing carbon dioxide levels and climate change. Other benefits, particularly the use of residual, byproduct, and waste materials to produce energy, reduces the amount of environmental waste caused by agriculture, industrial manufacturing, and other human activities. Fertilizers can be produced from the solid end products of biomass energy conversion. With respect to environmental concerns associated with biomass energy, like any land-intensive human activity, purposefully grown energy crop production can adversely affect natural land and soil conditions in a

given geography. One way this occurs is through the removal of nutrients from the land by the harvesting of plant material that would otherwise decompose and recycle nutrients back into the soil. The removal of vegetative matter, particularly with livestock and agricultural machinery, can affect soil texture and compaction, which can impact the ability of soils to absorb water. Without vegetation, where plant roots help to keep soil in place, large tracts of land become susceptible to rapid erosion.

In addition to the effects on land, erosion deposits particulate matter into rivers, lakes, and other bodies of water, which can become a problem to wildlife. The resident soil fauna can also be adversely affected through heavy harvesting, as well as pesticide and herbicide usage. As with other forms of fuel, biomass energy can produce carbon dioxide emissions if burned to produce heat or electricity. Large carbon sinks, such as forests, act to store carbon, which reduces the amount of carbon dioxide in the atmosphere. In addition to increasing atmospheric carbon dioxide and other greenhouse gases in combustion, the large-scale harvesting of such carbon sinks reduces the capacity for and prevents future carbon sequestering, the process that removes carbon from the atmosphere and stores it into a carbon sink, which can contribute to the greenhouse effect.

—*Josef Nguyen*

Further Reading

Barber, James. "Biological Solar Energy." *Energy ... Beyond Oil*. Ed. Fraser Armstrong and Katherine Blundell, 137-55. Oxford UP, 2007.

Bilgili, Faik, and Ilhan Ozturk. "Biomass Energy and Economic Growth Nexus in G7 Countries: Evidence from Dynamic Panel Data." *Renewable and Sustainable Energy Reviews*, vol. 49, 2015, pp. 132-38.

"Biomass Explained." *US Energy Information Administration*. Department of Energy, 17 July 2015. Web. 13 May 2016.

De Jong, Wiebren, and J. Ruud van Ommen, eds. *Biomass as a Sustainable Energy Source for the Future: Fundamentals of Conversion Processes*. Wiley, 2015.

Kartha, Sivan, and Eric D. Larson. *Bioenergy Primer: Modernised Biomass Energy for Sustainable Development*. United Nations Development Programme, 2000.

Lee, Sunggyu. "Energy from Biomass Conversion." *Handbook of Alternative Fuel Technologies*, edited by Sunggyu Lee, James G. Speight, and Sudarshan K. Loyalka. CRC, 2007.

Nunes, Leonel Jorge Ribeiro, Joao Carlos de Oliveira Matias, and Joao Paulo da Silva Catalao. *Torrefaction of Biomass for Energy Applications from Fundamentals to Industrial Scale*. Academic Press, 2018.

Silveira, Semida. *Bioenergy—Realizing the Potential*. Elsevier, 2005.

Strezov, Vladimir, and Hossain M. Answar, eds. *Renewable Energy Systems from Biomass: Efficiency, Innovation and Sustainability*. CRC Press, 2019.

BOILERS

FIELDS OF STUDY
Mechanical Engineering; Pipefitting and Boiler Trades; Thermodynamics

ABSTRACT
A boiler is a device used to heat a fluid, generally water, in order to provide space heating or process heat directly, or to generate electricity. There are various types of boilers but the general principal of their operation is the same.

KEY CONCEPTS

combustion zone: the region within a boiler at which fuel combustion is taking place

fluidized bed: a structure in which an inert carrier such as fine sand or alumina is suspended in a column of heated gas being blown through it; used for the combustion of powdered fuels or to conduct gas-phase chemical reactions

higher heating value (HHV): describes fluid gases used for heat transfer at temperatures above their combustion temperature

lower heating value (LHV): describes a fluid used for heat transfer at temperatures below the combustion temperature or boiling point of the fluid

BOILERS AND BOILER FLUIDS
A boiler is a device used to heat a fluid, generally water, in order to provide space heating or process heat directly, or to generate electricity. There are various types of boilers that vary in size and technical specification, depending upon their application, but the general principal of their operation is the same. A heat

source is used, typically the combustion of a fuel, to heat a fluid to, or near, its boiling point. Water is most often used as the working medium because it is cheap, readily available, and has suitable properties in terms of thermal capacity and boiling point. The resulting hot fluid or vapor is then circulated as required in an industrial process, heating device, or turbine. Unless the system is open (in which case the hot water or steam exits, for example as hot tap water), the hot fluid is returned at a cooler temperature to the boiler and recycled. When the heat source is a furnace used for the combustion of fuel, hot exhaust gases may or may not be used to preheat the water, which is related to the concept of lower- and higher-heating values (LHV and HHV) for fuels, corresponding to the fraction of their energy content that is actually exploited.

The heating value or energy value of a fuel is the amount of heat released during the combustion of a specified amount of fuel, and is expressed in terms of the LHV or HHV. The HHV is defined as the return of the combustion products to the pre-combustion temperature, including any vapor produced. The HHV is the same as the thermodynamic heat of combustion and takes into account the latent heat of evaporation/vaporization of water in the combustion products, which is the regained energy relating to the change of phase from gas to liquid. The LHV, on the other hand, neglects the latent heat of vaporization and therefore assumes that water leaves the system in the gaseous state. The distinction between the two is that this second definition assumes that the combustion products are all returned to the reference tem-

Victorian Railways J class boiler and firebox stored at the Seymour Railway Heritage Centre, Victoria, Australia. (Marcus Wong via Wikimedia Commons)

perature and the heat content from the condensing vapor is not considered useful.

TYPES OF BOILERS

Most boilers can be categorized into either fire-tube boilers, in which the hot gases from combustion flow through the inside of the tubes and heat is transferred to surrounding fluid; or water-tube boilers, in which the setup is reversed. The reason for the tubes is to maximize the surface area between the combustion zone and the fluid for heat transfer. Fire-tube boilers generally have lower initial investment and are more efficient, but are limited in terms of capacity and pressure. Water-tube boilers, on the other hand, can be built to any capacity and pressure, and have higher efficiencies. Other classifications of boilers include packaged boilers, in which the complete plant is en-

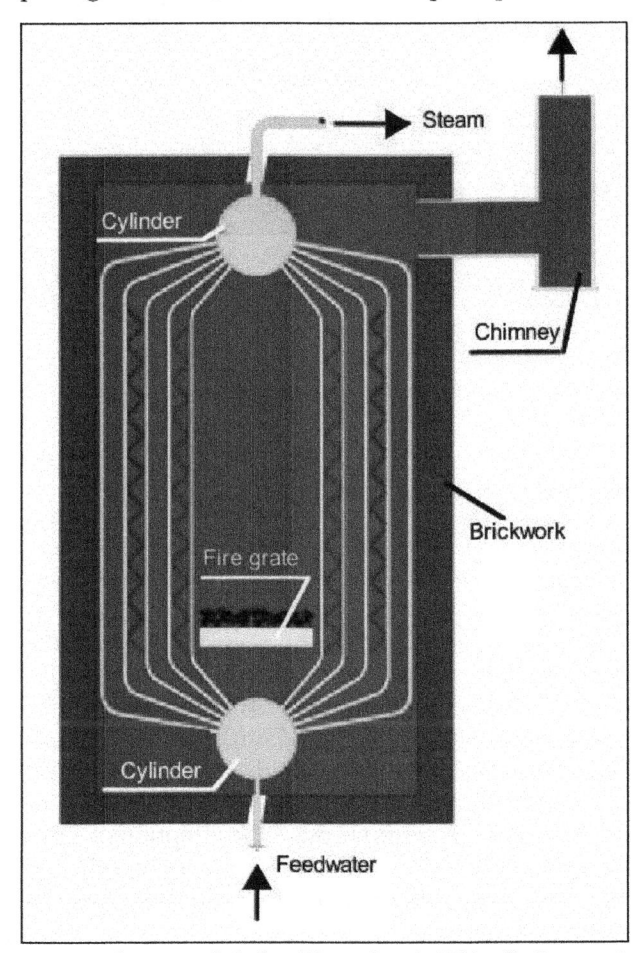

Diagram of a water-tube boiler. (Moonraker via Wikimedia Commons)

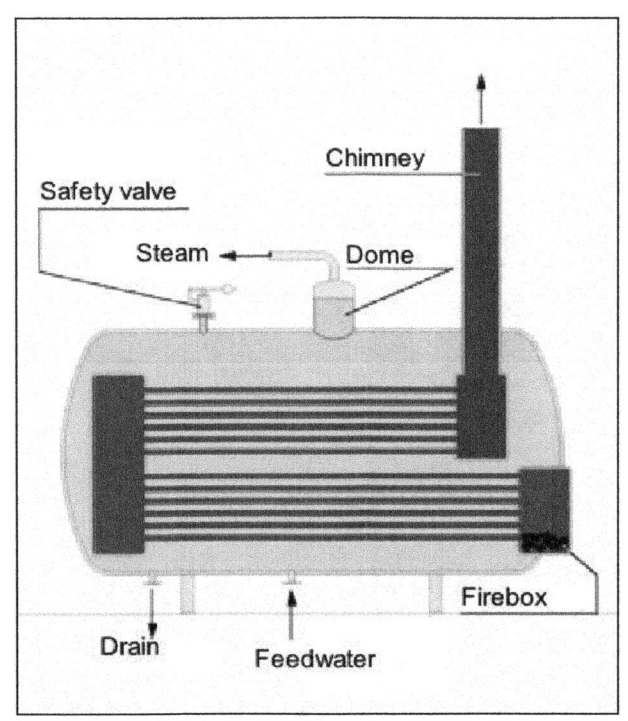

Diagram of a fire-tube boiler. (Old Moonraker via Wikimedia Commons)

closed in one modular unit, ready for connection onsite.

In addition, boilers can be categorized according to their method of combustion. A widespread type of boiler is the stoker boiler, in which coal is typically used as a fuel. One example is the traveling or chain grate stoker, in which a moving grate acts as a conveyor belt, carrying burning coal, while hot air is drawn through the grate to enable complete combustion. The coal should burn completely before passing off the end of the conveyor, which means that a precise control of conveyor speed, air flow, and coal feed rate is required. Another type of stoker boiler is the spreader stoker, in which coal is continuously fed in or "spread" into the furnace above a burning coal bed. Fine particles are burnt in suspension and larger particles fall to the bed to be burned. In this case, the combustion is very quick and this type of boiler responds very quickly to load fluctuations, hence it is common for some industrial applications.

Most modern coal-fired boilers use pulverized coal because the combustion is much more efficient than with larger lumps, due to the much larger surface area of a large number of fine particles compared to the

Industry boiler gas burner. (Vladdeep via iStock)

surface area of coal lumps. In fluidized bed combustion (FBC) boilers, the base of the combustion chamber is characterized by the fuel being in the fluid state, even though the fuels employed are solids such as coal and agricultural wastes. Small particles of sand are suspended in a stream of hot air, as when pulverized coal is fed into the furnace, which due to the high and uniform temperature of the bed combusts very quickly and efficiently. Generally, this approach is coupled with a conventional boiler at standard pressure, so that it is referred to as an atmospheric fluidized bed combustion (AFBC) boiler. If the combustion chamber is pressurized, however, which is similar to the FBC, the heat release rate can be increased and therefore the combustion efficiency increases. This is known as pressurized fluidized bed combustion (PFBC). Finally, some other variations include the waste heat boiler, which is used to recover waste heat, as from an industrial process, in order to provide another process with heat or to generate electricity.

For boilers raising steam for electricity generation, the main way of improving overall efficiency is to increase the turbine inlet temperature, which means higher temperatures and pressures. With inlet temperatures around 550°C (1022°F), they already operate at the current technological limit, in what is referred to as the supercritical region—in which the working fluid ceases to have distinct phases. This places extreme requirements on the materials for boiler construction and it is ultimately these materials, and their ability to withstand such high temperatures, that represent one of the main challenges for the future in this area. Through the adoption of alternative materials in steel alloys such as nickel, it is hoped that the efficiency of supercritical boilers can further be increased.

—*Russell McKenna*

Further Reading

Basu, Prabir, and Scott A. Fraser. *Circulating Fluidized Bed Boilers Design and Operations.* Butterworth-Heineman, 2013.

Ganapathy, V. *Steam Generators and Waste Heat Boilers for Process and Plant Engineers.* CRC Press, 2015.

Jawad, Maan H. *Stress in ASME Pressure Vessels, Boilers and Nuclear Components.* Wiley, 2018.

Kohan, A. *Boiler Operators Guide.* 4th ed. McGraw-Hill, 1997.

Rayaprolum, Kumar. *Boilers for Power and Process.* CRC Press, 2009.

Rogers, G., and Y. Mayhew. *Thermodynamics: Work and Heat Transfer.* 4th ed. Longmann Scientific, 1992.

University of Strathclyde, Glasgow. "Boiler Technology Adapted to Small Scale Heating." http://www.esru.strath.ac.uk/EandE/Web_sites/06-07/Biomass/HTML/boiler_technology.htm.

Breeder Reactors

FIELDS OF STUDY
Energy and Energy Resources; Nuclear Engineering; Physics

ABSTRACT
Properly managed and maintained breeder reactors produce useful energy, generate less waste than conventional light-water fission reactors, and reduce greenhouse gas emissions by reducing the use of fossil fuels.

KEY CONCEPTS
actinide: elements belonging to the actinide series in the periodic table, having atomic weights greater than that of actinium
breeder reactors: nuclear reactors designed to create more fuel than they use in the production of energy
fissionable: capable of undergoing nuclear fission, either spontaneously or by neutron bombardment
nuclear fission: the process in which an atomic nucleus is split into two smaller atomic nuclei

ADDING FUEL TO THE FUEL
Unlike normal nuclear fission reactors that use uranium-235 as their energy source, breeder reactors can make use of the much more abundant uranium-238 or thorium-232. Whereas a typical fission reactor uses only about 1 percent of the natural uranium-235 that starts its fuel cycle, a breeder reactor consumes a much larger percentage of the initial fissionable material. In addition, if the price of uranium exceeds a certain dollar value per kilogram, it becomes cost-efficient to reprocess the fuel so that almost all of the original fissionable material produces useful energy. Breeder reactors are designed to produce from 1 percent to more than 20 percent more fuel than they consume. The time required for a breeder reactor to generate enough material to fuel a second nuclear reactor is referred to as the doubling time; the typical doubling time targeted in power plant design is 10 years.

TYPES OF BREEDER REACTORS
Scientists have proposed two main types of breeder reactors: fast-breeder reactors and thermal breeder reactors. The fast-breeder reactor uses fast neutrons given off by fission reactions to breed more fuel from nonfissionable isotopes. The most common fast-breeding reaction produces fissionable plutonium-239 from nonfissionable uranium-238. The liquid metal fast-breeder reactor (LMFBR) breeds plutonium-239 and uses liquid metal, typically sodium, for cooling and for heat transfer to water to generate steam that turns a turbine to produce electricity. The thermal breeder reactor uses thorium-232 to produce fissile uranium-233 after neutron capture and beta decay.

Unlike fossil-fuel-powered plants, breeder reactors do not generate carbon dioxide or other greenhouse gas pollutants in their operation. Environmental concerns related to any type of breeder reactor include nuclear accidents that could emit radiation into the atmosphere and the difficulty of safely disposing of radioactive waste byproducts. In addition, for plutonium-based breeders a major concern is the possibility of the diversion of bred plutonium for nuclear weapon production. This concern can be addressed through the intermixing of actinide impurities with the plutonium; such impurities make little difference to reactor operation, but they make it extremely difficult for anyone to use the bred plutonium to manufacture a nuclear weapon.

BREEDERS AROUND THE WORLD
France has been the most prominent nation in the implementation of breeder reactors. The Superphénix

breeder reactor built on the Caspian Sea was used for power generation and desalination of seawater from 1985 to 1996. India has plans to build a large fleet of breeder reactors, including a large prototype LMFBR that uses a plutonium-uranium oxide mixture in the fuel rods. Russia, China, and Japan are also developing breeders. The Russian BN-600 fast-breeder reactor has experienced several sodium leaks and fires. If the sodium coolant in the central part of an LMFBR core were to overheat and bubble, core melting could accelerate in the event of an accident and release radioactive material into the environment. A strong reactor containment building and a reactor core designed so that the fuel rod bundles are interspersed within a depleted uranium blanket that surrounds the core could greatly decrease this effect.

—*Alvin K. Benson*

Further Reading
Bodansky, David. *Nuclear Energy: Principles, Practices, and Prospects*. 2nd ed. Springer, 2004.
Kessler, Günter. *Sustainable and Safe Nuclear Fission Energy: Technology and Safety of Fast and Thermal Breeder Reactors*. Springer, 2012.
Martin, Richard. *Super Fuel: Thorium-The Green Energy Source for the Future*. St. Martin's Press, 2012.
Mosey, David. *Reactor Accidents: Institutional Failure in the Nuclear Industry*. Nuclear Engineering International, 2006.
Muller, Richard A. *Physics for Future Presidents: The Science Behind the Headlines*. W.W. Norton, 2008.
Winterton, R.H.S. *Thermal Design of Nuclear Reactors*. Elsevier, 2014.

BUILDING ENVELOPE

FIELDS OF STUDY
Architecture; Construction Trades; Mechanical Engineering

ABSTRACT
The structures that separate the interior and exterior of a building are collectively known as the building envelope, sometimes also called the building enclosure. The building envelope protects internal spaces from excessive heat gain and moderates the need for mechanical cooling, as well as serving to prevent heat from escaping, thus reducing the indoor space heating required. Both function to limit the amount of energy required to maintain the space within at a proper temperature.

KEY CONCEPTS
dehumidification: the artificial removal of water vapor from the air inside of a building in order to maintain human comfort level and prevent damage due to condensation and mold growth
diurnal changes: quite literally, the difference between day and night
fenestration: the number, types, and sizes of the windows in a building
glazing: the type of glass used in window construction; single-glazed panes consist of a single sheet of glass; double-glazed panes consist of two sheets of glass separated by a layer of gas such as argon; triple-glazed panes consist of three sheets of glass with two gas-filled spaces separating them
humidification: the artificial addition of water vapor to the air inside of a building in order to maintain human comfort levels and avoid physical damage due to dehydration
seasonal changes: weather conditions according to the season of the year in a location

THE GREAT OUTDOORS AND INDOORS
The structures that separate the interior and exterior of a building are collectively known as the building envelope, sometimes also called the building enclosure. Essentially, this concept defines the distinction between interior, environmentally conditioned space, and the exterior environment. In hot climates, the building envelope protects internal spaces from excessive heat gain and moderates the need for mechanical cooling, whereas in cold climates it serves to prevent heat from escaping out of the building and helps to reduce the indoor space heating required. In both cases, the building envelope serves to moderate the amount of energy required to maintain the interior space at comfortable temperature levels. In addition to maintaining thermal comfort, a building envelope serves the functions of providing daylight to indoor spaces to limit the artificial lighting needed in the daytime, sound insulation of the interior from the external environment, and control of moisture or hu-

midity for maintaining indoor comfort. Structural integrity and aesthetics of a building also depend on the design and materials used in the construction of the building envelope.

The elements of a building envelope include walls, floors, roofs, doors, and windows directly in contact with the external atmosphere. Each of these elements is made up of assemblies of materials and individual components, which determine collectively the desired properties of the building envelop. Each assembly may consist of natural materials, such as stone and wood, or manufactured and highly processed materials, such as cement, steel, and specially engineered glass. For example, a building's fenestration (window structure) may be entirely of specialty glass such as a structural glass cladding with a supporting steel structure, or traditional windows that are made up of wood or metal framing with single-, double-, or triple-glazed panes. Thermal properties of each of these elements are aggregated to give a collective heat transfer coefficient, which indicates how well an assembly conducts heat, expressed as the U-value or U-factor of the window or cladding assembly. Sound insulation is also a collective property of the assembly, whereas the moisture-repelling capability depends primarily on the properties of the external surface and on the detail design of the assembly.

The main climatic (that is, diurnal and seasonal) changes that affect a building envelope are solar radiation, precipitation, and wind. These factors have an impact on both the external and the internal surfaces of the building envelope, although internal surfaces are affected mostly by seasonal changes because they are sheltered from daily climatic variations and are therefore subject to the moderating effects of artificial heating and cooling systems. Solar radiation in the daytime is responsible for heating the external surfaces of the building (facades), enabling them to radiate heat until arrival of cooler air in the evenings and night. The temperatures of the external facades thus fall to below the ambient temperature of the surrounding humid air, resulting in condensation on the facade sur-

The building envelope is all of the elements of the outer shell that maintain a dry, heated, or cooled indoor environment and facilitate its climate control, amidst such conditions such as this. (tab1962 via iStock)

faces. External building surfaces also directly receive moisture from the precipitation falling on them. Such water-soaked external surfaces may have the moisture driven to the internal layers of the envelope elements by the wind, causing interstitial condensation.

Repeated occurrences of external surface condensation and interstitial condensation, leading to internal dampness in the building envelope, can affect indoor air quality and hence internal comfort. Similarly, overheating of the building envelope will cause heat to be conducted to the internal surfaces of the envelope elements, causing overheating of the indoor spaces. Such weather-related effects can be managed by designing the elements of the building envelope to withstand the expected climatic effect on them. For example, adding insulation to the external walls prevents overheating of the internal spaces by preventing heat from flowing from the external heated surfaces to the interior. Similarly, designing an effective rain-screen cladding for the external walls can minimize the impact of precipitation on external walls, thus reducing moisture-related problems indoors. Mitigation of these effects requires energy-consuming appliances to provide heating, cooling, humidification, dehumidification and air exchange with the external atmosphere. The greater is the extent of mitigation required, therefore, the greater is the energy consumption profile of the building.

Design of the building envelope to mitigate extreme climatic effects is not a new concept and has been conducted by traditional societies through the ages. Even though the desired functions of the building envelope remain the same—to prevent overheating, dampness or moisture, and excessive cold in the internal spaces—the methods of achieving these goals through the design of the building envelope have significantly changed. The materials used in the construction of building envelope elements are now highly advanced and suited to today's high-rise and technically complex structures, in contrast to the natural and local materials used in traditional low-rise, passive architecture.

It is worthwhile to remember that achieving indoor comfort by mechanical means, such as air-conditioning and mechanical forced-air heating and ventilation, can overcome some weather-related problems. However, the energy used in mechanical systems is an expensive way of dealing with climate effects that can be completely eliminated, or at least reduced to a manageable level, through effective design of the building envelope. Passive environmental design of buildings not only helps to make healthier spaces to live and work but also reduces the carbon impact caused by the day-to-day operations of buildings.

—*Swati Ogale*

Further Reading

Dimitrov, Alexander V. *Energy Modeling and Computations in the Building Envelope.* CRC Press, 2016.

Durakovic, Benjamin. *PCM-Based Building Envelope Systems: Innovative Energy Solutions for Passive Design.* Springer, 2020.

Huang, Zujian. *Application of Bamboo in Building Envelope.* Springer, 2019.

Krarti, Moncef. *Weatherization and Energy Efficiency Improvement for Existing Homes.* CRC Press, 2012.

Lovell, Jenny. *Building Envelopes: An Integrated Approach.* Princeton Architectural Press, 2010.

National Institute of Building Sciences. "Building Envelope Design Guide and Whole Building Design Guide." http://www.wbdg.org/design/envelope.php.

C

Capacity (Electricity)

FIELDS OF STUDY
Electrical Engineering; Electrical Trades; Thermodynamics

ABSTRACT
Broadly defined, energy capacity is the amount of energy that can be stored or produced. The usage of the term capacity in this context commonly refers to the maximum electricity that can be supplied at a point in time. Energy capacities vary widely by region and technology. There are several important aspects of capacity, such as rated capacity, peak capacity, net capacity, declared net capacity, capacity factors, and battery capacity, which can be considered independently.

KEY CONCEPTS
gigawatt: 1,000,000,000 watts
kilowatt: 1,000 watts
kilowatt-hour: a continuous output or consumption capacity of 1,000 watts for a period of one hour
megawatt: 1,000,000 watts
terawatt: 1,000,000,000,000 watts
watt: named after Scottish engineer James Watt (1736-1819); one joule of energy per second, or the energy required to pass one ampere of electric current (or one coulomb of electrons) across a potential difference of one volt

PRODUCTION AND STORAGE OF ELECTRICAL ENERGY

Broadly defined, energy capacity is the amount of energy that can be stored or produced. The usage of the term capacity in this context commonly refers to the maximum electricity that can be supplied at a point in time. Energy capacities vary widely by region and technology. There are several important aspects of capacity, such as rated capacity, peak capacity, net capacity, declared net capacity, capacity factors, and battery capacity, which can be considered independently.

Rated capacity, also referred to as nameplate capacity, is the rate at which an energy generator can theoretically produce energy. The amount of energy produced can be derived by multiplying the capacity by the run time. For example, a generator with 1 kilowatt of capacity running for 10 hours will generate 10 kilowatt-hours of energy. In this example, the capacity is given in kilowatts, but different technologies rate their energy capacity in a variety of ways. For instance, the capacity may be recorded using pounds of steam per hour, millions of British thermal units per hour, boiler horsepower, or megawatts. Most electricity generators are standardized using the watt as the base unit. Since boilers produce heat energy instead of electricity, the capacity is rated at a specific temperature and pressure, and deviations from these conditions will result in an altered capacity. Similarly, other generators have rating conditions specific to the technology, such as fuel quality, wind speed, temperature, or incident light. It is important to understand what these conditions are, because most technologies will frequently produce less than their rated capacities in less than optimal conditions.

The capacity of an energy generator can vary widely. For instance, a single solar panel will have a nameplate capacity of a fraction of a kilowatt, whereas a large coal plant is more likely to have a capacity greater than 1 gigawatt. The coal power plant is therefore able to supply more than a million times as much energy per second than the solar panel. Currently, the largest-capacity power plant is the Three Gorges Dam in China, which has a cumulative capacity of 18.2 gigawatts.

Global electric capacity was 4,400 gigawatts in 2007. It is expected to grow by between 1 and 3 percent per year. This is consistent with the generation capacity growth between 1990 and 2007, which grew at 1.9 percent according to the U.S. Energy Information Administration. This suggests that global energy generation will increase by around 87 percent by 2035. The expected growth is primarily due to growth in the developing world. By 2035, the Organisation for Economic Co-operation and Development (OECD) countries are expected to increase from 46 to 61 percent of the energy supply. The largest capacities in 2007, by country, were the United States (995 GW), OECD Europe (836 gigawatts), China (716 gigawatts), Japan (279 gigawatts), Russia (221 gigawatts), and India (159 gigawatts). Until recently, when China took the lead, the United States had the largest electric capacity. In the United States, the total electric capacity consists primarily of natural gas, coal, nuclear, hydroelectric, and petroleum. Renewable energy sources make up only 5 percent of the total electricity-generating potential.

Historically, coal has provided the largest capacity of electricity to the world. In 2007, coal-fired power plants represented 42 percent of the global electricity supply. It is predicted to remain at this level in the near term, largely because of the two countries with the predicted highest electricity growth, China and India.

According to World Energy Outlook 2009, issued by the International Energy Agency, the global electric capacity in 2007 produced 19,756 terawatt-hours of electricity. This was expected to increase to 34,290 terawatt-hours by 2030. Of the 2007 total, 41.6 percent came from coal, 20.9 percent from natural gas, 15.6 percent from hydroelectric power, 13.8 percent from nuclear power, 5.7 percent from oil, and 2.4 percent from other sources.

TYPES OF POWER PLANTS

Some electricity sources have a higher installed capacity to produce the same amount of electricity as other types of energy. The difference between available power capacity and actual generated energy is due to the cost and availability of the different energy generation types, which must be managed according to the demand. Base-load power is cheaper and can be used almost continually. Peaking plants are used to supply more variable power needs, most often during the daytime, when industry, businesses, and air-conditioning are running. To be able to respond to variable electricity needs, peaking plants typically have a smaller capacity than base-load power plants or have stages of turbines that can be turned on and off to ratchet up and down the available capacity. These smaller capacity sizes allow for finer control over how much energy is produced.

Supply-side management requires having adequate peak power capacity, which may be easily switched on or off. A base-load power plant (such as a coal or nuclear plant) will have a much higher run time and therefore will produce more electricity per installed capacity in a year than will a peaking plant.

The advantages and disadvantages of small-capacity decentralized generation versus large-capacity centralized generators are prolific. Small-capacity plants can typically be located closer to the point of use, thereby lowering transmission and distribution losses and resulting in more widely dispersed environmental effects and less need for complex management patterns if most of their load can be used locally. However, they offer their own management challenges when incorporated into a larger energy grid. Large-capacity generators, which provide most electricity generation, often have difficulties in locating their facilities as a result of the "not in my backyard" (NIMBY) syndrome, and they are often located close to resources and farther from the urban environments where the electricity they generate is predominantly used. Large-capacity generators therefore suffer greater transmission and distribution losses. Large-capacity generators also have concentrated environmental effects.

Cumulative power plant capacities are required to meet the peak power loads. The declared net capacity is the net cumulative capacity minus the power needs of the power generator itself. This power will be available to meet the external demand.

CAPACITY FACTORS

A capacity factor is the actual energy produced by a generator over time, divided by the total possible energy that could be produced at maximum nameplate capacity. There are many reasons that the capacity factor is less than 1, including intermittency, decreases in efficiency, and downtime due to operations and maintenance. An example is a 500-megawatt base-load power plant with a capacity factor of 0.9; this means that 90 percent of the energy that could be produced by running this plant at 100 percent capacity is produced. In the course of a day, this power plant would produce:

$$500{,}000 \text{ kilowatts} \times 24 \text{ hours} \times 0.9 = 10{,}800{,}000 \text{ kilowatt-hours}$$

A more intermittent or peaking energy generator will have a much lower capacity factor. The same capacity of photovoltaic energy with a capacity factor of 0.2 will generate only:

$$500{,}000 \text{ kilowatts} \times 24 \text{ hours} \times 0.2 = 2{,}400{,}000 \text{ kilowatt-hours}$$

The capacity factor is the simplest way to calculate how much energy will be produced from a generator because it describes the aggregate effects of decreased efficiency, intermittency, and downtime necessary for operations and maintenance.

Battery capacity is the amount of stored charge. It is described differently from electricity generators. It is rated in ampere-hours (or amp-hours), at specific conditions. The amp-hour refers to the number of hours at which the battery can supply a rated current. Increasing battery capacity is an important field for the future. One barrier to clean energy deployment is the lack of adequate energy-storage mechanisms. Since many renewable energy technologies generate energy only under specific conditions—for example, when it is windy or sunny—storing the energy for later use is critical.

—*Madeline Tyson*

Further Reading

Casazza, Jack, and Frank Delea. *Understanding Electric Power Systems: An Overview of Technology, the Marketplace, and Government Regulations.* 2nd ed. IEEE Press/Wiley, 2010.

El-Sharkawi, Mohamed A. *Electric Energy: An Introduction.* 3rd ed. CRC Press, 2013.

Embassy of the People's Republic of China in the United States of America. "The Three-Gorges Project: A Brief Introduction." http://www.china-embassy.org/eng/zt/sxgc/t36502.htm.

International Energy Agency. *World Energy Outlook 2009.* International Energy Agency, 2009.

U.S. Energy Information Administration. Electricity Explained: Your Guide to Understanding Electricity. http://www.eia.doe.gov/energyexplained.

Zycher, Benjamin. *Renewable Electricity Generation: Economic Analysis and Outlook.* AEI Press, 2011.

Cellulosic Ethanol

FIELDS OF STUDY
Bioengineering; Fermentation Chemistry; Microbiology

ABSTRACT
Cellulosic ethanol, based on the hydrolysis of woody or fibrous biomass, is the second generation of biofuels, attractive because it uses the vegetable matter trash but not the food from crops such as corn or sugarcane and is more sustainable because it can be grown on degraded land.

KEY CONCEPTS

cellulose: a principal secondary product of photosynthesis, formed by the head-to-tail concatenation of glucose molecules that are the primary product of photosynthesis

first-generation ethanol: ethanol produced from food crop materials that would normally be used as food

hemicellulose: a secondary product of photosynthesis, formed by the incorporation of cellulose molecules and lignin molecules into the rigid molecular structure of hemicellulose

nonfood residues: materials such as woody stems, inedible leaves, corn cobs, etc., that are separated from food sources and discarded

second-generation ethanol: ethanol produced from food crop refuse and crops that are not used for food

ETHANOL FROM VEGETABLE WASTE PRODUCTS

Cellulosic ethanol is produced through the hydrolysis of woody or fibrous biomass. Feedstock for cellulosic biofuels is generally divided into two categories: dedicated energy crops (such as willow, switchgrass, and eucalyptus) and agricultural residues (the nonfood parts of crops, such as stems, husks, leaves, and forestry waste). As such, the ethanol itself is considered a carbon-neutral fuel, since the carbon it is produced from is taken directly from the atmosphere and only that amount can be returned to the atmosphere when the ethanol is burned as fuel.

Cellulosic ethanol is often referred to as a second-generation biofuel. Second-generation biofuels are more environmentally sustainable than first-generation biofuels because they can make use of abandoned and degraded land. First-generation biofuels are produced primarily from food crops (including sugarcane and vegetables such as corn), and there have been ongoing concerns about the displacement of food crops and competition over scarce resources, particularly the land and water used for the production of those food crops. Second-generation biofuels make use of residues from agricultural and forestry production and can thereby assist in creating additional revenue streams in rural areas without requiring increased use of arable land. Constraints occur in areas where rural populations are dependent on agricultural residues for animal fodder or domestic fuel.

PRODUCTION PROCESSES

Cellulosic materials, such as straw and wood, are much cheaper and often more readily available than starch- and sugar-based feedstocks. However, the fermentation of cellulosic raw material to ethanol requires complex and costly processes. Cellulosic material contains cellulose and hemicellulose, bound together by lignin, which makes the fermentation of ethanol difficult, because the lignin protects the cellulose from breakdown by enzymes; moreover, when the enzymes are able to reach the cellulose, they are hindered by the crystalline structure of the molecules. The process therefore involves a high-temperature pretreatment, followed by treatment with enzymes and a two-stage fermentation process.

Cellulosic ethanol is currently not economically cost-competitive with gasoline (a fossil fuel) or corn-based ethanol. More research is required to reduce the costs and improve the efficiency of commercial enzymes, including onsite production of enzymes as part of the ethanol production process. Currently, no large-scale cellulosic ethanol production facilities are operating or under construction. However, technological advancements in the processing could soon make cellulosic ethanol a more economically viable and sustainable option for expanded ethanol production.

GROWTH

Although second-generation biofuels are not yet commercially produced on a large scale, several pilot plants have been established. In 2007, the U.S. Department of Energy (DOE) invested $385 million for six biorefinery projects, which were expected eventually to produce more than 130 million gallons of cellulosic ethanol per year. Several demonstration plants for production on a commercial scale are under development in Denmark, France, and Europe. Projections published in the *World Energy Outlook 2009*'s "450 Scenario" estimated that biofuels could provide 9 percent of total transport fuel demand by 2030, which could grow to 26 percent by 2050. In 2050, second-generation biofuels are likely to make up 90 percent of all biofuels, with more than half of the production estimates to occur in countries that are not members of the Organisation for Economic Co-operation and Development (OECD), mainly China and India.

Bioreactor for cellulosic ethanol research. (Wikimedia Commons)

IMPACTS

Second-generation biofuels have potential to reduce impact on water quality significantly over their first-generation counterparts—largely because second-generation crops generally are more tolerant of dry weather and drought and can more easily be cultivated on degraded land. However, they may require more of this marginal land to achieve necessary levels of productivity. Furthermore, much like first-generation crops, advanced biofuel crops need to be sustainably cultivated to reduce soil erosion and depletion of natural resources.

More research is required to understand the impacts of cellulosic ethanol production on the hydrologic cycle. In some instances, the complete removal of agricultural residues for biofuel production may lead to erosion and nutrient runoff to water supplies, as well as soil impoverishment and concomitant requirements for additional nutrient application for future crops. As cellulosic bioenergy crops become more viable, policy instruments will be required, based on local conditions, to ensure sustainable production and reduce nonpoint pollution.

—*Aileen Anderson*

Further Reading

Buckeridge, Marcos Silveira, and Gustavo Henrique Goldman. *Routes to Cellulosic Ethanol*. Springer, 2011.

Eisentraut, Anselm. *Sustainable Production of Second-Generation Biofuels: Potential and Perspectives in Major Economies and Developing Countries*. International Energy Agency, 2010.

Elcock, D. *Baseline and Projected Water Demand Data for Energy and Competing Water Use Sectors*. U.S. Department of Energy, National Energy Technology Library, and Argonne National Laboratory, 2008.

Goettemoeller, Jeffery, and Adrian Goettemoeller. *Sustainable Ethanol: Biofuels, Biorefineries, Cellulosic Biomass, Flex-Fuel Vehicles, and Sustainable Farming for Energy Independence*. Prairie Oak, 2007.

International Science Panel on Renewable Energies. *Research and Development on Renewable Energies: A Preliminary Global Report on Biomass*. International Science Panel on Renewable Energies, 2009.

Intratec US. *Cellulosic Ethanol from Switchgrass Report Ethanol E81A Basic Cost Analysis*. Intratec, 2019.

Intratec US. *Cellulosic Ethanol from Wood Chips Report Ethanol E52A Basic Cost Analysis*. Intratec, 2019.

United Nations Environment Programme. *The Bioenergy and Water Nexus*. Oeko-Institution and IEA Bioenergy Task 43, 28 Nov. 2011.

Chernobyl

FIELDS OF STUDY
Environmental Engineering; Mechanical Engineer; Nuclear Technology

ABSTRACT
Chernobyl refers to both the nuclear power station and the local town in northern Ukraine, where a disastrous accident struck the reactor's Unit 4 on April 26, 1986. The Chernobyl nuclear accident, considered the worst in the history of nuclear technology, had far-reaching ramifications for the regulation and institutional protocols of the nuclear industry and radically altered the surrounding physical environment.

KEY CONCEPTS
involuntary park: an uninhabited wilderness area that was originally created and inhabited by humans, such as a "ghost town" that has been abandoned and reverted to wilderness

reactor core: the central component of a nuclear reactor, containing the fuel rods of refined uranium; the reactor core is constantly immersed in "heavy water" to absorb carry away heat and moderate the rate of nuclear fission within the fuel rods

sievert (Sv): the standard unit of radiation exposure that has replaced the rem, with 1 Sv being the equivalent of 100 rems; the sievert relates radiation dose and biological effectiveness

A DISASTER OF NUCLEAR PROPORTIONS
Chernobyl refers to both the nuclear power station and the local town in northern Ukraine, where a disastrous accident struck the reactor's Unit 4 on April 26, 1986. Experimentation in the reactor core caused a breach of its controls. The ensuing explosion blew off the containment lid and sent large amounts of radioactive material into the atmosphere, prompting immediate evacuation of the nearby worker settlement, Pripyat (Pryp'yat, ????'???), and adjacent highly contaminated areas. The exodus amounted to

116,000 evacuees in 1986 and another 230,000 in later years. Radioactive material rapidly spread to neighboring areas of Ukraine, Belarus, and the Russian Federation, each a part of the Soviet Union at the time of the disaster. In total, 50 tonnes of radioactive particles rained onto the immediate environs, northern and western Europe, and eventually, in trace amounts, North America.

NUCLEAR POLLUTION AND INVOLUNTARY PARKS

Human-made disasters and environmental pollution rarely have positive consequences for biological life-forms. At Chernobyl, however, the evacuation of a large resident population, along with the suspension of agricultural and other disturbance-inducing activities, helped create an unintended oasis for the diverse organisms of this region called Polesia (considered to be the origin of the eastern Slavs). The 30-kilometer (18.5-mile) exclusion zone experienced such tremendous transformation that it serves as the prime example of an involuntary park, defined as areas of high-quality wilderness preserved as an unintentional aftereffect of political, economic, or technological activities. Similar circumstances have formed around other nuclear facilities, such as the Hanford Nuclear Reservation in the United States.

The consequences of the nuclear reactor explosion were initially detrimental to the local biota. An entire forest, named the Red Forest, morphed from a mixed-conifer forest to a sea of dead, red-brown leaves and needles due to the radioactive fallout that billowed from the reactor. The extremely high doses of radiation in this ten-square-kilometer (four-square-mile) area prompted officials to bulldoze and bury this entire section of forest and soil under four feet of solidified sand. The effects of the radiation continue to manifest themselves in deformities, such

A typical scene from the evacuated town of Pripyat after Chernobyl. Radiation is expected to last 900 years. (Omer Serkin Bakir via iStock)

Chernobyl aerial view. (DeSid via iStock)

as the stunted conifers in this area. However, animals have fared well, with only minor reports of deformities such as partial albinism and a decline in barn swallows (which were probably more susceptible because of their migration patterns and pigment biochemistry) and some complications for ground-dwelling insects, whose habitat suffered greater exposure to radiation.

These complications remain limited compared to the broad biodiversity and habitat benefits provided by the exclusion zone and associated abandoned areas. At least 66 mammalian species inhabit the zone, while the diversity of flowers and other plant life rivals protected areas nearby. Rare, apex predators such as the Eurasian lynx now prowl the zone to prey on abundant roe deer. Wolves thrive on an ample supply of wild boar, while the Eurasian brown bear is thought at least to frequent the zone. Even endangered species such as Przewalski's horse, a wild horse nearly hunted to extinction in the mid-twentieth century, and vulnerable species such as the European wisent (or bison), now roam the forest and open areas of Chernobyl's exclusion zone.

CONSEQUENCES FOR THE HUMAN POPULATION

Across the surrounding exposure area-358,400 square kilometers (140,000 square miles) within three countries—approximately 4.9 million people were exposed to radiation. The evacuees, residents and emergency workers who were most directly exposed, received elevated doses of radiation amounting to exposures of 17 to 380 millisieverts. (Authorities consider doses above 100 millisieverts per year to impart a significantly higher risk of developing cancer.) More indirect pathways for exposure of the longer-lived radioactive elements have persisted. The majority of the surrounding population received, and continue to receive, doses of radiation roughly equivalent to background levels (1 millisievert per year). Wild game, forest products, and some agricultural products continue to show elevated radiation levels, due to the cycling of radioactive fallout by geochemical and biological processes.

Outside the local region, reports demonstrate connections to contaminated reindeer meat in northern Russia and Scandinavia as well as bioaccumulation of radioactive cesium in fish from lakes in Germany and Scandinavia. The International Atomic Energy Agency (IAEA) has advised continued monitoring and use of countermeasures to minimize these exposure pathways.

Public health continues to be of concern. The immediate disaster caused the death of two persons within hours and another 28 within four months from acute radiation sickness. The only link to longer-term radiation-induced disease thus far uncovered is the occurrence of thyroid cancer in 4,000 children, attributed to radioactive iodine released into the environment by the explosion. Some leafy vegetables produced in the area and milk from locally pastured cows were the primary conduits for the radioactive iodine to reach the population. Radioactive iodine will continue to affect the children in the area, who may be particularly susceptible given iodine deficiencies in the region. Although generally considered treatable, thyroid cancer resulted in nine deaths among those affected, and another 4,000 deaths can be expected among the hundreds of thousands directly exposed to the reactor explosion.

Exposure for many of the emergency workers occurred as they constructed a temporary sarcophagus in the six months following the disaster. International concern mounted regarding the integrity of the hastily constructed iron and concrete sarcophagus, threatened by moisture damage and high levels of radioactivity. To prevent a second release of nuclear material, the United States, the European Union, and Ukraine formulated the Shelter Implementation Plan to manage and fund the construction of a new shelter. Construction of the new shelter employed an innovative sliding arch method derived from a design competition in 1992, whereby large steel-armature arches are built in sequence and slid on rails into place over the original sarcophagus, minimizing the exposure of construction workers to radiation. Clad in protective panels, these sliding arches form a barrel vault over the entire reactor structure. Completion of the design was estimated to be completed in 2013, but was not actually completed until July, 2019. The arches are specified to last a minimum of 100 years and cost more than $1 billion.

The long-term stabilization and confinement plans remain less certain for the areas outside the nuclear power plant. Although cleanup crews have dug various trench and landfill-type deposits to bury large quantities of radioactive waste 0.3 to 9.3 miles (0.5 to 15 kilometers) from the plant, the emergency workers did not apply the necessary engineering barriers or study the hydrologic conditions around these deposits. Furthermore, only half of these deposits have been examined sufficiently to know their locations, quantities, and status. The IAEA continues to call for a properly integrated waste management plan for these outlying radioactive zones.

Because of the remaining hazards and uncertainties, an area called the exclusion zone surrounds the site out to 30 kilometers (18 miles). Access inside this zone is restricted primarily to workers, scientists, and land managers, who help police and monitor the zone. Paradoxically, the lack of human habitation, coupled with strict controls over access to this zone, has resulted in a rebound in the natural fauna and flora of the Polesian region. A web of ongoing studies continue to evaluate the ability of organisms and humans to survive radioactive conditions, types of countermeasures that should be applied to minimize exposure in nuclear accidents, and safeguards to prevent the disastrous consequences of nuclear breaches similar to Chernobyl.

—*Casey Lance Brown*

Further Reading

Baker, R.J., and R.K. Chesser. "The Chernobyl Nuclear Disaster and Subsequent Creation of a Wildlife Preserve." *Environmental Toxicology and Chemistry*, vol. 19, no. 5, 2000.

Higginbottom, Adam. *Midnight in Chernobyl: The Untold Story of the World's Greatest Nuclear Disaster*. Simon & Schuster, 2020.

International Atomic Energy Agency. "Thyroid Cancer Effects in Children." Aug. 2005. http://www.iaea.org/NewsCenter/Features/Chernobyl-15/thyroid.shtml.

Mara, Wil. *The Chernobyl Disaster: Legacy and Impact on the Future of Nuclear Energy*. Marshall Cavendish Corporation, 2011.

Mycio, Mary. *Wormwood Forest: A Natural History of Chernobyl*. Joseph Henry Press, 2005.

United Nations Chernobyl Forum Expert Group Environment. *Environmental Consequences of the Chernobyl Accident and Their Remediation*. Aug. 2005. http://www.iaea.org/NewsCenter/Focus/Chernobyl/pdfs/ege_report.pdf.

U.S. Nuclear Regulatory Commission. *Backgrounder on Chernobyl Nuclear Power Plant Accident*. April 2009. http://www.nrc.gov/reading-rm/doc-collections/fact-sheets/chernobyl-bg.html.

Yablokov, Alexey V., Vassily B. Nesterenko, and Alexey V. Nesterenko. *Chernobyl: Consequences of the Catastrophe for People and the Environment*. Blackwell, 2009.

CHINA SYNDROME (NUCLEAR MELTDOWN)

FIELDS OF STUDY
Electromechanical Engineering; Nuclear Chemistry; Thermodynamics

ABSTRACT
Nuclear scientists have used the term "China Syndrome" since the 1960s to describe the hypothetical consequences of a nuclear plant meltdown. The 1979 movie The China Syndrome *and the Three Mile Island accident greatly influenced American public opinion about the nuclear industry.*

KEY CONCEPTS
core material: the array of rods of refined uranium, or other fissionable elements, that are the source of energy in a nuclear reactor

fissionable materials: radioactive isotopes that undergo spontaneous nuclear fission, releasing various kinds of radiation and changing their elemental identity in the process

meltdown: the runaway release of energy from radioactive materials resulting in the production of heat well beyond that required to melt those materials

nuclear fission: the spontaneous splitting apart of the atomic nucleus of atoms of a radioactive isotope, resulting in the release of various kinds of radiation and changing the identity of the isotope to that of a lower element; for example, uranium is transmuted to lead through a nuclear fission process

TERM RECOGNITION
Although the term now refers primarily to the Earth-spanning influence of the Chinese economy, according to the 11th edition of *Merriam-Webster's Collegiate Dictionary*, the term "China Syndrome" refers to the geographically erroneous notion that the molten reactor contents of a nuclear facility in the United States could theoretically penetrate through Earth and reach China. As nuclear scientist Richard Muller noted, some scholars argued that, theoretically, the heat might be enough to let the fuel melt through the planet; some scholars argued that the meltdown would not move farther when it reached the center of the planet and gravity became negative; and other scholars believed that the fuel might only be able to melt down a few dozen meters.

The dictionary defines the term as it was first used in 1970. Historian of the nuclear age J. Samuel Walker believes that the term has been used by nuclear scientists since the 1960s. Historian of physics Spencer Weart has also noted that, soon after 1965, confidential warnings about the China Syndrome had been made by government safety researchers but gained little attention. The issue was later revisited by a group of students, scientists, and others who formed the Union of Concerned Scientists, based at the University of Boston. They found that the government spent a great deal of money studying how to develop nuclear facilities but little in studying how to manage possible emergencies. They publicized a report on this issue in March 1971. After that, the group received materials from unknown scientists about the risks of civil nuclear facilities; subsequently, it now focuses its agenda on the persistent issue of nuclear safety.

On December 12, 1971, Ralph Lapp—a Manhattan Project physicist who, with 76 other atomic scientists, in July 1945 had signed a petition urging President Harry Truman not to use the atomic bomb—used the term China Syndrome in his article "Thoughts on Nuclear Plumbing," which was published in the *New York Times* that day. Several sources then attributed to Lapp the coining of the term, and although that might not be true, he is among the first people who impressed the public with the concept, as well as the consequences of nuclear radiation.

HOLLYWOOD GETS INVOLVED
The popularization of the term is most likely attributable to the Burton Wohl's novel *The China Syndrome*, followed by the movie, both released in 1979. The movie, starring Jane Fonda, Jack Lemmon, and Mi-

The "China syndrome" refers to a hypothetical nuclear meltdown scenario so dire it would figuratively burn through the crust and body of the Earth until reaching the other side. (curraheeshutter via iStock)

chael Douglas, tells the story of an accident at a nuclear power plant: A television reporter (Fonda) and her cameraman (Douglas), while visiting a nuclear plant, witness it going through an emergency shutdown. The plant's shift supervisor, Jack Godell (Lemmon), gathers evidence that the plant is in severe danger of nuclear meltdown. Under threat from the company responsible for the faulty construction that caused the problem, Godell returns to the plant to find it at full power. Fearing nuclear meltdown, he forces everyone at gunpoint to leave the plant and insists on access to the media in order to forewarn the public of an imminent disaster, but the plant's managers bring law enforcement officers to the site, who shoot him before he can expose the situation. At the end of the film, a live television report on the plant is abruptly cut, and no one in the public was warned.

A REALISTIC REPRESENTATION OF THE NUCLEAR ENERGY INDUSTRY?

The movie was reviewed by John Dowling, a spokesperson for the nuclear industry on some issues. He adopted an antinuclear perspective, exaggerating the consequences of a possible nuclear plant accident and baselessly proposing a conspiracy theory that nuclear power plant managers were serving profit machines and were unconcerned about public safety. Dissatisfied with misleading statements, Dowling suggested that the movie should criticize the nuclear weapon competitions rather than civil nuclear development. Even in a popular magazine, movie reviewer David Denby could not buy the movie's portrayal of the nuclear plant management. A simple reason was that, only three days before the film was released, the Nuclear Regulatory Commission closed five nuclear power plants as unsafe in case of an earthquake. The decisions were made based on reports provided by the nuclear plants' managers themselves.

LIFE IMITATES ART
The Three Mile Island accident happened on March 28, 1979, 12 days after the movie's release, in a nuclear plant close to Harrisburg, Pennsylvania. Due to mechanical failure as well as human error, the nuclear reactor lost its coolant water and underwent a partial meltdown. When the shaken Pennsylvania governor announced the accident on television, the movie was in people's minds, although it had been filmed long before the incident. Subsequent investigation found that the plant officials did delay reporting the accident to the government and industry experts had downplayed the seriousness of the consequences of the accident. Investigators also found that an automatic emergency system installed in the nuclear facility successfully prevented a major disaster, such as the one that occurred in 1986 at Chernobyl in Ukraine, then part of the Soviet Union. The Chernobyl reactor underwent a complete meltdown, killing 31 immediately and unknown numbers in the following years as a result of radiation released into the environment. Tellingly, there is essentially no evidence that molten core material melted its way into the planet, although the nature of the event, the horrendous levels of radiation still present at the site, and the entombment method that was used to contain the remnant structures have completely prevented any investigation of such an event.

Although no one was hurt at Three Mile Island, the remarkable coincidence of its occurrence at almost the same time the movie was released amplified the influence of the China Syndrome concept. Public perception of the consequences of a nuclear meltdown soon turned American public opinion against the use of nuclear power. No new nuclear plant plan has been approved since then, although some environmentalists who have been against nuclear power in the past have now begun to advocate it as a clean energy resource that could mitigate the global climate change they see as a much more palpable threat to the planet and to human existence.

—*C.S. Eigenmann*

Further Reading
Denby, David. "More Heat Than Light." *New York Magazine*, 2 Apr. 1979.
Dowling, John. "The China Syndrome." *Bulletin of the Atomic Scientists*, vol. 35, no. 6, 1979.
Jorgensen, Timothy. J. *Strange Glow: The Story of Radiation*. Princeton UP, 2016.
Muller, R. *Physics for Future Presidents: The Science Behind the Headlines*. Norton, 2006.
Tomain, Joseph P. "Nuclear Futures." *Duke Environmental Law and Policy Forum*, vol. 15, no. 221, 2005.
Walker, J. Samuel. *Three Mile Island: A Nuclear Crisis in Historic Perspective*. U of California P, 2004.
Weart, Spencer R. *Nuclear Fear: A History of Images*. Harvard UP, 1988.

CLAUSIUS, RUDOLF

FIELDS OF STUDY
Physics; Physical Chemistry; Thermodynamics

ABSTRACT
A major figure in the fields of thermodynamics and kinetic theory, Rudolf Clausius is credited with improving the first law of thermodynamics that deals with the conservation of energy as developed by James Joule, and with articulating the second law of thermodynamics that concerns the transference of heat.

KEY CONCEPTS
caloric theory: an antiquated, and erroneous, view that heat was due to a "fluid" called "caloric" that existed within physical systems, such that temperature changes of the system were due to changes in the caloric content of the system
entropy: essentially, the energy of a system due to the temperature-dependent degree of randomness or chaos within a system
thermodynamics: the study of the nature and movement of heat

BACKGROUND
A German mathematical physicist, Rudolf Julius Emanuel Clausius was born in Koslin in the Prussian province of Pomerania on January 2, 1822. He is considered one of the 10 founders of thermodynamics, a field that developed over the course of a century of

scientific successes and failures that dealt with improving knowledge of energy, entropy, and absolute temperature. Clausius's father was both a minister and an educator, and Clausius began his education at his father's school. He entered the University of Berlin in 1840, majoring in history. However, he soon found that his true interests lay in science, and he graduated in 1844 with a degree in mathematics and physics. He received his PhD from Halle University in 1847 after completing his dissertation, *On Those Atmospheric Particles That Reflect Light*. Dealing with optical effects in the universe, the dissertation attempted to answer questions such as why the sky is blue and why shades of red are found in sunrises and sunsets.

During the Franco-Prussian War (1870-71), Clausius organized an ambulance corps and received an Iron Cross for his contributions to his country. In 1875, Clausius's wife, Adelheid Rimpham Clausius, died while giving birth to their sixth child; he then turned much of his attention to raising his family. He married Sophie Sack in 1866, and fathered a seventh child. Clausius's lecture "On the Energy Supplies of Nature and the Utilization of Them for the Benefit of Mankind" was published in book form in 1885. Clausius died on August 25, 1888, in Bonn.

SCIENCE

In 1850, Clausius became a professor at the Royal Artillery and Engineering School. That same year, he presented his first important paper on thermodynamics, analyzing the movement of energy during transformation processes. In that paper, Clausius was able to reconcile the pioneering work of Carnot and Émile Clapeyron (1799-1864), who had been heavily influenced by Carnot, with subsequent findings in the field of thermodynamics, which rejected the caloric theory that stipulated that heat was indestructible and could not be converted to work by any device. Clausius maintained that, rather than being indestructible, heat could be affected by both transmission and conversion. In 1855, Clausius transferred to Zurich Polytechnicum in Switzerland, but returned to Germany in 1867 to teach at Wurzburg University. He moved to Bonn University two years later and stayed there for the remainder of his life. In 1867, in *The Mechanical Theory of Heat*, Clausius articulated his two fundamental laws of mechanical heat: (1) The energy of the universe is constant; and (2) The entropy of the universe tends toward a maximum. Scottish physicist and mathematician William Thomson, Lord Kelvin, was working in the field at the same time as Clausius, and his work suggested to Clausius that he consider the dissipation of energy when forming his theories.

Clausius built on the works of other physicists, particularly that of Sadi Carnot (1796-1832), who is credited with pioneering the field of thermodynamics. Clausius was also able to build on the individual and collective works of James Joule (1818-89) and William Thomson, Lord Kelvin (1824-1907). Clausius succeeded in combining Carnot's understanding of the conversion of heat with Joule's findings on the conversion of heat. Clausius also used Carnot's work on the heat processes of steam engines to develop an understanding of what he called "entropy." The word

Rudolf Clausius. (Wikimedia Commons)

was derived from the Greek word "trope" (transformation) and was defined as a method of measuring disorder or randomness within a system. Clausius was concerned with two types of entropy. The first dealt with how heat is converted into work, and the second was concerned with how heat is transferred from higher to lower temperatures.

Clausius is best known for articulating in 1850 the second law of thermodynamics, which states that, on its own, heat cannot pass from a cold body to a hotter body. Clausius also improved on the scientific understanding of the first law of thermodynamics developed by James Joule, which states that changes in energy within a system can be measured by adding the heat in the system to the system itself and subtracting the labor involved by the system producing the heat.

In 1857, Clausius began working on the kinetic theory of gases and the theory of electrolysis. Between 1850 and 1865, he wrote nine memoirs on the mechanical theory of heat, including "A General Theory of Gases" (1856) and "On the Nature of the Movement Which We Call Heat" (1857). His findings were published in *The Mechanical Theory of Heat* in 1865. Clausius's work has had major applications for the contemporary field of industrial physics, a field that has contributed greatly to the development of technology in the contemporary world.

—Elizabeth Rholetter Purdy

Further Reading

Cardwell, Donald S.L. *From Watt to Clausius: The Rise of Thermodynamics in the Early Industrial Age.* Cornell UP, 1971.

Cropper, William H. *Great Physicists: The Life and Times of Leading Physicists from Galileo to Hawking.* Oxford UP, 2001.

Segrè, Emilio. *From Falling Bodies to Radio Waves: Classical Physicists and Their Discoveries.* W.H. Freeman, 1984.

Simonis, Doris, ed. "Clausius, Rudolf." *Scientists, Mathematicians, and Inventors: Lives and Legacies: An Encyclopedia of People Who Changed the World.* Oryx Press, 1999.

Climate and Weather

FIELDS OF STUDY
Climatology; Oceanography; Thermodynamics

ABSTRACT
Climate is the term for the average weather of a region, observed over a period of at least 30 years. Climatology is the scientific study of climate conditions, including temperature and precipitation of a defined climate zone, global atmospheric and oceanic currents. The driving force of all climate and weather conditions, and essentially all energy on Earth, is the energy that arrives from the Sun.

KEY CONCEPTS
anthropogenic: caused by or resulting from human activities

global conveyor belt: the planet-spanning system of interconnected surface and deep ocean currents that transports an immense amount of the energy received from the Sun and from Earth's interior

greenhouse gases: gases present in the atmosphere that absorb infrared (heat) radiation emitted from Earth's surface and then re-emit about two-thirds of that energy back into the atmosphere instead of into space

Hadley cells: a large, convective air circulation system driven by the influx of solar energy at the equator; the cells rise from the equatorial region, move toward the north and south poles, then sink to the surface to return toward the equatorial regions

thermohaline circulation: the global circulation of ocean currents driven by changes in thermal energy and salt content of the water

CLIMATE VS. WEATHER
Climatology comprises the recording of climate data (actual and historical), climate diagnostics regarding physical and chemical atmospheric processes, development of climate models, and forecasting consequences of anthropogenic impacts. Characteristic courses of one year describing temperature and precipitation are displayed in climate diagrams. Weather is based on similar data measurements but describes short-term events.

WEATHER
Meteorology is the scientific study of the atmosphere, investigating mainly the dynamics of weather events. Etymologically, the root of the term can be traced back to the Greek *meteoron*, referring to "above-Earth floating air and phenomena." As a branch of geosciences, meteorology studies physical and chemical atmo-

spheric events and phenomena, including interactions with the Earth's surface and solar radiation. Spatial-temporal studies range from microtur- bulences to general weather to climatology; the latter comprises more comprehensive interdisciplinary research.

The atmosphere is the most unstable and fastest-changing subsystem of the climate system, although it is relatively thin: The stratosphere reaches about 50 kilometers (31 miles) high, which is less than 1 percent of Earth's radius of 6,341 kilometers (3,963 miles). The atmospheric layers, from Earth's surface upward, are the troposphere, the stratosphere, the mesosphere, the ionosphere, and the exosphere. These layers can be classified with regard to their chemical, dynamic, thermal, and optical properties. Their vertical formation influences global weather and climate processes. With regard to meteorological research—which concerns mainly tropospheric and stratospheric events, because 99 percent of the entire atmospheric air mass is concentrated in these layers—temperature is the most important factor. Air density decreases with altitude. While at the surface a density of 1.225 kilograms per cubic meter can be measured, at the tropopause (the border between the troposphere and the stratosphere) air density is only 0.36 kilogram per cubic meter. Simultaneously, atmospheric pressure decreases from 1.013 hectopascals at the ground to about 200 hectopascals at the tropopause and only 1 hectopascal at the stratopause (the border with the mesosphere). Regardless of their other properties, the atmosphere's layers are defined by their temperature profiles, and both density and pressure change depending on weather conditions. In extreme cases, the tropopause at Earth's mid-latitudes is about 500 hectopascals.

The chemical composition of the atmosphere impacts its characteristic vertical temperature profile. Within the troposphere, from the surface up to the tropopause, temperature sinks from on average 15°C to minus 50°C, because the troposphere is heated primarily by absorbed solar radiation from the surface. Furthermore, reemitted thermal radiation is absorbed by greenhouse gases and thus is kept in the lower troposphere. In the stratosphere, temperature increases again, because solar radiation is partially absorbed by the ozone layer. Ozone results from photolysis of oxygen molecules because of very energetic ultraviolet radiation in the stratosphere. The higher temperatures in an upper atmospheric layer block cooler air from the lower troposphere from rising further.

Essential weather processes include evaporation, condensation, and atmospheric dynamics based on differences in temperature. Radiation energy not only heats the land surface but also causes evaporation: 90 percent of water vapor in the atmosphere is evaporated from oceans, rivers, and lakes. To evaporate one gram of water, 2.45 kilojoules of energy are required. Upon warming, humid surfaces form water vapor, rise up with warm air, condense into water

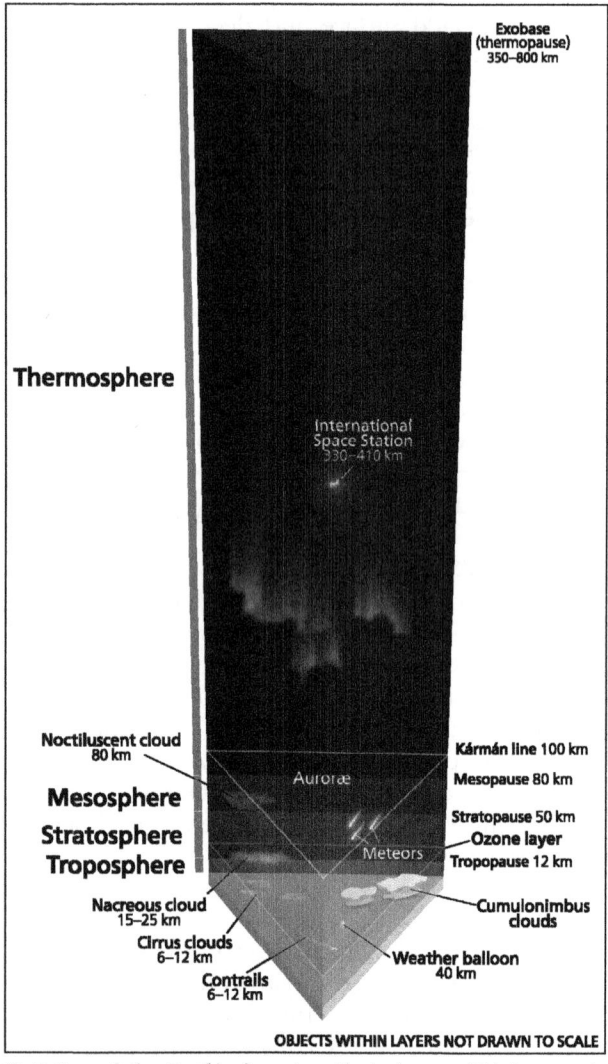

Depiction of the Earth's five atmospheric layers. (Kelvinsong via Wikimedia Commons)

upon cooling, and form clouds and precipitation. Converging air masses can produce heavy weather events, such as thunderstorms. Evaporation consumes energy, stored in water vapor as latent heat and released during condensation and rainfall.

SOLAR RADIATION

Solar radiation and the resulting atmospheric interactions determine the climate of Earth. Interactions include atmospheric heat storage, heat convection, global wind systems, and ocean currents, distributing heat and precipitation over the planet's surface. Altogether, the global climate system can be understood as a giant thermal power plant driven by solar energy. The radiation budget is the most important part of the Earth's energy budget. About 30 percent of the incoming solar radiation (that is, an Earth's albedo of 0.30) is reflected by clouds, air, and land surfaces, especially snow, to outer space. The remaining 70 percent is absorbed: about 20 percent by the atmosphere and 50 percent by land. The latter is reflected as thermal radiation and convection, heating the lower atmosphere.

Assuming that this energy would be completely reradiated into space and away from the planet, the average temperature on Earth would be minus 18°C; the average is actually 15°C. At minus 18°C, water would be frozen and life as we know it would not exist. Also, without the thermal regulation by the atmosphere and the oceans, temperature would fluctuate and would be much more dependent on solar radiation (on the Moon, with virtually no atmosphere, for example, the surface temperature fluctuates about 300 degrees).

Three factors influence the radiation balance. First, the amounts of incoming radiation can fluctuate because of changes in the sun's activity or alterations of Earth's orbit. Second, reflection can vary—for instance, because of aerosol particles released from volcanic eruptions. Third, concentrations of greenhouse gases change; again, volcanic eruptions can release carbon dioxide and sulfate compounds, for example. Earth's actual average temperature of about 15 degrees Celsius is possible thanks to the natural greenhouse effect of the atmosphere. At this point it is worth noting that the main components of the atmosphere—nitrogen (78.1 percent), oxygen (20.9 percent), and argon (0.93 percent)—do not contribute to

Ocean thermal energy conversion (OTEC) is a process or technology for producing energy by harnessing the temperature differences (thermal gradients) between ocean surface waters and deep ocean waters. OTEC systems using seawater as the working fluid can use the condensed water to produce desalinated water. (Vitafougue via Wikimedia Commons)

the greenhouse effect; instead, trace gases, such as water vapor, carbon dioxide, methane, and nitrous oxide, have a big impact on climate. These greenhouse gases reemit infrared radiation into the atmosphere as well as back toward Earth's surface. Furthermore, the trace gas ozone plays an important role for life on Earth. In the lower atmosphere, it acts as a green- house gas; in the stratosphere, it absorbs the hazard- ous ultraviolet radiation of the sun.

Local and regional climate conditions depend on many factors. Earth's surface may reflect an average 30 percent of solar radiation back into space, but regional conditions significantly affect local areas. When there is snow, reflected radiation may be as high as 40 to 90 percent; in deserts, 20 to 45 percent; in forests, 5 to 20 percent. Other factors that may affect reflected radiation include the wave angle of solar radiation reaching the surface as well as its intensity and duration, clouds and humidity, heat transport by wind, atmospheric layers, and so forth.

The particular wave angle of solar radiation at different latitudes causes warmer or colder temperatures and variable precipitation, resulting in the classification of distinct climate zones: the tropical rain climate zone, dry climate zone, temperate humid climate zone, cold-dry climate zone, boreal climate zone, polar climate zone, and mountain climate zone. Solar radiation and availability of water determine the plant cover essentially, and thus diverse habitats. Differences in temperature between cold and warm climate zones relate to congruent differences in air pressure, influenced by the Earth's rotation, leading to the atmospheric circulation system, which transports energy from tropical latitudes to higher latitudes. Differences in radiation between night and day and summer and winter are taken into account for descriptions of atmospheric circulation. The term *solar constant*, given as 1,368 watts per square meter (W/m²) and describing the amount of energy reaching the outer atmosphere from the sun, suggests a constant solar radiation. In fact, solar radiation has increased during the past 4 billion years by about 25 to 30 percent, and it also undergoes periodic fluctuations.

Clouds influence climate a great deal. Lower clouds have a cooling impact because their reflection of solar radiation is higher than their absorption, whereas higher ice clouds (cirrus clouds) have a warming impact due to their absorbing properties. Aerosol particles often function as condensation cores during cloud formation. They also have a cooling effect because they reemit solar radiation. Their climatological effects are most significant during and shortly after volcanic eruptions, when sulfate aerosols are blown up into the lower stratosphere and there absorb solar radiation, causing temperature increases at that altitude while temperature at the surface decreases.

RISE IN TEMPERATURE

The most remarkable climate feature during the last 1,000 years is the net increase in temperature by the end of the twentieth century, which is believed to have been caused by anthropogenic emissions of greenhouse gases. Apart from this peculiarity, global average temperatures fluctuated only about 0.5°C. Prior to the Industrial Revolution, climatic changes were caused only by volcanic activities or changes in solar radiation, and the latter had an impact of about 0.2-0.4°C. Volcanism may well have caused the Medieval Warm Period (between 950 and 1250), as well as the Little Ice Age (between 1550 and 1850). Studies of ice cores from Greenland and Antarctica demonstrated atmospheric impacts of volcanic eruptions. Sulfate concentrations in certain ice layers show evidence of the eruptions of Mount Tambora in 1815 and Mount Krakatoa in 1883, as well as another extremely strong eruption, in 1259. There were two significant maxima of solar radiation during the past 1,000 years, one during the early Middle Ages and the other during the twentieth century, while radiation was relatively low during the Little Ice Age. Its minimum occurred in the fifteenth century.

ATMOSPHERIC OSCILLATION AND WIND SYSTEMS

With respect to atmospheric dynamics of the last decades, consequences of global warming include a fortification of the North Atlantic Oscillation (NAO) and Arctic Oscillation (AO) and a reduction of planetary waves. This causes less warm and ozone-rich air to reach the polar stratosphere during winter months, resulting in what is called "climatic isolation" of the stratosphere. Similar trends can be described for the Antarctic Oscillation. Altogether, global warming amplifies ozone depletion and retards the regeneration of the ozone layer, even after the banning of chlorofluorocarbons in the late 1980s.

Any change in ozone concentrations has a direct impact on temperature in the stratosphere. Reduced ozone causes a cooling in the stratosphere, significantly above the poles. During 1969 and 1998, the temperature at the higher latitudes of the Southern Hemisphere decreased about 6°C in 18 kilometers (11 miles) altitude. The lower polar stratosphere even cooled between October and November 1985 by about 10°C. Such changes in temperature have also had an influence on atmospheric dynamics. For instance, a significant amplification of the polar vortex was observed to be combined with a retarded timing of its collapse about one month later. Because of this feedback effect, given reduced ozone concentrations additionally amplify the polar conditions, leading to further ozone depletion.

The amplification of the polar vortex in the spring of the Southern Hemisphere one to two months later results in a stronger tropospheric Antarctic Oscillation, again causing a cooling above Antarctica, with the exception of the west Antarctic Peninsula where strong west winds reduce polar cold air. This cooling is also a result of lower thermal radiation resulting from stratospheric ozone depletion. Similar trends have been described for polar latitudes of the Northern Hemisphere. Furthermore, the fortification of circulation around Antarctica causes less precipitation between latitudes 35 and 50 degrees south. Most alarming is the reduction of rainfall by 15 to 20 percent in southwest Australia during the last 50 years as a result of replacements of southward rain-laden cyclones. These examples demonstrate the influence of the stratosphere on climate.

At the equator, radiation is most intense; therefore, heating and evaporation at the equator are most intense too. Warm air expands, is less dense, and rises, thus transporting heat away from the surface, a process that is called "convection." As it rises, this air cools down, water vapor condenses, and heavy tropical rainfalls occur. This area of daily, heavy rainfall is called the "Intertropical Convergence Zone." The rising warm air leads to low pressure, causing suction of air from regions with higher air pressure. Equatorial air flows in this direction until it is chilled and sinks, inducing one of the global circulation systems, the Hadley cell, more popularly known as the trade winds (based on its surface characteristics). The result for climate is that the circulating air transports heat from the equator toward the poles.

Because the rotation of Earth redirects the circulation to the right, there are two Hadley cells, causing the northeast trade winds in the Northern Hemisphere and the southeast trade winds in the Southern Hemisphere. At the poles, similar circulations occur but are caused by reverse conditions: Cold air sinks above the poles and moves toward warmer regions, until it is warm enough to rise. The resulting polar cells are the reason for polar east winds. Between both systems lies a third, the Ferrel cells, where west winds are characteristic.

Ferrel cells, also referred to as midlatitude cells, do not comprise a real wind system; the west winds are more unsteady and stormier than the trades and the polar east winds are. They are much more a chaos of storms and weather systems, driven around the globe by jet streams from both sides of the west wind zones.

Wind systems distribute heat and evaporated humidity, including the absorbed latent heat of water vapor. They also determine precipitation and thus the availability of water. Without the wind systems, the tropical zone would be on average 15° warmer and polar regions would be 25° colder than they are now. The formation of rainfall is also influenced by the characteristics of Earth's surface. Where mountains block winds, air rises and it rains; conversely, in the rain shadows of mountains, deserts often develop.

OCEAN CURRENTS AND GLOBAL CONVEYORS
In the upper 3 meters (10 feet) of the oceans, there is as much energy absorbed as in the entire atmosphere. Thanks to the high heat, the storage capacity of water in the oceans has a balancing impact on climate, mitigating differences in temperature between winter and summer. Wind systems account for about 80 percent of the entire heat distribution; the remaining 20 percent is distributed by oceanic streaming systems. Because of the predominant wind directions and the distracting power of the rotation of the Earth, round vortices occur in the big ocean basins, transporting warm water away from the equator and cold water to the equator.

These surface currents are related to bottom currents; besides being driven by winds, they are moved by means of the sinking of dense, cold, very saline water in the polar sea, forming what has been termed a "global conveyor belt" that transports not only water but also massive amounts of thermal energy received from the Sun. Climate in the Northern Hemisphere is influenced by this global conveyor belt. Without it, climate in Europe would be much colder, comparable to that of Newfoundland. One important current is the Gulf Stream, which begins in the Gulf of Mexico and transports an estimated 1.3 billion megawatts of energy as North Atlantic current from the tropics northbound. Thanks to the Gulf Stream, the palm trees can grow as far north as Scotland.

The ocean conveyor belt is driven by trades that push water from West Africa to America. It leaves the Gulf of Mexico and moves through the Florida Straits, heating as Gulf Stream water along the U.S. east coast. West winds, Earth's rotation, and the cold Labrador Stream distract the current and lead it as the

North Atlantic current to Europe. Arriving at the European North Sea, the water is cooled; it also at this point has a high salinity due to evaporation. It sinks because of its increased density, which causes a suction so that the water keeps flowing. If winds are considered as a primary driving force, salinity and temperature determine the secondary driving force, called "thermohaline circulation." As a deep current, the water flows back into the southern Atlantic, restarting the circulation.

There are several oceanic regions where cold water sinks: in the Labrador Sea, the Weddell Sea, and the Ross Sea. Hence, periodic temperature fluctuations occur both in the Atlantic and in the Pacific, whose relations to the ocean currents and influences on climate are currently an object of further research.

—*Manja Leyk*

Further Reading
Aguado, Edward, and James E. Burt. *Understanding Weather and Climate.* 5th ed. Prentice Hall, 2009.
Barry, Roger Graham, and Richard J Chorley. *Atmosphere, Weather, and Climate.* Routledge, 2010.
Cook, Peter J. *Clean Energy, Climate and Carbon.* CSIRO Publishing, 2012.
Hodgson, Peter E. *Energy, the Environment and Climate Change.* Imperial College Press, 2010.
McIlveen, J.F.R. *Fundamentals of Weather and Climate.* Oxford UP, 2010.
Pittock, Jamie, Karen Hussey, and Stephen Dovers, eds. *Climate, Energy and Water: Managing Trade-offs, Seizing Opportunities.* Cambridge UP, 2015.
Stephenson, Michael. *Energy and Climate Change: An Introduction to Geological Controls, Interventions and Mitigations.* Elsevier, 2018.
van Aken, Hendrik M. *The Oceanic Thermohaline Circulation: An Introduction.* Springer, 2007.

Climate Neutrality

FIELDS OF STUDY
Climatology; Environmental Studies; Policy Studies

ABSTRACT
The first United Nations Conference on the Human Environment, held in June 1972 in Stockholm, Sweden, called for protection of the environment and natural resources across the world. The inconsiderate use of fossil fuels has caused greenhouse gas (GHG) emissions that have contributed to global warming. This is having a negative impact on human and other animal life, as well as causing a loss of valuable environmental resources. Climate neutral initiatives need to be implemented at both the individual and the organizational levels in order to protect the environment and mitigate climate change.

KEY CONCEPTS
carbon neutral: processes and activities that result in no net increase in atmospheric carbon
climate change: a shift in the global climate over a prolonged period of time, whether by natural or anthropogenic causes; often used interchangeably with "global warming"
climate neutral: processes and activities that have no net effect on the climate, whether globally or locally
global warming: the observed gradual increase over time of Earth's average atmospheric temperature, not counting seasonal variations; often used to mean "climate change"
greenhouse effect: the demonstrable phenomenon in which certain atmospheric gases absorb thermal energy radiating from Earth's surface level and return a significant portion of it back into the atmosphere instead of passing it directly out into space, resulting in a sustained higher-then-normal atmospheric temperature

ADDRESSING THE ISSUE OF CLIMATE CHANGE
The first United Nations Conference on the Human Environment was held in June 1972 in Stockholm, Sweden. It called for protection of the environment and natural resources across the world. Since then, there has been much debate over the impact of human activities on the environment, particularly regarding the degradation of natural resources (forests, agriculture, and water) and exploration for fossil fuels. Notably, the inconsiderate use of fossil fuels for industries and transportation has caused greenhouse gas (GHG) emissions that, many scientists agree, have led to global warming as a result of excessive CO_2 and other GHGs in the atmosphere, which exacerbate the greenhouse effect. This warming is having a negative impact on human and other animal life, as well as causing a loss of valuable environmental resources.

CLIMATE NEUTRALITY

The term *climate neutrality* refers to actions, commitments, and behaviors that do not result in GHG emissions that lead to global warming and diverse environmental hazards. Currently, human activities such as the use of home appliances, automobiles, and industrial technologies are resulting in different rates of GHG emission. The effort to achieve climate neutrality begins with taking stock of what we use and produce that is causing GHG emissions. Taking such an inventory can be an effective tool for estimating the amount of GHGs we are emitting and devising strategies to neutralize the effects of climate change. For instance, through measuring electricity consumption in kilowatt-hours, the amount and type of fuel used to heat water and warm the house, and how many miles we drive, fly, or ride in different vehicles, we can become aware of our impact on the climate and then determine what changes in habits would be required to make our activities more climate neutral.

On the individual level, climate neutrality means changing behaviors that otherwise contribute to GHG emissions. For instance, in our homes we can use light-emitting diode (LED) lightbulbs, which consume much less power and therefore reduce carbon emission; adopt solar panels to reduce overdependence on fossil-fuel-based electricity generation; switch off our computers and televisions instead of leaving them in standby mode; ride bicycles to avoid emissions from motor vehicles; and plant trees and other plants to absorb carbon dioxide.

On the business and corporate level, production of low-carbon-emitting goods and services is the goal. The low-carbon-emitting products will save the environment and can involve recycling products that otherwise would lead to additional manufacturing loads

Plenary session of the COP21 for the adoption of the Paris Accord, United Nations Climate Change Conference, 2015. (UN Climate change via Wikimedia Commons)

and the emission of new carbon into the global atmosphere. A climate neutral business program can consist of tracking, reducing, and offsetting business flights, which is usually the largest source of unavoidable emissions in a service-oriented business.

Climate neutral businesses can create a high level of customer satisfaction by advertising their environmentally friendly point of view and playing a role in environmental protection. Environmentally friendly business goods and services can increase profits through lower production costs and can boost sales by touting their environmentally friendly investments. However, climate neutral business solutions may require coaching and guidance in the design of innovative products.

Organizations can also play a vital role in advancing climate neutrality by participating virtually rather than physically in industry conferences and workshops. The virtual effort will save money and time and reduce carbon emissions globally by avoiding air travel. Different organizations are currently organizing online conferences and workshops as part of their effort to become climate neutral. The proper networking and merging of new climate neutral business ideas and their development are required. To some extent, business entities can fill their climate neutral roles through compliance with environmental protection laws and policies regarding manufacturing and by adhering to international standards (accredited by the International Standards

Organization, ISO) for carbon neutral certification. Compliance with environmental laws will support a company's sustainability as well as reduce carbon dioxide and other GHG emissions. Many countries and communities have pledged to become climate neutral. Some of them include Costa Rica, Denmark, Iceland, Norway, the Maldives, and New Zealand. These nations have started to use renewable sources of energy such as solar power, wind power, hydropower, and geothermal energy for producing electricity. New Zealand undertakes forest regeneration programs to offset the carbon emissions from its public sector.

CARBON OFFSETTING

Another strategy for becoming carbon neutral is to make use of carbon offsetting. Carbon offsetting involves balancing the GHG emissions from one source by reducing emissions from another source. High-GHG-emitting companies can use carbon offsetting, for example, by financing the planting of trees elsewhere. The offsets are measured in metric tons of carbon dioxide equivalent (CO_2e) but may involve the reduction of any of the six major GHGs: carbon dioxide (CO_2), methane (CH_4), nitrous oxide (N_2O), perfluorocarbons (PFCs), hydroflourocarbons (HFCs), and sulfur hexaflouride (SF_6). This is also termed a *carbon credit*, which an organization or country can buy from a lower-emitting entity through financing for green activities. The concept is discussed in the United Nations Framework Convention on Climate Change (UNFCCC) and is working to some extent.

Climate neutral actions are needed to reduce carbon and other GHG emissions into the atmosphere by means of offsetting or equivalent green initiatives. Moreover, developing climate neutral attitudes and behaviors will help save our environmental resources and promote sustainable global development.

—*Ronju Ahammad*

Further Reading

Bumpus, Adam G., James Tansey, Blas L. Pérez Henriquez, and Chukwumerije Okereke, eds. *Carbon Governance, Climate Change and Business Transformation*. Routledge, 2016.

"Climate Neutrality." http://www.firstclimate.com/climate-neutral-services/climate-neutrality.html.

Conlin, R. "On the Path to Climate Neutrality." *Yes*, vol. 10, 2010. http://www.yesmagazine.org/blogs/richard-conlin/on-the-path-to-climate-neutrality.

Delbeke, Jos, and Peter Vis, eds. *Towards a Climate-Neutral Europe: Curbing the Trend*. Routledge, 2019.

Hirsch, Tim. *A Case for Climate Neutrality: Case Studies on Moving Towards a Low Carbon Economy*. United Nations Environment Programme, 2009.

Hoffman, J.A. "Climate Change as a Cultural and Behavioral Issue: Addressing Barriers and Implementing Solutions." *Organizational Dynamics*, vol. 39, 2010, pp. 295-305.

Zhou, Shelley W.W. *Carbon Management for a Sustainable Environment*. Springer, 2020.

COAL AND ENERGY PRODUCTION

FIELDS OF STUDY
Geology; Mechanical Engineering; Mine Engineering

ABSTRACT
Coal has been one of the most widely used of the fossil fuels since the nineteenth century, driving industrial growth and modernization, although concerns about its environmental impacts slowed its growth in the twentieth and twenty-first centuries.

KEY CONCEPTS
acid mine drainage: water runoff from mines in which sulfur from certain minerals has reacted with the water to produce sulfuric acids that leach toxic heavy metals from other minerals present

carbon capture: trapping of carbon dioxide, hydrocarbons and other waste gases to prevent them from entering the atmosphere

Carboniferous Period: the period of geological history before the evolution of land-dwelling creatures, in which only plants existed on Earth's land surfaces

combustion: generally identified as "burning" with the formation of flames, but technically all oxidation reactions that produce carbon dioxide as a product of the reaction, with or without flames

fossil fuels: fuels derived from the remnants of once-living organisms, particularly coal, petroleum and natural gas (methane)

Industrial Revolution: the period beginning in the mid-1700s in which traditional animal and human power began to be replaced by steam power and mechanization

oxidation: any reaction in which a material combines with oxygen to produce oxides

COAL, BRIEFLY
Coal is a fossil fuel formed over centuries through the compaction and heating of decaying plant materials in oxygen-poor environments. It is mined from the earth through either surface or subsurface methods. Coal has been one of the most widely used fossil fuels since the Industrial Revolution of the nineteenth century; one of the leading fuels for electricity generation, it currently accounts for close to 30 percent of the world's primary energy production. In the United States it accounted for 33 percent in 2015, tied with natural gas. Coal is used for heating, generating electricity, and industrial processes such as the manufacture of coke and steel.

Coal's powering of the Industrial Revolution in Britain, Germany, and elsewhere led to its emergence as the driving force behind economic expansion well into the twentieth century. Usage rates declined mid-century as a result of environmental concerns and the development of cleaner energy sources. However, coal usage rates have undergone a resurgence since the late twentieth century because of the rising expense and energy security issues surrounding oil, the rising demand in developing countries such as China and India, and to a lesser degree the development of clean-coal technologies.

COAL ORIGINS
Coal formed when ancient plant materials accumulated in layers over the centuries in environments such as swamps, where they failed to decay completely due to lack of oxygen. Thermal processes in the anoxic environment decomposed the carbohydrate components of the plant matter, eliminating the hydrogen and oxygen as water and leaving only the carbon and mineral remnants of the plant materials. Centuries of heat and pressure converted this decaying organic material into the combustible, rock-like substance known as coal, which is mainly carbon. Scientists believe that most of Earth's coal was formed during the Carboniferous period, approximately 300 million years ago, with other large deposits forming during the Upper Cretaceous period, approximately 100 million years ago.

The dead vegetative material first forms into peat, which is soft and wood-like. Peat itself is burned as a fuel source, but its poor burning capacity and smoky

Peat excavation. (longtaildog via iStock)

A peat fire. (Wikimedia Commons)

output have limited its use, especially in modern times. If peat remains in the ground, layers of sediment gradually build on top of the peat, resulting in additional heat and pressure. This process slowly converts the peat into coal. Coal is most often found in lengthy horizontal layers of varying thickness known as seams or beds, which may lie at or near the surface or deep underground. There are various types of coal, depending largely on the amount of time it has remained in the ground. The three main classifications of coal are lignite, bituminous, and anthracite.

TYPES AND CLASSIFICATION OF COAL

The first type of coal formed is known as lignite coal, also known as brown coal; followed by bituminous, or soft, coal; and anthracite, or hard, coal. (Bituminous coal can be further divided to include the lower-value subbituminous coal.) The types are ranked in value, with lignite at the lowest part of the scale and anthracite at the highest. Lignite is the most abundant type of coal; has the highest moisture, oxygen, and ash contents; and has the lowest carbon content and heating value, or energy content. Anthracite is the least abundant type of coal; has the lowest moisture, oxygen, and ash contents; and has the highest carbon content and heating value, or energy content. Bituminous is the most common and widely used type of coal and falls in the middle of the scale.

Coal is classified in terms of heating value, with anthracite at the top of the scale, followed by bituminous, subbituminous, and lignite. The higher the percentage of pure carbon, the higher the heating value or rank of the coal. There is usually a direct correlation between age and pure carbon percentage. Anthracite tends to have a 90 percent or greater carbon content, bituminous ranges from 70 percent to 90 percent, and lignite from 40 to 70 percent. Coal may also be evaluated by grade, based on the amount of ash or sulfur content. Coal also contains a number of other minerals, chemicals, and trace elements that vary among different coal seams and even within individual coal seams. Examples include aluminum, zirconium, hydrogen, oxygen, nitrogen, and sulfur, as well as heavy metals such as mercury, arsenic, and uranium. These minerals are noncombustible and remain behind as ash or are released into the atmosphere as fly ash and other emissions when coal is burned.

Coal is defined as a fossil fuel, meaning it is an energy source of organic origins. Coal is one of the world's most common and commonly used fossil fuels, alongside oil and natural gas. It is the most carbon-intensive of the fossil fuels, releasing its energy through the combustion of carbon. Burning coal oxidizes the carbon in an exothermic reaction that releases a large amount of energy. Coal is considered a nonrenewable energy source because no more significant quantities of coal are expected to form in the human time-scale present or future, meaning that the world will eventually exhaust its supply of this energy resource through continued use.

COAL PRODUCTION

Although coal is classified as a nonrenewable energy resource, it exists in very large quantities throughout the world, including both proved and potential re-

Raw coke. (Stahlkocher via Wikimedia Commons)

serves. Proved coal reserves are spread over approximately 70 countries around the world and found in every continent. However, the United States, Russia, China, Australia, and India combined hold about two-thirds of global proved recoverable coal reserves, according to the World Energy Council in its Survey of Energy Resources 2010. Not all known resources are extractable with current technology; this adds to the uncertainty in the widely varying estimates of potential global coal resources. Estimates by the World Coal Institute indicate that current production rates and reserve estimates will allow for approximately 100 more years of coal production, far more than estimates for other fossil fuels. This figure could change with new discoveries and technological developments.

Coal is one of the most economical fuels, leading to its popularity since the nineteenth century. All types of coal are mined and utilized as fossil fuels; all types are used in the generation of electricity. Subbituminous and anthracite are also used for space heating. The leading coal producers are China— which alone accounts for about half of global production—the United States, India, the European Union, Australia, Russia, and Indonesia. Coal-mining operations exist in more than 60 countries across the globe, with output by the United States and China amounting to two-thirds of annual global production. Most of the coal produced is destined for the domestic market, with a small percentage sold on the international market. Coal export is more likely to dominate in less densely industrialized nations; in recent years Australia and Indonesia together have accounted for half of world coal exports.

The first step in the production of coal is exploration and discovery of coal seams, usually accomplished through geological mapping, geochemical and geophysical surveys, and exploratory drilling. Next, the extent and quality of the coal seam are determined, and a decision is made regarding whether or not it is technologically possible and economically feasible to begin mining operations. The two major types of coal mining are subsurface mining, also known as underground or deep mining; and surface mining, also known as strip, opencast, or open-cut mining. The decision is determined based on the geographical and geological features of the coal seam. Subsurface mining is dominant in global coal production overall, although surface mining is the most prevalent form in some leading coal production countries such as Australia and the United States. Subsurface mining is more dangerous but less environmentally damaging.

Surface or strip mining is utilized to obtain coal seams that lie close to the surface through the explosive breakup and removal of overlying sediment layers known as overburden. Miners then drill and recover the coal by mining out strips through the coal seam. Miners employ different types of surface mining based on the geography and geology of the individual coal seam and surrounding area.

Types of surface mining include contour mining, mountaintop removal, area mining, and open-pit mining. Equipment involved includes draglines, power shovels, and trucks for the removal of overburden and bucket-wheel excavators, conveyors, and trucks to move the coal. Finally, the coal is transported to a processing plant or primary consumer. Surface mining provides coal recovery rates at or above 90 percent of the total deposits.

Subsurface or underground mining is utilized to extract coal from seams that lie deep below the surface through the construction of tunnels into the earth. Different types of mines are constructed based on the depth and slope of the coal seam. Types of subsurface mines include drift mines, slope mines, and shaft mines. The main types of subsurface mining are room-and-pillar and longwall mining.

Room-and-pillar mining involves carving a series of rooms into the coal seam to allow for mining while leaving behind pillars of coal as roof supports. Longwall mining involves excavating all coal from a section of the seam known as a face. Miners use mechanical shearers to remove the coal and temporary hydraulic roof supports during the course of removal. Coal is brought to the surface through tracks, conveyor belts, coal trucks, or electric hoists.

Coal mining safety concerns include health risks such as lung diseases, the need for proper ventilation, and the potential for coal gas explosions, cave-ins, or flooding. In addition, inadequate management of mine "tailings" from underground operations can have disastrous consequences. In 1966, the collapse of a mountain of mine tailings from coal mining operations in Aberfan, Wales, buried much of the town below, destroying the school and killing some 46 adults and 156 schoolchildren as they sat at their desks. Explosions in mines are a concern because of the release of methane during subsurface mining. Although mining safety standards, worker training, and oversight have improved in modern times, mining remains a dangerous occupation, especially in countries with poor mining safety records, such as China.

Negative environmental impacts associated with coal mining include land and habitat destruction; soil erosion; dust, noise, water, air, and soil pollution; mine subsidence; and acid mine drainage. Mine subsidence occurs when subsurface coal mining results in the lowering of the surface ground level. Acid mine drainage results when rocks containing sulfur-bearing minerals react with water, forming an acidic runoff that dissolves heavy metals into the water. Measures to reduce or repair environmental impacts include project planning and monitoring, pollution control measures, and reclamation projects at abandoned mines.

COAL PROCESSING

After exploration and mining, the recovered coal then moves to the transportation and storage stage of production; these add to production expenses. Transportation of coal is most often by train or ship; other modes include trucks and coal slurry pipelines. Coal must be stored along the supply chain routes from producers to consumers, which can be dangerous if not done with care. Storage problems can include oxidization of improperly stored coal, and subsequent heat content and economic value losses, or the spontaneous combustion of coal piles due to heat buildup over time, which can result in uncontrolled fires or cause dangerous explosions.

The World Coal Association estimates that approximately 7 million people are directly employed in the coal industry. Indirect employment through related services and industries is also significant. In developing nations, many local mine enterprises result in the development of multi-use infrastructure, such as roads, water distribution, and communications systems. Coal exports provide countries with foreign hard currency as well as more cost-effective domestic energy sources and greater access to electricity than is typical in developing nations. Many have thus targeted coal as an important component of their strategic development programs, despite its environmental drawbacks.

USES OF COAL

Coal is a fossil fuel energy source that provides both a direct and an indirect source of fuel. Coal can be directly burned to provide heat or can be used as a secondary heat source, such as in the creation of steam to power engines. Humans have utilized coal-the "rocks that burn"-as an energy source throughout much of recorded history, with readily visible surface outcroppings providing the first preindustrial sources. Recorded uses of coal in China date back to the fourth century CE, and it had emerged as one of that nation's leading fuel sources by the eleventh century. Large-scale underground coal mining began later, around the thirteenth century.

The emergence of coal as an important domestic heating source in Europe beginning in the thirteenth century was driven in part by the destruction of wide swaths of forest for ship building and wood energy. When this traditional energy source was endangered, the resulting crisis drove the search for alternative energy resources. European nations such as Britain found an answer through the available supplies of coal. During the Middle Ages, coal powered such preindustrial applications as smelters and forges. By the fifteenth century, coal-burning stoves, furnaces, and fireplaces were heating homes.

The advent of the Industrial Revolution was in part powered by new developments in coal use technology. Britain led the way in the eighteenth century because of its large coal deposits and growing coal production rates. British inventor Thomas Newcomen introduced the first steam engine powered by coal in 1712. Another British inventor, James Watt, improved the Newcomen engine design to devise a revolutionary steam engine that was also sometimes coal-powered. The Newcomen, Watt, and other coal-powered engines were first used to pump water from mines and later powered factories, ships, and trains. Britain's abundant coal supplies and production, as well as its technological innovations, soon gave it the lead in energy-intensive manufacturing, notably in the textile and steelmaking industries. Britain gained military as well as economic advantages, as iron and steel were used in the manufacture of advanced weapons and armaments.

As a result of its industrial and domestic applications, coal had supplanted wood as a global energy source by the end of the nineteenth century and con-

Coal plant. (Drbouz via iStock)

tinued to grow into the first half of the twentieth century, although at a slower rate. Coal-burning stoves, furnaces, and fireplaces also continued to heat homes, businesses, and institutions. Coal also proved important in the manufacture of coke, a nearly pure carbon that results when soft coal is heated in the absence of air. Coke has a higher heat value than coal and produces carbon monoxide when heated, reducing iron oxides contained within ore to iron, making it vital to the steel industry as well as other chemical and metallurgical processes. The other major product of that reduction process, however, is carbon dioxide produced when carbon monoxide combines with the oxygen released from the iron oxides.

Coal also became essential in the generation of electricity. Coal burning provides the heat source necessary for creating pressurized steam, which is then used to power turbines. The turbines turn generators and produce electricity. Coal never dominated the automotive or other ground transportation industries—except for a period in rail transport—because coal-fired steam engines lacked the success of their gasoline- or diesel-powered counterparts.

Coal began to lose ground to oil and natural gas during the course of the twentieth century, although it remained a dominant source of energy for home heating and lighting. Coal consumption's downward trend accelerated after World War II. Natural gas provided a cleaner and better-smelling fuel source than coal distillates, while many industries began switching to relatively convenient and economical petroleum as their dominant fuel source. Key uses of oil and natural gas included the powering of transportation such as trains and automobiles, residential heating, and certain industrial processes. Nuclear, hydro, and renewable energy sources have also increased in use since the 1950s, although at a slower rate than the fossil fuels. Some renewable resources, such as wind power, provide a less expensive alternative to coal and other fossil fuels, are not limited by finite supplies, and continue to benefit from technology growth. Coal has maintained its dominance in other areas, though, such as electrical power generation.

COAL USE RESURGENCE

Coal enjoyed a resurgence in the mid- to late 1970s, during U.S. oil shortages resulting from an oil embargo imposed by the Organization of Petroleum Exporting Countries (OPEC) in 1973 and the Iranian Revolution in 1979. The shortages and effects of global politics on oil supply forced the concentration of oil for transportation industry needs, and sparked efforts to diversify energy sources in other areas, revitalizing interest in coal. Ongoing oil crises and politics, rising oil prices, and concerns over energy security have sustained the renewed interest in coal. Coal demand is expected to rise by more than 50 percent by the year 2030.

Coal currently provides nearly 30 percent of the world's primary energy supply, and coal usage is rising at a faster rate than that of any other fuel source. The electrical power generation industry is the largest modern consumer of coal as a fuel source; the burning of coal provides at least 42 percent of global electrical generation, according to the International Energy Association, and the electric power industry uses approximately 90 percent of the world's coal production. Much of the rising demand for coal comes from developing countries such as China, India, and South Africa, where it is used to increase electrification rates. As construction booms in such nations, other key industries utilize more coal, such as the concrete, iron, and steel manufacturers. The advent of cleaner combustion methods and other clean-coal technologies in the late twentieth century has provided some answer to the environmental objections over the pollution generated by coal burning—although much of these developments remain largely untested. Coal is expected to maintain its dominance in part because of its lower costs in comparison to other energy sources.

ENVIRONMENTAL CONCERNS AND THE DEVELOPMENT OF CLEAN-COAL TECHNOLOGIES

One of the biggest drawbacks to continued production and consumption of coal, besides its nonrenewable status as a fossil fuel, is its associated negative environmental impacts. Environmental problems associated with its use include soil, air, and water pollution, acid rain, and emissions of soot, sulfur oxides, nitrogen oxides, carbon monoxide, carbon dioxide, mercury, methane, cadmium, uranium, lead, and other particulates and trace elements. The fly ash remaining after coal combustion either escapes into the atmosphere as a pollutant or must be disposed of as a hazardous waste. These toxic emissions mean that coal is regarded as a "dirty fuel"; its emissions have been linked to respiratory and other diseases. The carbon dioxide and other "greenhouse gas" emissions produced by coal burning are also associated with climate change or global warming due to the greenhouse effect.

Historical uses of coal as a primary domestic heating source, coupled with growing factory use during the Industrial Revolution of the nineteenth century, often produced heavy black smog during periods of stagnant, humid weather that held pollutants close to ground level. Well-known examples included the 1952 Great London Smog and the 1948 Donora, Pennsylvania, smog. Modern urban, industrial areas have also seen a rise in pollution-related health issues and deaths, such as those documented in the international port city of Yokkaichi, Japan, resulting in victims' lawsuits against industrial ownership and more stringent government pollution regulations. Many nations have imposed environmental regulations designed to reduce carbon and other emissions. Ironically, when the COVID-19 pandemic shut down many of the factories that depended on electricity from coal-fired generation plants, as well as almost all road traffic, in 2020, many of the most air-polluted cities around the world found themselves with clean air for the first time in decades.

Recent environmental concerns have centered on the release of greenhouse gases such as carbon dioxide and methane, and their association with global warming. Coal is more carbon-intensive than other fossil fuels; it produces greater amounts of carbon emissions. International agreements such as the

United Nations Framework Convention on Climate Change and the Kyoto Protocol have targeted coal as a leading producer of carbon dioxide. Signatory countries have some incentive to reduce coal usage to meet required reductions in carbon dioxide and other greenhouse gas emissions. Some national governments have also moved toward cleaner and renewable energy resources in response to the growth of the green culture movement and consumer demand for sustainability, which have made it more difficult to mount coal-based projects and to expand the coal industry.

One area of initiative involves reducing or utilizing the dangerous waste products associated with coal production and use. Methane gas is created during coal formation and released during the coal-mining process. Although methane poses a danger because of its combustibility and classification as a greenhouse gas, it is also a potential source of natural gas energy that could be recovered from coal seams. Coal combustion generates waste in the form of the noncombustible minerals found naturally within coal as well as small amounts of carbon that are left behind. Processes have been developed to remove a number of these impurities before the coal is burned, improving the quality of the coal itself while reducing levels of waste and inefficiency at power-generating stations.

Two key techniques involve coal cleaning, also known as washing, and coal pulverization, to separate and remove the mineral and chemical materials often found within the coal and released as emissions during the burning process. Coal washing or pulverization techniques are standard in many developed countries but are lagging in many developing countries. Waste can be further reduced through high-efficiency coal combustion technologies. Potential uses for generated waste are also under consideration, with industrial uses in construction and civil engineering showing promise.

Another area features processes to convert solid coal into a liquid or gaseous state to produce alternate forms of coal fuel. The gasification process involves pulverizing coal and introducing it to steam combined with air or pure oxygen, which produces a reaction converting the coal into a mixture of gaseous hydrocarbons called "synthetic gas." The liquefaction process involves converting coal into a liquid form similar to that of petroleum. Liquefaction could allow for the use of coal as a motor vehicle fuel, which has proved impractical in the past and led to reliance on oil to fuel the transportation industry.

Other developments have focused on the removal of waste products from the smoke produced during the coal-burning process before it is released into the atmosphere. Key technologies for the reduction of harmful emissions include particulate control devices, various fluidized bed reactors, activated carbon injection, and flue gas desulfurization. Flue gas desulfurization (FGD) systems, known as scrubbers, work within the furnace or smokestack of a coal-fired plant. Traditionally, wet scrubbers have been employed, but the implementation of dry scrubbers utilizing the dry injection of compounds into the furnace is under experimentation.

CLEAN-COAL TECHNOLOGIES

One of the main environmental impacts of coal's use as fuel is the release of carbon emissions into the atmosphere. Clean-coal technologies present a promising new method of reducing carbon emissions and are currently undergoing intensive testing and development. These technologies are also known as carbon sequestration or carbon capture and geological storage (CGS). Clean-coal technologies would allow for the growth of coal usage while simultaneously reducing carbon emissions, and they have been cited as a critical technology by the Intergovernmental Panel on Climate Change. Although such technology would require additional costs for added equipment and energy and less efficient fuel usage, technological developments would reduce those costs over time.

There are numerous clean-coal projects in development or operation around the world and an ever-increasing amount of carbon in storage. The oil, gas, and chemical industries were the first to employ such technologies in the late twentieth century, including natural gas processing and coal gasification processes. The systems needed for commercial power stations have yet to be employed. Potential sites for use of clean-coal technologies include power stations utilizing fossil fuels as energy sources and the steel, aluminum, cement, and chemical manufacturing industries. Carbon capture processes with potential uses in the power generation industry include precombustion sequestration, postcombustion sequestration, and oxyfuel or oxyfiring.

CGS involves capturing the carbon dioxide produced during the burning of fossil fuels and storing it in deep geological formations rather than releasing it into the atmosphere. After the carbon is captured, it must next be transported to a final storage location. The main forms of transportation are via pipelines or ocean vessels. The third step in the process is the geological storage of the carbon. The carbon dioxide would be pumped underground, where the intense pressure would convert it into a liquid form. Deep underground reservoirs and marine storage sites under consideration include deep saline formations, depleted oil and gas fields, and coal seams that are not minable. Trapping mechanisms for keeping the carbon in place include structural storage, residual storage, dissolution storage, and mineral storage. Finally, storage sites would be mapped and closely monitored for leaks.

Challenges to overcome include the cost of developing and implementing clean-coal technologies. Clean-coal-burning technologies began to emerge in the late twentieth century in response to increased awareness of the environmental damage caused by traditional coal-burning methods. These technologies include chemical, physical, and biological methods and center on the reduction of pollutants released into the atmosphere through the coal-burning process. The World Coal Association, an industry group, asserts that the average new coal-fired electric generation plant emits about 40 percent less carbon dioxide than the average such plant did in the twentieth century. CGS is a promising avenue for the reduction of carbon dioxide emissions from the use of fossil fuels such as coal at an affordable cost. The use of these technologies currently ranges from full implementation to experimental trials.

—*Marcella Bush Trevino*

Further Reading

"Coal." International Energy Agency. OECD/IEA, 28 May 2015. Web. 13 May 2016.

"Coal & Electricity." *World Coal Association*. World Coal Association, 22 Oct. 2015. Web. 13 May 2016.

Evans, Robert L. "Clean Coal Processes." *Fueling Our Future: An Introduction to Sustainable Energy*. Cambridge UP, 2007.

Freese, Barbara. *Coal: A Human History*. Perseus, 2003.

Goodell, J. *Big Coal*. Houghton Mifflin, 2006.

Gorbaty, Martin L., John W. Larsen, and Irving Wender, eds. *Coal Science*. Academic, 1982.

International Energy Agency. *Coal Information 2010*. International Energy Agency, 2010.

Logan, Michael. *Coal*. Greenhaven, 2008.

Martin, Richard. *Coal Wars: The Future of Energy and the Fate of the Planet*. St. Martin's, 2015.

Miller, Bruce G. *Coal Energy Systems*. Elsevier, 2005.

Morris, Craig. *Energy Switch: Proven Solutions for a Renewable Future*. New Society, 2006.

Shepard, M. "Coal Technologies for a New Age." *EPRI Journal*, Jan.-Feb. 1988, pp. 4-17.

Cogeneration and Electricity Generation

FIELDS OF STUDY
Combustion Engineering; Power Generation Engineering; Thermodynamics

ABSTRACT
Cogeneration is a thermodynamically efficient use of fuel. When electricity is generated without the use of cogeneration techniques, some energy will be unused as waste heat. In cogeneration, this thermal energy is put to good use. Combined heat and power (CHP) systems are an energy-efficiency technology.

KEY CONCEPTS
biomass: material produced from the growth of plants, either as a crop byproduct or for direct use

cogeneration: processes that generate both electricity and recoverable heat

Second Law of Thermodynamics: heat moves only from a body of higher temperature to one of lower temperature

thermodynamics: the study of the movement and characteristic behavior of heat

COMBINED HEAT AND POWER
The conversion from heat to mechanical energy uses a thermodynamic cycle. Regardless of the thermodynamic cycle used, the second law of thermodynamics states that not all of heat can be turned into work. Cogeneration is the attempt to recover all or part of

the heat that must be downloaded from a thermal engine plant.

Cogeneration, or combined heat and power (CHP), integrates the production of usable heat and electricity in one single, highly efficient process. All power plants must emit a certain amount of heat during electricity generation. This can be released into the natural environment through cooling towers, as flue gas, or by other means. These efficient systems recover heat that normally would be wasted in an electricity generator and save the fuel that would otherwise be used to produce heat or steam in a separate unit. This contrasts with conventional ways of generating electricity, where vast amounts of heat are simply wasted.

The CHP process can be applied to both renewable and fossil fuels. CHP can use diverse fuels in the same boilers. This means that, as greener fuels like biomass (e.g., straw bales, waste-wood pellets, or certain specially grown crops) become available, they can be used in CHP plants, with an immediate reduction in carbon dioxide (CO_2) emissions and with the knowledge that these precious green fuels are also being used in the most efficient way possible.

The specific technologies employed and the efficiencies they achieve will vary, but in every situation CHP offers the capability to make more efficient and effective use of valuable primary energy resources. Today's CHP systems are based predominantly on existing, proven power generation technologies: steam

Hanasaari B, a coal-fired cogeneration power plant in Helsinki, Finland. (Matti Paavola via Wikimedia Commons)

turbines, gas turbines, and reciprocating engines used the world over to generate energy.

This adaptation of existing technology not only contributes to the relatively low cost of CHP but also ensures that it is a proven and reliable technology, capable of delivering an immediate impact in transforming our energy system.

The overall efficiency of CHP plants can reach in excess of 80 percent at the point of use. This compares with the efficiency of combined-cycle gas turbines (CCGTs); coal-fired plants fare less well, with an efficiency of around 38 percent. At this efficiency level, CHP systems, therefore, could effectively double the central electric system's average delivered fuel-use efficiency. Byproduct heat at moderate temperatures of 100°C to 180°C (212°F to 356°F) can also be used in absorption chillers for cooling. A plant producing electricity, heat, and cold is sometimes called a trigeneration, or more generally a polygeneration, plant. CHP captures some or all of the byproduct heat for heating purposes, either very close to the plant or as hot water for district heating with temperatures ranging from approximately 80° to 130°C. This is also called combined heat and power district heating (CHPDH). Small CHP plants are an example of decentralized energy.

Cogeneration plants typically arise near thermal users, because, given the high transmission losses, technically it is not simple or cost-effective to transmit heat over long distances.

ENVIRONMENTAL BENEFITS

Cogeneration can reduce fuel consumption without compromising the quality and reliability of the energy supply to consumers. As a result, it provides a cost-effective means of generating renewable energy, and using less fuel in the process reduces energy costs, too. It also offers an opportunity to secure cost-effective reductions in CO_2 emissions.

Because less fuel is combusted, greenhouse gas emissions, such as CO_2, as well as criteria air pollutants such as nitrogen oxides (NO_x) and sulfur dioxide (SO_2), are reduced. This technology reduces overall demand from centralized, such as large-scale coal or gas-fired, power stations, also reducing stress on the electricity grid. It also enhances the security of our energy supply and helps mitigate dependence on imported fuels.

CHP TECHNOLOGIES

CHP technologies can be divided into the following five types:
- Backpressure power plants, the simplest cogeneration power plants, in which CHP electricity and heat are generated in a steam turbine.
- Extraction condensing power plants, which generate only electricity; however, in an extraction condensing power plant, some part of the steam is extracted from the turbine to generate heat.
- Gas turbine heat recovery boiler power plants, wherein heat is generated with hot flue gases of the turbine. The fuel used in most cases is natural gas, oil, or a combination of these.
- Combined-cycle power plants, consisting of one or more gas turbines, heat recovery boilers, and a steam turbine; these have become quite common.
- Reciprocating engine power plants, wherein, instead of a gas turbine, a reciprocating engine, such as a diesel engine, can be combined with a heat recovery boiler, which in some applications supplies steam to a steam turbine to generate both electricity and heat.

The technology used by a cogeneration plant has to be planned with a specific and accurate analysis of electrical loads and heating requests by the user. The analysis should evaluate the maximum power demand and load curves (daily, monthly, and seasonally). This makes the application of a CHP system absolutely not generalizable; the choice of appropriate technology and the size have to be evaluated on a case-by-case basis.

FURTHER DEVELOPMENTS

In recent years, CHP has suffered from adverse market conditions such as increasing natural gas prices, which have reduced the cost-competitiveness of CHP; falling electricity prices, resulting from market liberalization and increased competition barriers to accessing national electricity grids to sell surplus electricity; and relatively high start-up costs.

According to the European Commission (2006), future CHP production capacity will largely be based on natural gas combined-cycle gas turbines and small gas turbines. These projections also show that the use of biomass is expected to emerge as a significant fuel for

CHP over the period to 2030 only if future ambitious renewable energy targets are reached.

Increasing the share of CHP will help countries to reduce greenhouse gas emissions and meet their commitments under the Kyoto Protocol of the United Nations Framework Convention on Climate Change.

—Guido Mattei

Further Reading

Amidpour, Majesh, and Mohammad Hasan Khoshgoftar Manesh. *Cogeneration and Polygeneration Systems.* Academic Press, 2021.

Bartnik, Ryszard, and Zbigniew Buryn. *Conversion of Coal-Fired Power Plants to Cogeneration and Combined-Cycle: Thermal and Economic Effectiveness.* Springer, 2011.

Breeze, Paul. *Combined Heat and Power.* Academic Press, 2018.

Combined Heat and Power Association. "History of the CHPA." http://www.chpa.co.uk.

European Commission. *Combining Heat and Power: Using Cogeneration to Improve Energy Efficiency in the European Union.* European Commission, Directorate-General for Energy and Transport, 2005.

European Environment Agency. "Climate Change and a Low-Carbon European Energy System." Report 1/2005. 29 June 2005. http://www.eea.europa.eu/publications/eea_report_2005_1.

European Environment Agency. "EN 20 Combined Heat and Power (CHP)." http://www.eea.europa.eu/data-and-maps/indicators/en-20-combined-heat-and.

Flin, David. *Cogeneration: A User's Guide.* The Institution of Engineering and Technology, 2010.

Kolanowski, Bernard F. *Small-Scale Cogeneration Handbook.* Fairmont, 2011.

United Kingdom, Department for the Environment, Food and Rural Affairs. "The Government's Strategy for Combined Heat and Power to 2010." http://www.lowcarbonoptions.net/resources/Policy-&-Measures/Policies-and-Measures-United-Kingdom/chp-strategy.pdf.

Cold Fusion

FIELDS OF STUDY
Catalysis; Nuclear Chemistry; Physical Chemistry

ABSTRACT
Cold fusion refers to a radiation-free form of fusion energy that, although much discussed, exists in theory only, yet has been posited to explain the results of various experiments at room temperature. Although disputed and controversial because of previous scandals associated with it following some exaggerated claims, the hypothesis continues to spark interest among scientists.

KEY CONCEPTS
acetone: 2-propanone, also known as dimethyl ketone
catalyst: a material that takes part in a reaction process, serving to lower the energy barrier to the reaction and so promote its occurrence, but is not itself consumed in the reaction
fusion: the joining together of two atomic nuclei to produce the nucleus of an atom of a heavier element
heavy water: water molecules in which the normal hydrogen atoms of H_2O have been replaced by the heavier isotope deuterium, producing D_2O

IN THEORY...
Cold fusion refers to a radiation-free form of fusion energy that, although much discussed, exists in theory only. The concept is not new among the scientific community, but it is one with which many scientists have been uncomfortable because of previous scandals associated with it.

Cold fusion theoretically provides a way to generate clean energy from uranium without incurring the harmful effects, including radiation, that come from traditional ways of generating a nuclear reaction. Although failed claims related to some researchers' investigations into cold fusion have brought disrepute to the notion, recent developments have once again made some scientists interested in cold fusion as a possible alternative energy source.

If cold fusion could prove to be practicable, or even possible, it could represent a limitless, cheap, and pollution-free source of energy. New and inexpensive measuring devices have reinvigorated the field, as these permit more laboratories to conduct cold fusion research. Additionally, recent findings indicate that cold fusion might occur naturally in certain bacteria, and some scientists have been excited by research that centered on a battery based on cold fusion. Despite this enthusiasm, however, a great deal of trepidation remains regarding cold fusion because of prior scandals and unsubstantiated claims regarding cold fusion discoveries and research.

EXPERIMENTAL EVIDENCE...MAYBE

Cold fusion received a great deal of media attention in 1989, when researchers Martin Fleishmann and Stanley Pons claimed they had achieved nuclear fusion at room temperature using a simple, inexpensive tabletop device. The small tabletop experiment involved electrolysis of heavy water on the surface of a palladium (Pd) electrode. Fleishmann and Pons hypothesized that the high compression ratio and mobility of deuterium that could be achieved within palladium metal using electrolysis might result in nuclear fusion. To investigate, they conducted electrolysis experiments using a palladium cathode and heavy water within a calorimeter, an insulated vessel designed to measure process heat.

Current was applied continuously for many weeks, with the heavy water being renewed at intervals. For most of the time, the power input to the cell was equal to the calculated power leaving the cell within measurement accuracy. They then decided to announce their findings. However, the wild excitement that greeted their news quickly dissipated when other scientists could not reproduce their results. Pons and Fleishmann eventually closed their laboratories and left for France to continue their research at a laboratory sponsored by Toyota Motor Corporation. That laboratory closed in 1998 without any success in achieving cold fusion, and the term itself fell into disrepute.

Immediately following 1990, attention given to cold fusion in the media and research conducted on the process dropped precipitously. Many in the scientific community believe that cold fusion is impossible, or at least highly improbable. Cold fusion depends on a nuclear reaction occurring within a crystal structure, which is unlikely, because all nuclei are positively charged and as a result they strongly repel each other. Without a catalyst such as a muon, very high kinetic energy generally is needed to overcome this repulsion. Skeptics of cold fusion assert that those calculations that demonstrate increased heat are based on faulty assumptions.

For example, Fleishmann and Pons assumed in their experiments that the efficiency of electrolysis is nearly 100 percent. Other researchers have asserted that if hydrogen and oxygen recombine to a significant extent within the calorimeter, electrolysis is less efficient and that this might account for the excess heat sometimes attributed to cold fusion. Due to skepticism regarding cold fusion, the U.S. Patent and Trademark Office has consistently rejected patent applications for apparatuses alleging cold fusion on the grounds that such devices do not work. Indeed, the U.S. Court of Appeals for the Federal Circuit, which hears many patent disputes, ruled in *In re Swartz* (232 F.3d 862, Fed. Cir. 2000) that a device alleging cold fusion can be rejected on its face as inoperable.

Since then, scientists' interest in the idea of cold fusion has revived, and examinations are being undertaken with the goal of alleviating the need for inexpensive energy sources. Many are looking for ways of becoming "greener" to help protect the environment from the damage that traditional sources of energy have caused. Cold fusion is envisioned as a way to help reduce the amount of pollution that comes from making energy, since it represents a clean energy source. Cold fusion's only by-product would be helium particles, which do not harm the environment. Cold fusion would also be much less expensive than producing the amount of oil people use on a daily basis.

ACHIEVING COLD FUSION...MAYBE

The most expensive element needed for cold fusion, heavy water, costs about $1,000 per kilogram retail, despite its ubiquitous nature. Heavy water is expensive because a great deal of energy is required to separate it from ordinary water and because demand is limited, so new separation technologies have not been developed. However, a huge amount of energy can be derived from heavy water from a fusion reaction. Even at $1,000 per kilogram, heavy water would be thousands of times less expensive than oil. In a heavy water cold fusion economy, a fraction of a percentage of the fuel would have to be recycled to keep the heavy water separation plants working, whereas today 7 percent of oil goes to refinery use and loss.

Proponents of cold fusion believe that, in order for the process to be successful, three conditions must be met. First, the cold fusion devices must be safe and nonpolluting to keep with the green goals that the process favors. Second, cold fusion generators, motors, heaters, and other devices must have high power density, so they can be roughly as compact as competing motors. Third and last, cold fusion must be transferable to a wide range of uses, from devices such as thermoelectric pacemaker batteries to automobile, marine, and aerospace engines.

While such benefits are enticing, little evidence supports the proposition. Indeed, most recent projects that have purported to demonstrate cold fusion have ended in scandal, recriminations, and sanctions. In 2002, for example, nuclear engineer Rusi Taleyarkhan claimed to have observed a statistically significant increase in nuclear emissions of products that he attributed to cold fusion, which he termed "bubble fusion." Taleyarkhan is a faculty member at Purdue University, prior to which he was a staff scientist at the Oak Ridge National Laboratory. He alleged that he had observed fusion reactions during acoustic cavitation experiments with chilled acetone bombarded with deuterium.

Although Taleyarkhan published several papers outlining his results, other scientists working independently were not able to replicate his work. After a series of disputes, Purdue began an investigation regarding Taleyarkhan's claims, which resulted in his being judged guilty of research misconduct for falsification of the research record. Despite this, the National Science Foundation provided Taleyarkhan with $185,000 in research funds to investigate bubble fusion.

The tantalizing potential benefits of cold fusion permit it to remain of interest to some, especially those who desire inexpensive, green energy. Despite ongoing research regarding cold fusion, however, little progress has been made. It appears that cold fusion will remain a tantalizing possibility for the foreseeable future.

—*Lisa Blagg*

Further Reading

Biberian, Jean-Paul, ed. *Cold Fusion: Advances in Condensed Matter Nuclear Science.* Elsevier, 2020.

Huizenga, J.R. *Cold Fusion: The Science Fiasco of the Century.* Oxford UP, 1994.

Kozima, Hideo. *The Science of the Cold Fusion Phenomenon: In Search of the Physics and Chemistry Behind Complex Experimental Data Sets.* Elsevier, 2006.

Mallove, E.J. *Fire and Ice: Searching for Truth Behind the Cold Fusion Furor.* John Wiley & Sons, 1991.

Sutton, Anthony C. *Cold Fusion: The Secret Energy Revolution.* Dauphin Publications, 2016.

Taubes, G. *Bad Science: The Short Life and Weird Times of Cold Fusion.* Random House, 1993.

Communication to Garner Support for Energy Development

FIELDS OF STUDY
Communications; Journalism; Social Sciences

ABSTRACT
Communication is critical for the energy industry if it is to garner public understanding and support. Scholars have long noted the barriers to this communication, such as the complexities of science, significant differences in the predispositions of scientists and most members of the public, and low public understanding of scientific and technological issues generally. The federal government establishes communication campaigns and other projects to improve energy efficiency nationwide.

KEY CONCEPTS

coercion: the use of punitive consequences for failing to comply with corporate and governmental directives

demonstration: typically, a small-scale project designed to show the features, characteristics and usefulness of a full-scale implementation

energy waste: unnecessary energy consumption and consumption beyond the immediate needs of the operations

greenhouse gas emissions: the emission of gases known to contribute to the "greenhouse effect" from power generation facilities and operations, particularly of carbon dioxide (CO_2), hydrocarbons, and other gases

offshore drilling: drilling of oil and gas wells in the waters overlying the continental shelves rather than on land

scientific terminology: words and phrases specific to various scientific fields and operations, not used in general parlance

COMMUNICATION IS KEY

Communicating to garner public understanding and support is critical for the development of science and technology, including energy. Scholars have long noted the barriers to this communication, such as the complexities of science, significant differences in the predispositions of scientists and most members of the public, and low public understanding of scientific and technological issues generally.

COMMUNICATING ENERGY TO THE PUBLIC

The fate of the energy companies, particularly those involving nuclear power and offshore drilling, is largely determined by public opinion. For example, whereas the nuclear energy industry's promise was thwarted by accidents at Three Mile Island in 1979 and Chernobyl in 1986, it is now experiencing a resurgence of interest as a source of energy with low greenhouse gas (GHG) emissions and hence a strategy for coping with global warming. As Baruch Fischoff points out, the most critical factor determining the future of nuclear energy is whether the industry can communicate effectively with the public.

Fischhoff proposes several ways in which the nuclear industry can gain public support through communication. The overall principle is that the nuclear industry needs to care about the public and bring genuine benefits with acceptable risks to society. Two-way communication based on empirical understanding of the public must be a strategic priority. In all activities, nuclear industry managers should avoid unwittingly sending out messages that contribute to a negative image of their business. The industry must pay attention to public perceptions of events and organizations related to the industry, particularly with regard to public health issues. Communication efforts and projects should be evaluated frequently, so that lessons from past experiences can be learned and solutions applied.

Several factors influence the effectiveness of energy communication to the public. A survey of 1,475 California adults by Juliet Carlisle and her associates shows that more than half of Californians believe that offshore drilling for oil is risky. Members of the public tend to believe messages that offshore drilling is risky regardless of the source of the message, but when the message says that offshore drilling is safe, the public tends to respond based on the source, with greater levels of belief if the source is an environmental organization scientist and less if the source is a governmental scientist; the public response is most skeptical when the source is an oil industry scientist.

PUBLIC PERCEPTION AND MEDIA COVERAGE OF ENERGY ISSUES

Americans' support for energy conservation is high, but their attitude toward the energy industry varies. Public support for nuclear energy and offshore drilling has generally been lower than for other types of energy production. Linda Pifer finds that antinuclear attitudes have been particularly strong among young adults. Moreover, from the 1960s to the 1980s, women were more likely to oppose nuclear power than men. A study finds that women show lower levels of knowledge about nuclear energy, tend to associate nuclear power with dangers, and are more likely to be antinuclear activists compared to men, while other studies find that nuclear knowledge is distributed equally between nuclear supporters and opponents. More science education leads to more favorable attitudes toward nuclear power. For example, Pifer found that a majority of those young adults who took at least four college science courses believed that the benefits of nuclear power outweighed its risks. Generally, in the United States, Democrats and liberals are more concerned with the risks of nuclear power and offshore drilling, whereas the Republicans and the conservatives are more likely to see benefits.

Some scholars have speculated that Americans' fear about nuclear power and offshore drilling stems from exaggeration of the risks on the part of media. Sharon Friedman and her associates' study do not find the media guilty, however. They analyzed the coverage of the Chernobyl accident by five major U.S. newspapers and three television networks and did not find exaggeration of the nuclear dangers, other than a large number of references to the need for stricter nuclear regulation in the United States.

However, although the coverage was balanced, with both pronuclear and antinuclear opinions, not enough background information was provided to put the accident in context. Among 394 articles and 45 newscasts devoted to the accident, less than 25 percent contained information about safety records, the history of nuclear accidents, and the current status of the nuclear industry; these were important to understand the accident and nuclear industry thoroughly.

COMMUNICATING ENERGY IN CAMPAIGNS

Communication campaigns are effective tools to increase public understanding of energy and inform the public about new energy technologies. Research suggests that successful campaigns should be planned based on social science theories, such as social learning, reasoned action, diffusion of innovation, and framing. These tools can show what made previous campaigns succeed and what should be expected from current campaigns. Benchmark evaluation of a campaign is also necessary, as it reveals existing audience knowledge, attitudes, and behavioral intentions and provides a basis for measuring campaign effects. Public engagement allows campaigns to be adjusted to achieve the best results and enhance public understanding of the information conveyed in the campaign. Finally, a clear and measurable objective helps to make the campaign focused.

One type of campaign often used by governmental agencies is demonstration. Demonstrations are usually used to promote a particular type of innovative technology. Although they are more expensive compared with alternative methods, such as communicating through mass media, they provide firsthand observation, experience, and measurements and attract bona fide users of the technologies. Susan Macey and Marilyn Brown have found that demonstrations cause changes in the long term but have few short-term influences. Identifying and engaging target audiences are essential. For instance, the Hood River Conservation Project, a $20 million resident conservation demonstration proposed by the Natural Resources Defense Council and funded by the Bonneville Power Administration, engaged many people within and outside those agencies in its design and evaluation plan development. The broad involvement led to a widespread use of the project's results. The Department of Energy has also sponsored demonstrations of catalytic distillation, an energy-efficient technology. Together with financial incentives, the demonstrations attracted a substantial number of petroleum companies, most of which adopted the new technologies. Sufficient publications, such as brochures, magazines, and articles, are also critical, because a successful demonstration needs to provide enough information to help people make the decision to adopt.

Energy companies also use demonstrations when they promote new technologies. In 1999, Southern California Edison, the primary electricity supplier in Southern California, initiated an "Emerging Technologies Showcase" to promote emerging energy-efficient technologies. The demonstrations were arranged in places where there was market potential. Most of the demonstrations provided information such as engineers' analyses of the advantages and disadvantages of the emerging technologies and the current technologies. The information was disseminated through on-site publications, the company's website, and business journals. Southern California Edison organized about 60 demonstrations and effectively opened a market for the emerging technologies.

OTHER GOVERNMENTAL ENERGY COMMUNICATION PROGRAMS

In addition to campaigns, governmental agencies have established other programs to improve public knowledge and efficient use of energy. The Department of Energy's Federal Energy Management Program (FEMP), for example, grants annual awards for energy and water management to individuals, organizations, groups, and agencies that have made significant contributions to efficient usage of energy and water resources. Pictures of award winners are published on the department's website. The Environmental Protection Agency (EPA) also has an Energy Star project, which has set voluntary international standards for energy-efficient equipment and is working with appliance producers to reduce standby power consumption. Governmental agencies also provide graphic information about energy efficiency on their websites and in their publications. For example, the EPA published a calculation that a hotel reducing 10 percent of its energy consumption would have the same financial benefit as selling an additional 930 rooms per year, and that a supermarket reducing 10 percent of its energy consumption could boost profit margins by 6 percent. The EPA also estimated that a desktop computer could save $25 to $75 annually if the power management function was turned on to allow it to enter low-power mode when it was idle. If all the power management functions in the monitors and the computers in the United States were activated, $1.5 billion could be saved.

U.S. presidents have also addressed energy issues. Jeffrey Peake and Matthew Eshbaugh-Soha found that presidents' national addresses on energy issues had the effect of increasing media coverage on those issues. In July 2001, President George W. Bush issued an executive order requiring all the federal agencies to buy equipment with low standby power functions.

BUSINESSES' INTERNAL ENERGY AWARENESS PROGRAMS

Businesses have also begun to educate their employees to use energy more effectively and thus reduce the amount of energy the businesses consume. In the past, workers have habitually left on lights, computers, and other electrical equipment overnight and on weekends. John Eggink calls this electrical waste. He is amazed that wasting electricity is allowed in many companies and even encouraged in some, deemed as a sign of financial success. He finds that electrical waste costs business billions of dollars each year. For all types of business, from plants to offices, more than 15 percent of the electrical bills is estimated to pay for wasted electric energy.

Increasing employees' awareness of energy consumption, energy costs, and environmental consequences and training them to use energy efficiently, through what Eggink calls energy awareness programs, is helping many companies to save hundreds of thousands of dollars. For example, a program of this type at Verizon cost less than $5,000 but saved 10 million kilowatt-hours and $750,000 in California alone. IBM estimated that it saved $17.8 million one year by encouraging employees to turn off idled electric equipment. Between 1991 and 2004, the UK telecommunications group BT saved $214 million because of its energy awareness program and another $564 million because of its efficient use of transport.

Finding out why people waste electricity is the foundation for a successful energy awareness program. Eggink finds that, because electricity is invisible, people do not naturally witness the energy lost. Although the wattage appears on the monthly utility bill, employees usually do not see the bill or may see it only long after the energy has been consumed. Showing the bill to everyone in a timely manner may be an effective way to increase awareness of energy usage and waste. Visualizing energy usage patterns with graphs is also helpful, allowing people to track the wasting of energy. Electricity generators usually charge higher rates during peak usage times, usually from noon to 7:00 p.m., to avoid the collapse of the grid. People unaware of these rate differences may schedule unnecessary usage of electricity during peak times and thereby create waste by incurring the higher rate, which could have been avoided had they scheduled usage at a different time. Therefore, it is important that energy users be made aware of peak periods.

Meanwhile, the complexity of the scientific terminology may dissuade people from understanding energy issues. Communicating in a relevant, personal, accurate, and easily understood manner and using terms such as *dollars* and *percentage* encourages understanding. Relating electricity to other nonrenewable energy resources, such as coal, oil, and natural gas, can also increase people's motivation to conserve. In fact, according to the U.S. Energy Information Administration's 2004 data, about 50 percent of the electricity in the United States is generated from coal, 20 percent from nuclear power, and 18 percent from natural gas.

In energy awareness programs, managers' direct orders and mandates can be counterproductive and may well generate rebellious behavior. Eggink notes that education focusing on positive reasons to conserve energy and successes tends to reinforce employees' desirable behaviors. Therefore, in each call to save energy, the knowledge, reasons, and possible benefits of saving energy should be provided. Coercion may be more effective than education only where governmental regulations and laws, and attendant consequences for noncompliance, are concerned. Energy awareness programs should be persistent, with materials freshened frequently. Basic information is important: One plant saved $30,000 on electric bills in the first year simply by letting everyone know which switches controlled which lights by labeling the switches.

Knowledge is essential in energy awareness programs, because even highly educated people may subscribe to myths, such as that leaving lights and computers on saves more electricity than switching them on and off, or that switching them on and off shortens their life spans. Eggink also found that people believe computer screensavers save energy. The success of energy awareness programs, he concludes, depends on proper communication regarding energy, which should ensure that employees understand the issues, know the impact of their individual energy use, and know that they can easily make a difference.

COMMUNICATING DURING CRISES

Crisis communication is a necessary skill for the energy industry, because crises that involve energy companies often have tremendous impacts. Sonya Duhé finds that having a crisis communication plan in place is the first step in dealing with a crisis. Assuming that a crisis will never happen and having no plan ready is dangerous. Although it is impossible to make precise predictions about when a crisis might happen or what its nature might be, management can still anticipate what types of crises could occur. The crisis communication plan should include backup communication and operation systems, locations to convene the media, and answers for questions that members of the media can be expected to ask. Moreover, the plan needs to be tested to ensure that it works. Selecting the appropriate spokesperson is also very important. During crises, researchers argue, a top manager or the chief executive officer should be the spokesperson. If necessary, a team of spokespersons may be formed to make sure that no miscommunication will occur and to amplify the consequences of the crisis.

Duhé has also provided some rules of crisis communication. Preparation is number one. Since there is no time to prepare after the crisis has occurred, failure can be guaranteed if there is no preparation beforehand. The preparation includes knowledge on the topic and the subject, such as the consequences of an oil spill or a nuclear meltdown. If there is no answer available, the spokesperson should inform the media that the answer is unknown but will be found soon. In response to a question, a spokesperson should never say "No comment," which will be interpreted as deceptive and will suggest that the company has something to hide. If the question is a "What if?" (hypothetical) question, it is fine to refuse to answer, unless the situation in the assumption is very likely to happen immediately. Agreeing and assuming that journalists will keep comments "off the record" is dangerous. Most of the time, such comments will have negative consequences; the job of journalists is to elicit information and tell the public. Using visuals, such as graphs and videos, will help journalists to understand the topic and produce their stories. Those visual elements will also help the public to absorb information about the crisis.

During the crisis of a public energy shortage, governments also need to communicate with the public—immediately and directly, usually through mass media, regarding their understanding and expectations about how and when the problem will be solved. Governments have tried to use celebrities, such as movie stars and comedians, as well as other attention-getting methods, to enhance the effectiveness of crisis communication.

For example, the governments of New Zealand, Norway, and Brazil have used pictures of empty reservoirs to inform people about their energy crises. The state of Arizona has asked people to raise room temperature a few degrees and reset the timers of pool pumps when utility equipment was not working. The Swedish government asked people to lower thermostats and postpone unnecessary electrical consumption to cope with a daylong electricity shortage one winter. These communication methods not only show that the government is in charge and taking appropriate steps to alleviate the problem but also demonstrate to the citizenry that they can take individual action to mitigate the crisis.

—*C.S. Eigenmann*

Further Reading

Carlisle, Juliet E., et al. "The Public's Trust in Scientific Claims Regarding Offshore Oil Drilling." *Public Understanding of Science*, vol. 19, no. 5, 2010.

Duhé, Sonya. "Crisis Communication." *Encyclopedia of Science and Technology Communication*, edited by S.H. Priest. Sage, 2010.

Eggink, John. *Managing Energy Costs: A Behavioral and Non-Technical Approach.* Fairmont Press, 2007.

Fischhoff, Baruch. "The Nuclear Energy Industry's Communication Problem." *Bulletin of the Atomic Scientists*, 11 Feb. 2009.

Leggett, M., and M. Finlay. "Science, Story, and Image: A New Approach to Crossing the Communication Barrier Proposed by Scientific Jargon." Public Understanding of Science, vol. 10, no. 2, 2001.

Macey, Susan M., and Marilyn A Brown. "Demonstrations as a Policy Instrument with Energy Technology Examples." *Science Communication*, vol. 11, no. 3, 1990.

Pifer, Linda K. "The Development of Young American Adults' Attitudes About the Risks Associated with Nuclear Power." *Public Understanding of Science*, vol. 5, no. 2, 1995.

Computers and Energy Use

FIELDS OF STUDY

Electrical Engineering; Electronics Technology; Social Sciences

ABSTRACT

The emergence of computing since the mid-twentieth century has had multiple consequences for the consumption, management, and development of energy resources. Although traditional wisdom may assert that computers are essentially electronic products that consume energy, a comprehensive analysis of computing's overall energy impact is far more difficult to calculate.

KEY CONCEPTS

analytical engine: a programmable mechanical calculating device proposed by Joseph Babbage but never actually constructed, as a refined and advanced version of his difference engine

data center: a large array of memory storage devices and server units in which information is stored and served out to remote users as a centralized service rather than on individual private machines, for example, Microsoft Cloud, Google Dropbox, etc.

difference engine: a mechanical calculating device proposed and partly constructed by Joseph Babbage in 1822, for carrying out reliable calculations of life expectancies

digital computer: an electronic programmable calculating device that uses Boolean logic to manipulate patterns of electronic "bits" ("high" and "low" voltage signals) for the performance of various functions

Moore's law: an empirically derived law based on the observation that the ability to etch ever smaller transistor structures on silicon-based semiconductor surfaces advances at such a rate that the computing power of the resulting devices doubles approximately every two years until the physical size limit of available space will be attained, at which point no further advancement is physically possible

network society: the social culture of computing systems connected to centralized and distributed network systems

processing speed: the rate at which a particular central processing unit is capable of performing calculations and data manipulations

THE RISE OF COMPUTER TECHNOLOGY

Throughout the second half of the twentieth century, the proportion of national energy supplies devoted to computers rose consistently. Changing trends in the way computers are produced and used has meant that consistent trends or predictions for the future are difficult. The emergence of computing since the mid-twentieth century has had multiple consequences for the consumption, management, and development of energy resources. Although traditional wisdom may assert that computers are essentially electronic products that consume energy, a comprehensive analysis of computing's overall energy impact is far more difficult to calculate.

Computers draw energy when used and represent considerable energy expenditure in their creation; however, media theorists and social theorists would also point out that the rise of the "network society" is changing human behavior, with potentially profound implications for energy use. Commentators such as Lev Manovich and Manuel Castells have argued that the computerization of culture is in the process of radically changing the ways in which society is organized.

Simple examples of this can be found in the way in which computer-based satellite navigation technology can lead to mass savings (or increases) in energy use as driving patterns are changed. Similarly, the increased possibilities for video conferencing and emailing have allowed for the possibility of great savings in the energy-intensive transportation of people. At a more direct level, technologies are now emerging that allow home owners to monitor and track their own energy use. From smart meters to smart appliances, the trend in energy markets globally is for an integration of computer network technologies with consumer appliances. In this case it is not simply human behaviors that computing affects: The behaviors of appliances themselves start to change as they are connected to smart grids that operate to balance supply and demand. This facilitates the adoption of intermittent renewable energy technologies that previously were not feasible on energy grids that required a consistently reliable flow of power to meet variable demand. Thus, social scientists would point out that while computer technologies have the potential to change behavior and save energy, it is extremely difficult to determine the savings versus the additional costs.

Calculating devices such as the abacus date back to at least 2400 BCE and used essentially no energy since they were operated by hand manipulation. Following the Industrial Revolution, the computer, like many other tools, was conceived as a mechanical device. Even the complex mechanism of the millennia-old "Antikythera device" found in the remnants of a shipwreck in the Mediterranean Sea, and now recognized as an intricate computing device, used only the energy provided by a human operator. Charles Babbage's difference engine (1822) and subsequent analytical engine (1837) are widely held to be the first major attempts to develop what can be regarded as a computer in the modern sense. Although he never finished it, Babbage initially conceived of the difference engine as a steam-powered device, famously protesting the tedium of manual calculations with the proclamation, "I wish to God these calculations had been executed by steam." Because Babbage's machines were never realized, it was not until the development of electronic digital computation that the modern era of computing began in earnest and significant quantities of energy were used for computing.

TWENTIETH-CENTURY ENERGY USE
Like their unrealized predecessors, twentieth-century computers were conceived as very large industrial devices. This meant that while they often consumed considerable quantities of energy in their operation, there were also few devices in operation relative to other industrial processes. In other words, early computers consumed a great deal of energy but were few in number and so had negligible impact on the power consumption of their host nations. However, the development of microchip technology led to the possibility and development of personal computers. In a reversal of the previous logic, the production of personal computers led to great savings in the energy consumption of individual machines but was followed by the exponential growth in number of devices. Over time, this has meant that computers have come to consume an ever-increasing proportion of many nations' energy budgets.

For the duration of computing history, machines have been widely governed by Moore's law, which states that the transistor density (and therefore the capacity of the computer's central processing unit, or CPU) doubles every two years at no increase in cost. In recent years, industry analysts have suggested that a new trend may be emerging in which energy consumption is as important a driver of research and development as is processing speed. With the move toward mobile computing, the manufacturing industry is now searching for energy efficiencies in order to circumvent the limitations of battery technologies. At the same time, much attention has been turned to the energy costs of the internet.

While personal computers may become increasingly efficient, the server farms used as the backbone of the internet are growing in quantity and energy demand. In response to a study undertaken in 2009, many British and American newspapers ran headlines claiming that performing two Google searches from a desktop computer can generate about the same amount of carbon dioxide as boiling a kettle. This sparked a public debate about the energy and carbon costs of the internet in which Google protested that such headlines were misleading and inaccurate. Google pointed out that the precise energy and carbon costs of internet usage have not been agreed upon by experts. Nevertheless, the story highlighted the fact that the internet is not an energy-neutral construct in the minds of the public.

Since then, energy usage has risen to the top of the information technology (IT) industry's agenda, with both Google and Microsoft investing significantly in both energy monitoring and renewable energy development projects. Headline-grabbing initiatives such as Google's patent for seawater-cooled, offshore server farms that generate their power natively from wind and wave power have served to highlight the broader significance of energy to the IT industry.

Concerns over the energy consumption of power-hungry data centers have led to the formation of an international corporate consortium known as the Green Grid. The consortium seeks to unify and organize a set of global industry standards that will help improve the energy efficiency of IT hardware and networks. In response to these concerns, some companies (notably IBM) have been developing water-cooling methods for their hardware that could divert energy for reuse in buildings and increase the energy efficiency of data farms. Other initiatives include the location of server farms in Iceland, where machines can be powered with renewable thermal energy and cooled with locally available freezing water. Similarly, the Scottish government has suggested that Scotland could become a wind-powered server hub for the United Kingdom and Europe.

Environmental campaigner and earth scientist James Lovelock has argued that computer use could be a positive way to reduce energy and therefore carbon emissions. Lovelock argues that computer use is a relatively low carbon activity compared with forms of entertainment, such as tourism, which involve energy-intensive processes of transportation. This argument, however, does not take into account the relationship between personal computers and the larger global computer networks to which they are connected. With this very problem in mind, analyst Jonathan Koomey has tried to quantify the energy consumption of server infrastructures around the world.

A look at the future of computers and their energy and carbon use suggests that the current shift toward cloud computing will further complicate the picture. While personal computers have traditionally stored their data locally on machines owned by users, the trend with mobile computing is increasingly toward the storage of personal data in "the cloud," that is, on remote servers whose location and the responsibility for whose maintenance are not concerns for the user. While such a shift will be more convenient for users, in many cases media analysts such as Sean Cubitt are beginning to question the wisdom of a process that has the potential to exponentially increase dependence and demand on remote, energy-intensive server farms. With this in mind, it seems likely that in the future the debate regarding energy and computing will be balanced between both personal devices and large-scale industrial network nodes reminiscent of the 1940s, 1950s, and 1960s.

—*Leon Gurevitch*

Further Reading

Cubitt, Sean, et al. "Does Cloud Computing Have a Silver Lining?" *Media Culture and Society*, vol. 32, no. 6, Nov. 2010.

Fen, Wu-Chun. *Green Computing: Large-Scale Energy Efficiency*. CRC Press, 2011.

The Green Grid. "About the Green Grid." http://www.thegreengrid.org/about-the-green-grid.

Kempos, Chris, David H. Wolpert, and Peter F. Stadler. *The Energetics of Computing in Life and Machines*. Santa Fe Institute Press, 2018.

Koomey, Jonathan G. "Estimating Total Power Consumption by Servers in the U.S. and the World." Final Report. Released 15 Feb. 2007, http://blogs.business2.com/greenwombat/files/serverpoweruscomplete-v3.pdf.

Lovelock, James. *The Revenge of Gaia: Earth's Climate Crisis and the Fate of Humanity*. Basic Books, 2006.

Manovich, Lev. *Software Takes Command*. 20 Nov. 2008, http://softwarestudies.com/softbook/manovich_softbook_11_20_2008.pdf.

Munir, Kashif. *Cloud Computing Technologies for Green Enterprises*. IGI Global, 2018.

Pierson, Jean-Marc, ed. *Large-Scale Distributed Systems and Energy Efficiency: A Holistic View*. Wiley, 2015.

Sui, Daniel, and David Rejeski. "Environmental Impacts of the Emerging Digital Economy: The E-for-Environment E-Commerce?" *Environmental Management*, vol. 29, no. 2, 2002, pp. 155-63.

The Times. "Revealed: The Environmental Impact of Google Searches." http://technology.timesonline.co.uk/tol/news/tech_and_web/article5489134.ece.

CORN ETHANOL

FIELDS OF STUDY
Biochemistry; Fermentation Science; Microbiology

ABSTRACT
The United States is the leading producer of corn ethanol, a corn-derived alcohol fuel. While corn ethanol can be used as an alternative to fossil fuels, the steady increase in its popularity owes more to its use as a gasoline additive. The process of manufacturing the fuel yields numerous commercial byproducts.

KEY CONCEPTS

biofuels: fuels derived from recently living organic matter

distillation: separation of the components of a solution by heating to the boiling point boiling and condensation of the resulting vapors of the individual components

E rating: a designation of the percent ethanol content by volume in an ethanol-blended gasoline

fermentation: digestion of available sugars by yeast or other bacteria with ethyl alcohol and carbon dioxide being the usual products

gasohol: specifically refers to gasoline-ethanol blends containing less than 25 percent ethanol by volume

pH: a rating of the degree of acidity of a water solution, according to the equation $pH = -\log[H^+]$

ETHANOL

Corn ethanol is produced through the fermentation, chemical processing, and distillation of corn biomass, and is the most common type of ethanol fuel produced in the United States, which is the world's leading producer of ethanol fuel (accounting for about 62 percent of global production in 2011). Ethanol is ethyl alcohol, the same alcohol found in alcoholic beverages, though because of legal and tax restrictions, ethanol sold for purposes other than consumption is adulterated with methanol (wood alcohol, which is highly toxic) or additives that make it too bitter to willingly ingest. Though fuel is its most common use, ethanol can also be used as a chemical feedstock for producing organic compounds, as an antiseptic (it is the key ingredient in most hand sanitizers and has led to the unhealthy practice of drinking sanitizer by individuals of different age groups), and as a solvent.

ETHANOL FUELS

Most ethanol fuel is used as motor fuel, in the form of fuel mixtures of gasoline with low levels of ethanol. The term *gasohol* is sometimes used for mixtures of less than 25 percent ethanol (the amount used in Brazil, the second-largest ethanol producer). A common ratio in the United States is 10 percent, commonly called E10; most retail motor fuel in the United States is E10. Blends with greater proportions of ethanol are not commonly used in the United States because only recent model cars are certified by the Environmental Protection Agency to use them without damage; a study on the safety of E15 remains incomplete, and as of a 2011 ruling, E15 could only be sold accompanied with a warning that it can only be used in flex-fuel vehicles and passenger vehicles with a model year of 2001 or later. E15 is expected to supplant E10 in the United States as older cars fall out of use. E10 is also used in Thailand and Finland, and has become more available in Australia. In the United States, E70 and E75 are reduced-ethanol versions of E85 sold during the winter months to avoid cold starting problems, a practice also followed in Sweden. E85 is the standard fuel for flex-fuel vehicles, which have internal combustion engines designed to run on more than one fuel; there are 10 million such cars, light trucks, and motorcycles in the United States, and there are another 17 million worldwide. (Popular models include the flex-fuel Chevrolet Impala and the flex-fuel Ford Escape.) Because it is created from an agricultural crop, corn ethanol is a renewable energy source. Because it is produced domestically, it is less subject to price volatility than fossil fuels.

PRODUCTION OF CORN ETHANOL

To produce corn ethanol, the corn is first milled. Wet milling steeps the corn in diluted sulfuric acid for one to two days, which separates the grain into its fiber, gluten, starch, and oil. Wet milling creates a number of corn products. Corn oil is sold for cooking and other purposes. The fiber is sold as corn gluten feed for livestock. The steeping liquid can be used to deice streets. The remaining water and starch can be processed either into corn syrup, modified corn starch, or corn ethanol. The process for turning the water and starch produced by wet milling into corn ethanol is substantially similar to the dry milling process. Dry milling grinds corn kernels into flour, which is then mixed with water and enzymes that convert the starch to dextrose (sugar). Yeast is added to convert the sugar to ethanol, just as in the process of natural fermentation or the fermenting of corn for bourbon or other whiskey.

The addition of ammonia balances the pH levels of the mix in order to maintain the ideal environment for the yeast and maximize fermentation. Left to its own devices, fermentation will stall when the environment becomes too acidic or too alcoholic for the yeast to thrive, even if there is remaining sugar to convert. Byproducts of wet milling include the carbon dioxide released by fermentation, which is used to manufacture dry ice or to carbonate soft drinks, and a nutrient-rich mash of solids that is processed into feed for livestock. Most American corn ethanol is made by the dry milling process, which can be performed fairly simply.

DEMAND PROS AND CONS

In recent years, more and more corn has been diverted to the production of ethanol in response to the increased demand for biofuels, a demand driven by consumer concerns with environmental impacts, government incentives, and the volatility of fossil fuel prices. Corn ethanol has many considerable environmental and economic impacts, including its carbon intensity and its impact on the food supply. The use of arable land to produce biofuel was one of the factors in the food commodity price shocks of the early twenty-first century. The increased demand for corn also drove up the price of the crop in Mexico, where it is the staple grain. In addition to considering other crops as feedstock for ethanol—one of those often mentioned is algae, because its cultivation would not replace arable land already in use—there have been suggestions that the cellulosic remnants from crop processing could be used to produce ethanol, in order to produce it as a byproduct of agricultural activity that already exists, rather than replacing that activity.

Corn ethanol production has been strongly supported by subsidies and other government incentives. The Energy Policy Act of 2005 mandated the use of ethanol blends, guaranteeing demand for corn ethanol; despite this guaranteed demand, corn crops (regardless of end use) receive the largest share of the United States' generous agricultural subsidies, about a third of the total. Until 2012, ethanol producers re-

A typical ethanol plant in West Burlington, Iowa. (Wikimedia Commons)

ceived a $0.45 per gallon subsidy on top of that crop subsidy. A tariff on ethanol imports maintains domestic producers' advantage. Today, only about 20 percent of American corn is grown for human food, and even that figure is inflated by the subsidy-driven reliance on corn syrup as a sweetener in the food industry, while ethanol has recently surpassed animal feed as the primary use of corn crops. Although environmentalists who support the use of ethanol have touted it as just a transitional fuel to reduce reliance on fossil fuels before adopting an even greener energy, market factors suggest that the profitability of corn ethanol, and the safety net provided by subsidies to mitigate risk, will make it difficult to change.

—*Bill Kte'pi*

Further Reading

Goettemoeller, Jeffrey. *Sustainable Ethanol: Biofuels, Biorefineries, Cellulosic Biomass, Flex-Fuel Vehicles, and Sustainable Farming for Energy Independence*. Prairie Oak Publications, 2007.

Hill, Jason, et al. "Environmental, Economic, and Energetic Costs and Benefits of Biodiesel and Ethanol Biofuels." *Proceedings of the National Academy of Sciences*, vol. 103, no. 30, 2006.

Lang, Susan S. "Cornell Ecologist's Study Finds That Producing Ethanol and Biodiesel from Corn and Other Crops Is Not Worth the Energy." *Cornell University News Service*, 2005. http://www.news.cornell.edu/stories/july05/ethanol.toocostly.ssl.html.

Pimentel, David. "Corn Ethanol as Energy." *Harvard International Review*, vol. 31, no. 2, 2009.

Pimentel, David. "Ethanol Fuels: Energy Balance, Economics, and Environmental Impacts Are Negative." *Natural Resources Research*, vol. 12, no. 2, June 2003.

Yacobucci, Brent D. "Fuel Ethanol: Background and Public Policy Issues." *Congressional Research Service*, 3 Mar. 2006. http://www.policyarchive.org/handle/10207/bitstreams/2757.pdf.

D

Dams

FIELDS OF STUDY
Environmental Engineering; Geology; Hydraulic Engineering

ABSTRACT
A dam is an artificial facility that is constructed in the path of a flowing stream or river for the purpose of storing water. Dams are designed for a number of purposes, including conservation and irrigation, flood control, hydroelectric power generation, navigation, and recreation; most major dams have been constructed to serve more than one of these purposes.

KEY CONCEPTS
coffer dam: a temporary structure designed to divert water flow around the location of work being carried in the construction of a dam
compressive stress: stress on a dam structure and its surrounding terrain by the weight of water being held behind the dam
piping: an intrusion of water into and through the structure of a dam, forming a slow leak that may develop into serious erosion and the failure of the dam
probable maximum precipitation (PMP): the maximum amount of rainfall or snowmelt that is likely to be experienced by a particular dam serving a defined watershed
siltation: the deposition over time of waterborne silt, rocks and debris carried into a dam reservoir
watershed: the land area in which the natural drainage enters a specific river or the rivers within that area

BACKGROUND
A dam is an artificial facility that is constructed in the path of a flowing stream or river for the purpose of storing water. Historically, dams are the oldest means of controlling the flow of water in a stream. The primary function of most dams is to smooth out or regulate flows downstream of the dam. Generally, dams are permanent structures. In some cases, however, temporary structures may be constructed to divert flows, as from a construction site. Coffer dams are used in this regard—not to store water but to keep a construction area free of water. Such temporary structures are designed with a higher assumed risk than are permanent dams. Dams date (in recorded history) to about 2600 to 2900 BCE.

USES OF DAMS
Dams are primarily used for four purposes: conservation, navigation, flood control, and generation of hydroelectric power. A fifth but somewhat less important use is recreation. Conservation purposes include water supply (including irrigation) and low-flow augmentation (to achieve wastewater dilution requirements). Flood-control objectives dictate that a dam's reservoir be as empty as possible so that any excess water from the watershed (the area upstream of the dam that sheds water to it) can be detained or retained in the reservoir to reduce any potential flood-related damage of property and/or loss of life. A single dam can be used for all of the foregoing purposes. In that case, the reservoir is called a multipurpose reservoir. Rarely can the cost of a major dam be justified for one purpose.

TYPES OF DAMS
Dams are classified according to the type of material used in their construction and/or the structural principles applied in their design (for structural integrity and stability). Generally, dams are constructed with concrete or earthen materials readily available at the construction site. In some cases, the nature of the construction site with regard to underlying rock formation, climate, topography, and sometimes the width and load-carrying capacity of the valley in which the dam is constructed significantly influences the type of

dam selected. Common types include earthfill or earthen dams, rockfill dams, and concrete dams, which include gravity dams (their structural stability depends on the weight of the concrete), arch dams, and buttress dams. Buttress dams are further categorized as flat-slab (also called Ambursen dams after Nils Ambursen, who built the first of this kind in the United States in 1903), multiple-arch, and massive-head dams.

Arch dams are designed to take advantage of the load-bearing abutments in the valley or gorge where the dam is constructed. The structure is designed so that all the loading is transmitted to the abutments. Therefore, these abutments must be rock of high structural integrity and strength capable of sustaining substantial thrust loading with little displacement. Arch dams are typically thinner than gravity dams and are constructed from reinforced concrete. Arch dams (as the name implies) are curved to ensure that compressive stresses are maintained throughout the dam. The basic buttress dam consists of a sloping slab supported by buttresses at intervals over the length of the reservoir.

Earthen (or earthfill) dams are constructed as earth embankments. To avoid the destructive effects of seepage through the dam (especially "piping," a slow leak that develops into destructive erosion through the core of the dam), an impervious core is constructed to prevent seepage. For example, the use of compacted clay core is common. Rockfill dams are

The Seebe Dam near Exshaw in Alberta, Canada is used to produce hydroelectricity. (James Gabbert via iStock)

like earthfill dams but use crushed rock as fill material. The impermeable core is usually constructed using concrete.

Dams can be further classified as major (with a storage capacity of more than 60 million cubic meters), intermediate (with a storage capacity between 1 and 60 million cubic meters), and minor (with a storage capacity of less than 1 million cubic meters). Major dams are designed to handle the probable maximum precipitation (PMP). PMP is the estimated maximum precipitation depth for a given duration which is possible over a particular geographical region at a certain time of year. Intermediate dams are designed to withstand the flood from the most extreme rainfall event considered to be characteristic of its watershed or basin. Floods that occur every fifty to one hundred years are used in the design of minor dams.

Dams associated with hydropower plants can be categorized by the height of the water surface at the plant intake above the tailwater (the water surface at the discharge end of the hydropower plant). This vertical difference in height is called the "head." Three categories exist: low (head between 2 and 20 meters), medium (head between 20 and 150 meters), and high (head above 150 meters).

SIZING OF DAMS

The useful or service storage volume of dams is determined by an analysis of streamflows occurring at the proposed dam site and of the expected average releases or demand flows from the reservoir. Several methods have been developed, ranging from the Rippl method (attributed to Wenzel Rippl in 1883, also called the stretched-thread method) to optimal reservoir sizing schemes which employ more sophisticated operations research techniques. Rippl's method is simply a mass diagram analysis technique. Another less cumbersome method for estimating the active (service) storage of a dam is the sequent peak method. Reservoir storage can be divided into three components: flood storage capacity for flood damage mitigation, dead storage needed for sediment storage, and the active storage for the regulation of streamflow and water supply.

RESERVOIR BENEFITS AND COSTS

The construction of a dam can be justified only if it is cost-effective. That is, the total benefits resulting from its construction must be greater than the direct or indirect costs incurred in its construction and operation. Benefits accrue from hydropower sales, water supply and flow augmentation, recreation activities, and flood damage mitigation. Costs (or disbenefits) include net loss of streamflow because of evaporation, loss of water to seepage, the inundation of areas upstream from the dam, destruction of aquatic habitat, and the prevention of migration of fish in the stream, to mention a few.

SILTATION

One of the factors that can shorten the useful life of a dam is the unavoidable siltation that will occur during the projected life (service period) of a reservoir. In the design of any reservoir, the portion of the reservoir earmarked or set aside for reservoir siltation is the part of the reservoir storage referred to as dead storage. Siltation occurs in the reservoir as the sediment load carried by flow entering the reservoir is trapped in the reservoir because of decreased flow velocities within the reservoir. This siltation can lead to other problems, especially if organic debris is carried in the sediment load. Because organic material will degrade, typically resulting in the depletion of dissolved oxygen in the stored water, water quality can easily be affected by such loads. In addition, because the sediment load in the water discharged from a reservoir may be decreased because of siltation, unnaturally clarified water in the channel downstream may lead to above-normal erosion of the downstream streambed. Further, sediment deposition and buildup in a dam can submerge and choke benthic communities (bottom animal and plant life), thereby changing the character of reservoir bottom plants.

In order to control the level of siltation in the reservoir, it may be necessary to undertake programs to reduce bank erosion. Activities which lead to increased sediment load, including construction activities, must be minimized and their effects carefully monitored. Bank stabilization schemes are very important first-line defense measures against the reduction of reservoir capacity by siltation.

HYDROPOWER PRODUCTION

Hydroelectricity is produced when flow from a reservoir is passed through a turbine. A turbine is the direct reverse of a pump. In the case of a pump, mechanical energy is converted to fluid energy. For example, in a typical pumping station, a pump is supplied with electric power that is converted to mechanical energy—usually to turn a motor. The mechanical energy is then converted to energy which is imparted to the fluid being pumped, resulting in increased fluid energy.

A turbine is used to generate hydroelectricity in a directly opposite manner. In this case, water from the reservoir travels through special piping called penstocks and impinges on the turbine wheels (which can be rather large—up to 5 meters in diameter), causing them to spin at very high rotational speeds. A significant proportion of hydropower installations are used for "peaking." Peaking is the practice of using hydropower plants to supply additional electric power during peak load periods. Because hydropower plants can be easily brought online in an electric power grid (in contrast to fossil-fuel powered plants), they are often used in this manner. Hence, hydropower plants are used in most cases as part of an overall power supply grid.

There are two types of hydropower plants: storage and pumped storage. In a storage hydropower plant water flows only in one direction. In contrast, in a pumped-storage plant, water flow is bidirectional. In pumped-storage facilities, power is generated during peak load periods, and during off-peak (low load) periods water flow direction is reversed—water is pumped from the tailwater pool (downstream) to the headwater pool (the upstream reservoir). This is economically feasible only because the price of energy is elastic and time dependent. During peak load periods, energy is relatively expensive, so the use of a hydropower plant to meet demand requirements is cost-effective. During off-peak periods, it is economical to pump water upstream, and water is stored (or re-stored) as potential energy. The amount of hydroelectricity generated for a unit flow of water is directly proportional to the volumetric rate of flow, the hydraulic head (approximately the difference between the surface water elevation at the headwater pool and the surface water elevation at the tailwater pool), and the mechanical efficiency of the hydropower plant (turbines).

Researchers continue to explore the most cost-effective way to incorporate hydropower production in the operation of multipurpose reservoir systems. Many operational research techniques have been reported in the literature. In general, mathematical formulations (models) that describe important interactions and constraints necessary to model and evaluate operational objectives are developed and solved by efficient methods. The value of such work lies in the fact that substantial benefits could be reaped from efficient use of existing reservoirs (dams) in contrast to the capital-intensive construction of new ones.

Hydropower production is a nonconsumptive use of water. This means that the water that passes through the turbines can be used (without undergoing any treatment) for other purposes. This nonconsumptive nature is one of the most attractive aspects of hydropower production; hydropower generation does not result in the significant degradation of water quality. However, problems with hydropower production do exist, most notably dissolved oxygen reduction and the adverse effects on aquatic life sensitive to changes in dissolved oxygen. Highly sensitive marine species which require a relatively stable aquatic system may suffer shock and undue stress from the drawdown of reservoir pool levels. Furthermore, pool elevation changes (swings) to sustain hydropower production may result in the significant reduction of recreational benefits. On the other hand, smaller downstream flows (during reservoir filling times and low water-demand periods) may cause water quality to degrade by increasing the temperature of water and pollutant concentrations.

Significant temperature increases can be devastating to some species. For example, it is known that even slight temperature increases affect trout and salmon. The optimum temperature range for salmonids is about 6°C to 13°C. Adult fish will die when the water temperature exceeds 28°C, while the juveniles will die when the temperature exceeds 22°C. Unfavorable thermal conditions may discourage fish migration and cause the death of marine life because increasing temperatures delay or postpone the migration of adult fish, encouraging or promoting the development of fungus and other disease organisms. Eventually the balance of the ecosystem is modified, since predator or competitive species are favored, adversely affecting the salmon or trout population.

OTHER ECOLOGICAL EFFECTS

With the construction of a dam, several permanent or temporary changes may occur to an ecosystem. These include changes attributable to the pool of water behind the dam, such as temperature increases because the water is relatively stagnant. In addition, salts and hydrogen sulfide could accumulate. These altered conditions could lead to the decimation of sensitive stream-type organisms, stressed marine life, diseases or disablement, and displacement of native marine and aquatic organisms. Increased temperature of the reservoir pool could also lead to increased evaporation and other ecological problems. In delta areas, the lowering of the rate of river flow (and volume) by upstream dams may result in saltwater intrusion.

—Emmanuel U. Nzewi

Further Reading

Evans, Stephen G., Reginald L. Hermanns, Alexander Strom, and Gabriele Scarascia-Mugnozza, eds. *Natural and Artificial Rockslide Dams.* Springer, 2011.

Fell, Robin, Patrick MacGregor, David Stapledon, Graeme Bell, and Mark Foster. *Geotechnical Engineering of Dams.* 2nd ed. CRC Press, 2015.

Hager, Willi H., Anton J. Schleiss, Robert M. Boes, and Michael Pfister. *Hydraulic Engineering of Dams.* CRC Press, 2021.

Leslie, Jacques. *Deep Water: The Epic Struggle Over Dams, Displaced People, and the Environment.* Farrar, Straus and Giroux, 2005.

Mays, Larry W. *Water Resources Engineering.* John Wiley & Sons, 2005.

Scudder, Thayer. *The Future of Large Dams: Dealing with Social, Environmental, Institutional, and Political Costs.* Earthscan, 2005.

Tanchev, Ljubomir. *Dams and Appurtenant Hydraulic Structures.* 2nd ed. CRC Press, 2014.

Wang, Jun-Jie. *Hydraulic Fracturing in Earth-Rock Fill Dams.* Wiley, 2014.

Daylighting

FIELDS OF STUDY
Architecture

ABSTRACT
Daylight refers to the visible spectrum of radiation received from the sun, and daylighting is used to indicate the capture and use of this visible spectrum for illuminating spaces inside buildings and underground.

KEY CONCEPTS

daylighting: interior lighting achieved by capturing and directing sunlight to the interior spaces of buildings

diffuser: a device designed to scatter incoming sunlight evenly to reduce its intensity and harshness on the eyes

facade: the side of a wall, whether interior or exterior, that faces the outside; interior facades are used to capture heat energy and diffuse light entering from the exterior facade

Fresnel lens: a type of lens that uses a series of concentric edges to refract light towards a central focal point

light shelves: essentially a series of baffles that reflect incoming light within a light tube to produce diffuse light as its output

light tube, solar tube: a device built into the roof or walls of a building to capture and direct sunlight to the interior of the building

CAPTURING SUNLIGHT

The capture of visible light for use in interior spaces, in its simplest form, is achieved by allowing it to penetrate into a building through side windows, roof windows, and other areas of glazing, such as glass doors and curtain walling. Energy savings are accrued from daylighting, as a result of the reduced use of electrical lighting in the daytime as well as reduced demand for space heating in cold climates resulting from passive solar heating.

Daylight or natural light is important for maintaining physiological processes in the human body. The hormone melatonin, which induces sleep, is stimulated by darkness and secreted by the pineal gland in the brain, and it is at its highest levels in the dark hours of the night. The normal day-night cycle sees melatonin levels rise toward the evening and night and decline in the early morning hours, which enables humans to wake up from their night's sleep. In polar regions, the length of day varies according to the time of the year; the day increases up to 24 hours in the summer months and decreases up to zero hours in the winter months. It is now known that seasonal affective disorder (SAD), also called winter depression, occurs among the populations of some countries in the polar regions, which have long nights for five to six months of the year. Those affected by this disorder

Daylighting is the practice of placing windows, skylights, other openings, and reflective surfaces so that sunlight (direct or indirect) can provide effective internal lighting. (Julie Anne Birch via iStock)

show symptoms of depression, oversleeping, and related social problems. One of the recommended treatments for this condition is light therapy, whereby the patient is exposed to a bright white, full spectrum light for prescribed durations to simulate daylight in order to suppress the secretion of melatonin in alignment with a more normal pattern and level.

Thus, the extent of daylight received on Earth varies according to latitude: Locations close to the equator have roughly 12 hours of day and 12 hours of night all year round, while in the northern and southern regions from the tropics of Cancer and Capricorn, the total daylight received is much lower than that received in equatorial regions. Moreover, daylight varies according to local weather conditions and cloud cover. Therefore, designing buildings to maximize opportunities for illuminating living and working spaces by daylighting is a critical requirement in places where daylight is scarce for the reasons of geographical location, climatic patterns, or both.

The strategies used to maximize daylighting in buildings are varied and serve different requirements. Perimeter daylighting zones, to capture daylight through optimally glazed facades, are used in large buildings where the majority of spaces needing daylight are based along the perimeter. The core, which is cut off from the outside environment, generally contains spaces that do not require direct daylight, such as conference and audiovisual rooms. Even there, however, advanced methods of capturing daylight can be used; these take advantage of reflective surfaces to deliver daylight by multiple reflections into areas that otherwise are completely cut off from the outdoor environment.

One such device is called a light tube or solar tube, which has a transparent receptor on the roof to admit light into highly reflective flexible tubing. The diffuser at the end of the tube is located in the space to be illuminated. These tubes can also have directional collectors, reflectors, or even Fresnel lens devices that assist in collecting additional directional light down the tube. In regions where window shading is essential to keep out the glare from the bright tropical or equatorial sun, light shelves are used to bring diffused natural light into indoor spaces and eliminate overheating and glare. These light shelves are placed in positions such that the sunlight incident on their surfaces is reflected inside the building spaces onto white or light-colored walls or ceilings. The light is then reflected back into the room from the white surfaces as a uniformly diffused natural light without heat.

In equatorial regions, factory buildings were traditionally designed with north-facing roof windows, since the light received from this direction is without glare and suitable for direct illumination without the need for window shading or glare control. Buildings in colder regions typically have large glazed areas along the sun's path to maximize the daylight received, whereas buildings in hot regions in the south, southeast, or southwest have heavily shaded windows or deep verandas to take advantage of air circulation and diffused daylight but protect living and working spaces from direct sunlight. Thus, strategies for daylighting differ according to the geographical locations of buildings as well as the functions that they are intended to serve.

—*Swati Ogale*

Further Reading
Bainbridge, David, and Ken Haggard. *Passive Solar Architecture: Heating, Cooling, Ventilation, Daylighting, and More Using Natural Flows*. Chelsea Green, 2011.
Konis, Kyle, and Stephen Selkowitz. *Effective Daylighting with High-Performance Facades: Emerging Design Practices*. Springer, 2017.
National Institute of Building Sciences. "Building Envelope Design Guide and Whole Building Design Guide." http://www.wbdg.org/design/envelope.php.
Schiller, Mark. *Simplified Design of Building Lighting*. John Wiley and Sons, 1997.
Tregenza, Peter, and Michael Wilson. *Daylighting Architecture and Lighting Design*. Routledge, 2011.

Demand-Side Management (DSM)

FIELDS OF STUDY
Energy Policy Studies; Resource Management

ABSTRACT
Demand-side management is a method of managing resources, such as energy, that emphasizes changing demand for those resources by encouraging customers to modify their levels and patterns of use (i.e., their individual demand).

KEY CONCEPTS
bottom-up approach: a management approach that begins from the position of the end user with regard to needs and adjusts the higher levels of management to minimize costs and maximize efficiency
peak period: the period during the day in which demand for a resource such as electricity is highest
real-time pricing: the cost to the consumer of a resource such as electricity adjusted continually according to the time of day the use occurs
smart meters: meters that continuously monitor use of a resource by the consumer and relay that data to the utility company in real time
time-of-use pricing: the cost to the consumer of a resource such as electricity according to the set rates for the time of day the use occurs

MANAGEMENT FROM THE CONSUMERS' POINT OF VIEW
Demand-side management (DSM) is sometimes called a "bottom-up approach" to management because it focuses on modifying the behavior of individual customers who are assumed to prefer to obtain services at lower cost by removing barriers (such as lack of information, lack of capital, or misplaced incentives) to enable them to make choices that will further this self-interest. In the field of utilities management, implementation of DSM often includes differential pricing that encourages customers to use less energy at peak hours and may also include components such as education and personalized feedback concerning energy use.

U.S. interest in DSM as a means of lowering demands for energy began in the 1970s, when the Arab oil embargo of 1973 and then the Iranian Revolution in 1979 abruptly raised the cost of energy and increased interest in conservation. DSM was seen as a way to reduce energy demand and thus avoid the expense of building new power plants or expanding existing plants. A number of programs were tried, including time-varying rates for industrial customers and time-of-use pricing for residential customers, to encourage people to use less electricity at peak hours, but because these programs were primarily crisis-driven and aimed at immediate results, little hard monitoring or evaluation was performed to see which programs were most effective. In the first half of the 1980s, interest shifted toward using a variety of means, including conservation, load management, and strategic electrification, and in the second half of the decade experiments with real-time pricing (RTP) were also tried.

In the early 1990s, DSM activity rapidly expanded in some states, and in 1993, 447 utility companies had DSM programs. There was also a new focus on measuring the environmental benefits of DSM programs. However, in the mid-1990s, faced with competition from independent power producers, many traditional vertically integrated utilities reduced or dropped their DSM expenditures as a cost-cutting effort in the face of this new competition.

In the late 1990s, utility regulators added a "public goods charge" on the sale of electricity by distribution utilities (rather than the power providers) to cover DSM expenditures. DSM programs at this time were often implemented by third-party energy service companies (ESCOs). Price spikes in wholesale power markets in 2000 again increased interest in DSM programs, and the 2001 power crisis in California raised interest in pricing reform as well.

Several principles can be drawn from the history of DSM in the United States. One is that DSM programs often reduce utility revenues; hence, utilities are reluctant to implement them unless they are assured of cost recovery. Another is that appropriate pricing, reflecting marginal costs, is necessary if DSM programs are to succeed in the long term. A third is that cash rebates should be used with care, primarily to encourage people to participate in programs at their beginning rather than as a long-term strategy.

TYPES OF DSM PROGRAMS

There are two principal types of DSM programs relevant to utilities: load curtailment programs and dynamic pricing programs. Both are aimed at reducing electrical demand at peak periods, because the utility must maintain the ability to produce sufficient electricity at the period of peak demand even if it is not needed for most of the time of operation. If the peak demand can be lowered, less capacity is required and thus costs are reduced.

Load curtailment programs pay customers for reducing their use of energy during peak periods. This may be in the form of an up-front incentive payment (the more traditional model) or a pay-for-performance incentive payment (more typical of new DSM programs). Up-front incentive programs may include direct load control of appliances such as water heaters and air conditioners for residential customers and rates that can be interrupted for industrial and commercial customers. Pay-for-performance incentive programs, also known as demand bidding or buyback programs, may include specified payments for each unit of electrical load curtailed during peak periods. Although both types of programs offer economic benefits to consumers, historically speaking they have been a tough sell in the United States, perhaps because electrical rates are generally low as compared to many other countries and thus savings involved seem not to be worth the effort.

Several types of dynamic pricing programs are possible, carrying different amounts of risk to customers and utilities. Time-of-use pricing (TOU) involves set rates that are lower in off-peak periods and higher in peak periods and may also include an intermediate rate for shoulder periods. Critical peak pricing (CPP) is similar to TOU but includes a much higher critical peak price, which may be imposed with minimal warning (perhaps even the same day) for a fixed number of days each year. Extreme day pricing (EDP) is similar to CPP, but the higher price is imposed for 24 hours of a given day, and such days are not known until 24 hours in advance. Extreme day CPP (ED-CPP) is similar to CPP but applies the critical peak and lower off-peak prices only on extreme days. Real-time pricing (RPT) allows rates to vary hourly or subhourly, with notification of rate changes a day or an hour ahead.

—*Sarah Boslaugh*

Further Reading

Charles River Associates. *Primer on Demand-Side Management with an Emphasis on Price-Responsive*

Programs. CRA D06090. World Bank, 2005. http://siteresources.worldbank.org/INTENERGY/Resources/PrimeronDemand-SideManagement.pdf.

Faruqui, Ahmad, Ryan Hledik, Sam Newell, and Johannes Pfeifenberger. "The Power of Five Percent: How Dynamic Pricing Can Save $35 Billion in Electricity Costs." The Brattle Group Discussion Paper, 16 May 2007. http://www.brattle.com/_documents/UploadLibrary/Upload574.pdf.

Gabbar, Hassam A., ed. *Energy Conservation in Residential, Commercial, and Industrial Facilities.* John Wiley & Sons, 2018.

Galbraith, Kate. "Why Is a Utility Paying Customers?" *New York Times,* 23 Jan. 2010.

Jamasb, Tooraj, and Michael G. Pollitt. *The Future of Electricity Demand: Customers, Citizens and Loads.* Cambridge UP, 2011.

McLean-Conner, Penni. *Energy Efficiency: Principles and Practices.* PennWell, 2009.

Torriti, Jacopo. *Peak Energy Demand and Demand Side Response.* Routledge, 2016.

DIESEL, RUDOLF

FIELDS OF STUDY
Combustion Science; Mechanical Engineering; Thermodynamics

ABSTRACT
Rudolf Diesel was a German inventor and theorist. He studied the theories of French physicist Sadi Carnot, learning the principles of modern internal combustion. His studies convinced Diesel that he could build an engine four times as efficient as the steam engine by injecting fuel into an engine with compressed air heated by compression to a temperature high enough to ignite the fuel. This is the principle of all Diesel engines in use today.

KEY CONCEPTS
cetane number: a classification of fuel oil combustibility relative to pure cetane (hexadecane, a C-16 hydrocarbon); analogous to the octane number for gasoline fuels relating gasoline combustibility to pure iso-octane (a C-8 hydrocarbon)

compression ignition: compressing a gas causes its temperature to increase; in a Diesel engine the temperature increase of air due to compression is high enough to ignite the fuel

four-stroke engine: an engine requiring four piston strokes for a single fuel combustion; one stroke draws in air, a second compresses the air, fuel is injected and ignited to drive the piston through the third stroke, and the fourth stroke pushes out the combustion gases. All four strokes are timed to the opening and closing of intake and exhaust valves.

fractional distillate: petroleum (unrefined oil) is a complex mixture of different compounds; fractionation of the mixture by distillation separates the components according to their respective boiling points. Each portion according to boiling point range is a fractional distillate

BACKGROUND
Rudolf Christian Karl Diesel was born March 18, 1858, in Paris. His father was a leather craftsman and his mother was a governess/language tutor. Diesel earned admission at age 12 to Paris's best school, the *École Primaire Superieure.* Classed as enemy aliens in France, Diesel's parents moved to London during the Franco-Prussian War of 1870, with Diesel himself moving to Augsburg and entered the Royal County Trade School. He then enrolled at the newly opened Industrial School of Augsburg. A scholarship enabled him to attend Munich's *Technische Hochschule* (Technical High School), where he studied engineering and graduated in 1879 with the highest examination scores on record. Also, at Royal Bavarian he met and worked with Carl von Linde, inventor of refrigeration and gas separation.

Diesel worked for Sulzer Brothers at Winterthur, Switzerland, for two years before he returned to Paris as a refrigeration engineer and manager at Linde Refrigeration Enterprises. Working for Linde, Diesel filed patents in France and Germany in the 1880s.

Diesel married in 1883 and had three children. On September 30, 1913, while crossing the English Channel to consult with the British Admiralty, Diesel disappeared at sea, and his body was never recovered.

A portrait of Rudolf Christian Karl Diesel and an illustration of the engine he invented appeared on a German stamp to commemorate the 100th anniversary of his invention.

SCIENCE AND TECHNOLOGY

Diesel studied the theories of French physicist Sadi Carnot, learning the principles of modern internal combustion. His studies convinced Diesel that he could build an engine four times as efficient as the steam engine by injecting fuel into an engine with compressed air in a ratio up to 25 to 1. The compression level would heat the air to 537°C (almost 1,000°F), sufficient to ignite fuel without plugs or other ignition systems. He opened his first shop in 1885 in Paris to work on his engine. Work on the engine continued when he moved to the Berlin office of Linde Enterprises, working on thermodynamics and fuel efficiency.

The Diesel engine is a compression ignition engine with fuel ignition through sudden exposure to high temperature and pressure, rather than through a separate source such as a spark plug. Diesel worked on different designs for over a decade before receiving the 1892 patent for one running on the cheapest available fuel, powdered coal. He published *Theory and Construction of a Rational Heat-Engine to Replace the Steam Engine and Combustion Engines Known Today* in 1892, seven years after Karl Benz received the patent for the first automobile. The publication presented the core concepts of the diesel engine. He patented the engine in Germany in February 1892. After receiving the patent and financial assistance from the Krupp brothers and Augsburg-Nuremberg Company, he began building the prototype, which was ready to test in July 1893. The prototype used powdered coal injected by compressed air. Previously, Diesel had worked with compressed ammonia. The first prototype stood 10 feet tall, resembled steam engines of the era, and achieved compression of 8100 kPa (80 atmospheres), but required outside power. It demonstrated the feasibility of forced air ignition, but it exploded and almost killed Diesel. Hospitalized for months, he thereafter had health and vision problems.

Seven months later, on February 17, 1894, he ran a 13-horsepower single-piston engine for a minute. The engine used high compression of fuel to ignite it, eliminating the spark plug used in the internal combustion engine. He modified the second prototype over four years. By 1897, he had a working 25-horsepower, four-stroke, single-vertical-cylinder compression engine. This prototype had such a strong enough resemblance to Herbert Akroyd Stuart's engine of 1890 that for some years they were in patent disagreements and arguments over whose came first. When Diesel prevailed, the engine came to carry his name. Diesel engines were relatively simple and as efficient as he had anticipated, but he allowed manufacture of only heavy stationary engines. He initially used coal dust because it was cheapest. By 1897, kerosene replaced powdered coal, and the engine presented at the 1900 *Exposition Universelle* ran on peanut oil.

After three additional years of development, production began, with the first engine at a U.S. brewery. Licenses from firms in several countries made Diesel a millionaire. The first Diesels were unsuitable for automobiles, but popular for locomotives and ships. The Diesel engine propelled cars, trucks, motorcycles, boats, and locomotives. Packard used diesel motors in airplanes from 1927, and in 1928, Charles Lindbergh flew a Stinson SM1B with a Packard Diesel. Diesels powered aircraft that set nonstop flight records in the late 1920s and early 1930s. The *Hindenburg* had four

Rudolf Diesel. (Wikimedia Commons)

16-cylinder Diesels, each capable of 850 horsepower for sustained cruising, and bursts of 1200 horsepower. The first Diesel automobile trip took place in 1930, 1,280 kilometers (800 miles) from Indianapolis to New York City. In 1931, Dave Evans drove a Cummins Diesel equipped car the entire Indianapolis 500 without a pit stop.

Diesel's engine runs on a specific fractional distillate of fuel oil, usually from petroleum. It also handles variants derived from other than petroleum, including biomass to liquid, gas to liquid, or biodiesel. Diesel is about 15 percent denser than regular gasoline distillate, simpler to refine, usually cheaper to purchase (but fluctuations in the market, as when cold weather increases demand for heating oil refined the same way as diesel, cause diesel to be more expensive). Diesel-powered vehicles normally have better gas mileage than gasoline-powered vehicles because diesel has greater energy content and the diesel engine is more efficient. The difference is up to 40 percent better mileage than gasoline. At the same time, diesel emits 15 percent more greenhouse gases than a comparable amount of gasoline, reducing the advantage. Diesel fuel also contains sulfur and, in the United States, a lower cetane number (which measures ignition quality), making for poorer cold weather performance.

Diesel's 1892 invention was a variant of the hot bulb engine. The modern diesel engine retains Diesel's compression ignition, but incorporates cold-fuel injection (pumps raise fuel to extremely high pressures that activate fuel injectors), rather than compressed air, which heats the fuel slightly (thus hot-fuel injection). Hot fuel remains the preference for large marine engines and other low-speed, high-load conditions with poorer quality fuels. Many also incorporate "glow plugs" to provide a hot spot for fuel ignition on cold start-ups. Diesel engines became markedly cleaner in 2007 after manufacturers shut down production for two years to convert to lower sulfur diesel. The Diesel remains highly efficient due to high compression and high output.

—*John H. Barnhill*

Further Reading

Bain, Don. "We Should Have Celebrated Rudolph Diesel's Birthday Yesterday." *Torque News*, 19 Mar. 2012, http://www.torquenews.com/397/we-should-have-celebrated-rudolph-diesel%e2%80%99s-birthday-yesterday.

Madehow.com. "Rudolf Christian Karl Diesel Biography (1858-1913)." http://www.madehow.com/inventorbios/11/Rudolf-Christian-Karl-Diesel.html.

Moon, John Frederick. *Rudolf Diesel and the Diesel Engine*. Priory Press, 1974.

"Obituary, Rudolf Diesel." *Power*, vol. 38, no. 18, 1913, p. 628. http://books.google.com/books?id=b70fAQAAMAAJ&pg=PA628&dq=rudolf+diesel+explosion&hl=en&sa=X&ei=52GZT97TMcbs2QXQ8uClBw&ved=0CGMQ6AEwCA#v=onepage&q=rudolf%20diesel%20explosion&f=false.

Speedace.info. "Rudolp (sic) Diesel and Diesel Oil." http://speedace.info/diesel.htm.

E

EINSTEIN, ALBERT

FIELDS OF STUDY
Nuclear Energy; Nuclear Physics; Quantum Electrodynamics

ABSTRACT
Albert Einstein was a leading theoretical physicist whose theories of relativity and gravitation overturned Newtonian physics. He also pioneered quantum theory and his work was integral to the development of atomic energy.

KEY CONCEPTS
Brownian movement: a phenomenon observed and described by Robert Brown (1773-1858) in which very small particles suspended in a still fluid "jiggle" without an apparent cause

chain reaction: in a nuclear fission chain reaction, a subatomic particle ejected from one nucleus strikes another nucleus, causing it to undergo fission and release another particle, the number of ejected particles increasing geometrically as the chain reaction "runs away"

nuclear fission: the splitting apart of the atomic nucleus whether spontaneously or by impact with other particles

photoelectric effect: the phenomenon in which light striking the surface of a material causes its surface atoms to emit electrons that can be harvested as an electric current

quantum (pl. quanta): the discrete "particle" of energy that characterizes a particular wavelength of light

wave-particle duality: the principle that light propagates as if it is a wave but interacts with matter as if it is composed of particles

BACKGROUND
Albert Einstein was born in Ulm, Germany, and at age six weeks moved to Munich, where he remained until he entered school. His family moved to Italy, and Albert entered school in Aarau, Switzerland. His academic performance was less than stellar, and existing report cards indicate that he was not expected to have any significant career. He developed a strong curiosity about science and also learned to play violin and piano. At the Swiss Federal Polytechnical School in Zurich, where he trained from 1896 to 1900 to teach mathematics and physics, he met Mileva Maric, a fellow student who graduated with him in 1900. Einstein married Maric in 1903, and the couple had two sons and a daughter. The two divorced in 1919, the year Einstein married his cousin, Elsa Lowenthal (d. 1936).

He renounced German citizenship during this period and was stateless until becoming Swiss in 1901,

Albert Einstein. (Wikimedia Commons)

97

the year he took a clerkship at the patent office. He received his doctorate in 1905. He remained at the patent office until 1908, moving to professional positions in Prague, Zurich, and then Berlin, where he became director of the Kaiser Wilhelm Institute for Physics in Berlin in 1913 and a German citizen in 1914. From 1914 to 1933 Einstein served as director of the Kaiser Wilhelm Physical Institute. He was awarded the 1921 Nobel Prize in Physics, but following the rise to power of Adolph Hitler he was removed from that position because he was Jewish as well as an outspoken pacifist and socialist. He once again renounced his German citizenship, moved to the United States, and from 1934 on lectured at Princeton University, becoming a full professor in 1940. Einstein was offered the Presidency of Israel in 1952, but declined in order to continue his theoretical work. The great man died in his sleep at Princeton in 1955.

THE FUNDAMENTAL SCIENCE OF ENERGY
In 1905, Einstein produced four papers that revolutionized conceptions of energy, matter, motion, time, and space. Each of the papers that Einstein contributed to *Annalen der Physik* (Annals of Physics) in 1905 became the basis for a new branch of physics. His idea that the universe is made up of quanta, tiny particles of energy and matter, altered the definition of reality and explained the photoelectric effect as due to light particles. Einstein proved the existence of atoms in describing a new way of counting and measuring atoms or molecules in a defined space and in explaining Brownian movement as molecular kinetic energy. Three months after the paper on quanta, he published the article on special relativity, contradicting his quanta theory by postulating light as waves. He finished the year by extending special relativity to tie energy and matter in $E=mc^2$.

SPECIAL THEORY
Einstein developed his special theory of relativity to make electromagnetic field laws compatible with Newtonian mechanics. Neither space nor time is absolute, contrary to Isaac Newton's universal laws that mandated absolute space and time independent of motion. Einstein argued that motion is measurable only in relation to the observer. The special theory of relativity showed that time was relative, altering conceptions of time, space, motion, mass, and gravitation.

The special theory states that energy equals mass times the speed of light squared ($E=mc^2$). Matter and energy are thus interchangeable; one can become the other because they are simply two aspects of a single thing. Previously, according to Antoine Lavoisier's eighteenth century theory of conservation of matter, the assumption was that matter is constant and can only change form. In the nineteenth century, the same conservation was established for energy—it can take many forms, but cannot be created or destroyed. Then, Einstein postulated that conservation applied across the two aspects of the single construct, matter/energy. The switch from one form to another was $E=mc^2$. The magnitude of the speed of light squared is 9×10^{16} m^2sec^{-2}. Thus, a tiny amount of matter comes from a massive amount of energy, and a tiny amount of matter transforms into massive amounts of energy.

BROWNIAN MOVEMENT
Einstein's work on making statistical mechanics consistent with quantum theory led to his explanation of Brownian movement in molecules. He developed calculations to demonstrate how movement of molecules can cause Brownian motion in liquids. Brownian movement is named for Scottish plant expert Robert Brown, and refers to the microscopic jiggling motion observed in small particles suspended in a still fluid. By explaining movement of minute particles floating in gas or liquid, Einstein provided a strong argument of proof for the existence of atoms.

PHOTON THEORY/QUANTUM PHYSICS
Einstein used Max Planck's quantum theory to create the photon theory of light and explain the photoelectric effect. Planck explained that black bodies take all light energy they receive, reflecting none, because light is not continuous waves, but streams of discrete tiny particles, of energy called "quanta." Light particles strike metal and free electrons; waves cannot achieve that photoelectric effect. Einstein demonstrated that quanta striking atoms forced the atoms to release electrons, substantiating quantum theory and enabling the creation of the photoelectric cell. His work on quanta earned Einstein the 1921 Nobel Prize. He continued to work on quantum theory after his migration to the United States, and his development of the quantum theory of a monoatomic gas and

other theories were invaluable. Planck and Einstein were the founders of quantum physics.

GENERAL THEORY

In the 1910s, Einstein realized that his special theory of relativity would provide a theory of gravitation. Einstein submitted his theory of gravitation, the general theory of relativity, to the Prussian Academy of Sciences in 1915 and the *Annalen der Physik* in March 1916. The general theory of relativity shows that matter and energy shape space and time. Because energy and matter are interchangeable, although photons have no mass, light behaves like mass, for instance, bending in response to solar gravitation. Stars appear in different positions when behind the sun than they do when the sun is in another position relative to them. Arthur Eddington measured the deflection of starlight by the sun during a 1919 solar eclipse, confirming Einstein's general theory of relativity.

In 1917, Einstein used general relativity to model behavior on a full universal scale, becoming the first to publish a work of cosmology, the study of the behavior of a universe in its entirety. In 1919, he anticipated the uncertainty principle by indicating that quanta having both wave and particle natures might make impossible any definitive link of effects with cause. Gravity is what we feel by taking the most direct path through curved space/time. The general theory of relativity remains basic to the understanding of the universe.

NUCLEAR POWER

Splitting a uranium atom releases 2 million times the energy generated by breaking a carbon-hydrogen bond in wood, oil, or coal. Traditional combustion is a reaction that occurs only between electrons in the outermost atomic shells of the reacting atoms. But no less than 99.98 percent of the mass of an atom is in the nucleus. To liberate the massive amount of energy from the nucleus requires the breaking up and conversion of the atom. Einstein was among those who doubted the feasibility of splitting an atom. In his 1905 article on special relativity, Einstein didn't initially grasp the implications of converting mass to energy by other than physicochemical means, but he indicated that the heat of radium might cause the conversion of a small amount of the radium to energy.

By the late 1930s, other scientists had demonstrated the feasibility, and Einstein was among those who recognized the potential energy available through splitting an unstable uranium atom. In 1939, the longtime pacifist was convinced of the need to use force to prevent Hitler's Germany from developing an atomic bomb and was instrumental in convincing Franklin Roosevelt to begin production of the atomic bomb (Manhattan Project), technology enabled by Einstein's theories. The Manhattan Project exploded three weapons—a plutonium implosion bomb in New Mexico, an enriched uranium bomb over Hiroshima, and a second plutonium bomb over Nagasaki. Both detonations over Japan took place in 1945, well after Germany had been defeated, and both killing hundreds of thousands of people, mostly women and children, in an instant, including Allied POWs being held near those cities. The aftereffects of radiation poisoning from those bombs are still being experienced by residents more than 75 years later. The project began in 1939 and cost $23 billion in 2007 dollars.

Einstein did not work on the Manhattan Project itself. After the war, he became a strong advocate of nuclear disarmament. He declined the presidency of Israel in 1952, but collaborated in the creation of Hebrew University in Jerusalem and was active in the World Government Movement. He was Time magazine's Person of the Century.

OTHER CONTRIBUTIONS

Einstein shifted direction in the 1920s, focusing on development of unified field theories. He spent decades pursuing a mathematical link between gravitation and electromagnetism, but he never found the sought-after unified theory. In 1925, he built on ideas of Satyendra Nath Bose that postulated a new state of matter, the Bose-Einstein condensate, to go along with solid, liquid, and gas. The condensate was created at exceptionally low temperatures only in 1995.

His honorary doctorates came from many European and American universities in science, philosophy, and medicine. He lectured throughout the world in the 1920s and became a fellow of many of the leading scientific academies. Awards include the Copley Medal of the Royal Society of London (1925) and the Franklin Medal of the Franklin Institute (1935). Einstein's theoretical writings include *Special Theory of Relativity* (1905), *Relativity* (English translations, 1920 and 1950), *General Theory of Relativity* (1916), *Investigations on Theory of Brownian Movement* (1926), and *The Evolution of Physics* (1938.) He also wrote nonscientific works such as *About Zionism* (1930), *Why War?* (1933), *My Philosophy* (1934), and *Out of My Later Years* (1950). His theories of relativity and gravitation advanced physics significantly beyond Newtonian theory and provided a paradigm shift in science and philosophy.

—*John H. Barnhill*

Further Reading
Gamow, George. *Thirty Years That Shook Physics: The Story of Quantum Theory.* Anchor Books, 1966.
Nobel Foundation. "The Nobel Prize in Physics 1921: Albert Einstein." http://nobelprize.org/nobel_prizes/physics/laureates/1921/einstein-bio.html.
Nothiger, Andreas. "Einstein Theory." Hyperhistory, http://www.hyperhistory.com/online_n2/History_n2/index_n2/einstein_theory.html.
Pais, Abraham. *"Subtle Is the Lord": The Science and the Life of Albert Einstein.* Oxford UP, 1982.
Rigden, John S. *Einstein 1905: The Standard of Greatness* Harvard UP, 2005.
Tucker, William. "Understanding E=mc²." 21 Oct. 2009, http://www.energytribune.com/articles.cfm/2469/Understanding-E-=-mc2.
Vujovic, Ljubo. "Albert Einstein (1879-1955)." Tesla Memorial Society of New York, http://www.teslasociety.com/einstein.htm.

Electric Grids

FIELDS OF STUDY
Electrical Engineering; Environmental Science; Resource Management

ABSTRACT
Electric grids are networks that connect electric generation plants to individual customers. Recent developments in alternative energy, involving wind and solar plants in remote areas, have increased interest in long-distance transmission networks and new grid technologies.

KEY CONCEPTS
microgrid: an electric grid system that serves a small area such as a single town

smart grid: an electric grid system with built-in digital monitoring and control functions

step-down: reduction, or stepping down, of a high voltage to a lower voltage by the use of a transformer

step-up: the increase, or stepping up, of a relatively low voltage to a higher voltage by the use of a transformer

transformer: a constructed electrical device that uses the principle of induction to induce a voltage in a secondary coil by the current in a primary coil with a different number of "turns" of wire; a higher number of turns on the secondary coil results in induction of a higher voltage, while a lower voltage

transmission network: an electricity delivery system originating at a generating station, passing through any number of user nodes and interconnections, and terminating at the generation site to complete the electrical circuit

TRANSMISSION NETWORKS
Electricity transmission is in general considered to be the bulk transfer of electrical power or energy. The lines that connect the bulk transmitters of power (generating stations) to their substations are typically called a grid. The wires or connections from substations to consumers are typically called a power distribution network. There are many types of electric grids, including direct current (DC) grids, alternating current (AC) grids, international grids, hyper grids, smart grids, and microgrids. Grids vary depending on the type of power being transmitted.

In grid systems there must be some power to put into the grid at the beginning of the process. Generating plants produce electricity using different means including coal, natural gas, nuclear, hydropower, renewable energy, and petroleum. In general, electricity moves through the grid in stages as it is: (1) generated, (2) stepped up, (3) transmitted, (4) stepped down, and (5) delivered to the end-user consumer.

Usually, transmission lines transmit electricity in the 110 kV range. This may be done with overhead wires or underground wires. In order to move electricity from one region to another there is often a system in place to match up the various systems of energy. This is typically referred to as an HVDC power transmission system (HVDC stands for High Voltage Direct Current). An HVDC system has the ability to adjust so that different systems can work together.

AC AND DC POWER

There are two main types of power: DC power (DC stands for direct current), and AC power (AC stands for alternating current). With DC power the charge goes in one direction only, not both. AC power alternates its direction of flow. In North American AC utility circuits, it changes direction approximately 60 times every second. DC current is most commonly thought of in the sense of batteries; AC current is most commonly thought of as the wall outlets in a house. DC power does not go as far with as much strength as AC power does. This is important for electrical grids, as this difference is central to the discussion of centralized grids or decentralized grids. Decentralized grids most often refer to their power usage and generation as localized power generation or localized grids. Centralized power is more typically the big power generating plant producing power that is sold and used throughout a region of the country.

AC grid systems are presently the common systems of choice. AC grid systems are used by a majority of countries to a greater or lesser extent. DC grid systems are broken down into two sets of systems: (1) HVDC, which is used fairly universally as a bridge or interpreter helping to match AC grid systems together, and low voltage DC systems. Low voltage DC systems are either on or off of the main grid of a country. In many cases a low voltage grid exists as either a remnant of earlier times, when DC power was used more extensively in grid systems than today. Another case is in remote or industrial settings. For example, in Africa and in Italy and other countries, remote places like mining towns may be on their own town-wide low voltage DC grid. Finally, the issue with DC is that since it costs more to transmit it great distances, it is usually kept close to its area of use. In many cases equipment or appliances used on a daily basis operate on DC, and when plugged into an AC outlet, there is a rectifier or inverter changing the cycle of power so the DC equipment can use it.

The problem of how to combine DC grids into an AC grid system or switch back to DC systems has been revived because of increased generation of alternative energy, which is often DC, not AC. In some countries like India, Denmark, Norway, and Germany they have already begun to make DC grids. These countries are working on either integration with their existing AC grid or parallel use of a DC grid with the existing grid infrastructure.

THE U.S. GRID

U.S. grid system is made up of basically three regional interconnection areas or in another sense four basic categories of voltage. The categories are similar to the three regional interconnection areas plus an interpreter, which is the fourth voltage level. The regional interconnections or voltages carried are: (1)

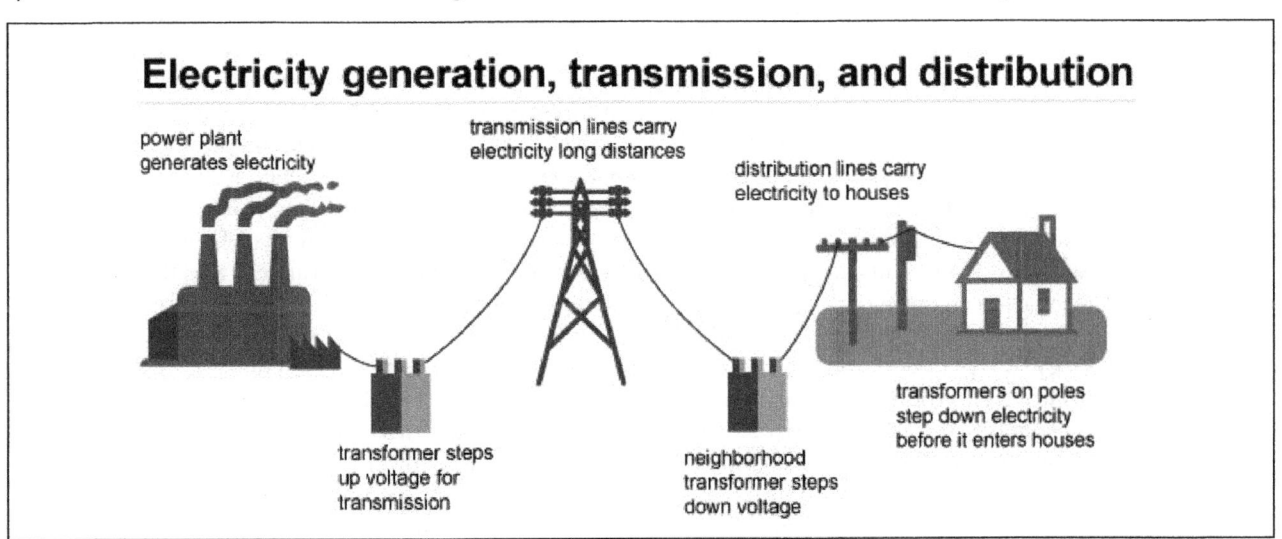

Diagram of an electric power system. (Adapted from the National Energy Development Project)

345-499kV, (2) 500-699kV, and (3) 700-799kV, all of these being AC current, and then 1,000 kV (HVDC serves as the "interpreter," able to change the speed or rate of transfer when needed). The three regional interconnection grid areas are called a "wide area synchronous grid." Each regional interconnection area interconnects AC power operating at the same frequency and phase within its area or region. These regions are overseen by various reliability councils and the members of these reliability councils are in part utility companies, power producers, and distributors.

Once the generating plant has produced electricity, it is prepared for transmission by going to a generating station step-up transformer to increase the voltage for better transmission. Once the power is packaged, it is transmitted into transmission lines at a voltage within one of the three regional voltage ranges. When it reaches its destination, since it is carrying so much energy it needs to be stepped down through a substation step-down transformer so that it may be distributed to users with much smaller voltage needs (for example, American homes with 120V outlets). These smaller distributed networks are on an AC grid.

SMART GRIDS
The term *smart grid* refers to the addition of types of artificial intelligence to electric grid systems. These software programs track, read, compute, and determine the most optimal ways and means for grid systems to work. Smart grids are important in that they can make determinations without manual intervention. They must be protected from corruption or viral invasion; this area is typically called "cybersecurity." Since grids are now being made smart we must be able to identify if grids are connecting to each other through the interpreter (HVDC) or if other grids are interfering. In the area of international or hyper grids these identifications and protections are going to be critical to how the smart grid works.

The U.S. smart grid was first established under Title XIII of the Energy Independence and Security Act of 2007 (EISA). According to the U.S. Department of Energy's original estimates the smart grid was supposed to cost $165 billion over 20 years. There are infrastructure costs at all levels, such as changing from mechanical meters on houses to digital "smart" meters linked wirelessly or through some other means.

The benefit is in better service, fewer outages, less leakage of power resources, and more. The consumer and the vendor also have better control over their expenses. The challenges involve problems such as that reported by the U.S. National Energy Technology Laboratory, which found that while 60 percent of transmission systems have supervisory control and data acquisition systems (SCADA) that send data from a sensor to a central computer, only 2 percent of the local distribution network had SCADA as of January 2009.

MICROGRIDS
Microgrids are created when like or dissimilar renewable energy resources are tied together before being attached to the grid, either through corporate or community efforts. Resources may be tied to the grid through a common access point, or through a common corporate access point. In the case of a common energy system in which a community runs on solar power and wind power, the need to buy energy from the main grid may be removed or substantially lowered. In this case the main reason for being tied to the grid is to purchase power either individually or through an association like an energy cooperative when power is needed. These configurations draw on energy systems, but if the houses have more energy being produced by solar and wind than they can consume, the users are creating cheap energy for the utility system to buy and then resell. Sometimes a loose confederation of users allow a company to manage all of their energy issues, and all usage and data goes to that entity first and then it reports that data to the utility companies attached to the grid.

Renewable energy sources of power are typically DC. Homes with renewable energy sources usually have either a DC power-using home or an AC home with an inverter (which converts DC power to AC) between the power source (solar cells and the house) and its point of energy use. This is important for the grid as a whole because renewable energy cannot usually be put directly into the grid. The U.S. FERC (Federal Energy Regulatory Commission) controls licensing for people or entities wishing to attach to the U.S. grid to give the grid power.

Grid management systems, sensors, and flexibility are paramount in the new smart grid and microgrid world. Microgrids are useful in rural and remote areas where laying transmission lines is a challenge for the local or regional power companies, especially in rugged terrain or in disaster areas. Microgrids will, however, have to meet some form of standard as continually regulated by FERC. Nonetheless, there are some real advantages to them. For example, in a microgrid or local power distribution scenario the whole grid or region is less vulnerable to one large catastrophic event. Hospitals, police, and military may be on localized microgrids, a configuration that makes it harder for a negative event to affect the whole of the grid, or even a significant portion of it depending upon how the microgrids are constructed. Microgrids would most likely be DC-based since they tend to be in the low power generation cycles and this would also free up current power generation for other more immediate high-power needs and use.

—Richard Brent

Further Reading

Berger, Lars T., and Krzysztof Iniewski. *Smart Grid Applications, Communications, and Security*. John Wiley & Sons, 2012.

D'Andrade, Brian W., ed. *The Power Grid: Smart, Secure, Green and Reliable*. Academic Press, 2017.

De La Rosa, Francisco C. *Harmonics, Power Systems and Smart Grids*. 2nd ed. CRC Press, 2015.

Digital Communities. "21st-Century Smart Grids Update: U.S. Electric Grid." http://www.digitalcommunities.com/templates/gov_print_article?id=99859074.

Global Energy Network Institute. "The International Grid." http://www.geni.org/globalenergy/library/technical-articles/transmission/new-scientist/the-international-grid/index.shtml.

Pappu, Vijay, Marco Carvalho, and Panos M. Pardalos, eds. *Optimization and Security Challenges in Smart Power Grids*. Springer, 2013.

Singh, Chanan, Panida Jirutitijaroen, and Joydeep Mitra. *Electric Power Grid Reliability Evaluation*. IEEE/Wiley, 2019.

Sorebo, Gilbert N., and Michael C. Echols. *Smart Grid Security: An End-to-End View of Security in the New Electrical Grid*. CRC Press, 2011.

U.S. Energy Information Administration. "Power Transactions & Interconnected Networks." http://www.eia.doe.gov/cneaf/electricity/page/prim2/chapter7.html.

Electric Potential

FIELDS OF STUDY
Classical Mechanics; Electronics; Electromagnetism

ABSTRACT
Electric charge is a property of atomic particles. As charged particles move, they generate electric currents. By properly storing these currents, electric potential energy can be stored in devices such as batteries and capacitors later to be released to serve a purpose. The relations that affect electric charges are described in this article.

KEY CONCEPTS

conductor: a material that has a low resistance to electric charges, allowing them to move through it easily

coulomb: the basic unit of charge in the International System of Units (SI)

current: the movement of electric charges from one place to another

insulator: a material that has a high resistance to electric charges, preventing them from moving through it easily

joule: the SI derived unit of energy, equal to one kilogram-square meter per second squared (kg·m^2/s^2)

voltage: the work done per unit charge when moving a charge against an electric field

work: the use of energy to move an object over a distance by means of the application of force.

ELECTRIC POTENTIAL ENERGY

As a rainstorm gathers in the atmosphere, a dance and motion of charges begins to take place. As the storm gets stronger, it creates faster updraft winds. These updraft winds pick up water droplets and raise them high in the clouds. At these high altitudes, the temperatures are extremely low, and the water freezes to form ice. As more water droplets are carried up by the updrafts, they start to collide with some of these ice particles. During these collisions, electrons are taken away from the ice. The electrons remain lower in the cloud. Eventually enough electrons build up to produce a lightning strike. These electrons travel through the air, which is typically an insulator. However, when enough electrons build up, they can break through the insulator, turning it into a conductor.

Electric Potential

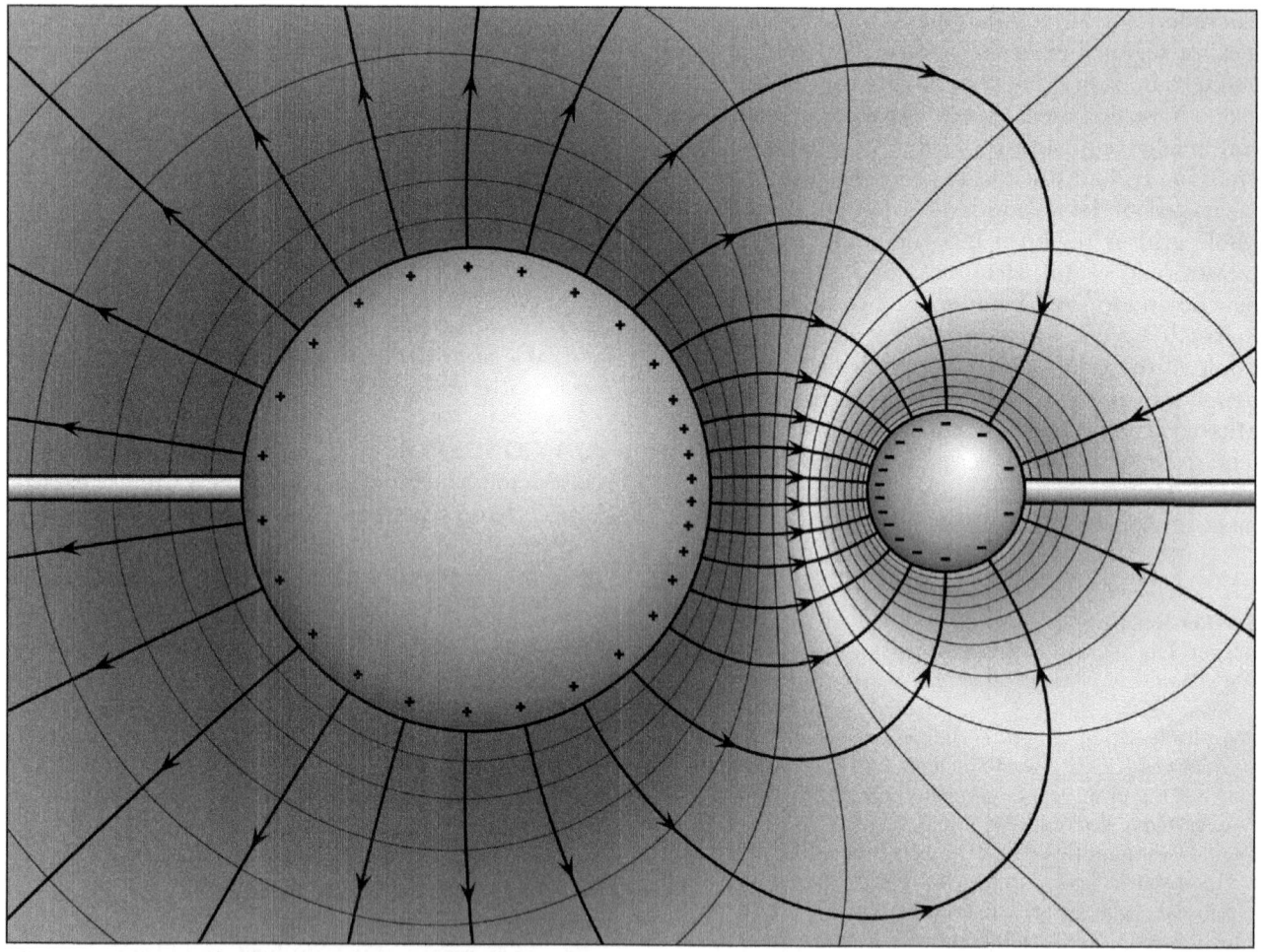

Electric potential around two oppositely charged conducting spheres. Purple represents the highest potential, yellow zero, and cyan the lowest potential. The electric field lines are shown leaving perpendicularly to the surface of each sphere. (Geek3 via Wikimedia Commons)

During the buildup of charges at the base of the cloud, an interesting effect takes place. Electrons at the base of the cloud start attracting positive particles on the ground. This interaction builds up the electric potential energy in the system.

The electric potential energy between the clouds and ground is due to the configuration of electrons and protons, which are treated as point charges, or idealized dimensionless charged particles. Electric potential is defined as electric potential energy per unit charge, or the electric potential energy of a single point charge at any point in an electric field. It is equal to the amount of work that would be necessary to carry the charge to that point when moving against the electric field.

CALCULATING ELECTRIC POTENTIAL

The electric potential energy (U_e) between two point charges is a function of the electrostatic constant (k), also called Coloumb's constant; the values of the individual point charges ($Q1$ and $Q2$); and the distance between the two charges (d), measured in meters (m):

$$U_e = k(Q2 - Q1)/d$$

The electrostatic constant is measured in newton-square meters per coulomb squared ($N \cdot m^2/C^2$) and is equal to 8.99×10^9 $N \cdot m^2/C^2$. Any physical energy is measured in the International System of Units (SI) unit of energy, the joule (J). The charges are measured in SI units of coulombs (C). In a system of more than two charges, the total electric potential energy is the sum of the electric potential energy of each pair of charges.

The equation to calculate the electric potential (V) of one of the two point charges is similar to the equation for electric potential energy. It, too, is a function of the charge amount (Q), the electrostatic constant (k), and the distance to the charge (d):

$$Ue = V = kQ/d$$

Electric potential is measured in volts (V), an SI derived unit named in honor of the Italian physicist Alessandro Volta (1745-1827). This equation calculates the electric potential generated by one point charge. Electric potential is simply electric potential energy per unit charge. Therefore, it can also be defined as electric potential energy (Ue) per unit of charge (Q):

$$V = Ue/Q$$

When dealing with accumulations of charges, a different technique must be used. Imagine two horizontal charged plates separated by a certain distance. The charges in these plates create a uniform electric field between them. Finding the electric potential using the equation above for point charges would take a great deal of time, because the number of individual charges can be very large. Instead, we turn to the electric field. The electric potential at any point in a uniform electric field with a strength of E newtons per coulomb (N/C), generated by two plates separated by a given distance (d), is:

$$V = Ed$$

On its own, the electric potential of a single point charge is not a meaningful quantity. However, the difference in electric potential between two points (ΔV) is a very useful quantity. This quantity, called potential difference, is more commonly known as voltage. It is the amount of work, measured in units of electric potential (i.e., volts), that would be necessary to move a charge between the two points in the opposite direction of the electric field.

Electric potential due to a uniform electric field generated by parallel plates is something that affects the world daily. Electronic devices are made with a collection of smaller circuit parts. One of these parts is a capacitor, or a pair of parallel conductive plates that store electric potential energy between them. The electric potential energy (Ue) stored by a single capacitor in a circuit is a function of the capacitance (C) of the capacitor, measured in farads (F), and the potential difference or voltage (ΔV) between the two plates.

CURRENTS AND CIRCUITS

Although charged particles run through them, circuits do not work using individual charges. They require an electrical current, a series of charges moving as a function of time. In a circuit there is a voltage that is supplied by a power source, for example a battery. The voltage supplied by the battery depends on the amount of resistance and the current allowed by the resistance. This relation is given by Ohm's law, which states that the voltage (ΔV) in a circuit is a function of current (I), measured in amperes (A), and the resistance (R), measured in ohms (Ω):

$$\Delta V = IR$$

Solving this equation for current shows that it is equal to voltage divided by resistance:

$$I = \Delta V/R$$

Resistors work by slowing down or stopping the electrons from moving through a circuit. That is, the more resistance, the less current will flow. A good example of how resistors dissipate potential energy is an incandescent lightbulb. They are made out of a filament that has very high resistance properties. As electrons flow through the filament, they are slowed down or stopped by the filament. The electric potential energy the electrons carry cannot just disappear. It changes form into thermal energy, which is emitted as light and heat.

The energy in an electrical circuit is directly related to the work done by the charges in the circuit, the current. Work is the energy used by a force to move an object over a distance. A simple circuit can be made using a battery, which provides the voltage, and one resistor. The work done by the system (W) is a function of the voltage supplied by the battery (ΔV), the current in the circuit (I), and the amount of time the current stays flowing (t):

$$W = \Delta V I t$$

In this same circuit energy is being dissipated by the resistor. There is a limited amount of energy in the battery. Eventually the battery will stop working. The resistor has slowly used up all the energy. The power (P) at which the resistor works and dissipates energy is a function of the voltage (ΔV), the current (I), the resistance (R). There are a number of versions of the power equation that can be used to calculate the power dissipated by the resistor.

CHARGING CIRCUITS

Hearts beat because of an electrical discharge that makes the heart contract, thereby pushing blood into the arteries. Unfortunately, like most circuits, the circuit that makes the heart beat sometimes fails. It can fail in many ways. Sometimes it gets out of control. Sometimes it stops working altogether. In each of these cases, scientists have developed a device that can help the heart beat normally again. Pacemakers and defibrillators make use of capacitors. They supply a charge or current that gets stored in a capacitor. When the capacitor stores as much charge as it can, no more charge flows through this circuit. As soon as the pacemaker detects that the heart has stopped beating or is beating at different pattern, it releases the charge. At this point it begins to charge the capacitor again. In many ways these capacitors work like batteries that supply the shock needed to keep the heart beating.

—Angel G. Fuentes

Further Reading

"Electric Potential." *Encyclopaedia Britannica*. Encyclopaedia Britannica, 3 Apr. 2014. Web. 19 June 2015.

"Electric Potential Energy." *Khan Academy*. Khan Acad., n.d. Web. 19 June 2015.

"Introduction to Circuits and Ohm's Law." *Khan Academy*. Khan Acad., n.d. Web. 19 June 2015.

Giambattista, Alan, Betty McCarthy Richardson, and Robert C. Richardson. *Physics*. 3rd ed. McGraw, 2016.

Nave, Carl R. "Electric Potential Energy." *HyperPhysics*. Georgia State U, n.d. Web. 19 June 2015.

Young, Hugh D., Philip W. Adams, and Raymond J. Chastain. *Sears & Zemansky's College Physics*. 10th ed. Addison, 2016.

ELECTRICITY ENERGY TRANSMISSION, SECONDARY

FIELDS OF STUDY
Electrical Engineering; Electrical Trades

ABSTRACT
Electrical energy (commonly called "electricity"), though a secondary form of power, remains the most important energy. Production and consumption of electricity involves generation, transmission, and distribution.

KEY CONCEPTS

secondary energy: energy that is derived from a primary source rather than being a source itself

transformer: an electrical device that converts input electricity at a particular voltage to output electricity at a different voltage that is either higher or lower than that of the input electricity

transmission loss: the dissipation of electricity as heat due to the friction of electrons moving through the transmission lines

THE MOST IMPORTANT SECONDARY ENERGY

Electricity is not a primary energy source in and of itself but is rather a means of transferring energy from a source to a use in the form of work. Central to the use of electricity, therefore, is its transmission, along with its production and consumption. Benjamin Franklin first discovered that, to generate electricity or for electrical energy to be at all, there must be a conductor or a circuit that will enable the transfer of the energy. Electrical energy is said only to occur when electric charges are moving or changing position from one element or object to another. After electricity is produced at power plants, it has to get to the customers who use the electricity.

Today, most of the world's cities are crisscrossed with transmission lines that transfer electricity through transformers developed by George Westinghouse. The transformer has allowed electricity to be efficiently transmitted over both short and long distances.

WHAT IS ELECTRIC TRANSMISSION?

Electric transmission of energy, otherwise known as high-voltage electric transmission, is generally seen as mass movement or bulk transfer of electricity through transmission lines, from power-generating plants to substations, the latter usually situated near population centers. In the United Kingdom, electric transmission lines are interconnected and referred to as the "national grid"; such networks are known as "power grids" in the United States. North America has three major grids: the Western Interconnection, the Eastern Interconnection, and the Texas Interconnection (operated by the Electric Reliability Council of Texas, ERCOT).

Electricity is transmitted at high voltages (110 kilovolts or greater) to reduce the energy lost in long-distance transmission. It is important to note, however, that because electrical energy for the most part cannot be stored, a more sophisticated system of control is vital to ensure balance of supply and demand. Failure to do this could result not only in national electricity failure but also in regional blackouts (such as those that occurred in 1965, 1977, and 2003 in the northeastern United States).

ELECTRIC TRANSMISSION SYSTEMS

There are three types of electric transmission systems: long-distance, subtransmission, and merchant.

Long-distance transmission of electricity, across thousands of miles, is cheap and efficient, with costs of $0.005 to $0.02 per kilowatt-hour (compared to annual averaged large-producer costs of $0.01 to $0.025 per kilowatt-hour, retail rates upwards of $0.10 per kilowatt-hour, and much more for instantaneous supply at unpredicted, highest-demand moments). Thus, it can be less expensive to use distant suppliers than it is to use local sources; for example, New York City buys much of its electricity from Canada. Multiple lo-

A series of electrical transmission towers. (YinYang via iStock)

cal sources, however—even if more expensive and infrequently used—can make the transmission grid more tolerant of weather and other disasters that can disconnect distant suppliers. When electricity travels long distances, it is better to have it at higher voltages; electricity can be transferred more efficiently at high voltages. Transmission lines are made of copper or aluminum in long, thick cables; these are materials that have low electrical resistance. The lower the resistance, the less heat is generated during transmission and hence the less energy is lost along the way. High-voltage transmission lines carry electricity long distances to a substation. The power lines go into substations near businesses, factories, and homes. There, transformers change the very high-voltage electricity back into lower-voltage electricity that can be used for the appliances and devices at these locations.

Subtransmission is part of an electric power transmission system that runs at relatively lower voltages. It is uneconomical to connect all distribution substations to the high main transmission voltage, because the equipment is larger and more expensive. Typically, only larger substations connect with this high voltage. It is stepped down and sent to smaller substations in towns and neighborhoods. Subtransmission circuits are usually arranged in loops so that a single line failure does not cut off service to a large number of customers for more than a short time. Although subtransmission circuits are usually carried on overhead lines, buried cables may be used in urban areas.

Merchant transmission refers to an arrangement in which a third party constructs and operates electric transmission lines through the franchise area of an unrelated utility. Advocates of merchant transmission claim that this activity creates competition to construct the most efficient and lowest-cost additions to the transmission grid. Merchant transmission projects typically involve direct current (DC) lines because it is easier to limit flows to paying customers. Merchant transmission remains a work in progress, so there few examples, but some include the Neptune RTS transmission line from Sayreville, New Jersey and New York; ITC Holdings' transmission system in the American Midwest; Path 15 in California; and Bass link between Tasmania and Victoria in Australia.

As C.L. DeMarco has observed, in electric transmission systems much relatively new technology has reached reasonable maturity since the start of the twenty-first century. The future seems bright as these new technologies are now waiting in the wings to offer greater capability to utilize grid resources to satisfy our electricity needs.

—*Soji Oyeranmi*

Further Reading

American Society of Civil Engineers. *Guidelines for Electrical Transmission Line Structural Loading*. ASCE, 2020.

Chudnovsky, Bella. *Electrical Power Transmission and Distribution: Aging and Life Extension Techniques*. CRC Press, 2013.

Conejo, Antonio J., Luis Baringo, S. Jala Kazampour, and Afzal S. Siddiqui. *Investment in Electricity Generation and Transmission: Decision Making Under Uncertainty*. Springer, 2016.

DeMarco, C.L. "Trends in Electrical Transmission and Distribution Technology." *Transmission and Distribution Technology*, 1 Feb. 2006. http://www.mrec.org/pubs/06%20MREC%20DeMarco%20PSERC.pdf.

Energy Quest. "The Energy Story." http://www.energyquest.ca.gov/story/chapter07.html.

Reilly, Helen. *Connecting the Country: New Zealand's National Grid, 1886-2007*. Steele Roberts, 2008.

Embodied Energy

FIELDS OF STUDY
Accounting; Materials Science; Thermodynamics

ABSTRACT
The energy that is used in the entire process of producing a material, a product, or a service is known as embodied energy.

KEY CONCEPTS

carbon footprint: a measure of the amount of atmospheric carbon dioxide a product or service represents

cradle-to-grave: a term representing the span of different stages of a particular material, product or service from the moment of its inception to its final disposition; similar appropriate terms may be used to indicate different stages

embodied energy: an accounting of the energy consumed in the existence of a material, product or service through its cradle-to-grave life span

emery: a contraction of embodied energy used to indicate the embodied energy in equivalent solar energy

energy burden: term used in the case of fuel and energy products and services to indicate the embodied energy cost of the product or service relative to the energy to be derived from the product or service

ENERGY ALL THE WAY ALONG

Embodied energy can represent the energy used up to produce a construction material, a household appliance, a consumer good, or a less tangible service, even an energy fuel itself. It is one measure of an environmental impact arising from the processing of the material or product or the activities involved in delivering the service. In the case of an aluminum window frame, for example, the embodied energy would be all the energy used to mine, extract, refine, and process the materials constituting the frame; manufacture the frame; transport and install the frame to an end user; and decommission the product at the end of its use.

The values of embodied energy are often expressed in megajoules per kilogram (MJ/kg) or the equivalent and are a comparative measure of the energy used to provide the product, compared to available alternative products. The embodied energy bears no relationship to the calorific value of a finished product or its ability to provide energy as a fuel. Embodied energy is, rather, an accounting concept to describe a material's, product's, or service's consumption of energy during its process of being transformed from its initial raw state to finished article. Embodied energy may be considered to be the energy use or energy debt that a material, product, or service has consumed during its life.

Different life-cycle assessments are used for the calculation of embodied energy, depending on the boundary conditions that are relevant to the study of the material, product, or service. The life cycle starts from birth (cradle) and ends at an appropriate accounting gateway, such as the factory, the point of use (site), or at death (grave). These life-cycle assessment periods may be cradle-to-factory, cradle-to-site, or cradle-to grave, depending on what position in the life cycle the energy consumption is truncated.

Embodied energy figures should be used as a relative comparator rather than an absolute figure, as they are variable across different studies. Embodied energy methodologies vary and may consider process, input-output, or hybrid methods, and both vertical and horizontal system boundaries and assumptions are an important attribute in determining the results and applying data from embodied energy tables or spreadsheets.

Analyses are dependent on time period, geographical location, boundary conditions, assumptions, methodologies, and the rigor of the life-cycle assessment. Different periods of time have different energy technology availability, and different cultures have different energy uses to sustain their cultures. Studies may be related to a particular country (such as Australia, Canada, the United Kingdom, or the United

The main methods of embodied energy accounting as they are used today grew out of Wassily Leontief's input-output model and are called Input-Output Embodied Energy analysis. (Wikimedia Commons)

States), and transportation energy may be an important factor (e.g., stone from Brazil shipped to Europe may have greater embodied energy than indigenous, locally produced stone hewn and used in Europe).

The assessment of human services in support of materials, products, and services is also an area of debate and uncertainty. A product manufactured in a wealthy society by workers with a high standard of living may include a proportion of energy used to support the lifestyle. For example, whether workers travel to and from work by a motor vehicle, with excessive fuel consumption, or by bicycle may need to be taken into account. One would expect greater embodied energy in a service provided by energy-intensive societies. Some materials, such as aluminum, may have increasing proportions of recycled material that dramatically cut embodied energy; hence, knowledge of the proportion of recycled versus raw material is necessary when comparing studies.

In the special case in which the material, product, or service is itself an energy technology system (such as oil, gas, coal, nuclear, wind, hydro, tidal, solar, photovoltaic, or biofuel), the embodied energy used to extract, refine, process, and transport the energy can be expressed as an energy burden on the resulting fuel. The gross energy of the energy product is reduced by the embodied energy to give the residual net energy available. For example, where it takes 2 buckets of coal to power the technology to mine 9 buckets of coal, there is an energy burden (embodied energy) of 2 buckets of coal and a net energy resulting of 7 buckets of coal.

When the coal is so deep or difficult to extract that 3 buckets' worth of embodied energy are required to deliver 2 buckets of coal, the net energy is minus 1 bucket, a negative return that suggests the venture is no longer a sensible energy investment. This net energy accounting may be clear when a single energy resource is considered but is more difficult when numerous energy source types subsidize the energy system.

Some analysts are concerned that different energy resources are not easily aggregated using heat equivalence, economic price-derived factors, or physical thermodynamic analysis, as these aggregations lose information about the quality of an energy resource. Howard T. Odum promoted a system of energy and ecological analyses in terms of solar energy equivalents using transformity factors to bring all energies used to a common solar base, measured in a unit he termed emjoules. He used a special term emergy (with an m) to denote a special accounting of embodied energy using the embodied solar energy equivalent. International Standard 13602-1 attempts to describe a set of rules by which input and output and boundary conditions can be prescribed for technical energy systems.

With increasing interest in climate change and environmental concerns, there is a move to measure embodied energy in terms of the amount of carbon dioxide (CO_2) released in the use of the embodied energy. Measures of embodied energy are cited in terms of carbon released into the atmosphere (in kilograms of carbon dioxide per kilogram). This is analogous to the material's, product's, or service's carbon footprint over its lifetime.

—*Gilbert Valentine*

Further Reading

Benjamin, David S. *Embodied Energy and Design: Making Architecture Between Metrics and Narratives.* Columbia UP, 2017.

Cleveland C. "Energy Quality, Net Energy, and the Coming Energy Transition." http://www.oilcrisis.com/cleveland/EnergyQuality NetEnergyComingTransition.pdf.

Costanza R. "Embodied Energy and Economic Valuation." *Science*, vol. 210, no. 4475, 12 Dec. 1980.

Crawford, Robert H. *Licycle Assessment in the Built Environment.* Spon Press, 2011.

Dixit, M.K., J.L. Fernandez-Solis, S. Lavy, and C.H. Culp. "Identification of Parameters for Embodied Energy Measurement: A Literature Review." *Energy and Buildings*, vol. 42, 2010.

Hammond, G., and C. Jones. "University of Bath, Inventory of Carbon and Energy." http://www.bath.ac.uk/mech-eng/sert/embodied/.

International Standards Organization. "Technical Energy Systems: Methods for Analysis, Part 1: General." ISO 13602-1 2002. http://www.iso-standard.org/13626.html.

Odum, Howard T. *Environmental Accounting: Energy and Environmental Decision Making.* John Wiley and Sons, 1996.

Pomponi, Francesco, Catherine de Wolf, and Alice Moncaster, eds. *Embodied Carbon in Buildings: Measurement, Management and Mitigation.* Springer, 2018.

Spreng, Daniel T. *Net-Energy Analysis and the Energy Requirements of Energy Systems.* Praeger, 1988.

Energy and Power

FIELDS OF STUDY
Classical Mechanics; Electronics; Electromagnetism

ABSTRACT
There are many different ways to measure the amount of energy used to perform a task. Some involve knowing the force used and the resulting velocity. Others use the work done and the time it took to do that work. When dealing with circuits, there is a third way of calculating the energy used and the power expended.

KEY CONCEPTS
displacement: the difference between the initial position of an object and its final position, regardless of its path

force: the result of an interaction between two objects that changes the pattern of motion of the objects. A force can be a pull or a push.

joule: the base unit of energy, equal to one kilogram-meter squared per second squared (kg·m^2/s^2)

kilowatt-hour: a unit for measuring electricity consumption, equal to one thousand watts of power consumed over one hour, or 3.6×10^6 joules

newton: the base unit of force, equal to one kilogram-meter per second squared (kg·m/s^2)

potential energy: the energy that is stored in objects and can be converted to other forms of energy, such as kinetic energy

watt: the base unit of power, equal to one joule per second (J/s)

work: the energy used by a force to move an object over a given distance

FROM ENERGY TO POWER
As a person lifts an object, he or she is increasing the object's gravitational potential energy. That energy is being transferred to the object at different amounts per second, as the force a person expends when lifting an object is not constant. When the object is lifted, work is being done on the object. The work done on the object, like the force expended, is also not constant. The amount of work done per unit of time is what physicists call "power." Whether it takes a person three seconds or 10 minutes to lift a box one meter above the ground, the amount of work done is the same; it is the power that is different. In terms of the amount of energy spent per second in each case, the person who took 10 minutes to lift the box spent energy at a much lower rate. The person developed less power, which is why it took the person longer to lift the box. The power spent is equal to the total work done divided by the time it took to do that work.

Because work is a measurement of the difference in energy, in the International System of Units (SI), it is measured in joules (J). This means that power is measured in units of joules per second (J/s), or watts (W). Named after Scottish engineer James Watt (1736-1819), one watt represents the amount of work done in joules over a period of time measured in seconds. Power is also the amount of radiant energy produced by a light source. For instance, a hundred-watt lightbulb produces one hundred joules of energy each second.

CALCULATING POWER
Power (*P*) is a function of work done (*W*) over a given period of time (*t*):

$$P = W/t$$

Therefore, if a washing machine does 9×10^5 joules of work per load, and one load takes 1,800 seconds to complete, then the power spent by the washing machine in one load is calculated as follows:

$$P = (9 \times 10^5)/1800 = 500 \ W$$

This definition can be mathematically expanded by considering the definitions of work and energy themselves. Work is the difference in energy between a starting point and an end point. It depends on the force (*F*) acting on an object and the displacement (*d*) of the object. This relationship is given by the equation:

$$W = Fd$$

Substituting this value into the original equation for power produces the equation:

$$W = Fd/t$$

Remembering that velocity (*v*) is equal to distance over time, this equation can be simplified:

$$P = Fv$$

The SI unit of force is the newton (N). The energy spent to exert one newton of force over a distance of one meter is equal to one joule.

There are many ways to calculate power; which method to use depends on the type of problem. For example, imagine a hundred-kilogram man walking into an elevator. The elevator takes the man up to the second floor, a displacement of ten meters. It takes the elevator five seconds to do this. The force the elevator has to overcome is the weight of the man himself. Force is equal to the product of mass (m) and acceleration (a):

$$F = ma$$

Therefore, the weight of an object is also equal to its mass times its acceleration—specifically, the acceleration due to gravity (g), which on Earth is 9.8 meters per second squared (m/s2). This is the rate at which the velocity of an object in freefall will increase. Given this information, the power required for the elevator to lift the man is calculated as follows:

$$P = W/t = Fd/t = mad/t = (100)(9.8)(10)/5 = 19.6\ W$$

As seen in this example, power depends directly on the distance the object moves, the time it takes to do so, and the force, which itself depends on the mass of the object. Thus, accurate measurements of these quantities must be taken in order to obtain correct measurements of power.

These examples deal with the power spent against a specific kind of potential energy, known as gravitational potential energy. Other forms of potential energy exist, such as electrical potential energy. To calculate electrical potential energy, one must use a different equation of power. The power provided by batteries for circuits is the product of the current (I) that flows from the batteries, measured in amperes (A), and the voltage (V) supplied by the batteries:

$$P = IV$$

Note that one volt (V), the base unit of electric potential difference, is equal to one watt per ampere. If a six-volt battery supplies a current of 0.002 ampere into a circuit, then the power provided by the battery is calculated as follows:

$$P = (0.002\ A)(6\ V)$$

$$P = (0.002\ A)(6\ W/A)$$

$$P = 0.012\ W$$

MEASUREMENTS OF POWER

Knowing the amount of energy used by appliances, machines, and other devices in homes is of great importance to people and companies. Power companies install wattmeters in their customers' homes to measure the amount of energy consumed by a household. This amount can then be quoted to the consumer in units of kilowatt-hours (kWh). A kilowatt-hour is used to measure energy consumption per unit time, such as the total energy used during a last billing cycle. One kilowatt-hour is equal to 3.6×10^6 joules.

Other types of measuring devices are employed when measuring the power produced by machines. Dynamometers are used to measure the power produced by engines. These devices measure the rotational speed and the torque produced by the engine to calculate the power the engine can achieve.

ENERGY VERSUS POWER

The words "energy" and "power" are commonly used to mean the same thing. People quote the energy of lightbulbs they buy by incorrectly quoting the power. These quantities are very much related, yet they represent different things. While energy is the ability of an object to perform work, power is the rate of energy use. Energy consumption is a subject that affects every society. When buying LED lightbulbs, for instance, people have noticed that they have lower power ratings. An LED bulb with a power rating of just 5 watts can be as bright as a hundred-watt incandescent lightbulb because the LED uses most of this power to produce visible light rather than heat. Hundred-watt incandescent bulbs use a smaller proportion of that power to produce visible light; the rest is used to heat the element. Accurately measuring the energy spent and how fast it is being spent affects everything from the amount of power produced by utility companies to horsepower in cars.

—*Angel G. Fuentes, MS*

Further Reading

"Electrical Power." *BBC Bitesize*. BBC, n.d. Web. 5 May 2015.

Giambattista, Alan, and Betty McCarthy Richardson. *Physics*. 2nd ed. McGraw, 2010.

Khan, Sal. "Work and Energy (Part 2)." *Khan Academy*. Khan Acad., 2015. Web. 30 Apr. 2015.

Nave, Carl R. "Work." *HyperPhysics*. Georgia State U, 2012. Web. 5 May 2015.

Santo Pietro, David. "Power." *Khan Academy*. Khan Acad., 2015. Web. 5 May 2015.

Young, Hugh D., Philip W. Adams, and Raymond J. Chastain. *Sears & Zemansky's College Physics*. 10th ed. Pearson, 2016.

Energy Conservation

FIELDS OF STUDY

Electrical engineering; Environmental Sciences; Mechanical Engineering

ABSTRACT

While the distribution and utilization patterns of energy resource consumption has been justified, it is unsustainable for future generations. Energy consumption is expected to double in industrialized countries and triple in developing countries by the year 2050.

KEY CONCEPTS

alternative energy sources: refers to energy sources that do not rely on unsustainable fossil fuels (petroleum, coal, natural gas) but are entirely renewable (wind power, solar photovoltaics, small-scale hydropower, etc.)

LED lightbulbs: lightbulbs that use high-intensity light-emitting diodes (LEDs) rather than incandescent tungsten filaments or mercury vapor as the light source; LEDS typically use 3 to 5 watts of power to produce the same light output as a 100-watt incandescent bulb, and last thousands of times longer

solar photovoltaics: solid state devices that rely on sunlight and the photoelectric effect to produce an electrical current output

sustainability: refers to processes that can be carried on for generations of time, typically based on entirely renewable resources

UNSUSTAINABLE PATTERNS

Widespread availability of energy at an affordable cost is one of the most important factors in economic progress and improving the quality of life, but can result in wasteful energy use that in turn leads to environmental degradation. Energy conservation provides a range of resources that can simultaneously minimize energy consumption while keeping it widely available and affordable. More recently, energy conservation has been harnessed to reduce emissions of greenhouse gases and other pollutants, and individuals such as Amory Lovins have widely (and successfully) promoted conservation as an alternative to the continued construction of new power plants.

Although thought of as a recent development, energy conservation has a long history. In ancient Greece and Rome, scarce and expensive fuel resources led architects and engineers to design both efficient buildings and heating systems. These advance-

Occupancy sensors can conserve energy by turning off appliances in unoccupied rooms. (Z22 via Wikimedia Commons)

ments were largely lost after the fall of the Roman Empire, but were rediscovered during the European Renaissance, when improved chimneys and stoves were reintroduced. As early as 1618, German writer Franz Kessler wrote *Holzsparkunst* (Saving Wood), which promoted efficient heating systems in order to save Germany from deforestation. While the Germans learned this lesson, the French (and others) did not, and the introduction of efficient stoves into Paris in the eighteenth century simply made heat available to a large population, which led to rapid deforestation of the French countryside. Benjamin Franklin saw this same pattern in America and invented a more efficient stove to avoid a similar deforestation in America (which happened anyway).

The dilemma of conservation is that by reducing the energy required to do a specific task, the cost will also be reduced and people will use more of it. One of the best examples of this is the introduction of vehicle fuel economy standards in the United States, which made cars more efficient, but as a result, Americans increased the numbers of miles they drove by 151 percent over the next 25 years, five times more than the growth in population.

One of the more effective and widespread conservation methods employed in recent years is the promulgation of efficiency standards for a wide range of appliances as well as buildings. Refrigerators, air conditioners, furnaces, and lightbulbs have become much more efficient, and building codes in many areas now mandate that new and renovated buildings must be very efficient.

Energy conservation has been aided by the introduction of smart meter and energy control systems, which allow even residential consumers to monitor and control their energy consumption. As is the case with most energy conservation measures, they are more likely to be adopted by consumers with the financial resources to do so, although governments and nongovernmental organizations (NGOs) have become increasingly active in reducing energy use in low-income areas of developed nations and throughout developing countries. The introduction of more efficient appliances, low-power computers, and other technology has also made it possible to introduce them into areas depending on limited renewable resources such as solar photovoltaics (PVs).

Increasing and rampant energy consumption is polluting the global atmospheric environment through the emission of excessive carbon compounds, while it is gradually limiting access to energy for a substantial amount of the world's population. Energy conservation is thus needed to reduce consumption of limited energy resources and to encourage the use of energy-efficient products in our daily life in such a way that would reduce environmental trade-offs and ensure sustainable use.

AVAILABILITY OF ENERGY RESOURCES

Energy resources are found in renewable and nonrenewable forms. Nonrenewable energy resources are most commonly consumed and include fossil fuels such as oil, natural gas, and coal. Nonrenewable energy resources are extensively converted into other forms to generate electricity. Although nonrenewable energy resources have multiple uses in daily life and their consumption is expected to increase, their future quantities are limited. Notably, fossil-fuel-based energy consumption has already given rise to environmental protection concerns, calls for energy conservation, and searches for alternative energy resources.

The largest amount of fossil-fuel-based energy is consumed in the developed parts of the world, but only 30 percent of global population lives there. It has been well documented that since the industrial era of the late 1800s, greenhouse gas (GHG) emissions have increased exponentially due to fossil fuel combustion in developed nations. Of particular concern is the influx of carbon dioxide (CO_2) into the atmosphere, which has risen to more than 450 ppm from the normal level of 250 ppm in the preindustrial period. Due to excessive CO_2 emission into the atmosphere, global warming has already increased and the resulting weather variability has heavily impacted social and economic conditions in many poor countries. With regard to climate change, energy consumption issues have been the focal point driving the quest for alternative energy sources.

DEVELOPING CONSERVATION MEASURES

Energy conservation and the concomitant development of alternative energy sources can contribute to both social and economic development. Energy conservation can most importantly reduce the need for finite fossil-fuel-based energy products and thus reduce GHG emissions into the atmosphere. Reduced energy consumption is likely to result in lower per capita demand for energy as well as lowered industrial production costs. The resulting energy savings will lead in turn to lowered costs associated with the investment, supply, and distribution of energy to meet the increased demands driven by population growth.

Just as humans have driven energy consumption, humans have the capacity to drive energy conservation. Energy is required in the accomplishment of our daily work. In general terms, energy is the ability to do work, for instance to move objects, light our homes and cities, and power our factory machines. Energy conservation should thus start from our daily lives and surroundings.

Individuals can reduce their daily energy consumption and adopt low-carbon lifestyles through changing their behavior and product usage. Households have a significant role in energy efficiency because of their diverse demands for energy use and household activities with respect to their diverse social and economic situations. Households use both direct (natural gas, electricity, biomass, and other fossil fuels) and indirect (production, transportation, and disposal of goods and services) energy.

Energy conservation efforts most commonly include the following steps (which are not exhaustive): walking, riding a bicycle, or riding mass transit instead of driving automobiles, which emit 60 percent of the air pollution in cities; using LED lightbulbs, which can reduce energy consumption 20 to 30 times and more as compared to incandescent lightbulbs; air-drying clothes instead of using a clothes dryer; and recycling newspapers, aluminum cans, and plastic bottles to reduce future energy costs for manufacturing the same products.

METHODS OF CONSERVATION

Energy conservation also involves the type of appliances we use in our households and offices. Different appliances have varied energy consumption rates and consumers should act proactively to reduce their energy consumption. Examples include raising or lowering the water or air heater or air-conditioning thermostat, switching off the lights when leaving a room, not using the standby mode of electronic appliances like computers, sealing air leaks around windows and doors to reduce excessive heating and cooling, planting trees to block sunlight near windows to reduce associated cooling costs, cleaning or changing air filters in air heating and cooling systems to improve efficiency, and the use of energy-efficient appliances.

Many factors influence energy conservation at the individual and household levels as well as within energy-intensive business units. One key issue is a lack of awareness regarding the selection and use of energy-efficient products and services. Awareness campaigns, such as the distribution of information booklets and broadcast programs and the marketing of energy-efficient products may increase people's conscious decisions to adopt products that facilitate energy conservation. People's awareness of their energy use and their self-motivation in terms of environmental concerns can determine their consumption patterns. People select energy products with respect to their economy, comfort, and context. Increased energy saving awareness can motivate people toward reduced consumption and the preference of alternative energy sources. In this regard, energy policies must encourage the adoption of efficient products.

The core idea of energy conservation is to use energy more efficiently. Energy efficiency is determined by lifestyle, but also by access to more efficient products. Energy-efficient products include those that consume less energy, are not purely produced from fossil fuels, are recyclable, and are driven by alternative energy sources. For example, electricity is usually generated from finite, nonrenewable fossil fuels, but there is a huge potential for the production of electricity from renewable solar, wind, and hydro sources.

Alternative energy sources such as solar energy are efficient means for energy conservation and the reduction of energy consumption and are directly accessible resources in the world around us. While fossil-fuel-based electricity generation is not equally available in developing and least-developed countries, there is great potential for the adoption of solar-based home electricity systems. To some extent, promotion of solar panel-based electrification systems would improve the social and economic conditions of rural populations in those countries that seldom have access to fossil-fuel-based energy resources while reducing over-reliance on biomass fuel or the outside importation and exploration of fossil fuel resources.

Energy conservation is a global issue but conservation roles and responsibilities vary among nations. Developed nations need to play the vital role in conserving energy, as these nations consume the highest percentage of energy resources due to greater access and to technological and financial investment in energy exploration. Residents in developed nations use a plethora of appliances in their daily life, which are powered by electricity generated from a variety of mostly nonrenewable energy resources. There is also an upsurge in demand for oil for fuel combustion to support the growing urban traffic of those nations. Citizens of highly energy-dependent countries need to adopt more energy-efficient products to reduce their consumption levels. At the same time, energy conservation efforts must be accompanied by the development of alternative renewable energy technology, such as solar power, wind power, and hydropower. Some developed nations have already adopted alternative energy sources—such as nuclear energy in the United States and Sweden, and wind power in Denmark and the Netherlands—to generate electricity and boost industrial economic growth.

Demand for energy is projected to at least remain constant to maintain growth in developed countries, and demand will rise in developing and least-developed countries to drive economic growth. The role and demand of energy-deficit countries regarding energy conservation must not be downplayed. Raising economic growth in these areas will require sufficient access to energy. Renewable alternative energy sources have already showed promise. The access of rural people to energy resources is vital to accelerate the national development of poor countries, which may be viable if alternative energy can be provided in remote areas. Many initiatives are already underway to promote alternative energy sources in Asian and African countries. For example, a few private agencies are providing solar panels to rural residents with no access to national electric grids in Bangladesh, while a group of people formed a cooperative to use solar panels for their businesses in Namibia.

The adoption of alternative energy sources and new technologies, however, are not yet cost-effective nor easily accessible for most remote impoverished populations, and little government funding is in place to support such efforts. Human life will not be sustainable, however, without energy conservation efforts focused on environmental protection and global socioeconomic realities.

—*Ronju Ahammad*

Further Reading

Bynyard, Peter. "Pursuing the Energy Efficiency-Conservation Path." *Energy Policy*, 1988.

"Communications on Energy: Household Energy Conservation." *Energy Policy*, 1984.

Green, John A.S., ed. *Aluminum Recycling and Processing for Energy Conservation and Sustainability*. ASM International, 2007.

Koerth-Baker, Maggie. *Before the Lights Go Out: Conquering the Energy Crisis Before It Conquers Us*. John Wiley & Sons, 2012.

Kumar, Anil, Om Prakash, and Prashant Singh. Chauhan. *Energy Management: Conservation and Audits*. CRC Press, 2021.

Lovins, Amory. "Energy Strategy: The Road Not Taken?" *Foreign Affairs*, Oct. 1976.

Maher, Neil M. *Nature's New Deal: The Civilian Conservation Corps and the Roots of the American Environmental Movement*. Oxford UP, 2009.

Sheldrick, Bill, and Sally Macgill. "Local Energy Conservation Initiatives in the UK: Their Nature and Achievements." *Energy Policy*, 1988.

Thumann, Albert. *Plant Engineers and Managers Guide to Energy Conservation*. 10th ed., River Publishers, 2020.

ENERGY INDEPENDENCE AND SECURITY ACT OF 2007

FIELDS OF STUDY

Government Affairs; Policy Studies; Transportation Engineering

ABSTRACT

The Energy Independence and Security Act (EISA) of 2007 is a comprehensive U.S. energy law comprising a variety of innovative measures to address energy security.

KEY CONCEPTS

CAFE standards: Corporate Average Fuel Economy standards, the minimum average fuel economies all automobile manufacturers are to achieve, designated by class and type of vehicle

cellulosic ethanol: ethanol produced from degradation and fermentation of cellulose-based plant matter

RFS: Renewable Fuel Standard, the minimum amount of renewable fuels to be produced and used in transportation fuel blends

BACKGROUND

In December 2007, U.S. president George W. Bush signed into law the Energy Independence and Security Act (EISA). It comprises a variety of innovative measures to address energy security and, to a lesser degree, climate change. Most prominently, the law directed U.S. automobile manufacturers to significantly increase the Corporate Average Fuel Economy (CAFE) standards for several vehicle classes.

Since the turn of the century, rising energy prices and growing concerns over climate change had invigorated the debate over a new comprehensive energy policy in the United States. While addressing climate policy gathered pace only slowly in Congress, energy security was addressed more proactively. After several years of debate, Congress finally agreed on the Energy Policy Act of 2005. The law was negotiated while the Republican Party held the majority in both the House and the Senate and basically followed President Bush's focus on increasing domestic production of fossil fuels and advancing nuclear energy technology. The EISA of 2007 deviated from this previous course, introducing significant changes to energy policy only two years later. The EISA was initiated after the midterm elections of November 2006 by the new Democratic majority in Congress.

The House version of the bill included, among other measures, a variety of tax provisions as well as a renewable portfolio standard (RPS). The former, totaling about $21 billion, were designed to repeal tax subsidies for oil and gas companies and promote renewable energy sources. The latter would have obliged electricity-producing companies to derive a certain percentage of their electricity supply from renewable sources. The RPS and most of the envisaged tax provisions, however, were not included in the final version of the bill because of the resistance of Senate Republicans and the threat of a veto by the president.

The final bill did, nevertheless, include several new provisions aimed at decreasing U.S. demand for nonrenewable sources of energy. Most important, the EISA of 2007 modified the U.S. CAFE standards in several ways. CAFE standards had originally been established as a reaction to the oil crisis of 1973. In 2007, the standard was 27.5 miles per gallon (mpg) for passenger cars and 22.2 mpg for light trucks. The distinction between passenger cars and light trucks had created a loophole for a class of vehicles designated as sport utility vehicles (SUVs) and minivans, which had been designed so they were regulated under the light truck standard.

Furthermore, some SUVs were not regulated at all, because they exceeded the light truck upper gross vehicle weight limit of 8,500 pounds. The growing popularity of such vehicles had actually decreased the overall average U.S. fuel economy in the years preceding the legislation. The EISA aimed at narrowing the SUV loophole. It stipulated that the combined fuel economy average for passenger cars and light trucks be raised to 35 mpg by 2020. Moreover, it directed the administration to regulate, for the first time, work trucks and commercial medium-duty or heavy-duty on-highway vehicles.

The EISA also substantially increased the renewable fuel standard (RFS), compared to the provisions included in the Energy Policy Act of 2005. Following passage of the 2007 law, the minimum amount of renewable fuels to be used in the transportation sector was set at 9 billion gallons in 2008 and was to be increased to 36 billion gallons by 2022. This standard is to be met, to an increasing degree, by advanced biofuels such as cellulosic ethanol. Although such advanced fuels comprised only a small fraction of the overall standard in 2008, the 2022 level will have to include 21 billion gallons of such fuels. Moreover, the act prescribes specific minimum life-cycle reductions in greenhouse gas (GHG) emissions from renewable fuels. These minimum reductions are calculated as the difference from traditional fossil fuels' emissions and range from 20 percent for traditional biofuels to 60 percent for cellulosic ethanol. This latter provision was included in the final version of the bill despite the disapproval of the president.

Several other titles in EISA cover other areas of energy efficiency. The act establishes appliance and lighting efficiency standards. It sets a goal for achieving zero-net-energy use for commercial buildings by 2025. Several provisions target the energy efficiency of federal agencies and raising consumer awareness.

Moreover, efforts to promote carbon capture and sequestration (CCS) technologies, designed to store energy-related greenhouse gases, are expanded.

—Jörn Richert

Further Reading

Attenberg, Roger H., ed. *Global Energy Security*. Nova Science Publishers, 2009.

Heath, Garvin A. *Life Cycle Assessment of the Energy Independence and Security Act of 2007*. National Renewable Energy Laboratory, 2009.

Kaplan, Stan Mark. *Smart Grid: Modernizing Electric Power Transmission and Distribution; Energy Independence, Storage and Security; Energy Independence and Security Act of 2007 (EISA); Improving Electrical Grid Efficiency, Communication, Reliability, and Resiliency; Integrating New and Renewable Energy Sources*. TheCapitol.Net, 2009.

Sissine, Fred, coordinator. "Energy Independence and Security Act of 2007: A Summary of Major Provisions, CRS Report for Congress." 21 Dec. 2007. http://energy.senate.gov/public/_files/R1342941.pdf.

ENERGY INTENSITY

FIELDS OF STUDY
Economics; Process Engineering

ABSTRACT
Energy intensity is an economic measure of the amount of energy consumed per amount of monetary value produced, and it often declines over time as a result of technological progress.

KEY CONCEPTS
embodied energy: the amount of energy represented by a specific product or resource throughout its lifetime, from the initial step of its development to its ultimate disposal

emissions intensity: the amount of pollution emitted relative to the amount of energy used for the production of a certain amount of economic value

energy efficiency: the value of a product or resource relative to the amount of energy that is consumed in its production

energy intensity: the amount of energy consumed by the production of a product or resource relative to its economic value

ENERGY AND ECONOMIC VALUE

Energy intensity measures the amount of energy that is consumed to produce a particular value of goods and services in the economy. It is most easily thought of as the cost of turning energy into money. It allows countries to calculate and compare their energy use over time, even as the size of their economy changes. Many factors affect a country's energy intensity, although in most countries intensity has declined over time, mostly because of technology. Energy intensity is related to several other energy terms, such as energy efficiency and emissions intensity.

Energy intensity is often expressed as units of energy required to produce some goods and services, divided by the dollar value of that same amount of goods and services. By using the same units of measurement over time, this expression of energy intensity lets people calculate whether it is taking more, less, or the same amount of energy to produce the same value of economic activity from year to year. When energy intensity decreases over time (the number that expresses energy intensity becomes smaller), the amount of energy it takes to produce economic value is decreasing, or the cost of turning energy into money is going down. The reverse is true if energy intensity increases over time.

Simply producing fewer goods or services will not necessarily decrease energy intensity. Total energy consumed may go down, but the number of goods and services (and their economic value) will also go down, and energy intensity is a function of both. To decrease energy intensity, the amount of energy used must decrease more than the value of the goods and services produced by that energy falls. Conversely, the value of goods and services produced must increase by more than the amount of energy used rises.

Because the prices of the same goods and services can change over time in response to factors such as inflation, prices are often scaled to a reference year. Thus, only the change in energy required to produce those products changes, making energy intensity easier to calculate and compare over time.

FACTORS AFFECTING ENERGY INTENSITY

Energy intensity often decreases over time as a result of new technologies that allow the same value of products and services to be made with less energy. The energy savings can happen directly, as when recycling

GDP per unit of energy use

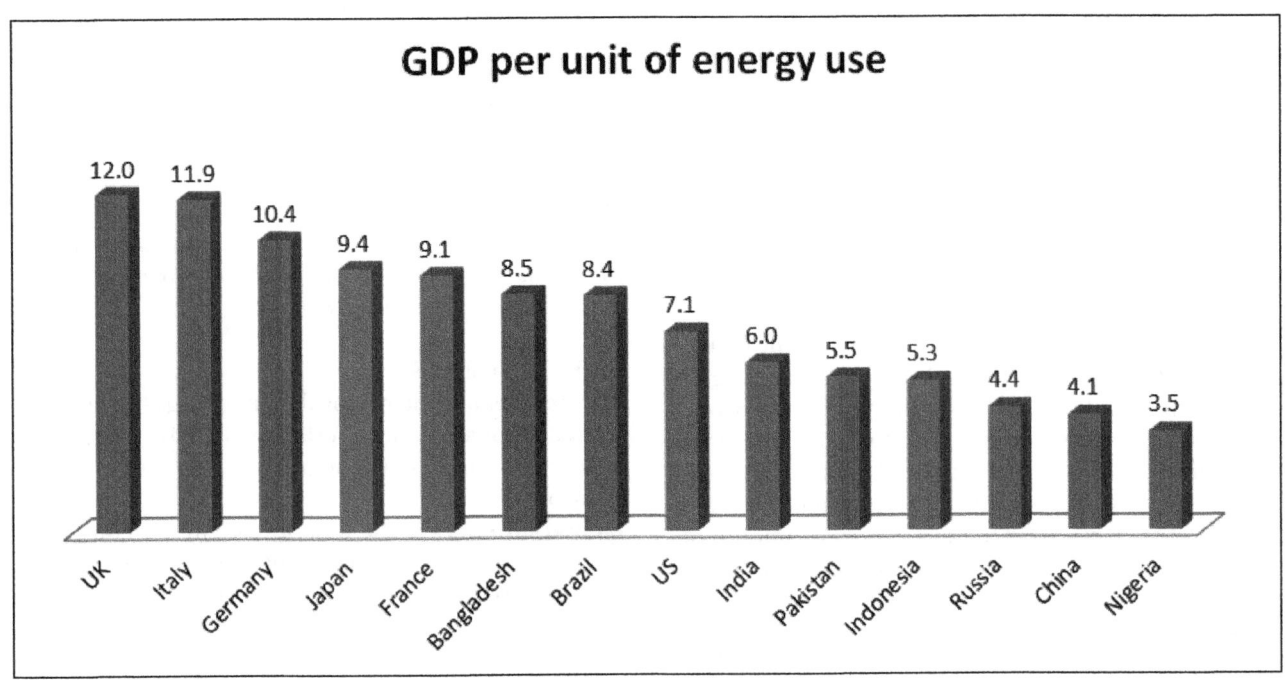

UK	12.0
Italy	11.9
Germany	10.4
Japan	9.4
France	9.1
Bangladesh	8.5
Brazil	8.4
US	7.1
India	6.0
Pakistan	5.5
Indonesia	5.3
Russia	4.4
China	4.1
Nigeria	3.5

The World Bank: purchasing power parity (PPP) $ per kg of oil equivalent (2011). (Wikimedia Common)s

allows an aluminum can to be turned back into another aluminum can, avoiding the energy required to mine and process raw aluminum from the ground. It can also happen indirectly, as when a lawyer uses the Internet to provide legal advice remotely, rather than having to use the energy of flying to a client's location, or when a better air conditioner allows factory workers to comfortably assemble automobiles all day while using less energy to cool themselves. Energy intensity can also change in response to changes in behavior, such as switching to riding a bicycle to work instead of driving an automobile.

All else being equal, people usually prefer lower energy intensity to higher, since this means less energy must be consumed to produce the same amount of economic activity. Many factors affect how energy intensive a country is. Countries with large economies may have lower energy intensity, because they have more money to invest in technologies, such as more efficient air conditioners, that reduce energy use. Countries with more policies that encourage energy-saving activities, such as recycling, may also have lower energy intensities.

However, countries with particularly hot or cold climates often have higher energy intensities, regardless of the size of their economies, because of the greater amount of energy required to cool and heat their buildings. Other factors that may influence a country's energy intensity include the types of energy sources available (some are more expensive to use than others); the size of the country, its population density, and the distances between population centers (which influence energy spent on traveling); the fraction of energy-intensive industries (such as manufacturing) that make up the economy; and whether energy costs are subsidized by the government (which tends to promote less energy savings).

As countries became more industrialized, their energy intensity tended to increase, because energy-intensive industries like manufacturing replaced less energy-intensive industries, such as subsistence agriculture. Later, as less intensive industries, such as financial services, replaced manufacturing and as countries invested more to invent and use energy-saving technologies, intensity tended to decrease. Different countries have experienced (or are experiencing) this evolution at different times, and therefore energy intensity has tended to vary widely from country to country. Today, because of relatively free trade in goods and services, converging patterns of consumption, and the spread of energy-saving technologies, energy intensity varies much less across countries, and the difference continues to narrow.

RELATED TERMS

Energy efficiency can be thought of as the inverse of energy intensity. It can be defined as the amount of product made divided by the amount of energy required to make it. For example, if the fuel efficiency of a car is described in miles per gallon (miles traveled for every one gallon of gasoline consumed), the equivalent fuel intensity of a car might be described in gallons per mile (gallons of fuel consumed for every one mile traveled) or dollars spent on fuel per mile traveled. Energy intensity is also related to *emissions intensity*, the amount of pollution emitted when a certain amount of energy is used to make a certain amount of economic value. It is also sometimes confused with *embodied energy*, which is the amount of energy consumed in the lifetime of a particular product from its inception to its final disposal, instead of a particular amount of economic value. It differs from energy intensity as used in physics, which describes the average flow of energy over time.

—*Kyle Gracey*

Further Reading

Cao, Jing. *China's Energy Intensity: Past Performance and Future Implications*. Economy and Environment Program for Southeast Asia, 2010.

Hu, Zhaoguang, and Zheng Hu. *Electricity Economics: Production Functions with Electricity*. Springer, 2013.

Kruger, Paul. *Alternative Energy Resources: The Quest for Sustainable Energy*. John Wiley, 2006.

Madureira, Nuno Luis. *Key Concepts in Energy*. Springer, 2014.

Ortiz, David, and Jerry Sollinger. *E-Vision 2002: Shaping Our Future by Reducing Energy Intensity in the U.S. Economy*. RAND, 2003.

United Nations, Department of Economic and Social Affairs, Division for Sustainable Development. "Intensity of Energy Use." http://www.un.org/esa/sustdev/natlinfo/indicators/isdms2001/isd-ms2001economicB.htm.

U.S. Energy Information Administration. "International Energy Statistics: Energy Intensity." http://www.eia.doe.gov/cfapps/ipdbproject/IEDIndex3.cfm?tid=92&pid=46&aid=2.

Energy Payback

FIELDS OF STUDY
Architecture; Economics; Electrical Engineering

ABSTRACT
Energy payback provides a calculation of quantity and time frame required to repay the embodied energy represented in a site of further energy production.

KEY CONCEPTS

decommission: the removal of an energy-production facility, or any other production facility, from active service, and may include its demolition and disposal

embodied energy: the amount of energy consumed in all aspects of the lifetime of a product or resource

energy payback: a time-based determination of the amount of energy produced by a facility that is equal to the embodied energy of the facility

energy return on investment: a calculation of the overall amount of energy produced by a facility relative to the energy invested in its operations

lifetime: the period of existence of a product, resource or facility from its original idea and design to its final stage of disposal

photovoltaics: electronic devices that produce electrical current by the photoelectron effect, in which incident sunlight causes surface atoms to eject electrons

TO RECOVER EMBODIED ENERGY

Energy payback is generally regarded as the time it takes for an energy-generating unit to produce the same amount of energy originally used to produce the unit, plus the energy it takes to maintain the unit throughout its lifetime, plus the energy it will take to decommission the unit. This "embodied energy" calculation usually accounts for the amount of energy exerted to start and end energy production, thereby revealing the amount of time expected to achieve positive yield. As such, energy payback calculations include both energy use numbers and temporal information.

ENERGY PAYBACK OF PHOTOVOLTAICS VS. FOSSIL ENERGY SOURCES

Often examined, energy payback calculation is used in the assessment of photovoltaic (PV) energy units such as solar cells. The U.S. Department of Energy has a history of assessing the advantages of PVs in terms of energy payback, given that PVs create zero pollution, produce zero greenhouse gases, and do not produce unsustainable outputs, as does the burning of fossil fuels. PVs seem attractive for commercial and individual power production for these reasons. However, costs for PV development have been outperformed by fossil fuel source costs, which has led to continued scrutiny of PVs in relation to the length of time it takes to recoup higher costs for solar power production.

This time is slightly longer than it is for fossil fuel energy sources, because of the latter's extant infrastructure resources, widespread access, and consistent fuel delivery. In the future, it is presumed, fuel and production resource scarcity may increase fossil fuel payback, given increased energy expenditures to attain resources in the production and maintenance portions of the energy payback calculation.

In recent years, the interest of energy companies in PVs has increased because of the sustainable nature of solar energy; moreover, the energy payback has improved for many types of PVs. In current assessments, energy payback calculations make it more likely that an energy-payback-sensitive consumer will choose to install thin-film PVs over multicrystalline units, given that thin-film units can produce electricity at similar or improved levels while using less production energy. The latter's payback time is shorter. Thin-film panels require less energy to produce the film and substrate materials they use for solar conversion to electricity. Multicrystalline units require much more energy to purify and crystallize silicon, their primary module and conversion material. Frame structures are also more energy intensive for the crystalline units than for the thin-film versions, as heavier frames are required to secure and stabilize the heavier silicon.

ENERGY PAYBACK AND WIND ENERGY

Wind energy production has also been assessed for energy payback in terms of unit propellers, turbines, foundations, and gears. Wind turbine life-cycle analysis yields the description and energy quantities of manufacturing turbine components and installing rotors in safe and secure structures. This analysis includes metals, plastics, glass, concrete, and chemicals. Such analysis demonstrates energy payback across domestic and international wind turbine units ranging from 1.3 to almost 8 months. Such a time line indicates that wind units, when placed in locations with high and predictably frequent wind velocity, pay back their production, usage, and decommission costs fairly quickly.

As with any energy production genre, externally limiting factors must be assessed for the sake of output calculations. In the previous renewable energy examples, sunlight and wind intensity and frequency are included as a portion of the output summary. Energy payback calculations amplify the need to consider the severity of these output qualifiers.

OTHER CONCEPTS IN ENERGY PAYBACK CALCULATIONS

There are many related concepts that can help clarify energy payback. One such concept is the idea of breaking even, or the question of whether or not an energy-producing entity provides at least as much energy as was used to produce the entity in the first place. The quantity of original energy used becomes a kind of "line in the sand" measurement against which production energy is measured. If an energy-producing unit will "cross the line," it is deemed to be at least breaking even in terms of its energy production. Energy payback is slightly more complex than the concept of breaking even, given that the former accounts for the time it takes to break even in terms of energy production contrasted to original energy expenditures. In fact, many energy analysts refer to the concept as the "energy payback time line" because of this important feature.

A second related concept, energy return on investment (EROI), is usually represented in terms of an energy-generating facility's capacity to go beyond breaking even in social, political, or economic terms. The calculation for EROI is usually represented as a divisible term: The numerator is calculated using an energy facility's nominal capacity multiplied by average load and the facility's lifetime; the denominator is calculated as the energy used to build and decommission the facility plus the multiplied outcome of named capacity, average load, portion of energy output used to maintain the facility, and facility lifetime. Of note here, EROI can be increased according to this model by reducing a facility's production costs, increasing load, increasing facility lifetime, and reducing maintenance energy usage. The outcome of EROI reveals the investment value for the energy produced, and such a value holds weight for energy consumers and municipalities in terms of development and investment.

"Extended EROI" and "ensemble EROI" calculations build on both the energy payback and the EROI concepts. The extended EROI model continues to calculate energy returned to users but places in the denominator energy used for acquisition, delivery, and usage of energy. The ensemble EROI model provides a calculation to assess payback and overage in terms of additional unit creation on the part of wide-scale and long-term energy-producing facilities with nonlinear or asymmetrical additions.

—*Curt Gilstrap*

Further Reading
Cleveland, Cutler. "Net Energy from Oil and Gas Extraction in the United States, 1954-1997." *Energy*, vol. 30, 2005.
Costanza, Robert, and Cutler Cleveland. "Net Energy/Full Cost Accounting: A Framework for Evaluating Energy Options and Climate Change Strategies." http://summits.ncat.org/docs/EROI.pdf.
Hall, Charles, et al. "What Is the Minimum EROI That a Sustainable Society Must Have?" *Energies* 2 (2009).
Harvey, L.D. Danny. *A Handbook on Low-Energy Buildings and District-Energy Systems: Fundamentals, Techniques and Examples*. Earthscan, 2006.
Kessides, Ioannis, and David Wade. "World Bank Working Research Paper: Toward a Sustainable Global Energy Supply Infrastructure." http://ideas.repec.org/p/wbk/wbrwps/5539.html.
Kurokawa, Kosuke, Keichi Komoto, Peter van der Vleuten, and David Faman, eds. *Energy from the Desert: Practical Proposals for Very Large-Scale Photovoltaic Systems*. Earthscan, 2007.
Lee, Yuh-Ming, and Yun-Ern Tzeng. "Development and Life-Cycle Inventory Analysis of Wind Energy in Taiwan." *Journal of Energy Engineering*, vol. 134, no. 53, 2008.
Tiwari, G.M., and Arvind Tiwari Shyam. *Handbook of Solar Energy: Theory, Analysis and Applications*. Springer, 2016.
Turner, John. "A Realizable Renewable Energy Future." *Science*, vol. 285, 1999.

U.S. Department of Energy. "What Is the Energy Payback for PV?" http://www.nrel.gov/docs/fy04osti/35489.pdf.

Energy Policy

FIELDS OF STUDY
Governmental Affairs; Policy Studies; Social Sciences

ABSTRACT
Energy policy allows governments to influence levels of supply and demand for energy, the composition of the energy sector, impacts from energy consumption and production, and energy markets.

KEY CONCEPTS
bilateral policy: meaning "two sides," a policy agreement involving two nations or governing bodies
cap-and-trade system: one in which a maximum amount of emissions is assessed for an industry and each contributor of emissions is assessed the maximum amount of emissions allowed (the "cap"). Contributors who do not emit that amount are allowed to sell or trade (the "trade") their remaining emissions allowance to contributors who may have to exceed their allotment in order to maintain the overall total within the specified limits of the policy
domestic policy: a policy established and applicable only within that countries national jurisdiction
fiscal policy: short-term policy agreements established based on annual monetary consideration of energy
multilateral policy: meaning "many sides," a policy agreement involving three or more nations or governing bodies
regulatory policy: long-term policy agreements established in regard to regulation of resource management and energy use

DEFINING PURPOSES, GOALS, AND PROTECTIONS
The energy policy of a decision-making body or government, whether local, state, national, regional, or international, is a broad course of action comprising a portfolio of strategies and objectives to guide the system of production and use of energy. Governments utilize a suite of energy policies in seeking to achieve

different goals in both the short term and the long term. Governments can use fiscal and regulatory policies to control short-term prices of energy and thereby consumer behavior; while using longer-term strategies, such as investments in research and development or in infrastructure, to develop new energy technologies, diversify fuel mixes, and transform energy systems to minimize environmental and public health impacts. High-priority concerns that many energy policies around the world seek to address include access to energy, energy security, and energy sustainability.

THE NEED FOR POLICY
Concentrated sources of energy—in liquid, solid, and gaseous forms of hydrocarbon chains—are important inputs for economic activity and one of the basic drivers of human civilization. Since the mid-1700s and throughout the Industrial Revolution, human societies have increasingly utilized hydrocarbon-based fossil fuels for personal and industrial transport, electricity, and agricultural purposes. Growing dependence on fossil fuels, which as of 2008 accounted for approximately three-fourths of global energy use and 25 percent of global energy-related carbon dioxide (CO_2) emissions, is one of the major factors driving the need for coordinated energy policy at all levels of government. Many of these policies seek to ensure that fossil fuel production remains reliable and sufficient for economic development and steady growth. Maintaining the safety and sustainability of energy production and consumption has also become an increasing concern over the past several decades, as global consumption of energy has nearly doubled: nuclear energy, coal, and offshore and unconventional sources of fossil fuels, such as oil sands and shale rock, have been increasingly exploited in order to meet growing global energy demand, coupled with growing concern over possible impacts on public health and environmental health.

ENERGY SECURITY
Growing demand, price volatility, supply disruptions in the wake of extreme weather events and other natural disasters, and political interventions all contribute to concern over national energy security. These concerns can be exacerbated by a perceived overreliance on imports, such as experienced by the United States, Japan, and the European Union (EU), which currently import more than half of their total energy. In order to ensure consistent energy supplies, governments can articulate multiple and often complementary policies.

One option is to increase domestic production of energy, which could entail (1) taking advantage of indigenous renewable energy through development and deployment of solar, wind, hydro, geothermal, and other alternative energy technologies; (2) supporting expansion of domestic fossil fuel recovery, potentially from previously uneconomic or unconventional sources; or (3) developing surplus production capacity, in the event that refinery capacity limits domestic supplies. For example, Iceland has managed to produce more than 80 percent of its energy domestically, largely through renewable geothermal and hydro technologies. Although this domestic production is made possible largely by Iceland's optimal geological characteristics and glaciers, Iceland's government has helped to stimulate private-sector investment in renewable energy development, hydrogen-powered transportation, and domestic industry to reduce imports. Conversely, increasing domestic drilling for oil has been an important part of the U.S. discussion on energy security policy over the past several years, ultimately becoming a component of the U.S. strategy for reducing reliance on oil imports by one-third over 10 years.

Governments can also work to reduce dependence on both foreign and domestic sources of energy by increasing energy efficiency. For example, the EU set a target to improve economy-wide energy efficiency by 20 percent by 2020. China also established a goal of decreasing energy intensity of gross domestic product (GDP) by 20 percent over the 2005-10 period.

Another option for increasing energy security is to diversify sources of energy. This includes utilizing an array of energy feedstocks and production technologies, such as renewables, nuclear energy, biofuels, natural gas, oil, and coal, as well as avoiding the consolidation of energy imports around a single or select few trading partners. For example, the United States made diversification of its energy supplies a priority in its national security strategy of 2010 and in the president's Blueprint for a Secure Energy Future of 2011. Denmark has prioritized its diversification of energy sources since the 1970s; in order to achieve

Energy Policy   PRINCIPLES OF ENERGY

Russia is a key oil and gas supplier to Europe. (Wikimedia Commons)

124

this goal, the Danish government has invested heavily in renewable energy (especially wind power), combined heat and power (CHP), increased oil and natural gas drilling in the North Sea, and reliable energy trading relationships with neighboring Scandinavian countries and Germany.

In the event that a country is highly dependent on a single trading partner or region for much of its energy needs, it becomes a political priority to develop political or economic agreements to ensure a steady flow of energy imports. For example, the EU imports more than a quarter of its oil and gas from Russia. After experiencing a series of gas cutoffs in 2007, the EU and Russia developed an agreement to avoid future supply crises.

Governments can also opt to develop strategic or auxiliary reserves of oil and gas or public stocks. While not a long-term solution to supply disruptions, such a program can still head off the immediate effects of a sudden disruption in normal energy supply. Other emergency preparedness strategies, such as demand restraint and fuel rationing, can play smaller but significant roles in maintaining energy security in the event of unexpected supply disruption.

OVERSEAS ENERGY INVESTMENTS
Countries not only seek to expand domestic energy supply, through the aforementioned strategies, but also invest in energy development overseas, particularly in developing countries where there is growing demand. Developed countries as well as fast-growing, resource-poor countries may undertake such foreign investment strategies in response to the imperatives of increasing energy access and facilitating economic growth.

Governments also support the expansion of energy infrastructure as part of development aid policies through multilateral bodies like the World Bank and their development finance institutions. For example, the Overseas Private Investment Corporation (an agency of the U.S. government) established the Global Renewable Energy Fund in 2008 to invest in renewable energy projects in emerging markets.

ENERGY ACCESS
Domestic, multilateral, or bilateral energy investments can address the policy objective of increasing energy access in developing countries. The International Energy Agency (IEA) estimates that as of 2011 more than one-fifth of humanity remained without access to electricity, having to rely on less concentrated and more polluting forms of energy, such as wood, charcoal, and other biomass. Increasing the access of these communities to modern energy grids can help improve livelihoods and public health.

Multilateral funding mechanisms such as the World Bank have provided billions of dollars in loans over the past several decades to developing countries such as Mozambique, Kazakhstan, Nepal, and Vietnam for renewable energy and energy access projects, largely in rural areas. Coal plants, while controversial for their significant greenhouse gas (GHG) emissions, are also supported in certain areas when there are not economically viable low-GHG options for improving energy access. World Bank funding, for example, comes from donor countries that sit on the board of directors and determine which energy projects meet the objectives of World Bank funding, including reducing poverty and increasing sustainable economic growth.

REDUCING ENVIRONMENTAL IMPACTS
Environmental impacts can result from each step along the life cycle of energy production and consumption. Biological communities and wildlife habitats can be degraded from fossil fuel mining and drilling. Air and water pollution in the form of GHGs, nitrogen oxides, sulfur dioxide, particulate matter, mercury, and other heavy metals are emitted during fuel production and combustion. A significant proportion of energy policies is devoted to addressing, mitigating, or minimizing the emissions and other impacts from the energy sector. Strategies to achieve these objectives include requiring energy producers to use the best available technologies (BATs) in their operations and other pollution control requirements; setting emission standards and limits to ambient pollution concentration; requiring permits for pollution-emitting facilities; taxing or pricing the environmental externalities of energy consumption; and requiring environmental impact assessments prior to any new energy extraction or production facility construction.

Requiring energy companies to carry out environmental impact assessments (EIAs) is a preventive step for minimizing the impact of new energy recovery projects. In carrying out an EIA, the company must investigate all potential impacts of an individual project before it is undertaken. Degradation of wildlife habitat, stress on water supplies, air pollution and water pollution, noise, and other disruptive activities during construction or operation of the facility, devaluation of surrounding properties, and risk of malfunctions or disasters that could have an impact on nearby populations are all variables that must be accounted for. Normally, the energy company must demonstrate in the EIA how it plans to address or minimize identified potential impacts and submit the document to a regulatory entity for review. The regulator can determine whether the EIA has been appropriately broad and detailed in its analysis and whether the proposed actions for addressing the impacts are sufficient.

BAT requirements state that the most advanced technologies must be used to limit pollution from energy production. The U.S. Clean Air Act, originally passed in 1963, refers to best available control technology (BACT) for reducing the emissions of air pollutants, including particulate matter, carbon monoxide, sulfur oxides, nitrogen oxides, and ground-level ozone. Setting explicit limits on ambient pollution levels in a local area is a related pollution reduction strategy, and the two polices can be complementary.

In combination with these policies, regulatory authorities can also require operating permits for pollution-emitting entities, which would contain information on what kinds of pollutants are released from the facility, the rate of emission, and required mitigation strategies. The permitting process is useful if facilities are regulated under multiple statutes in order to clarify the array of regulatory requirements.

Setting a price on the emissions from energy production, commonly referred to now as a 'carbon tax' in reference to carbon dioxide emissions, has become a more extensively discussed option for regulating environmental impacts. One of the first examples of such a policy was the U.S. sulfur dioxide cap-and-trade system implemented in 1990. This was highly successful in reducing U.S. sulfur dioxide emissions by more than 50 percent from 1990 to 2002. In later years, many governments have discussed applying the emissions pricing scheme to carbon dioxide emissions from energy combustion, and some regions have even implemented such a scheme. In 2015 President Barack Obama and the Environmental Protection Agency (EPA) established the Clean Power Plan for reducing carbon emissions from power plants. In February 2016 the U.S. Supreme Court stayed implementation of the plan pending judicial review.

SUSTAINABLE ENERGY SYSTEMS

Policymakers not only want to incrementally reduce the environmental impacts from existing energy use and infrastructure, as described above; they also aim, over the long term, to develop energy systems that contribute near-zero emissions and that depend on renewable feedstocks, as opposed to exhaustible supplies of oil, coal, and natural gas. Policies meant to increase the market penetration of renewable technologies, increase energy efficiency, and utilize lower-impact transportation technologies and modes can facilitate the transition toward sustainable energy systems.

Renewable energy technologies can be promoted through a suite of policy strategies, including renewable energy portfolio standards, renewable energy goals, feed-in tariffs, cap-and-trade schemes, carbon taxes, and various grants and subsidies. For example, the EU has a goal of renewable technologies providing 20 percent of total energy consumption and 33 percent of electricity consumption by 2020. In order to meet this goal, the EU has put additional policies in place, including a cap-and-trade system for certain economic sectors, which began in 2005, and multiple national-level feed-in tariff programs and other aggressive funding schemes. These have made the EU the largest renewable energy consumer in the world.

Increasing energy efficiency is also a key component of transitioning to a sustainable energy system, both to reduce use of fossil fuels in the near to medium term and to partially offset rising global energy demand over the coming decades. A country's average energy efficiency can be measured in terms of amount of energy required for each unit of GDP generated. An energy efficiency policy can seek to reduce this number over time. For example, China established a target in 2005 to reduce energy consumed per unit of GDP by 20 percent over five years, a target that China came close to meeting through significant engagement with and oversight of local governments and industries. Other, smaller-scale energy efficiency policies can set efficiency standards for specific buildings, appliances, vehicles (referred to as "fuel efficiency standards"), and economic activities.

Transportation accounts for nearly a third of all energy consumed globally and is one of the only sectors in which energy use continues to grow in developed countries. More than 95 percent of the transport sector is powered by fossil fuels. To achieve sustainable transport, countries have implemented policies to make vehicles more efficient, to increase the amount of renewable fuels (namely biofuels) used in personal vehicles, to invest in research on alternative vehicles and battery technologies, and to increase use of public buses, trains, and other low-emission modes of transport, such as bicycles.

To reduce GHG emissions and increase efficiency of international shipping and air transport, discussions have been under way at the United Nations Framework Convention on Climate Change (UNFCCC), the International Maritime Organization (IMO), and the International Civil Aviation Organization (ICAO) to regulate these emissions. International agreements are already in place to reduce air and water pollutants from shipping under the International Convention for the Prevention of Pollution from Ships (MARPOL) and the Convention on the Prevention of Marine Pollution by Dumping of Wastes and Other Matter (the London Convention), among others.

MAXIMIZING SAFETY

Energy recovery and production activities have long been associated with health and safety risks, particularly for oil and gas drilling, coal mining, and nuclear energy. In recent history, a number of energy-related disasters have occurred that had severe impacts on surrounding economic activities and ecosystems.

During 2010 and 2011, multiple incidents encouraged policymakers to revisit their energy health and safety policies: the BP Deepwater Horizon oil spill in the Gulf of Mexico; the Chilean coal mining incident in which 33 miners were trapped in a collapsed mine for 69 days; the Fukushima nuclear reactor meltdown in Japan; and concern about groundwater contamination from shale gas drilling, or fracking, throughout the United States. Policy responses to these events varied. In response to the issue of the toxic fluids generated by shale gas drilling, the Pennsylvania governor ordered that all gas-drilling wastewater cease to be treated at public drinking water treatment facilities. After the Deepwater Horizon oil spill, the United States imposed a temporary moratorium on deepwater offshore drilling. Longer-term policy changes include dividing regulatory authority for offshore drilling activities and strengthening safety and emergency procedure standards. Following the Fukushima disaster, Germany shut down its oldest nuclear plants and reversed its plans to reduce the rate of phaseout of nuclear power. China suspended new plant approvals in order to develop new nuclear safety standards.

In addition to these large-scale disasters, improving public health is also an argument for regulating the air and water emissions from the normal daily operations of energy production facilities, the policy strategies for which are described above.

HARMONIZING MARKETS

One strategy for reducing transaction costs of energy distribution is to harmonize energy regulations across regions. This can allow for transfer of greater amounts of energy and streamlined production processes with fewer steps of monitoring and regulation. Fractured regulation creates barriers to trade and competition, thereby increasing inefficiency and reducing the flexibility of energy markets to meet demand.

For example, different countries and even different states have distinct standards for gasoline quality, such as sulfur content. A refinery may either produce according to a single sulfur standard, in which case its market is reduced to those areas of the corresponding standard, or it can match the standard of the most aggressive regulation, which is also costly. Harmonized national or global sulfur standards would allow for each gasoline producer to maximize its market potential and minimize the need for multiple stages of monitoring across regional borders as gasoline travels from producer to consumer.

The EU has been working for more than a decade, with partial success, to harmonize its internal energy policies and market rules in order to ease energy trade, especially of natural gas and electricity, within the region and to allow market access to more suppliers. EU policymakers seek to open the EU energy market fully to internal and external competition while harmonizing fuel quality standards across member states, which the EU government argues create a barrier to competition.

The Common Market for Eastern and Southern Africa (COMESA) also developed a policy framework for regional energy market harmonization in 2007, largely in order to increase efficiency of regional energy production, increase consumer access to energy supplies, and stimulate greater energy investment.

LIBERALIZATION AND STATE OWNERSHIP

Governments can control many aspects of the energy sector in their countries, from direct ownership of oil companies to capping the price of electricity to regulating the electricity suppliers from which consumers can purchase their power. Governments can exert different levels of influence over energy-sector development, depending on how much of the sector is regulated or state-owned as opposed to liberalized. State-controlled or national oil companies control more than three-quarters of global oil reserves, which confers on them significant political influence and the ability to control global energy supplies and prices. The IEA has estimated that governments provide more than $550 billion annually to companies and consumers.

Conversely, small private companies are more characteristic of the renewable energy sector, which has begun to achieve economies of scale over the past decade in many parts of the world, albeit with significant levels of subsidization (though a fraction of fossil fuel subsidies).

RESEARCH AND DEVELOPMENT

Policies to support research and development (R&D) of energy technologies can be used by governments to achieve multiple goals; publicly funded research, alongside private research, can increase the efficiency of mature technologies, expand the applications of technologies, develop technologies to mitigate impacts from energy production and consumption, and develop new technologies with lower costs and impacts.

Many of the U.S. government energy research programs were first undertaken in the 1970s, largely in response to the Arab oil embargo of 1973. Concern over security of conventional energy supplies spurred governments to devote funds to developing alternatives to oil-dependent technologies and synthetic alternatives to oil. Global spending on energy R&D has declined from a peak of more than $20 billion in the early 1980s to about $15 billion at the end of the twenty-first century's first decade.

Governments are increasingly supporting carbon capture and sequestration (CCS) research, lower-cost batteries and fuel cells, advanced biofuels, and electrical grid improvements such as "pressurized-water reactor" upgrades, among other initiatives for reducing environmental impacts, reducing dependence on oil, and diversifying the energy mix. Government R&D support is particularly important for technologies that are at early stages of development and that therefore receive low levels of investment from private firms. As government funding advances the state of technology, private-sector confidence can grow such that the later stages of development and distribution are taken over by firms.

INTERNATIONAL ENERGY POLICY

Much of the discussion so far has been devoted to energy policy at the national level, yet meaningful energy policy is also created at other levels of governance, from the local to the international scale.

Much of international policy comes from the United Nations in the form of various international agreements to which a predetermined number of countries must agree for them to enter into force and otherwise be effective. International energy policy has evolved out of global initiatives to develop norms of behavior and international standards related to the law of the sea, transboundary impacts, safety, and climate change.

While it does not explicitly address energy policy, the United Nations Framework Convention on Climate Change (UNFCCC), established in 1992, is the international institution tasked with developing international climate policy. UNFCCC parties reached agreement in 2010 to an objective of limiting global temperature rise to 2°C above preindustrial levels by 2050. Meeting this goal requires a significant ramp-up of renewable technology deployment in both developed countries and developing countries as part of an overall effort to change the development paradigm away from increasing reliance on lower-cost, higher-carbon fuels such as coal.

The United Nations Convention on the Law of the Sea (UNCLOS), which entered into force in 1994, contains sections that explicitly address energy recovery in the global commons of the high seas. Under

UNCLOS, coastal states are responsible for preventing, reducing, and controlling marine pollution from offshore oil and gas activities. The convention also provides some guidance on energy exploration and exploitation rights in the high seas.

The Espoo Convention on Environmental Impact Assessments, which entered into force in 1997, was developed by European countries and the United States to address transboundary environmental pollution. The Strategic Environmental Assessment (SEA) Protocol to the Espoo Convention requires that all plans and programs undertaken by a country that have a potential for transboundary impacts must carry out an SEA. The protocol is applicable to offshore oil and gas recovery.

Smaller multilateral bodies can also develop energy policy, both legally binding and nonbinding. For example, the Arctic Council, an intergovernmental forum comprising the eight Arctic nations with some additional observer countries and organizations, developed a set of nonbinding guidelines for offshore oil and gas activities, which advises energy companies on minimizing the impacts of hydrocarbon recovery on other Arctic sectors.

Conversely, under the Convention for the Protection of the Marine Environment of the North-East Atlantic (OSPAR, named for the cities in which it was developed, Oslo, Norway, and Paris, France), the decisions made by parties are legally binding. Since 1977, OSPAR parties have adopted nineteen decisions and recommendations with regard to the offshore oil and gas industry in the northeast Atlantic.

The United Nations Conference on Climate change was held in Paris from November 30, 2015 to December 12, 2015, during which 186 countries published their action plans for reducing GHG emissions with the goal of limiting global warming by 2100 to below 2 degrees Celsius (3.6 degrees Fahrenheit). On Earth Day, April 22, 2016, 175 nations signed the nonbinding Paris agreement on climate change.

STATE AND LOCAL POLICIES

States, cities, towns, and local communities can also implement energy policies. States and local governments can seek to diversify energy sources, improve energy efficiency in buildings and public transport, implement renewable energy portfolio standards, regulate GHGs, and undertake educational or outreach campaigns to influence consumer behavior.

For example, California passed state legislation to regulate GHG emissions in 2006 through an emissions trading scheme. As described above, climate policy has significant implications for the energy sector. California has also implemented a low carbon fuel standard (LCFS), which requires transport fuel producers to meet certain quotas of less GHG-intensive fuels, including natural gas, biofuels, electricity, and hydrogen.

The city of Berlin, Germany, exemplifies an outreach campaign through its public transportation promotional advertising. The city has even established a "personals" website where passengers who have met briefly on the train and would be interested in getting to know each other can post a comment in search of their counterpart, making public transportation more of a social and interactive experience.

Local communities can work with nongovernmental organizations (NGOs) or foreign governments to improve energy access and install more efficient and cleaner technologies to improve public health. For example, the U.S. government launched its Global Alliance for Clean Cookstoves initiative in 2010; the program seeks to supply communities in need with high-efficiency cookstoves, in partnership with private partners, NGOs, and several government agencies. This reduces the dependence of local women on wood, crop residue, and dung, the burning of which can cause high levels of local air pollution and breathing ailments.

—*Emily McGlynn*

Further Reading

European Commission. "Communication from the Commission to the European Parliament, the Council, the European Economic and Social Committee and the Committee of the Regions: Energy 2020, A Strategy for Competitive, Sustainable, and Secure Energy."

EUR-Lex. European Union, 1998-2016. Web. 16 May. 2016. http://eur-lex.europa.eu/LexUriServ/LexUriServ.do?uri=CELEX:52010DC0639:EN:NOT.

Foulds, Chris, and Rosie Robison, eds. *Advancing Energy Policy. Lessons on the Integration of Social Sciences and Humanities*. Palgrave/MacMillan, 2018.

International Energy Agency. *Comparative Study on Rural Electrification Policies in Emerging Economies: Keys to Successful Policies*. International Energy Agency, 2010.

Müller-Kraenner, Sascha. *Energy Security*. Earthscan, 2008.

Philibert, Cédric. *Interactions of Policies for Renewable Energy and Climate*. International Energy Agency, 2011. Digital file. http://www.iea.org/papers/2011/interactions_policies.pdf.

Taner, Tolga, ed. *Energy Policy*. IntechOpen, 2020.

United Nations. "List of Parties That Signed the Paris Agreement on 22 April." *United Nations: Sustainable Development Goals: 17 Goals to Transform Our World*. UN, Apr. 2016. Web. 16 May. 2016.

United Nations Conference on Climate Change. "More Details about the Agreement." United Nations Conference on Climate Change. COP21, n.d. Web. 16 May. 2016.

U.S. Environmental Protection Agency. "Clean Power Plan: Clean Power Plan for Existing Power Plants." *EPA*. EPA, 11 Feb. 2016. Web. 16 May. 2016.

ENERGY POVERTY

FIELDS OF STUDY
Environmental Management; Policy Studies; Social Sciences

ABSTRACT
Energy poverty is a global problem resulting from the lack of access to energy sources because of either unavailability or cost. Energy poverty is most serious in developing countries, but it can also occur in industrialized countries.

KEY CONCEPTS
electrification: the provision of a secure and adequate electrical supply system in a particular area or region

electrification rate: the portion of an area or region that has been provided with secure and adequate access to electricity relative to the total area requiring such access

energy poverty: a state of the absence of or inadequate access to a secure and adequate supply of electricity and fuel for household needs such as cooking, hygiene, heating, etc., typically in undeveloped and underdeveloped areas of the world

fuel poverty: a state of the inability to afford to pay the cost for household utilities and fuel needs, typically in well-developed regions of the world

THE PROBLEM: LACK OF ACCESS
Energy poverty can be defined as lack of access to modern, clean, and safe energy sources to meet basic needs, such as lighting, cooking, boiling water, heating, cooling, earning a living, and using information and communication media. In addition, the term *energy poverty* is used to describe the unavailability of energy sources and the inability to afford energy sources because of their relative or absolute costs. Most people affected by energy poverty live in the rural areas of developing countries. However, increasing energy costs resulting from rising energy demand worldwide and anticipated scarcity of fossil energy sources are making access to energy a growing challenge in industrialized countries as well.

ENERGY, POVERTY, AND DEVELOPMENT
Available energy services fail to meet the needs of the poor. In most developing nations, only a minority of the population use commercial energy in the same way people in the industrialized world do. The overall electrification rates are low, and traditional biomass (firewood, agricultural residues, and animal dung) remains a major fuel source. According to the International Energy Agency (IEA) estimates, worldwide, 1.1 billion people lacked access to electricity as of 2016, and 2.8 billion relied on direct combustion of traditional biomass, kerosene, or coal to meet their basic energy needs. The projections for 2030 suggest that these numbers will largely persist. It is expected that by 2030, 674 million people will still lack access to electricity and, regardless of efforts to substitute traditional biomass as an energy source, 2.3 billion people will rely on traditional biomass. Geographically, most of these people live in sub-Saharan Africa and South Asia. In South Asia, the situation is expected to improve somewhat because of falling birth rates, infrastructure investments, and economic development; in sub-Saharan Africa, however, there is less hope for rapid improvement.

THE REALITY
People living in energy poverty are affected in all aspects of daily life, including health, education, work, access to water, access to information, and gender-related issues. Across borders, the social groups most affected by energy poverty are women and children. Because they are usually in charge of collecting fire-

wood, they also suffer most from the indoor air pollution caused by the direct combustion of biomass. According to the World Health Organization (WHO), health hazards caused by indoor air pollution were responsible for 3.8 million deaths in 2016, of which 90 percent occurred in the developing world; taking this into account, the death rate related to energy poverty is higher than from malaria (445,000) and tuberculosis (1.6 million).

Energy poverty affects not only current living conditions but also prospects for social and economic development in developing nations. Multiple and complex interrelations exist between the access to modern energy services and social and economic progress. Direct correlations can be found between the proportion of the population living below the poverty line and a country's electrification levels and electricity consumption. Accordingly, the lack of access to efficient, affordable, and clean energy sources interferes with efforts to reduce poverty.

THE METRICS

In order to measure the qualitative and quantitative progress a country makes in the fight against energy poverty and to better understand the role that energy plays in human development, the IEA introduced the Energy Development Index (EDI) in 2004. The EDI is composed of four indicators, each of which captures a specific aspect of potential energy poverty in developing countries: per capita commercial energy consumption, per capita electricity consumption in the residential sector, share of modern fuels in total residential sector energy use, and share of population with access to electricity.

The costs of fighting energy poverty were estimated by the IEA in 2017. To achieve universal access to electricity and clean cooking facilities by 2030, an additional investment of $786 billion would be required. Although the amount is manageable, most governments in developing countries are unable to provide sufficient funds to improve and expand modern energy services. It is also unlikely that the needed investment in rural areas will be financed by the private sector, as the purchasing power of the rural population is too low to attract government interest. Therefore, joint international efforts will be necessary to provide knowledge and funds to fight energy poverty.

For a long time, many donor countries neglected energy as a development factor. Between 1973 and 2005, only 5 percent of foreign aid was spent on energy projects; however, the situation is changing. Since the relevance of energy for social and economic development has been internationally perceived and energy security has become a central topic worldwide, energy is back on the foreign aid agenda, and the proportion of aid for energy projects is increasing.

Possible instruments in the fight against energy poverty are renewable energy and energy efficiency measures. Renewable energy technologies can directly affect livelihoods in remote areas by introducing decentralized sources of energy that can provide access to modern energy services without expensive grid extensions. Energy efficiency can help countries to channel economic growth in an environmentally sustainable way.

SUSTAINABLE DEVELOPMENT GOALS

Without improvements in the energy sector, poverty alleviation is unlikely to be achieved. Consequently, the topic of energy has to be integrated into strategies to fight poverty. The most important international agreement on poverty reduction is the Sustainable Development Goals (SDGs). These grew out of the United Nations Millennium Declaration of 2000, which articulated eight timebound antipoverty objectives known as the Millennium Development Goals (MDGs). Although energy was not directly addressed in any of the original eight MDGs, the successor SDGs target universal access to modern energy by 2030 and acknowledge the interplay between societal problems, such as the lack of energy access and poor health outcomes.

The relevance of energy services to reducing poverty was first explicitly acknowledged at the World Summit on Sustainable Development in Johannesburg, South Africa, in 2002, officially recognizing a critical link between energy access, sustainable development, environmental concerns, and natural resource use. It is important to integrate energy topics into future poverty alleviation strategies and policies, but it will be even more critical to transfer this knowledge into practice.

INDUSTRIALIZED COUNTRIES

Energy poverty in industrialized countries is also referred to as "fuel poverty." In contrast to energy poverty in developing countries, fuel poverty in the developed world is the result of strictly economic causes.

The leading research on energy poverty in developed countries has been conducted in the United Kingdom, were the problem has been known since the early 1990s. The term "energy poverty" is applied to households where fuel costs surpass the national median and their net income would be below the country's poverty line. In addition to space heating, energy expenses for water heating, lighting, electric appliances, and cooking are taken into consideration to define energy poverty.

Until recently, energy poverty has rarely been seen as an issue in most industrialized nations. With an increasing number of households struggling to pay for their energy needs, however, the subject has garnered more attention. Today, energy poverty exists in many industrialized countries with different social systems, from the United States to the United Kingdom to Romania. Energy poverty results from a combination of factors in these nations, including rising energy prices, the poor energy efficiency of older properties, and increasing proportions of the population with low household incomes. Efforts to increase energy efficiency and promote renewable energy use have begun to overlap with those to close the energy poverty gap.

—*Julia C. Pfaff*

Further Reading

Akli, Michaël, Patrick Bayer, S.P. Harish. and Johannes Urpelainen. *Escaping the Energy Poverty Trap: When and How Governments Power the Lives of the Poor.* MIT Press, 2018.

International Energy Agency. Energy Access Outlook 2017: From Poverty to Prosperity—World Energy Outlook Special Report. OECD/IEA, 2017, www.iea.org/publications/freepublications/publication/WEO2017SpecialReport_EnergyAccessOutlook.pdf. Accessed 9 Oct. 2018.

Modi, V., S. Mcdade, D. Lallement, and J. Saghir. Energy Services for the Millennium Development Goals: Achieving the Millennium Development Goals. World Bank and United Nations Development Programme, 2005. http://www.unmillenniumproject.org/documents/MP_Energy_Low_Res.pdf.

Nalule, Victoria R. *Energy Poverty and Access Challenges in Sub-Saharan Africa.* Palgrave MacMillan, 2019.

"9 Out of 10 People Worldwide Breathe Polluted Air, but More Countries Are Taking Action." World Health Organization, 2 May 2018, www.who.int/news-room/detail/02-05-2018-9-out-of-10-people-worldwide-breathe-polluted-air-but-more-countries-are-taking-action. Accessed 9 Oct. 2018.

Poor People's Energy Outlook. Practical Action, policy.practicalaction.org/policy-themes/energy/poor-peoples-energy-outlook.

Simcock, Neil, Harriet Thomson, Saska Petrova, and Stefan Bouzarovski, eds. *Energy Poverty and Vulnerability: A Global Perspective.* Routledge, 2018.

United Kingdom, Department for Business, Energy & Industrial Strategy. *Annual Fuel Poverty Statistics Report, 2018 (2016 Data).* Crown, 2018, assets.publishing.service.gov.uk/government/uploads/system/uploads/attachment_data/file/719106/Fuel_Poverty_Statistics_Report_2018.pdf.

World Bank. Renewable Energy Development: The Role of the World Bank Group. World Bank, 2004.

Energy Storage Techniques

FIELDS OF STUDY

Chemical Engineering; Electrical Engineering; Hydraulic Engineering

ABSTRACT

Energy storage involves converting energy into a different form for use at a later time and is becoming much more relevant as less-predictable renewable electricity capacity is installed.

KEY CONCEPTS

hydropower: power generated by the use of moving water

primary battery: a battery that cannot be recharged once its energy has been drained

pumped storage hydroelectric plant: a hydroelectric plant that uses some of its generated electricity to pump water back into the reservoir for reuse

secondary battery: a battery that is designed to be recharged by replenishing the energy that has been removed from it

thermochemical heat storage: the use of excess or available energy such as solar radiation to drive the chemical synthesis of a product that can be used at a later time in a reaction that releases heat

tidal energy: the energy of oceanic tides, derived from the gravitational influence of the Sun and Moon

ENERGY TYPES AND ENERGY STORES

The term *energy storage* refers to the temporary storage of energy for use at a later time, usually because the supply and demand for a particular energy carrier do not coincide. There are various types of energy storage according to the type of energy carrier, the time for which they can store, the speed at which they can be loaded and unloaded, as well as various other nontechnical, economic, and environmental characteristics.

Nature's own stores include different types of energy that are converted, either naturally or through human intervention, into other types of energy. The best-known example is the Sun, which continuously converts massive amounts of nuclear potential energy into electromagnetic radiation in the form of light and heat, and is the source of almost all energy on Earth—only tidal and geothermal energy derive at least partly from other sources. Another example is fossil fuels such as oil, gas, and coal, which are widely combusted by humans to extract the chemical energy stored in them and convert it to another useful form, such as heat or power. Nuclear power can also be added to this list, as it is based on utilizing the potential energy within unstable atoms, which is released as ionizing radiation when these atoms radioactively decay. Finally, there is a great deal of gravitational potential energy stored in the water cycle on Earth; as water flows downhill, this potential energy is converted to kinetic energy, which can be used to generate electricity, as in the case of hydropower.

ELECTRICITY

The versatility of electricity, which is manifested through its many applications, is contrasted with the disadvantage that this energy carrier cannot be stored, at least not in its original form. Hence, electricity generation systems have traditionally evolved to generate precisely as much electricity as is currently needed: in other words, they are led by demand. Power systems based on a large number of centralized power stations generating electricity for transmission across national and/or international networks have therefore evolved to enable rapid adjustment of the electricity generated to the load at any given time.

Artist's 3D rendering of Hydrogen energy storage unit. (Petmal via iStock)

If the difference between supply and demand in the electricity network at any one time exceeds very small tolerances, the network can become unstable due to the fact that the frequency of the network has to be adjusted to account for the discrepancy. In most European countries, most electrical appliances are designed to operate at a frequency of 50 hertz (Hz), such that only small deviations away from this value may result in damage to these appliances, as well as to the network. In North America and other regions, the operating frequency of the electrical system is 60 Hz.

The large growth in electricity generation from renewable resources over the past few decades, such as in Germany and Spain, has meant that the whole electricity system is no longer as controllable as it once was. This problem is expected to worsen in the future through the continued development of renewable energy and the growth in electric vehicles, which frequently need charging for extended periods. Although there are models for short-term wind forecasting, for example, the supply of electricity from renewable energy cannot be precisely predicted, especially over large time frames. Apart from measures to turn off power plants such as wind turbines, renewable electricity supply cannot easily be controlled. This means that there is a renewed interest in storage devices for electricity in order to balance supply and demand and thus account for the fact that, on some days, large amounts of electricity are generated when they are not needed (i.e., supply exceeds demand), and vice versa.

ELECTRICITY STORAGE

Available devices for storage of electricity generally convert it into potential energy. The most ubiquitous electricity storage devices used for these purposes, such as network balancing, are pumped storage hydroelectric plants. In these plants, excess electricity is used to pump water up from a basin into a reservoir (i.e., convert it into gravitational potential energy) until a time when electricity is needed again and the water flows out of the reservoirs to turn turbines and generate electricity.

The overall conversion efficiency is typically 80 percent, depending upon the turbine and type of plant. A similar principle is applied to pressurized air storage devices, which exploit the thermodynamic properties of gases to compress air, which is then later decompressed through a turbine. This technology is known as compressed air energy storage (CAES), and can be used in conjunction with gas turbine plants in order to improve the overall generation efficiency. The principle of this technology, which relies on underground caverns to store the compressed air, means that the location of any power plant wishing to exploit this concept is crucial.

Another form of device for storing electrical energy is the battery, which converts stored chemical energy into electricity, and vice versa. Primary batteries can only be used once, whereas reusable or rechargeable "secondary" batteries can be recharged and thus used in both directions. There are a wide variety of rechargeable battery types, which vary greatly in terms of their capacity, charging/discharging time, and deterioration over time (i.e., number of recharge cycles).

Yet another type of energy storage device is a flywheel, which is a large rotating mass designed to store (rotational) kinetic energy. When the flywheel is accelerated up to a certain speed, its large mass means that it has a large momentum (actually moment of inertia), so provided there is little friction on the shaft upon which the wheel is mounted, the flywheel will store this kinetic energy as it continues to rotate. Due to its large mass, the flywheel tends to resist changes in its rotational velocity; hence, they are used in applications where the demand for energy is continuous, but the supply is discontinuous—such as in a vehicle, to temporarily store the rotational energy from the drive shaft when the clutch is disengaged and the vehicle comes to a stop. They are also used for applications where a rapid discharge of the stored energy is required, such as in wind-up toys.

Finally, another type of energy store is that for storing heat energy. There are various types of thermal storage devices, including: sensible heat storage, whereby the store actually heats up, for example an insulated hot water tank; latent heat storage, which exploits a change in phase of a material, such as when salt is employed as a storage medium for solar thermal power plants; and finally thermochemical heat storage, whereby heat is stored in a chemical (endothermic) reaction, which can later be reversed (an exothermic reaction) to release the heat. In general, the latter approach is not constrained to heat storage, but is also applied in fuel cells, for example in order to generate and store electricity, typically in the form of hydrogen. In fact, the vision of a future energy system based largely on hydrogen as a chemical energy store is shared by many scientists.

—Russell McKenna

Further Reading

Alami, Abdul Hai. *Mechanical Energy Storage for Renewable and Sustainable Energy Resources.* Springer, 2020.

Baxter, R. *Energy Storage: A Non-Technical Guide.* PenWell Corporation, 2006.

Breeze, Paul. *Power System Energy Storage Technologies.* Academic Press, 2018.

Cabeza, Luisa F., ed. *Advances in Thermal Energy Storage Systems: Methods and Applications.* 2nd ed. Woodhead Publishing, 2020.

Energy.gov. "Energy Storage." http://www.oe.energy.gov/storage.htm.

Huggins, R.A. *Energy Storage.* Springer, 2010.

Inage, Shin-Ichi. "Prospects for Large-Scale Energy Storage in Decarbonised Power Grids." International Energy Agency Working Paper, http://www.iea.org/papers/2009/energy_storage.pdf.

International Energy Agency. "Technology Agreements: Energy Storage." 2011. http://www.iea.org/techno/iaresults.asp?id_ia=13.

International Energy Agency, Energy Technology Network. "Energy Conservation Through Energy Storage." 2012. http://www.iea-eces.org.

Kalaiselvam, S., and R. Parameshwaram. *Thermal Energy Storage Technologies for Sustainability: Systems Design, Assessment and Applications.* Elsevier, 2014.

Kularatna, Nihal. *Energy Storage Devices for Electronic Systems: Rechargeable Batteries and Supercapacitors.* Academic Press, 2015.

Roth, Kurt, and James Brodrick. "Seasonal Energy Storage." *ASHRAE Journal,* Jan. 2009.

Rufer, Alfred. *Energy Storage Systems and Components.* CRC Press, 2018.

Sørensen, Bent, ed. *Solar Energy Storage.* Academic Press, 2015.

ENTHALPY

FIELDS OF STUDY
Chemical Engineering; HVAC Trades; Mechanical Engineering; Thermodynamics

ABSTRACT
Enthalpy is a measure of the total energy of a thermodynamic system, including the energy required to create the system or body and the amount of energy needed to displace its environment and establish its volume and pressure in order to make room for the system or body.

KEY CONCEPTS
enthalpy: the energy of a system expressed as heat

entropy: essentially, the energy of a system due to the temperature-dependent degree of randomness or chaos within a system

free energy: also Gibbs free energy, the useful work obtainable from a thermodynamic system at constant temperature and volume

mechanical theory of heat: the theory that heat is produced by mechanical movement of atoms and molecules within materials

thermodynamics: the study of the active nature and movement of heat energy

WHAT ON EARTH IS EN'-THAL-PY?

Enthalpy is defined by *Merriam-Webster's Collegiate Dictionary* as "the sum of the internal energy of a body or system and the product of its volume multiplied by the pressure," with "internal energy" referring to that energy required to create the system or body and the "product of its volume multiplied by the pressure" describing the amount of energy needed to displace its environment and establish its volume and pressure in order to make room for the system or body.

Enthalpy is one of the properties of a system that often baffles students when they encounter this subject in their studies of chemistry and physics. Its initial discovery is often attributed to American Josiah Willard Gibbs, the first professor of mathematical physics at Yale University. Gibbs not only taught and lectured at Yale but also was a student there and was one of the first to be granted a doctorate from that institution. Gibbs was interested in how molecules respond to changes in states of energy and eventually defined a new property to represent this value, which he called the "free energy of the substance."

Gibbs observed that when energy is added to matter, either the randomness of the matter changes on a molecular level or heat is added or removed from the substance. The changes in the disorder of the substance were defined to be *entropy*, and the changes in heat were defined to be *enthalpy*. Gibbs was the first to be recognized for realizing that the heat energy of a substance and the atomic level of order were related and intertwined. Gibbs's studies of energy and thermodynamic equilibrium are summed up in the equation that bears his name. The Gibbs free energy equation he theorized and proved is:

$$G = H - T \times S \text{ and } \Delta G = \Delta H - (T \times \Delta S)$$

where H = enthalpy, T = temperature, S = entropy, Δ = the difference in two states or change in two states, and G = Gibbs free energy.

If Gibbs free energy neither enters nor exits a control volume, then G = 0 and the system is at thermodynamic equilibrium. In this specific case, the Gibbs free energy formula becomes H = T × S, or ΔH = (T × ΔS), and mathematically this defines H or entropy as the multiple of temperature change and disorder change of the substance inside the control volume. Simply stated, entropy is the measure of the change in atomic or molecular disorder, and/or the change in temperature, when heat energy is added. The addition of heat tends to provide the energy needed to loosen chemical bonds, and this is one of many mechanisms whereby chaos or entropy increases. This concept also means that ordered structures such as crystals will tend to start to break apart and/or increase in temperature from the addition of heat.

The addition of heat to water offers a good example of how entropy works. When water is heated it will increase in temperature and expand unless constricted. If it is constricted and the volume is held constant, then not only the temperature but also the pressure will rise. If enough heat is added, the liquid structure will become a gas, which has a much more random structure than the liquid. The opposite applies if heat is removed from water. In this case, the water gets cooler and will even form ordered crystals (ice). If one can return the water to the exact same temperature and chemical randomness when heat is removed and added (i.e., the process is 100 percent reversible), then this is called an adiabatic process. Mathematically, an adiabatic process is represented by the entropy being held constant when heat (enthalpy) is added or subtracted and only temperature changes with heat addition. This is only true in mathematics, however; in the real world, some randomness is always changed and entropy is never truly constant.

Using the modified Gibbs free energy formula, with G = 0 as a starting point, the formula can be further modified to represent a greater understanding of enthalpy. Since the randomness of the substance in a control volume can be changed or controlled by constraining either the pressure or the volume itself, T × S can be broken down into:

$$H = T \times S = E + P \times V$$

where E = internal energy, P = pressure, and V = volume.

This formula mathematically shows that if heat energy is added or removed, the substance studied can change in pressure, if volume and internal energy are held constant; conversely, the substance can change in volume if pressure and internal energy are held constant. For example, this can be observed in everyday life when water expands when heated in an open pot. It also can be observed that the same water contracts when heat is removed. Another common observation is that if water is held in a closed container and heated, the water will increase in pressure. The formula H = E + P × V represents that adding heat will make substances expand and pressurize and, conversely, contract and cool if heat is removed. The formula also demonstrates that it is a change in enthalpy that is driving the internal energy, pressure, and volume changes.

The definition of "work" is to move a force through a distance, so "pressure work" can be defined as P × V. Substituting this in the equation above, the relationship for enthalpy and work can be shown as H = E + W, where W = work. This equation is helpful to scientists and engineers, who use this relationship to estimate the work that can be done by adding heat. Engineers commonly use enthalpy when determining the amount of heat added or removed by heating, ventilation, and air-conditioning (HVAC) equipment, and they also use it in the power industry to see how much heat energy must be added to water to make steam for Rankine-cycle power plants. Enthalpy is expressed in many units. Some of the common forms are the British thermal unit (Btu), the joule, and the calorie. A Btu was originally defined as the heat input to raise the temperature of one pound of water by one

Fahrenheit degree. A joule is the amount of energy exerted, or work done, when a force of one newton is applied over a displacement of one meter. A calorie is the amount of heat energy required to raise the temperature of one gram of water by 1 Celsius degree.

—Damon Eric Woodson

Further Reading

American Institute of Physics. "Josiah Willard Gibbs, 1839-1903." http://www.aip.org/history/gap/Gibbs/Gibbs.html.

Bodner Research Web. "Gibbs Free Energy." http://chemed.chem.purdue.edu/genchem/topicreview/bp/ch21/gibbs.php.

Chopey, Nicholas P., et al. *Handbook of Chemical Engineering*. McGraw-Hill, 1984.

Kondepudi, Dilip, and Ilya Prigogine. *Modern Thermodynamics from Heat Engines to Dissipative Structures*. 2nd ed. Wiley, 2015.

National Aeronautics and Space Administration, Glenn Research Center. "Enthalpy." http://www.grc.nasa.gov/WWW/K-12/airplane/enthalpy.html.

Parris, George. *Thermodynamics: Heat Capacity, Enthalpy, Entropy and Free Energy*. 3rd ed. Independent, 2019.

Sherwood, Dennis, and Paul Dalby. *Modern Thermodynamics for Chemists and Biochemists*. Oxford UP, 2018.

Entropy

FIELDS OF STUDY
HVAC Engineering and Trades; Thermodynamics

ABSTRACT
Entropy is a key concept in thermodynamics and is an indicator of the amount of energy in a system that is no longer available for additional work; it measures the amount of dispersion of heat in a thermodynamic system.

KEY CONCEPTS

energy: a term describing the innate potential of a system to do "work" on that system and its surroundings

equilibrium: the state of a dynamic system in which no net change can occur without the input of energy from an external source

negentropy: meaning "negative entropy," as occurs when a random collection of different components is assembled into an ordered whole structure, as is the case when a machine is assembled from its individual component parts

steady state: often used to mean "equilibrium" when referring to dynamic systems, but a system in a steady state is not necessarily a dynamic system

thermodynamics: the science of energy as the relationship between enthalpy (heat energy), entropy (the dissipation of heat energy) and the work that can be done by a thermodynamic system

THE CONCEPT OF ENTROPY

Entropy is a key concept in thermodynamics. It was first coined and utilized in the second law of thermodynamics (the entropy law) by Rudolf Clausius in the 1850s. It is an indicator of the amount of energy that is no longer available for additional work. This indicator measures the amount of dispersion of heat in a thermodynamic system, which is often considered also to be a measure of the amount of disorder, or randomness, in a system. Although the first law of thermodynamics is time-invariant—that is, the amount of energy is a constant—the second law reveals time as the change in entropy. Entropy, then, is essential for understanding energy flow throughout the universe (both for the universe of all known mass and space and for a specific "universe," or system and its surroundings) and therefore is fundamental to numerous chemical, biological, mechanical, and physical processes.

WHAT BECOMES OF THE ENERGY?

The second law of thermodynamics, formulated by Clausius in the 1850s, governs the transformation of energy among its various forms. The law states that as energy is used in work, the amount of available, or free, energy decreases. The energy does not disappear; it is conserved, as the first law of thermodynamics stipulates; instead, the energy becomes degraded and cannot perform as much work as in its pre-work, energy-concentrated state. Clausius used the term *entropy* for the value of the degradation, or disorder, that occurs. As free (available) energy is transformed, there is a change from order toward disorder in the system; for energy, this manifests itself as the decrease in the amount of energy available to perform work. For example, as the energy in coal is transformed into electricity, some of the energy in the coal is transformed into a less intense state, heat.

The second law also reveals the irreversibility of energy flow and, even more abstract, of time. Unless obstructed by a barrier such as a chemical bond, energy tends to dissipate from high-potential to low-potential states, but the reverse does not occur—the amount of potential work that can be completed by an energy source does not increase as it is used. As free energy is transformed, the amount of entropy increases. Wood in a stove can be burned to release heat that will warm a home; however, the heat will dissipate from the warmer stove to the (assumed) cooler room. The temperature of the air in the room will move toward equilibrium with the surrounding environment without additional wood—an increase in the available energy—placed in the stove.

The remains of the burned wood do not have much, if any, energy available for work. On a larger scale, when energy flow stops, evolution ceases. Until that moment, time can be measured by assessing the amount of entropy; the presence of more entropy indicates a progression in time. However, entropy does not increase steadily, as specific events such as a forest fire can speed the pace of entropy's change and others can slow it down.

ENERGY, ENTROPY, AND WORK

Entropy multiplied by temperature measures the amount of energy that has been degraded from a state in which it is able to perform work to one in which it is not. This occurs as heat is exchanged between two systems. This principle was the basis of Sadi Carnot's study of the physics of steam engines; heat is transferred along the thermal gradient from the hot part of the engine to the cool part, and the energy available for future work decreases while the entropy increases. In order for the engine to continue running, more free energy would need to be added.

In equation form, entropy is calculated as $S = Q/T$, where S stands for entropy, Q for heat exchanged between a system and its surroundings, and T is temperature. As coal is burned, it produces heat; in other words, the energy in coal is transformed from a higher-intensity and lower-entropy level to a lower-intensity and higher-entropy level. For the system, the amount of entropy is a function of the free energy and temperature. $E = F + TS$, where E is the total energy, F is the free energy, T is the temperature, and S is the entropy. Restated for entropy, the equation becomes: $S = (E - F)/T$.

In a closed system, entropy continues to increase as heat disperses until it reaches its maximum at an equilibrium, or steady state. At that point, there is no free energy and the system is in disorder. There is no change, and evolution ceases. Until that moment, time can be measured by assessing the amount of entropy; the presence of more entropy indicates a progression in time. However, entropy does not increase steadily; as mentioned earlier, specific events, such as a forest fire, can speed the pace of entropy's change and others can slow it down.

Creating order in a system, measured as negative entropy (negentropy), requires the transformation of free energy. For example, building a car from many parts requires human or machine labor, both of which require energy to power the work. Order is measured in the form of a car, but during the assembly of the car, disorder is measured from the heat produced as food is transformed in the body of the human or electricity is used to power a robotic machine.

—*Kirk S. Lawrence*

Further Reading

Atkins, Peter. *Four Laws That Drive the Universe.* Oxford UP, 2007.

Ben-Naim, Arieh. *Entropy Demystified: The Second Law Reduced to Plain Common Sense with Seven Simulated Games.* Expanded ed. World Scientific, 2008.

Chaisson, Eric J. *Cosmic Evolution: The Rise of Complexity in Nature.* Harvard UP, 2001.

Lemons, Don S. *A Student's Guide to Entropy.* Cambridge UP, 2013.

Müller, Ingo. *A History of Thermodynamics: The Doctrine of Energy and Entropy.* Springer, 2007.

Smil, Vaclav. *Energy in Nature and Society: General Energetics of Complex Systems.* MIT Press, 2008.

Thess, André. *The Entropy Principle: Thermodynamics for the Unsatisfied.* Springer, 2011.

EXERGY

FIELDS OF STUDY

Ecology; Mechanical Engineering; Social Sciences; Thermodynamics

ABSTRACT

In thermodynamics, the exergy of a system is the amount of energy available for use. Exergy is sometimes also called ideal work, reversible work, available energy (or simply availability), or exergic energy.

KEY CONCEPTS

exergy: a concept describing the amount of usable work that can be reasonably obtained from a thermodynamic system

heat engine: any device that employs a thermodynamic system of cycle to provide the power of its operation

perpetual motion machine: a hypothetical thermodynamic device that produces as much energy or work as that consumed in its operation

ENERGY AND HEAT ENGINES

The study of exergy goes hand in hand with the history of thermodynamics. In the nineteenth century, French engineer Sadi Carnot considered the steam engines then in use and put forth the proposition that there is a limit to the possible improvements to steam engines, insofar as any heat engine (of which the steam engine is a subtype) works by transferring energy from a warm area to a cool one and converting some of the energy to mechanical work. Carnot described an ideal engine, now known as the Carnot engine, which transfers heat back and forth (i.e., it is reversible), such that energy exits the system as work and enters the system as heat energy. Carnot's theorem says that no engine operating between two heat reservoirs can be more efficient than this Carnot engine; it would be a perpetual motion machine, which physics says is impossible. Furthermore, all reversible engines operating between the same heat reservoirs are equally efficient.

The consequence of Carnot's theorem is that maximum efficiency is possible only if no new entropy is created. The attempt to articulate the upper bound of an engine's potential work resulted both in the concept of exergy and in the formulation of the second law of thermodynamics, which says that over time, differences (in temperature, pressure, and chemical potential) in an isolated system will achieve equilibrium.

American physicist and chemist Josiah Willard Gibbs formalized the physics of exergy in 1873 as part of his work synthesizing the extant body of knowledge of thermochemistry, although the term *exergy*—from the Greek *ex ergon*, "from work"—was not coined until 1956 (which is why so many variant terms remain in use to discuss this old concept with a young name). For

instance, in the chemicals industry, exergy has historically been called available work or availability.

EXERGY IN CONTEXTS

Exergy is an important concept in the design of power plants, the usefulness of which needs to be quantified in terms of the amount of usable energy they will produce under given conditions. Scientists working in ecological economics or systems ecology examine the impact of human activity on the environment by performing exergy-cost analyses. Policy possibilities will be compared, such as between methods A and B of creating an economic good, in order to determine which method consumes the most exergy. The answer to that question relies on a standard set of reference conditions; the operations of a hypothetical power plant, for instance, will be determined in a hypothetical set of conditions of a particular temperature, pressure, and atmospheric composition. The actual operations of the plant will vary and may be more or less efficient on very hot or cold, or very dry or humid, days—but this standard reference state is used first, serving the same purpose as Carnot's hypothetical engine.

Exergy must also be considered in the study of sustainable energy sources, such as tidal energy and solar power, and the impacts of nonrenewable resources such as fossil fuels. Various methods of power generation or energy use can be compared in terms of their effect on the exergy of Earth's natural resources.

—*Bill Kte'pi*

Further Reading

Aziz, Muhammad, ed. *Exergy and Its Application-Toward Green Energy Production and Sustainable Environment.* InTech Open, 2019.

Dincer, Ibrahim, and Marc A. Rosen. *Exergy: Energy Environment and Sustainable Development.* 2nd ed. Elsevier, 2013.

Mirskiy, Anton G. *Thermochemistry and Advances in Chemistry Research.* Nova Science, 2009.

Perrot, Pierre. *A to Z of Thermodynamics.* Oxford UP, 1998.

Szargut, J. *Exergy Method: Technical and Ecological Applications.* WIT Press, 2005.

EXTERNAL COMBUSTION ENGINE

FIELDS OF STUDY
Electrical Engineering; Mechanical Engineering; Steam Power Engineering; Thermodynamics

ABSTRACT
External combustion engines are characterized by an external combustion process followed by heat transfer into the working fluid within the engine, and are widely employed in the form of steam engines for electricity generation, as well as in Stirling engines for less centralized applications.

KEY CONCEPTS
Carnot limit: the upper limit for the efficiency of a heat engine; depends on the temperature difference between the working fluid and the surroundings; the higher this difference, the higher the theoretical maximum energetic efficiency

closed system: one in which the working fluid (steam) is condensed back to liquid water and returned for reuse rather than being vented away

heat exchanger: a structure that separates a heat source from a working fluid but allows heat to transfer from one to the other

open system: one in which the working fluid is not recaptured for reuse and must be replenished in order to maintain operation of the system; the opposite of a closed system

Rankine cycle: thermodynamic cycle employed in a steam engine involves four main stages: pumping of water, heat addition, work production, and heat removal

EXTERNAL HEAT SOURCE
An external combustion engine (EC engine) is a heat engine where, in contrast to an internal combustion engine (IC engine), the combustion takes place outside of the engine. Thus, the working fluid in the engine is externally heated through a heat exchanger or the wall of the engine. Most of the world's electricity generation occurs in fossil fuel or nuclear plants that raise steam to drive a turbine, that is, in an EC engine. Other examples of EC engines include the Stirling engine, which uses gas as the working fluid, or the Organic Rankine Cycle, which uses steam as the working fluid and is particularly well suited to electricity generation from low temperature heat.

Depending upon whether the working fluid operates in one or two phases, liquid or gas, or liquid and gas, the system may be referred to as single or dual phase. The EC engine may further be categorized depending upon whether the thermodynamic cycle is open or closed, and hence whether the working fluid is replenished or remains within the system: A steam engine using water is typically closed, as is a Stirling engine using gas. The upper limit for the efficiency of a heat engine is known as the Carnot limit, and depends on the temperature difference between the working fluid and the surroundings; the higher this difference, the higher the theoretical maximum energetic efficiency.

STEAM ENGINES
The most widespread type of EC engine is the steam engine, which is used the world over in centralized electricity generation plants, which are fueled by fossil fuels or nuclear power, and in some cases biomass. The history of the steam engine goes back thousands of years, but it wasn't until the Industrial Revolution in the eighteenth century that the engines found widespread application for providing motive power, after James Watt made significant improvements to a previous design, including the addition of a separate condenser. The thermodynamic cycle employed in a steam engine is the Rankine cycle, named after the Scottish polymath, Professor William Rankine, which involves four main stages: pumping of water, heat addition, work production, and heat removal. The components of the steam engine that carry out these stages are the pump, boiler, turbine, and cooling towers, if present, respectively. In cooling towers, the excess heat in the form of steam is condensed to water vapor, and this heat is transferred to water in the cooling cycle (not part of the closed-loop Rankine cycle) before being released to the environment. The steam can be exhausted directly to the environment, as in the case of a steam locomotive, but this is very inefficient because of the adverse effect on overall efficiency, due to the fact that the temperature difference between working fluid and environment is not maxi-

Model Stirling engine, with external heat from a spirit lamp (bottom right) applied to the outside of the glass displacer cylinder. (Zephyris via Wikimedia Commons)

mized. In the case that cooling towers are present, the condensed hot water is pumped back to the boiler to begin the cycle again. In a small boiler for domestic use, the heat is either directly released, or used to preheat the boiler water as in condensing boilers.

Whether or not the removed heat is used, such as in centralized combined heat and power (CHP) applications with district heat networks, has a massive effect on the overall efficiency of the plant. While pure electrical efficiencies of about 40 percent may be achieved in coal-fired electricity-only plants, in CHP plants the overall efficiency is around 80 percent. The thermodynamic price paid for this higher overall efficiency is a reduction in electrical efficiency, however, because some of the steam is extracted before passing through the turbine. Because this excess heat is generally not used for local heating, so that centralized electricity generation has average overall efficiencies lower than 50 percent on a global scale, this is an area in which there are large potential improvements in overall system efficiencies. In some European countries, particularly Scandinavian countries, the proportion of electricity generation from CHP plants is very high—in Denmark over 50 percent, for example—such that the overall efficiency of the electricity system is also high.

STIRLING ENGINES

One of the best known and probably most widely employed EC engine other than the steam engine, the Stirling engine was invented in 1816 by Robert Stirling, who at that time demonstrated the first closed-cycle air engine. It was only in the latter half of the twentieth century that the term "Stirling engine" was universally applied to this kind of heat engine. In this device, the cyclical compression and heating, followed by expansion and cooling, of gas results in the conversion of heat energy to work output. The thermal efficiency of a typical Stirling engine for domestic applications ranges between 15 and 30 percent, but care needs to be taken in interpreting this value, given that a large proportion of the remaining energy input is available, and is therefore used in such an application, making the overall efficiency around 85 percent. A diesel or gasoline engine may have a higher overall electrical efficiency approaching 40 percent, but will not be much more efficient overall, perhaps reaching 90 percent. Hence, EC engines typically have a lower power to heat ratio, so are suited to CHP applications with substantial heat loads.

Due to the external combustion process, the Stirling engine is also very quiet in operation, which makes it useful for applications where this is an advantage, such as in small combined heat and power (CHP) applications for households, where the unit may sit in the kitchen or a utility room. It is also very versatile, because it can be applied in any application with an external heat source, and is very reliable compared to IC engines. The latter means that the Stirling engine has lower maintenance costs, but these are somewhat offset by the higher investment compared to IC engines for household CHP applications. Another application for Stirling engines is in concentrating solar power plants, whereby solar radiation is focused with a parabolic dish onto a point at the input to the heat engine. Hence, the solar radiation serves as a heat input for the Stirling engine, which can be used to generate electricity. This application is especially useful for decentralized power generation, in locations where access to the electricity network or other renewable energy sources might be limited.

ADVANTAGES AND DISADVANTAGES

On the other hand, there are several disadvantages of EC engines compared to IC engines. First, the latter are smaller and therefore lighter, due to the fact that the combustion, heat transfer, and work transfer all occur inside the same chamber. This size difference is only minor when comparing, for example, a Stirling engine with a gasoline engine; but when comparing a steam engine and an IC engine for the same application, such as providing motive power for a vehicle, the difference becomes much more significant. The large external boilers required for a steam engine mean that the whole engine requires a large amount of space, even for applications requiring only a small amount of power. Another disadvantage of EC engines is the relatively long time it takes for them to start, compared to IC engines that can be started almost instantly.

Steam engines are most suited to operation in the steady state, which is most efficient and therefore economical at their design point. Hence, large steam engines, such as those used in power plants, take a long time to run up and run down; operating them away from this optimal load makes them less efficient. In addition, one large disadvantage of steam engines is the very high pressures at which they operate, which poses a risk of explosion. Ways in which a boiler can fail include overpressurization, overheating due to insufficient water and/or flow rate, or steam leakage from the boiler and/or pipe work. Typically, steam engines have devices to account for these risks, including an emergency valve that allows a maximum pressure to build up inside the boiler. Furthermore, lead plugs may be employed in the boiler crown, so that if the pressure and temperature reaches an excessive level, these plugs melt and thus automatically allow some of the steam to be released and the pressure to dissipate.

Further Reading

Beith, Robert, ed. *Small and Micro Combined Heat and Power (CHP) Systems. Advanced Design, Performance, Materials and Applications.* Woodhead Publishing, 2011.

Darlngton, Roy, and Keith Strong. *Stirling and Hot Air Engines: Designing and Building Experimental Model Stirling Engines.* Crowood Press, 2005.

Drbal, L., K. Westra, and P. Boston. *Power Plant Engineering.* Springer, 1996.

Eastop, T.D., and A. McConkey. *Applied Thermodynamics for Engineering Technologists.* 5th ed. Longman, 1993.

Rogers, G., and Y. Mayhew. *Thermodynamics: Work and Heat Transfer.* 4th ed. Longman Scientific, 1992.

F

FARADAY, MICHAEL

FIELDS OF STUDY
Electrical Engineering; Electrochemistry; Physics

ABSTRACT
Michael Faraday was a leading British chemist and physicist who laid the foundation for many discoveries in the fields of electromagnetism and electrochemistry, essential to the study and understanding of "energy."

KEY CONCEPTS
diamagnetism: a small repulsive force experienced in a magnetic field by all materials
electrolysis: chemical reactions driven by an electric current
electromagnetism: the generation of a magnetic field by an electrical current
experimentalist: a scientist who uses the observation of well-defined practical experiments to determine the apparent properties of processes and materials
induction: the generation of a magnetic field in a second material by the presence of a magnet; similarly, the generation of an electrical current in a conductor by a moving magnetic field
oxidation numbers: a formal method of defining the electronic state of atoms in compounds according to the number of electrons they appear to have gained (reduction) or lost (oxidation)

FARADAY'S CONTRIBUTIONS
Michael Faraday was a British chemist and physicist involved in the foundation of the fields of electromagnetism and electrochemistry. Although he came from humble origins and did not enjoy much advanced education, Faraday conducted studies that led to the discovery of diamagnetism, electromagnetism, induction, and the laws of electrolysis. As a chemist, Faraday discovered benzene and helped develop the system of oxidation numbers. He was very influential on later scientists, and Albert Einstein considered him a great source of inspiration. Considered by many to be one of the most adept experimentalists ever, Faraday shaped the sciences and remains one of the most honored advocates of the scientific method and means of inquiry.

BEGINNINGS
Faraday was born to a working-class family outside London on September 22, 1791. His father, Joseph, was a blacksmith and a member of the Glasites, a Protestant sect that required rigid adherence to biblical teachings. Faraday's formal education was minimal, and at the age of 14 he became an apprentice to a

Michael Faraday. (Wikimedia Commons)

145

local bookbinder, George Riebau. During his seven years working for Riebau, Faraday read many books about science and developed a fascination with electricity. He especially was impressed by the work of Jane Marcet, the author of a series of books that popularized science. Toward the end of his apprenticeship, Faraday began to attend a series of lectures given by Sir Humphry Davy of the Royal Society of London for Improving Natural Knowledge and the Royal Institution of Great Britain. Davy, an inventor and chemist, was the discoverer of chlorine and iodine and the creator of the Davy lamp; he was a popular public figure who was renowned for his lectures. When Faraday sent Davy notes he had compiled from his lectures, Davy replied immediately, engaging Faraday as his secretary. Davy later appointed Faraday as a chemical assistant at the Royal Institution and had him accompany Davy on a European lecture tour.

SCIENTIST
Once ensconced in Davy's laboratory, Faraday embarked on a series of experiments that greatly increased his scientific knowledge. Working under Davy, Faraday discovered two new chlorides of carbon, began rough experiments related to the diffusion of gases, and produced several novel types of glass designed for optical uses. Faraday's best-known work, however, had to do with electricity and electromagnetism. As early as 1812, Faraday had begun experimenting with the voltaic pile, an early form of a battery (named for its inventor, Alessandro Volta) that was composed of zinc and copper discs. After Danish chemist Hans Christian Ørsted had discovered electromagnetism in 1821, Faraday worked, unsuccessfully, with fellow Briton William Hyde Wollaston to build an electric motor. Although his early attempts with Wollaston were unsuccessful, Faraday continued to work on his own to develop what he termed an "electromagnetic rotation," a wire in a pool of mercury that would rotate around a magnet to which it was attached. Faraday published this work, which is the foundation of modern electromagnetic experimentation, but failed to acknowledge his previous work with Davy and Wollaston, creating hard feelings between him and the others and resulting in Faraday's reassignment within the Royal Society, which prevented his further work on electromagnetic topics for several years.

Although his experiments exploring electromagnetic themes began again, after Davy's death in 1829 Faraday felt better able to devote more time to the subject. Faraday's work on induction came after he took an iron ring and wrapped two insulated coils of wire around it. When he passed a current through one coil, Faraday found that a momentary current was induced in the other coil, a phenomenon he termed mutual induction. Faraday later found that when he moved a magnet through a loop of wire, this caused an electric current to flow through the wire. This event also occurred if the loop was moved over a stationary magnet, thus proving that a changing magnetic field produces an electric field, a relationship later modelled mathematically as Faraday's law by James Clerk Maxwell. Faraday's law became one of the four Maxwell equations, which have evolved into what is today known as field theory.

In 1845, Faraday wrote about a phenomenon he termed diamagnetism, which refers to the weak repulsion from a magnetic field that many materials demonstrate. Work Faraday conducted in 1862 with a spectroscope to search for changes of spectral lines by an applied magnetic field was later praised by Pieter Zeeman, who was awarded the 1902 Nobel Prize in Physics. Zeeman was able to make discoveries about light in the presence of a static magnetic field using better equipment than Faraday had available. Faraday also continued to work on static electricity, using an experiment with an ice pail to show that a charge resided only on the exterior of a charged conductor and that this exterior charge had no effect on material enclosed within the conductor because the external electrical field causes the charges to rearrange, canceling the internal fields. This shielding effect is known as a Faraday cage.

Although a brilliant experimental scientist, Faraday lacked the mathematical background to explain the electromagnetic phenomena he discovered. James Clerk Maxwell, a Scottish physicist, took the work of Faraday and others and consolidated these into a series of equations that lie at the base of modern theories of electromagnetism. In 1885, Maxwell published *A Dynamical Theory of the Electromagnetic Field*, which served as a unifying basis for much of the subsequent work done in physics.

Faraday was appointed the inaugural Fullerian Professor of Chemistry by the Royal Institution in

1833, a position he held until his death on August 25, 1867. During his lifetime, Faraday declined both a knighthood and the opportunity to serve as president of the Royal Society. Nevertheless, he remains one of the most celebrated of all scientists, and numerous roads, parks, and other public installations bear his name.

—Sean G. O'Keefe

Further Reading

Forbes, Nancy, and Basil Mahon. *Faraday, Maxwell and the Electromagnetic Field: How Two Men Revolutionized Physics.* Prometheus Books, 2014.

Hirshfeld, A.W. *The Electric Life of Michael Faraday.* Bloomsbury Publishing, 2009.

James, Frank A.J.L. *Michael Faraday. A Very Short Introduction.* Oxford UP, 2010.

Lindley, D. *Degrees Kelvin: A Tale of Genius and Invention.* Joseph Henry Press, 2004.

Russell, C.A. *Michael Faraday: Physics and Faith.* Oxford UP, 2000.

Flex-Fuel Vehicles (FFV)

FIELDS OF STUDY

Automotive Engineering; Electronics Engineering and Technology; Thermodynamics

ABSTRACT

A flexible-fuel, or flex-fuel, vehicle has an engine that runs on more than one fuel. Such vehicles play an important role in increasing the use of alternative sources of energy. The manufacture of flexible-fuel (also known as flex-fuel or flexi-fuel) vehicles started in the United States early in the 1990s after the approval of the Alternative Motor Fuels Act of 1988.

KEY CONCEPTS

bifuel vehicle: a vehicle that can operate using either one of two entirely different types of fuel rather than a mixture of two fuels

compression ratio: the ratio of the volume of an engine cylinder when fully open to its volume when fully closed; for example, an engine cylinder with a maximum volume of 1 liter and a minimum volume of 100 mL has a compression ratio of 10:1

hybrid vehicle: a vehicle that has both an electric motor and a secondary internal combustion engine

multifuel vehicle: a vehicle that can operate using either one of three or more entirely different types of fuel rather than a mixture of fuels

ONE ENGINE, MORE THAN ONE FUEL

The manufacture of flexible-fuel (also known as flex-fuel or flexi-fuel) vehicles started in the United States early in the 1990s after the approval of the Alternative Motor Fuels Act of 1988, which stimulated the development of vehicles that can use mixtures of ethanol and gasoline as part of a strategy to reduce dependency on petroleum sources. This policy's aim was to decrease emissions that contribute to global warming and minimize economic risks represented by the Organization of Petroleum Exporting Countries (OPEC).

The flex technology consists of an internal combustion engine designed to identify and work with mixtures of gasoline and ethanol or methanol. It is capable of running on any proportion of these fuels by using electronic sensors that detect the fuel composition, sending this information to an electronic control module that adjusts the ignition point, fuel injection time, and the opening and closing of valves. To prevent corrosion, the fuel distribution system is made of stainless steel and nickel.

Ethanol and gasoline require different engine compression ratios, so a flex engine adopts an intermediate ratio that is optimal for neither one nor the other. The result is a small loss of efficiency when compared to engines designed exclusively for pure fuel. In addition, ethanol has a lower energy yield. To consumers, the biggest advantage of having a flex-fuel engine is the freedom of choice between two fuels, which allows the consumer to take into account price, quality, and environmental considerations.

By 2015, Brazil was the leading country in flex-fuel use, having the biggest fleet of flex-fuel cars and motorcycles, which can burn fuels ranging from pure ethanol to pure gasoline in varying proportions. Cars sold in the United States can function on pure gasoline or E15, a mixture of gasoline and ethanol up to a maximum of 85 percent ethanol; ignition of pure ethanol engines is compromised in low temperatures. The U.S. Environmental Protection Agency approved E15 use in light-duty internal-combustion vehicles

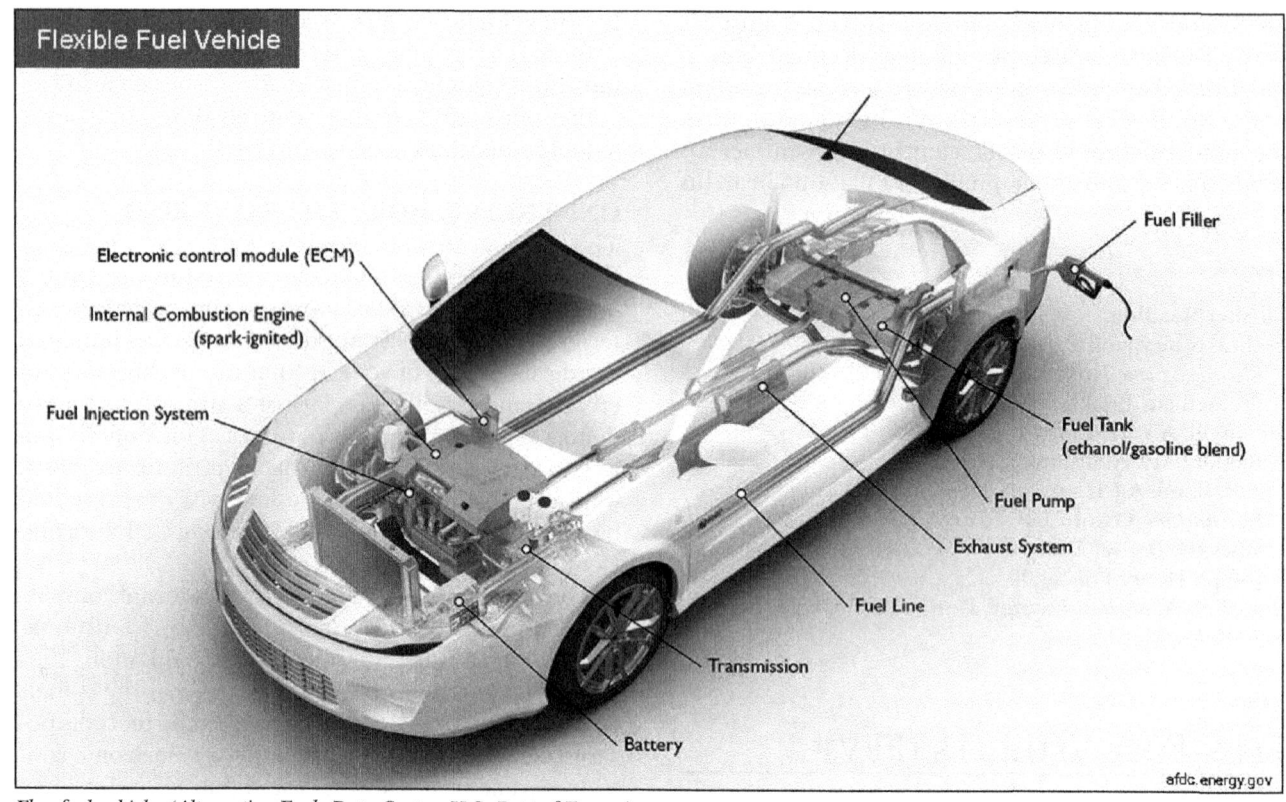

Flex-fuel vehicle. (Alternative Fuels Data Centre, U.S. Dept of Energy)

produced after 2001, although most were not optimized for it. As of 2018, nearly all automotive fuel sold at gas stations contains at least some ethanol, with E10 being the most common mixture, and the precise composition varies both by region and season.

In countries where flex-fuel vehicles are in common use, various forms of governmental support have encouraged industries to produce and costumers to buy such automobiles. Tax exemptions, reduced registration charges, reduced road taxes, and even (in Sweden) free parking are among the incentives to use flex fuels. Similarly, taxing fossil fuels can incentivize the use of flex fuels and alternative fuels.

HYBRID AND ELECTRIC VEHICLES

A hybrid vehicle uses electricity as its alternative source of power and requires a combustion engine in order to charge its battery, whereas a flex-fuel vehicle uses gas or a biofuel, mainly ethanol, as its alternative source. A hybrid car equipped with a flex-fuel engine could result in high fuel economy and reduction of automotive emissions.

The electric car has no combustion engine and relies entirely on electricity to recharge its battery through a plug-in system.

BIFUEL AND MULTIFUEL

Sometimes the term *bifuel* is used interchangeably with *flex-fuel*, but technically the two terms have different meanings. The former refers to the use of two fuels stored in different tanks with an engine that runs on only one fuel at a time, such as a bifuel vehicle using Compressed Natural Gas (CNG) and gasoline; these cars switch from one to the other fuel manually. On the other hand, flex-fuel vehicles operate on one fuel—whether blended or pure—stored in only one tank. The engine detects and works automatically with any mixture.

A multifuel vehicle works with three or more fuels and is generally a flex-fuel car equipped with a CNG system whose engine, therefore, can run using natural gas, gasoline, or ethanol.

ADDITIONAL USES OF FLEX-FUEL TECHNOLOGY

Flex-fuel technology is also being developed and applied in aviation and power generation in order to reduce greenhouse gas emissions and create an economical alternative for fueling. Again, Brazil has led the way: Inspired by the Embraer Ipanema agricultural aircraft, which uses pure ethanol (E100), engineers sought to develop an engine fitted with special components and electronic management software that will make it compatible with aviation gasoline and ethanol mixed in any proportion. In 2018, the standards-setting organization ASTM International approved the use of ethanol in commercially producing synthetic paraffinic kerosene, an aviation biofuel.

The use of the technology is also being applied in a thermoelectric plant at Juiz de Fora, supplying electricity to half a million inhabitants. It is the world's first power plant to operate on natural gas and ethanol, as Petrobras and General Electric outfitted a 43.5-megawatt gas turbine generator system to operate on ethanol by changing its combustion chamber and nozzles and installing peripheral equipment to receive, store, and move liquid fuel.

—*Rodrigo Marcussi Fiatikoski*

Further Reading

Bastian-Pinto, Carlos, et al. "Valuing the Switching Flexibility of the Ethanol-Gas Flex Fuel Car." *Annals of Operations Research*, vol. 176, 2010.

"Ethanol Benefits and Considerations." *Alternative Fuels Data Center*, Energy Efficiency & Renewable Energy, U.S. Department of Energy, 28 Mar. 2018, www.afdc.energy.gov/fuels/ethanol_benefits.html. Accessed 10 Oct. 2018.

Ferreira, Alex Luiz, et al. "Flex Cars and the Alcohol Price." *Energy Economics*, vol. 31, 2009.

Goettemoeller, Jeffrey, and Adrian Goettemoeller. *Sustainable Ethanol: Biofuels, Biorefineries, Cellulosic Biomass, Flex-Fuel Vehicles, and Sustainable Farming for Energy Independence*. Prairie Oak, 2007.

Kamimura, Arlindo, et al. "The Effect of Flex Fuel Vehicles in the Brazilian Light Road Transportation." *Energy Policy*, vol. 36, 2008.

Scragg, Alan. *Biofuels: Production, Application, and Development*. CABI, 2009.

U.S. Environmental Protection Agency. *E85 and Flex Fuel Vehicles*. U.S. Environmental Protection Agency, Office of Transportation and Air Quality, 2010.

Worldwatch Institute. *Biofuels for Transport: Global Potential and Implications for Energy and Agriculture*. Earthscan, 2007.

Zurschmeide, Jeff. "The Truth about Ethanol in Gasoline." *Digital Trends*, 6 Aug. 2016, www.digitaltrends.com/cars/the-truth-about-ethanol-in-your-gas. Accessed 10 Oct. 2018.

FLYWHEELS

FIELDS OF STUDY

Electrical Engineering; Mechanical Engineering; Physics

ABSTRACT

A flywheel is a heavy rotating wheel, the key component of a flywheel energy storage system, and a method of storing energy as rotational energy. They are used as alternatives to batteries or where energy use can be made more efficient by harnessing rotational energy from energy that is already being exerted.

KEY CONCEPTS

centrifugal force: an apparent force produced by the rotation of a mass about a central axis; it appears to act perpendicular to the axis of rotation in opposition to the centripetal force

centripetal force: a real force produced by the rotation of a mass about a central axis; it is formally the tension between the individual particles of the mass and the central axis proportional to the rate of rotation of the mass

drag: essentially, friction between moving surfaces and the surrounding air resulting in the generation of heat

gyroscopic force: a reactive force generated by a rotating mass to resist displacement from its original axis of rotation, consistent with conservation of angular momentum

magnetic bearing: a type of bearing that uses magnetic repulsion rather than physical contact to stabilize the rotation of an axle

rotational speed: simply, the rate at which a rotating mass travels about its central axis; for all particles in the mass the angular rotation speed is equal, but the physical distance traveled by individual particles about the axis increases directly with radius

tensile strength: the extent to which a material is able to resist forces acting in opposite directions without deformation and failure; a flywheel rotating at such a rate that the centrifugal force exceeds the centripetal force will fail and fly apart

A WHEEL THAT FLIES AROUND A CENTRAL AXIS

A flywheel is a mechanical device used to store rotational energy in an amount proportional to the square of its rotational speed. A flywheel energy storage system works because of the principle of conservation of energy. Rotational speed and stored energy are increased by applying torque to the flywheel, while the flywheel releases its energy by applying torque to a mechanical load. The common toy car that is dragged backwards and then released, propelled by the energy stored by the initial rotation of the wheels, is a familiar example of a simple flywheel system.

A component called a "flywheel" is an integral part of the electric ignition system of internal combustion engines. That application, however, is not intended to store energy for extended use. Rather, it is a large diameter heavy steel plate with a gear-tooth perimeter designed to engage the small spider gear of the electric starter motor, which drives it to turn the complete engine and initiate the firing sequence of the cylinders. The inertial mass of that flywheel serves to aid in smoothing the initial rotation by counteracting the force of the first few cylinder firings until all cylinders are operating, and to dampen the rundown of the engine when it is turned off. It does not store energy for further use.

Flywheels were a key component in James Watt's steam engine design, and some of the flywheels used in the first generation of those steam engines remain in use today. Although flywheel energy storage systems using mechanical energy to accelerate the flywheel are in development, current systems use electricity. Apart from the superconductors used to power magnetic bearings, if applicable, flywheels are not as affected by ambient temperature or temperature change as battery systems are, and have a much longer working lifespan.

DESIGN

Flywheel energy storage (FES) systems usually consist of a vacuum chamber (in order to reduce friction, or "drag"), a combination motor/generator that accelerates the flywheel and generates electricity, and a heavy steel rotor suspended on ball bearings inside the vacuum chamber. More sophisticated systems may use magnetic bearings to further reduce friction, though powering them may not be economically efficient. The advantage of a flywheel system is that it provides continuous energy, even when the output of the energy source is discontinuous or erratic. Energy is collected over a long period of time and released in a short spike, thus allowing the energy release to greatly exceed the ability of the energy source. Since 2001, FES systems have been able to provide continuous power with a discharge rate faster than batteries of equivalent storage capacity. Furthermore, flywheel

The G2 Flywheel Module. (Wikimedia Commons)

maintenance is half as expensive as traditional battery system maintenance—even less than that when magnetic bearings are used. They are increasingly used as part of the integrated power design of large data centers. FES systems have also long been used in research and development laboratories where circuit breakers and other electrical equipment are tested.

APPLICATIONS

FES systems have numerous applications in transportation. Formula One race cars sometimes use flywheels to recover energy from the drive train during braking, which is then re-deployed during acceleration. While the purpose of this application is to improve acceleration, and the economic efficiency of the system is considered by different criteria in competitive motor sports than in the design of passenger vehicles, a similar system is used in electric vehicles to increase fuel efficiency or to provide faster acceleration than the power source would otherwise be able to provide. Flywheel systems have been proposed to replace chemical batteries in electric vehicles; a small number of vehicle designs before the current generation of electric and hybrid vehicles incorporated flywheels into their designs, including flywheel-powered Swiss buses in the 1950s.

Flywheels have also been used in rail systems, especially electric rail systems, to provide boosts to power or to provide power when there is an interruption in the electrical supply. Since 2010, the London Midland train operator has operated two flywheel-powered railcars. New York's Long Island Rail Road (LIRR) has begun a pilot project to explore the possibility of using flywheels lineside to generate some of the electricity used by the LIRR's electric trains, to improve their acceleration and recover electricity during braking. Elsewhere in New York, Beacon Power opened a 20-megawatt flywheel energy storage plant in Stephentown, where power is purchased at off-peak hours and stored with fewer carbon emissions than traditional plants.

LIMITATIONS

The biggest limitations to flywheel systems are the rotor's tensile strength and the energy storage time. The tensile strength of the rotor determines how fast the rotor can rotate, and therefore the system's capacity; if it is exceeded, the flywheel can shatter in an explosion, the risk of which requires a containment vessel which increases the system's mass and cost. Because of the rotation of the planet, flywheels change orientation over time, which is resisted by the flywheel's gyroscopic forces exerted against the bearings. The resulting increase in friction results in a loss of energy of as much as 50 percent over two hours, when mechanical bearings are used, though careful design can reduce that amount by half. Magnetic bearings are much more efficient, losing only a few percent of their energy. Even so, until magnetic bearings are more cost-efficient, this limitation is one reason why flywheels are best used in applications where energy won't need to be stored for long periods of time, and why they are well-suited for public transit applications, where they are in near constant use.

—*Bill Kte'pi*

Further Reading

Bhandari, V.B. *Design of Machine Elements*. 3rd ed. Tata/McGraw-Hill, 2010.

Bitterly, J.G. "Flywheel Technology: Past, Present, and 21st Century Projections." *Aerospace and Electronic Systems Magazine*, vol. 13, no. 8, Aug. 1998.

Breeze, Paul. *Power System Energy Storage Technologies*. Academic Press, 2018.

Hebner, R. "Flywheel Batteries Come Around Again." *Spectrum*, vol. 39, no. 4, Apr. 2002.

Khan, B.H. *Non-Conventional Energy Resources*. Tata/McGraw-Hill, 2006.

Larminie, James, and John Lowry. *Electric Vehicle Technology Explained*. Wiley, 2003.

Leclercq, Ludovic, Benoit Robyns, and Jean-Michel Grave. "Control Based on Fuzzy Logic of a Flywheel Energy Storage System Associated with Wind and Diesel Generators." *Mathematics and Computers in Simulation*, vol. 63, no. 3-5, Nov. 2003.

National Aeronautics and Space Administration. *Metallic Rotor Sizing and Performance Model for Flywheel Systems*. NASA, 2019.

Rufer, Alfred. *Energy Storage Systems and Components*. CRC Press, 2018.

Fossil Fuels

FIELDS OF STUDY

Geology; Petroleum Engineering; Mine Engineering; Refinery Operation

ABSTRACT
Fossil fuels play a major role in providing the world's energy supply. Fossil fuels are mineral fuels formed from the fossilized remains of dead plants and animals, compressed and heated over hundreds of millions of years, and are found within the top layer of Earth's crust.

KEY CONCEPTS
British thermal unit (Btu): defined as the amount of heat required to raise the temperature of one pound of water by one Fahrenheit degree

caprock: an overlying layer of dense rock that prevents petroleum and natural gas from escaping from where it has accumulated, effectively trapping it in the more porous rock below

decomposition: the breakdown into its molecular components of once-living organic matter, whether of animal or plant origin

directional drilling: a deep drilling method in which the direction the drill bit cuts and travels through rock can be controlled from the surface

fossilization: the process by which once-living biological matter, whether of plant or animal origin, is converted to a mineral form

HEAT, PRESSURE, AND TIME
There are three types of fossil fuels: coal, petroleum, and natural gas. These energy sources are extremely important, because they are used in generating the majority of the world's electricity supply and providing transportation fuels, accounting for approximately 85 percent of the world's total energy use. Sometimes referred to as mineral fuels, fossil fuels are generally thought of as nonrenewable energy sources, because they have been formed over hundreds of millions of years through the process of decomposition of prehistoric plants and animals, thus accounting for their being called "fossil fuels." There is a limited quantity of fossil fuels available in the world.

Due in part to old advertising campaigns by large oil companies and because fossil animal remains have at times been found in tar deposits, many people think that dinosaur remains have created the fossil fuels that are used today. But in fact, most of the fossil fuels found today were formed millions, even billions of years before the first dinosaurs. These living things that died hundreds of millions of years ago, bacteria that were the only living things on the planet for at least a billion years before dinosaurs and the plant matter of the great swamps that existed for millions of years, slowly accumulating their detritus over all of that time, decomposed and over time became buried under thousands of feet of rock, sand, and mud successively layering on top of the decomposing organic matter. Over time, this matter was exposed to heat and pressure in Earth's crust and became fossilized, forming the fossil fuels known and used today. The time it took to form the different types of fossil fuels depended on the combination of different kinds of animal and plant materials that were present and the environmental conditions, such as temperature and pressure, to which they were subjected while they were decomposing.

Some decomposing materials were once covered by water formations that eventually dried up, leaving additional sediments from the preexisting oceans or lakes. After these waters receded, the combination of heat, pressure, and bacteria began the process of compacting and compressing these organisms under layer upon layer of silt. Oil and natural gas began to form in many of these areas as a thick liquid which, buried at greater depths underground, began to "cook" under heat and pressure. It is in places where large, dense rock formations blocked the eventual seeping up to the surface of these liquids or gases that many of the oil and natural gas reserves being mined today have been found, hidden under these so-called caprocks. Coal has been formed through a similar process, also taking place over hundreds of millions of years, although the primary ingredient for coal is the accumulation of partially decayed and decomposing trees and land plants such as ferns. Over time, the organic material began to decompose, and with increased pressure and high temperatures, the layers of sediment become solidified into rocky material.

COAL
Coal—"rocks that burn"—is believed to have been used at least as early as some 3,000 years ago by the Chinese to smelt copper in northeastern China. It was later used by the Romans for heating, and around the beginning of the 1700s the English began using coal in place of wood charcoal once they discovered that coal burns more cleanly and produces more heat. When the steam engine was invented in the 1800s, it was designed to run on steam produced by burning

coal, and with the rise of the Industrial Revolution, huge amounts of coal were extracted to power the boilers for steam-driven ships and trains. Fueled by coal, these modes of transport became commonplace. Today, coal still supplies a significant share of the world's energy demand, although some nations have sought to reduce its use due to its polluting qualities.

Of all the fossil fuels, coal is the most abundant, especially in the United States, which has an estimated quarter of all global coal reserves. The United States is one of the largest exporters of coal in the world. Coal predominantly consists of carbon and its energy content can range anywhere from 5,000 to 15,000 British thermal units (Btu) per pound. Four types of coal, sometimes referred to as "ranks," are mined today:

- Lignite, with the lowest energy content of the four types, is also the most commonly found in the world, characterized by its dark brown, blackish color and its relatively soft texture, sometimes exhibiting the original wood form.
- Subbituminous coal, which ranks just above lignite, is slightly more potent as a source of energy when burned and is characterized by its dull black color.
- Bituminous coal, which produces more energy than the previous two types, is sometimes referred to as "soft" coal.
- Anthracite coal, which gives off the most energy of all the four ranks, is also the rarest and hardest type of all.

Coal is used for electricity generation and as a heating fuel in industrial and manufacturing industries, particularly for the production of steel. Burning coal releases toxins such as mercury, barium, zinc, lead, chromium, arsenic, nickel, ammonia, and hydrochloric and sulfuric acids into the atmosphere. Products made from coal, in addition to the heat generated through burning it, are used in many different products, such as methanol and "coal oil." These components of coal can be isolated and used to make medicines, fibers, plastics, and fertilizers. For instance, steel industries use coal to make coke by cooking it in furnaces to drive off impurities that reduce its heating value; the burning of coke, which gives off extremely high temperatures, is necessary for iron ore to be smelted in order to make steel. The high temperature given off by burning coke is critical in steel production, as it gives steel the needed flexibility and strength for its later uses in construction and manufacturing.

Coal is extracted either by surface mining or by deep underground mining. The process of surface mining is relatively simple, less expensive than deep mining, and takes place in generally flat areas. The extraction process begins with the removal of topsoil and leveling of the surface so that the coal bed is exposed. Once the area is cleared, the coal is dug out and transported from the pit. The removed topsoil is then sometimes replaced in the final land-reclamation stage.

If the coal beds are too deep below the surface or the landscape surrounding them too hilly or mountainous, then underground mining techniques are used. Deep mining was once a dangerous and risky task, but nowadays, with the advent of automated digging, many underground coal mines are highly mechanized. Automated mining has meant that the digging of shafts and tunnels is done by long cutting machines that continuously move along the tunnels to cut and extract the coal. Once loosened or removed from the walls, the coal is lifted to the surface by means of a conveyor, reducing the health and safety risks for workers.

PETROLEUM

It is believed that the Chinese used petroleum as far back as 2000 BCE to heat their homes and burn in lamps for light. The extraction of "crude oil" (as petroleum is often called) from coal and shale began in earnest in the mid-nineteenth century, when its value first began to be realized. Since then, the exploration, knowledge, and use of crude oil have developed rapidly, especially with the advent of the automobile in the early twentieth century. This invention, run by an internal combustion engine, demanded gasoline to run its engine, and by the 1950s oil had become one of the most commonly used energy sources all over the world, mostly for transportation.

Oil stores the greatest amount of energy and is more powerful than any other fossil fuel. It is a combination of hydrocarbons, oxygen, nitrogen, and sulfur formed over hundreds of millions of years. In the twentieth century, the world relied heavily on the abundance and relatively cheap cost of oil to fuel global economic development. The majority of petroleum is used to produce gasoline or as a fuel source to be burned in combustion engines, and the transportation and industrial sectors, which account for the majority of energy use in general, especially depend on petroleum and other liquid fuels. Petroleum is also used to produce diesel fuel, heating oil, propane, jet fuel, and as a petrochemical feedstock (or raw material input) for the manufacture of other chemicals and a broad variety of common products, such as plastics, fertilizers, rubber, washing liquids, and tires.

Oil reserves exist in many parts of the world, but the largest concentrations, and thus biggest oil-producing countries, are in the Middle East, Russia, and North America. Some of the largest producers are the United States, Saudi Arabia, Russia, Canada, China, Iraq, the United Arab Emirates, and Brazil.

Petroleum, or crude oil, is a black or yellow liquid substance found in large quantities underground in areas called reservoirs. Contrary to what most people believe, petroleum does not sit in large pools underground, waiting to be extracted, but as tiny drops found in small spaces or pores that are embedded under high pressure in rock formations, visible only with the use of a microscope. Oil extraction is an expensive task, as much information is needed about a reservoir before drilling can begin. First, scientists can use sound waves to analyze the type of rocks below ground and whether or not they hold oil. It is important to have a good understanding of the pores in the rocks, how easily and quickly the oil droplets can be moved through these pores, and where spaces in the rock formations are located, so that drilling can be more precisely targeted. Another way of determining if the rocks have oil is to drill a preliminary well and remove some rocks in order to examine them for oil droplets.

Once it is confirmed that a reservoir has the right characteristics to be drilled, a production well is constructed. When the drill first hits the reservoir, the weight of millions of tons of rocks and millions of years of built-up heat and pressure may force some oil quickly up to the surface. This creates what is commonly thought of when people discover oil reservoirs: a fountain of oil that comes spurting up and shoots into the air (although today oil producers have installed pressure-control equipment on the wells to prevent this waste from happening).

Petroleum reservoirs also exist deep below the ocean, so offshore platforms are built from which to drill for the oil. These platforms, also known as oil rigs, float off the coast and can vary in size, depending on the characteristics of the targeted reservoir. The platform must hold all the necessary equipment for deep-sea drilling to take place, including pressure-control equipment to prevent oil from gushing up and out into the water when the well is first tapped.

After a well has been drilled, there are two stages of oil production. The first, called primary production, can last from a few days to years. The oil is forced up through the pores in the rocks or in the fractures in the rock formation by pressure that has naturally built up under the surface. Oil flows easily up from the reservoir through the wells in this stage.

After the pressure begins to drop, the secondary stage begins: a process called secondary recovery, which uses an additional well and steam or water to force up the oil remaining in the reservoir. Natural gas can be found in conjunction with oil, and when the petroleum is pumped out, the natural gas can be separated and is sometimes pumped back into the reservoir to increase the pressure and maintain the flow of oil. Tertiary, or enhanced, recovery methods may also be used when it is economical to maximize production with more involved techniques.

The different forms of oil recovery include the following:

- Thermal recovery, a process that uses steam injectors to thin the oil so that it will flow up more easily from the reservoir;
- Gas injections, which increase the pressure in the reservoir to force up the oil and increase the speed at which it flows through the well; and
- Chemical injections, which help to ease the flow of oil droplets along the surface, reducing friction and thereby speeding up the extraction process through the well.

Alternative methods such as horizontal drilling and hydraulic fracturing (fracking) can also be used to improve yields from less traditional petroleum reserves such as shale oil. These techniques became increasingly common in the twenty-first century, greatly changing the global oil industry. However, they generated much controversy for their negative environmental impacts.

Once the petroleum has been extracted from a well, it must be cleaned, transformed, and refined into a form that can be used to produce the fuels and chemicals to be made from it. Impurities such as sulfur, nitrogen, and heavy metals are removed during the refining process, and a distillation process is used to separate the different hydrocarbons, separating the crude oil into liquids and vapors, such as gasoline, based on the boiling points and weights of the various hydrocarbons of which oil is composed. The oil is processed through big furnaces, where the lightest of the components, such as propane, butane, and petrochemicals used to manufacture plastics and fabrics, turn to vapor and easily rise to the top and later condense into liquids. The second lightest fractions are gasoline, kerosene, and diesel fuel. Finally, the heaviest components, used to make home heating oil or shipping fuel, remain on the bottom. This separation process allows for the creation of the various products and fuels that are petroleum-based. Once refined, the different fuels are distributed via pipelines to additional refineries or storage facilities from which they can be shipped elsewhere.

NATURAL GAS

Natural gas, typically consisting of methane and ethane, is an abundant energy resource, the least carbon-intensive of all the fossil fuels, and it is either released from coal or found mixed with oil. It is easy to transport and use, and the demand for natural gas has been growing steadily since its first use in the early 1800s to light street lamps. Its production and consumption are therefore expected to rise even further in the future.

The Greek historian Plutarch was probably the first to "discover" natural gas, as noted in his writings on the "eternal flames" in what is today known as Iraq. Although originally thought to be a useless gas, natural gas wells were drilled by the early nineteenth century, and by the mid-nineteenth century, Robert Bunsen had invented a way to mix natural gas with air and burn it in what came to be known as the Bunsen burner. Natural gas was then used to generate heat for cooking and heating buildings. With the advent of better engineering and technology for metal piping, natural gas distribution and a network of pipelines have rapidly expanded, creating a large market for its use in industrial and commercial heating, electricity generation, and cooking all over the world. Natural gas can also be compressed and used for fuel in transportation.

Natural gas reserves are found all over the world, from Africa and Asia to the United States and Latin America. The majority of reserves are concentrated in the Middle East (Iran, Qatar, Saudi Arabia, and the United Arab Emirates), Europe, and Russia, with some slightly smaller reserves in Asia (approximately 16 percent of the world's reserves), and the Americas (around 8 percent). As of 2018, the United States was the biggest producer of natural gas, followed by Russia. There are estimates that many large reserves still exist, including deposits of methane hydrate found on the cold sea floors, but despite recent technological advancements, accurately estimating the amount remaining is a challenge.

Natural gas is currently used in a number of ways. Electricity can be generated with natural gas, and in many countries highly efficient, combined turbine and heat-recovery systems have been developed to produce electricity from both the turbine and a heat and steam generator. Besides its importance as a reliable and efficient energy source, natural gas is used to produce many household goods, such as medicines, fertilizers, plastics, and paint. Propane, used for barbecues and as a household fuel source in rural areas, is yet another example of a product of natural gas. Natural gas is often thought of as smelling like rotten eggs, but actually it is an odorless, colorless gas. The smell is a chemical added by gas companies as a safety measure so that it can be smelled if it is leaking from a pipe.

Natural gas wells are drilled underground to extract the reserves found in porous rock. The difficulty and cost of producing natural gas depend on where it is found, as some reserves require more complex extraction processes. Natural gas reserves are found in shale rock formations, coal seams, and sandstone beds. Fractures in rock formations through which the gas must be forced out can be opened by a process of cracking the rock in order to improve the flow of gas, or fluids can be pumped into the reservoirs to increase the pressure in the well. Methane gas can be found in coal beds and is a potentially valuable source of fuel, despite the risk it poses to miners. Sandstone is yet another source, although extraction from this rock can be especially costly, as the deposits are often trapped in very small pores or holes.

Natural gas production is in some ways similar to oil extraction; in fact, in the process of drilling for oil, natural gas is sometimes also found and can be extracted as well. Just as oil can freely flow to the surface, lightweight natural gas can easily be forced up to the surface, thanks to the pressure trapped in underground reservoirs. However, these easily tapped wells are not very common, and most use a pumping system that functions, similarly to those used in oil extraction, to force out the remaining gas trapped underground. Once natural gas is extracted from the reserves, it is processed so that liquid hydrocarbons, sulfur, carbon dioxide, and other components can be removed and the gas can be sent directly on to end users. Otherwise, the gas is piped to large, underground storage terminals until it is needed.

Liquefied natural gas (LNG) is another natural gas product. After the gas is extracted from the well, it is treated to remove impurities, and then, as the name implies, it is liquefied through a process of cooling and condensing. Once the LNG is made, it can be easily transported by tankers and other ships to where it is needed; after delivery, it is transformed back to its original, gaseous state. Then it can be piped and delivered, as a gas, to end users. However, the transport and conversion process for LNG is rather complex and can be very expensive.

ENVIRONMENTAL CONCERNS

Although they have become a key part of industrialized society, fossil fuels are often criticized by environmentalists. As nonrenewable resources, they are not considered part of a long-term sustainable development model. Burning fossil fuels also releases pollutants, including atmospheric carbon dioxide. This and other greenhouse gases contribute to global warming and other forms of anthropogenic (human-induced) climate change, regarded by scientists as a key threat to both human society and life as a whole. Other forms of pollution and environmental damage attributed to fossil fuel use include acid rain, particulate air pollution, mountaintop removal from strip mining, and oil spills. Both terrestrial and aquatic ecosystems can be heavily disrupted by fossil fuel prospecting and extraction activity.

—*Cary Hendrickson*

Further Reading

Berners-Lee, Mike, and Duncan Clark. *The Burning Question: We Can't Burn Half the World's Oil, Coal, and Gas. So How Do We Quit?* Greystone, 2013.

Curley, Robert, ed. *Fossil Fuels: Energy Past, Present, and Future.* Britannica Educational, 2012.

Deffeyes, Kenneth S. *Hubbert's Peak: The Impending World Oil Shortage.* Princeton UP, 2008.

Duffield, John S. *Fuels Paradise: Seeking Energy Security in Europe, Japan, and the United States.* Johns Hopkins UP, 2015.

"Fossil." *Energy.gov*, U.S. Department of Energy, www.energy.gov/science-innovation/energy-sources/fossil. Accessed 18 Aug. 2020.

Heinberg, Richard, and Daniel Lerch, eds. *The Post Carbon Reader: Managing the 21st Century's Sustainability Crises.* Watershed, 2010.

International Energy Agency. *Natural Gas Information 2016.* Organisation for Economic Co-operation and Development, 2016.

Maugeri, Leonardo. *Beyond the Age of Oil: The Myths, Realities, and Future of Fossil Fuels and Their Alternatives.* Praeger, 2010.

Nunez, Christina. "Fossil Fuels, Explained." *National Geographic*, 2 Apr. 2019, www.nationalgeographic.com/environment/energy/reference/fossil-fuels/. Accessed 18 Aug. 2020.

Raymond, Martin S., and William L. Leffler. *Oil and Gas Production in Nontechnical Language.* PennWell, 2006.

Wenar, Leif. *Blood Oil: Tyrants, Violence, and the Rules That Run the World.* Oxford UP, 2016.

"World Energy Outlook 2019." *International Energy Agency*, Nov. 2019, www.iea.org/reports/world-energy-outlook-2019. Accessed 18 Aug. 2020.

FRANKLIN, BENJAMIN

FIELDS OF STUDY
History of Science; Physics

ABSTRACT
One of the founding fathers of the United States, Benjamin Franklin was almost as accomplished in his scientific work as in his statesmanship and diplomacy. In addition to his fame

as a printer and inventor, he was a key figure in the early history of the study of electricity, and developed much of its vocabulary.

KEY CONCEPTS

charge: indicates that something has a component of electricity in its fundamental composition, or to load components of electricity into an object. In modern physics, charge is a fundamental property of matter, with the electron being the fundamental unit of negative charge and the proton being the fundamental unit of positive charge. Subatomic particles were unknown in Franklin's time.

condense: to collect charge. In Franklin's time, electricity was thought to be a kind of fluid that could condense or concentrate, as when an object builds up a charge of static electricity. The modern electronic component called a "capacitor" acquires a build up of electrical charge as it functions, and was originally called a "condenser."

conductor: any material that will allow the passage of an electrical current, which all materials do to a greater or lesser degree

discharge: the removal of a build-up of electrical charge from an object

electrify: to provide electricity to something to enable it to function

FROM PUBLISHING TO SCIENCE

Benjamin Franklin was one of the founding fathers of the United States. Though the only elected office he held was a three-year term as the sixth president of Pennsylvania from 1785-88 (before the office was replaced with that of governor), he was an important political theorist and journalist in the years leading up to the American Revolution, the first ambassador to France (as well as Sweden) in the years after the revolution, and the first postmaster general, establishing the U.S. Post Office after years of running the colonial mail. He is as famous as a scientist as a statesman, and his scientific career was as varied as his political one: In addition to contributing to the history of electricity, he invented bifocals and the Franklin stove, and his public and intellectual work intersected in his formation of the country's first public lending library.

Born to a working-class family in Boston, Franklin was the eighth of 10 children of Josiah and Abiah Franklin; Josiah had an additional seven children from his late wife Anne. The family had enough money to send Franklin to the Boston Latin School for two years, after which he directed his own education through reading and became apprenticed at age 12 to his brother James, a printer. In 1726, Franklin left his apprenticeship without permission, at age 17. He worked in printing in both Philadelphia and London before settling down in Philadelphia for the long term in 1726. The following year, he founded the Junto (named for his mistaken belief that it meant "meeting" in Spanish). The Junto, also known as the Leather Apron Club, was an intellectual mutual improvement society that discussed political, business, and philosophical issues. Each meeting began with Franklin asking various questions to guide discussions, ranging from the success or failure of local businessmen to works the members had recently read. It was the first of many such societies in Philadelphia and the colonies, and helped to further the cause of the American intellectual community. It was because

Benjamin Franklin. (Wikimedia Commons)

of the Junto that Franklin thought of the subscription library.

He founded the Library Company of Philadelphia in 1731, which pooled funds from member dues in order to buy books that would be lent out to members to read. Franklin's intellectual endeavors and printing experience soon led him to publish the *Pennsylvania Gazette* in partnership with Poor Richard's *Almanack*, beginning in 1733. The papers provided him with a forum to share his views and develop his voice; many of the adages he wrote under the pseudonym Richard Saunders continue to be quoted and misquoted today. Publishing, especially the *Almanack*, made Franklin wealthy, affording him the time and resources to pursue his scientific work. He never patented his work; his publishing wealth meant that he didn't need to profit from it, but he was also committed to the scientific world and was afraid that patents restricted innovation.

"Benjamin Franklin Drawing Electricity From the Sky," painting by Benjamin West, circa 1816, oil on canvas, Philadelphia Museum of Art.

INNOVATIVE INVENTIONS

The Franklin stove or Pennsylvania fireplace, for instance, was invented in the early 1740s and was first sold in 1744. The cold was a perpetual problem in the northern colonies, and as they became more urban, the amount of wood used to keep fires going to provide sufficient heat became increasingly difficult to acquire. Franklin's invention was a more efficient stove, one that produced more heat and less smoke. A hollow baffle at the rear of the metal-lined fireplace, along with an inverted siphon, transferred more of the heat from the fire's hot fumes to the room's air while letting the smoke pass up the chimney.

In 1747, Franklin stated that electricity was not created, but collected, a notion prerequisite to his most famous experiment. Though remembered as an inventor and statesman, his theoretical and experimental work was easily as significant. He introduced key modern English-language terms for electricity concepts, including charge and discharge, electrify, conductor, condense, and armature, but it is worth remembering that these only applied to static electricity when they were coined as this was the only form of electricity that could be reproduced. He was one of the only Americans to be elected as a fellow of the Royal Society in England, and was given its Copley Medal.

In 1749, Franklin invented what was probably the first lightning rod, and certainly was the first to popularize it. Though architectural features that might serve the same purpose as lightning rods have been found from earlier eras, it is not clear that they were intended as lightning rods; even in Franklin's time and decades after his death, lightning rods were widely misunderstood, which makes it seem unlikely that a pre-Franklin civilization would have knowingly adopted them. Franklin's "lightning attractor" was a simple, pointed iron rod that attracted and grounded lightning strikes, protecting the structure to which they were attached (a building or ship at sea). Lightning rods spread in popularity, and became especially important in reducing urban fires as American cities became more densely populated in the early nineteenth century.

THE FAMOUS EXPERIMENT

In 1750, Franklin designed an experiment to demonstrate that lightning is a form of electricity, and performed a variation of the experiment two years later. The experiment used conductive rods to attract lightning to a Leyden jar, a then-recent invention named for the Dutch city of Leyden, where it was introduced. The jar was a primitive conductor that stored static electricity between two electrodes in a water-filled glass jar. Franklin originally intended to attach the rods to the steeple of a church then under construction, but when he tired of waiting for it to be ready, he adapted the idea to the experiment he actually performed, the famous "kite experiment."

Contrary to popular depictions, a bolt of lightning did not strike the key—there was probably no visible lightning during the experiment at all. A Leyden jar and conductive key were attached to a kite that was flown into storm clouds, where sparks were drawn to it and static electricity was successfully collected. The key became negatively charged, giving him a mild shock when he approached it. This confirmed his belief that static electricity was present in those clouds, and that lightning must be made of it.

The experiment was widely duplicated by others, and Franklin discredited the prevailing theory of "electrical fluid" in two types, vitreous and resinous. Instead, the "fluid" had a charge that was either positive or negative. He was the one to introduce the use of the "+" and "-" symbols to denote those charges, though he guessed incorrectly as to which charge was actually positive and which was negative. Building on his 1747 statement, he formulated the principle of the conservation of charge, which states that electric charge is neither created nor destroyed: The balance of positive and negative charge in the universe is a constant.

—*Bill Kte'pi*

Further Reading

Franklin, Benjamin. *Experiments and Observations on Electricity Made at Philadelphia in America*. 1769. Reprint. Cambridge UP, 2019.

Isaacson, Walter. *Benjamin Franklin: An American Life*. Simon & Schuster, 2004.

Schiffler, Michael Brian. *Draw the Lightning Down: Benjamin Franklin and Electrical Technology in the Age of Enlightenment*. U of California P, 2006.

FREQUENCY

FIELDS OF STUDY
Acoustics; Communication Technology; Harmonics; Physics

ABSTRACT
Physical and electromagnetic wave phenomena are characterized in part with regard to their frequency. The frequency of a wave is expressed as the number of cycles occurring in a specific amount of time. The time required for one cycle is the period of the wavelength. Phase relationships and interference are other wave phenomena.

KEY CONCEPTS
frequency: the number of complete waves or cycles that occur in one unit of time

harmonics: the study of the interaction of wave phenomena

hertz: a unit of frequency defined as one cycle per second

period: the length of time for one complete cycle of a wave or other cyclic property to occur

reciprocal: the inverse of a value, calculated as 1 divided by the value

sinusoidal: having a shape or pattern of behavior that can be described by a sine wave function.

speed: the distance traveled per unit of time

wavelength: the distance from any point in a wave to the identical point in the next wave, usually measured from crest to crest

CYCLIC PHENOMENA

The term "cycle" generally indicates something that goes around in a circle. In physics, "cycle" indicates that a specific property or function has a value that progresses through a succession of other values and returns to the starting value in a precise manner that repeats. Phenomena that exhibit this behavior are associated with either circular or sinusoidal wave motions and properties. Such motions can be described by the same math functions, the sine and cosine.

The sine and cosine functions are themselves simple ratios of the lengths of the two sides of a right triangle at one vertex. The radius (plural: radii) of the circle can be rotated about the center by any amount to form the corresponding angle. A vertical line to the

Modern frequency counter. (Coaster J via Wikimedia Commons)

point on the circumference where it meets the displaced radius forms a right triangle with a base that is proportionately shorter than the length of the radius. In this right triangle, the displaced radius forms the hypotenuse and the vertical height of the triangle is the opposite. (The "opposite" is the side of the right triangle that is opposite the angle formed at the center of the circle.) The base of the triangle is called the "adjacent." The sine of the angle formed by the base and the hypotenuse is just the ratio of the length of the opposite to that of the hypotenuse (i.e., the radius). Likewise, its cosine is the ratio of the length of the adjacent to that of the hypotenuse. The reciprocal values of the sine and cosine are called the "secant" and "cosecant," respectively.

As the radius rotates, the angle that it forms at the center changes continuously. The value of the sine also changes accordingly. A graph of this variation produces the sideways S-shaped curve that is recognized as a sine wave. The value of the cosine follows the same pattern but is shifted from the sine values. The cosine at any angle has the value of the sine of an angle that is greater by 90 degrees.

There are two methods of describing the amount of rotation about the center, or axis of rotation. In one, the amount is stated in degrees of rotation, with one full revolution totaling 360 degrees. The other measurement of angles is in radians (rad). One radian is the angle formed by two radii when the length of the circumference they mark off is equal to the radius of the circle. There are 2p radians in one complete revolution.

PROPERTIES OF CYCLIC PHENOMENA

All cyclic phenomena, whether wavelike or circular, share several characteristics. The primary feature of all of them is that their behaviors or values repeat in the same regular way. The number of times that the cycle of any particular phenomenon repeats in a specific amount of time is its frequency. The most common of these is revolutions per minute (rpm) for rotational movements of physical objects, and cycles per second (cps) for wavelike properties. The conventional unit for cps is hertz (Hz), in honor of Heinrich Hertz (1857-94), for his contributions to the physics of electromagnetism (EM). The term is most often used to refer to EM waves.

The duration of just one cycle of the phenomenon is the period of the cycle. The period is calculated simply as the reciprocal value of the frequency. For example, a wave frequency of 100 cps has a corresponding period of 1/100, or 0.01, seconds per cycle (spc). If that same wave is progressing, or propagating, at a speed of 100 meters per second (m/s), each cycle will have moved it through a distance of 1 meter. Since this corresponds to the distance covered by just one complete cycle of the wave, it is the specific wavelength. Wavelength is only used to describe phenomena that travel through space or time. It is not applied to rotational motions.

FREQUENCY, WAVELENGTH, AND SPEED

EM waves such as visible light all travel at the same speed—the speed of light. Physical waves, such as sound, travel at different speeds determined by the density of the medium. The speed of sound in air, for example, is 331 meters per second (about 1,086 feet per second) at 0°C (32°F), and 342 meters per second (about 1,122 feet per second) at 18°C (65°F). In water, the speed of sound is about 1,140 meters per second (about 3,740 feet per second). The greater the density of the medium is, the more it can transmit sound waves. Another factor that affects the transmission of both physical and EM waves is the relative motion of the wave source and the observer or receiver of the emitted waves. When the two move toward each other, the apparent frequency of the waves increases and the apparent wavelength decreases, but if they move apart, the apparent frequency decreases and the wavelength increases. This is known as the "Doppler effect." It accounts for the apparent changes to the sound of a passing train, as well as the red shift and blue shift in the light observed from distant stars.

For physical waves, speed, frequency, and wavelength are all related. For EM waves, however, the constant speed of light requires that the frequency and wavelength are related in a manner that maintains the constancy of the speed of light.

FREQUENCY AND PHASE

The fundamental character of a wave is determined by its frequency. All wave phenomena have neutral points called "nodes" at which the value of the amplitude is zero. When two identical waves are synchronized such that their nodes and amplitudes coincide precisely, they are in phase. Their amplitudes do not matter. At all other times, they are out of phase by a particular amount. For example, two waves might be said to be 10 degrees out of phase. The amplitudes of waves that are in phase add together as constructive interference to produce a wave with greater amplitude. When they are out of phase, such that their positive and negative amplitudes overlap, they add together as destructive interference to produce a wave with less amplitude and loss of harmonic characteristics.

Waves of different frequencies can never be in phase. Instead, nodes and amplitudes may coincide occasionally in such a way that a harmonic beat frequency may arise. This is most apparent when one frequency is a whole-number multiple of the other. In such cases, nodes occur consistently when a number of wavelengths of one frequency span the same period of time as a different number of wavelengths of the other frequency.

—*Richard M. Renneboog, MSc*

Further Reading

American Industrial Hygiene Association. *Radio-Frequency and Microwave Radiation*. 3rd ed. American Industrial Hygiene Association, 2004.

Chaichian, Masud, Hugo Perez Rojas, and Anca Tureanu. *Basic Concepts in Physics: From the Cosmos to Quarks*. Springer, 2014.

De Pree, Christopher G. *Physics Made Simple*. Three Rivers, 2013. Digital file.

Gilbert, P.U.P.A., and W. Haeberli. *Physics in the Arts*. Elsevier, 2008.

Gunther, Leon. *The Physics of Music and Color*. Springer, 2012.

Kumar, B.N. *Basic Physics for All*. University Press of America, 2009.

Serway, Raymond A., John W. Jewett, and Vahé Peroomian. *Physics for Scientists and Engineers with Modern Physics*. 9th ed. Brooks, 2014.

Fuel Cells and Energy Efficiency

FIELDS OF STUDY

Electrochemical Engineering; Materials Science; Physical Chemistry

ABSTRACT

Fuel cells generate electricity and heat by reacting a fuel such as hydrogen with an oxidant through electrolysis. Due to their high efficiencies, quiet operation, and ability to store energy, these devices have many applications, but have had limited uptake due to prohibitive costs.

Fuel Cells and Energy Efficiency

KEY CONCEPTS

anode: the relatively positive terminal of an electrochemical circuit; the anode receives electrons emitted from the cathode

cathode: the relatively negative terminal of an electrochemical circuit; the cathode emits electrons that are to be received by the anode

electrochemical: a chemical reaction system that functions through the transfer of electrons from one material (the cathode) to another (the anode)

electrolyte: a solution or intermediary substance that contains mobile ions, such as saltwater

oxidant: also termed oxidizing agent, the substance accepts electrons in an electrochemical reaction, and so oxidizes the other material in the reaction

WHAT A FUEL CELL DOES

Fuel cells are electrochemical cells that convert chemical energy into heat and electricity by reacting a fuel with an oxidant, typically through the process of electrolysis. One common example is the hydrogen-based fuel cell, which converts hydrogen and oxygen to heat, power, and water. Other fuel cells use other hydrocarbons or methanol as fuels. The idea for a fuel cell goes back to the first half of the nineteenth century, but it was not until the late 1950s that the devices were developed beyond the prototype stage, finding their first application in National Aeronautics and Space Administration's (NASA's) Project Gemini. At that time, fuel cells for other applications were developed, but none took off on a large scale.

The structure of all fuel cells is common in principle: an electrolyte layer is sandwiched between two electrodes, a positively charged anode and a negative cathode. The two electrodes are formed by the fuel and the oxidant respectively, such that their direct mixing is prevented, and the only type of interaction is through the transmission of electric (ionic) charge through the electrolyte. A catalyst at the anode may oxidize the fuel, producing a positively charged ion and a negative electron. Hence, as electrons cannot pass through the electrolyte, during the process of electrolysis the anode is "used up" in the sense that it contributes positively charged ions to the electrolyte, which are then carried to the cathode. The electrons move the other way around the circuit, that is, through the wire connecting anode and cathode, before reaching the positive ions at the cathode, and there reacting with another chemical, typically oxygen, to form water or carbon dioxide. An individual fuel cell produces a relatively small voltage, such that these cells are typically combined into "stacks" depending upon the application, whereby connection in series and parallel can be used to provide higher voltage and current, respectively.

TYPES OF FUEL CELLS

Based upon these basic principles of operation, various specific types of fuel cells have been developed. A common type of fuel cell is the polymer electrolyte membrane fuel cell (PEMFC), which uses a solid polymer as the electrolyte, runs on hydrogen and oxygen, and employs carbon electrodes with a platinum catalyst. The latter is required in order to separate the positive and negative ions issued from the anode, but this significantly adds to the system cost. The fast start-up time from running at low temperatures, and the associated durability from less wear on components, are contrasted by disadvantages in terms of high costs and the need for hydrogen as a fuel.

Alternative types of fuel cells, such as molten carbonate or solid oxide fuel cells (MCFC and SOFC, re-

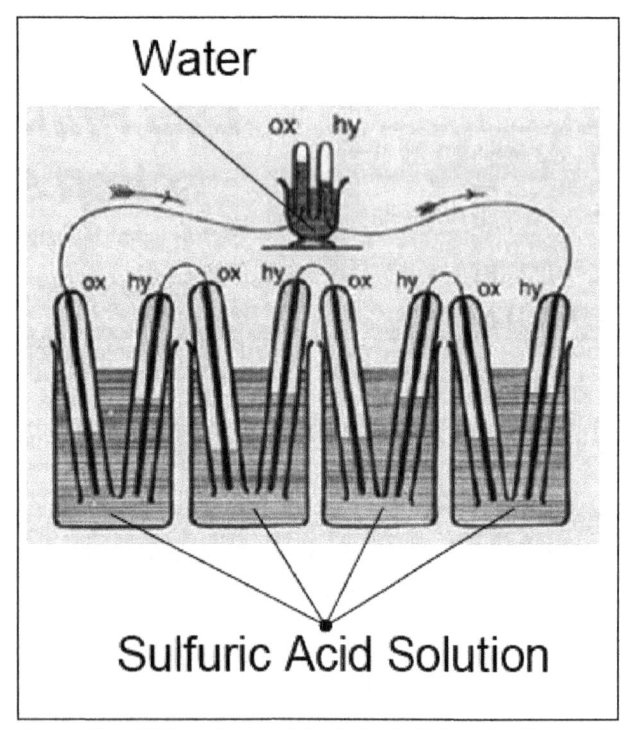

Sketch of Sir William Grove's 1839 fuel cell. (Wikimedia Commons)

spectively), which operate at significantly higher temperatures, have therefore been developed. These devices operate in the temperature range of 500°C-1,000°C, which eliminates the need for a precious metal catalyst and thereby reduces cost. Another advantage of the higher temperature is that, while these fuel cells still use hydrogen as a fuel, they can internally reform other fuels to hydrogen, such that it does not need to be externally produced, which can further reduce costs. Typically, these higher temperature fuel cells have higher overall efficiencies; however, precisely because of the high temperatures, they have longer startup times and suffer from greater wear and tear on the components.

EFFICIENCIES

Depending upon the type of fuel cell, the electrical efficiency varies between around 30 percent and 80 percent. Fuel cells also produce heat, and if this heat can be utilized, the overall system efficiency will increase. Also, where the hydrogen and oxygen for the fuel cell have to be produced, this needs to be considered when evaluating efficiencies. Unlike natural gas, for example, hydrogen and oxygen cannot be found in pure form in nature, and hence energy has to be invested to produce or separate these substances. If the whole supply chain efficiency for fuel cells is considered, it is therefore much lower than the upper value of 80 percent.

APPLICATIONS OF FUEL CELLS

Fuel cells have a wide range of applications, wherever heat and/or power is required, such as in vehicles to provide motive power, as mobile power devices, and as combined heat and power (CHP) devices for industrial or residential needs. Due to the lack of moving parts, fuel cells are highly efficient compared to other CHPs, but they are still relatively expensive compared to those other devices, which has limited their market penetration. Vehicle applications are not limited to road transport, but also include providing motive power for marine, aerospace, and industrial applications (such as forklift trucks), as well as light-duty vehicles (LDVs) and buses. Currently, there are no fuel-cell powered vehicles in mass production, as the high costs have confined their development to prototype vehicles. Due to the lack of moving parts and therefore the low maintenance required, their com-

Sir William Grove anticipated the general theory of the conservation of energy, and was a pioneer of fuel cell technology. (Wikimedia Commons)

pactness, and high power-to-weight ratio, fuel cells are very suited to remote applications, such as in space exploration or submarine warfare. The very low noise of fuel cells in operation is advantageous for the latter application.

Energy storage is another area where fuel cells can make a significant contribution: Because of their possible two-way operation, in generating heat and power in one direction, or converting (excess) electricity to chemical energy in the other, they have a key role to play in a future energy system dominated by fluctuating renewable electricity generation.

CHP applications for domestic use are generally more efficient than separate generation of heat (e.g., through a boiler) and electricity supply through the grid, which might have an overall efficiency of around 30 to 40 percent. However, there are other technologies for such micro-CHP applications in the range up to about 50 kilowatts of electrical power, which are significantly cheaper than fuel cells. Fuel cells have a heat-to-power ratio in the range of 0.3 to 1, which is much more favorable for domestic applications requiring more electricity than heat than do other CHP devices such as internal combustion engines. This means that, technically at least, they are more suited to cogeneration in buildings with only modest heat loads. Given that, primarily due to better insulation through retrofit and new build measures, the residential heat demand is expected to reduce significantly in the coming decades, the fuel cell could be a very attractive technology for such residential applications, if the costs can be reduced substantially.

One further barrier to the widespread adoption of fuel cells based on hydrogen, or any other fuel that is not widely available, is the lack of supply and distribution infrastructure for this fuel. Unlike gasoline, diesel, or electricity, little or no infrastructure exists for the widespread distribution of hydrogen, so there is a "chicken and egg" situation in which the energy companies that would develop the infrastructure, alongside or integrated into conventional fueling stations, have little incentive to do so because the fleet of vehicles using this fuel is so small that it would not be economically feasible. On the other hand, vehicle manufacturers have a similar disincentive to develop and produce fuel-cell powered vehicles on a large scale, exactly because there is a lack of infrastructure for this fuel. Neither side has so far been willing to take the substantial risk involved in such a venture; what is probably required is collaboration between both parties on a large scale, to guarantee a minimum capacity of vehicles and refueling points. There are some refueling stations for hydrogen, for example, in the United States or in European countries, but the number and total capacity is negligible compared to other conventional fuels. Notwithstanding these barriers, fuel cells are commercially available in many countries, such that applications in all areas have experienced growth.

Further future cost reductions combined with increasing costs for alternatives, such as fossil fuels, would normally enable these devices to become much more competitive over the coming years. However, the present focus on production of electric vehicles and the vast improvements in battery design that accompany it essentially guarantee that fuel cells are likely to remain a specialty niche application for many years to come, unless there is a significant breakthrough in development that renders fuel cells far more competitive than they are currently.

—*Russell McKenna*

Further Reading
Breeze, Paul. *Fuel Cells.* Academic Press, 2017.
Dincer, Ibrahim, and Osamah Siddiqui. *Ammonia Fuel Cells.* Elsevier, 2020.
Fergus, Jeffrey W., Rob Hui, Xianguo Li, David P. Wilkinson, and Jiujun Zhang, eds. *Solid Oxide Fuel Cells: Materials, Properties and Performance.* CRC Press, 2009.
Gou, Bei, Woon Ki Na, and Bill Diong. *Fuel Cells: Modeling, Control and Applications.* CRC Press, 2010.
Hoogers, G. *Fuel Cell Technology Handbook.* CRC Press, 2002.
Huggins, R.A. *Energy Storage.* Springer, 2010.
International Energy Agency. "IEA Advanced Fuel Cells: Implementing Agreement," 2012. http://www.ieafuelcell.com.
Sammes, N. *Fuel Cell Technology: Reaching Towards Commercialization.* Springer, 2006.
Sørensen, Bent. *Hydrogen and Fuel Cells: Emerging Technologies and Applications.* 2nd ed. Academic Press, 2012.

Fundamentals of Energy

FIELDS OF STUDY
Physics; Thermodynamics

ABSTRACT
Every source of energy is a manifestation of the four fundamental forces of physics: gravitational, electromagnetic, the weak, and the strong nuclear forces. The most important force for human-scale systems is the electromagnetic, which comes to Earth in the form of solar radiation and is stored as chemical energy by plants.

KEY CONCEPTS
differential heating: because of the nature of water, land surfaces heat up more quickly from sunlight than water areas do; land surfaces also give up that heat much faster than bodies of water do
electronic bonds: electrons in atoms occupy an orderly arrangement about the nucleus that is determined by specific energy relationships of the

nuclear components; in order to achieve the most stable arrangement within those parameters, atoms can accept, or share, an electron with another atom, which binds them together

exergy: an assessment of the amount of usable work that can be derived from an energy source

field: the manifestation through space of the potential of an electric or magnetic source

net energy ratio: the difference between the total energy that can be obtained from a source, and the energy expended to obtain the energy from that source

radioactive decay: the spontaneous splitting apart of an atomic nucleus with the concomitant release of subnuclear particles and high-energy electromagnetic radiation

semiconductive: describes a material that does not normally conduct and electrical current but that can be enabled to do so either electrically or by stimulation with electromagnetic frequencies

THE BASIC CONCEPTS OF ENERGY

We often think of energy as a resource that makes life easier. It allows us to heat and light our buildings, power our machines and appliances, and move ourselves and goods around the globe. This conception of energy is quite close to the actual yet abstract definition of energy. Formally, energy is the capacity to do work, and it is conserved during any change in nature. Every time a living thing or matter changes in any way, energy flows from one location to another and changes from one form to another. All these changes—all physical interactions—result from four fundamental forces: the gravitational and electromagnetic forces and the strong and weak nuclear forces. These four fundamental forces are linked to the primary energy sources used to sustain and grow human civilization.

THE STANDARD MODEL

Particle physics describes all physical interactions, matter, and forces within the universe in terms of 16 elementary, subatomic particles. This theory is called the "Standard Model" of particle physics. Scientific work on the Standard Model is ongoing today, but it began with work by Sheldon Glashow, Steven Weinberg, Abdus Salam, and Peter Higgs in the 1960s.

The model divides the 16 elementary particles into three groups: quarks, leptons, and gauge bosons. The quarks come in three "flavor" pairs of positive and negatively charged particles, grouped according to their mass. The lightest are the up (u) and down (d), then the strange (s) and charm (c), and then the top (or sometimes truth, t) and bottom (or sometimes beauty, b) quarks. Within the nucleonic particles—the proton and neutron—quarks are bound in triplets, such that a proton is an uud triplet and a neutron is a ddu triplet. Quarks may also be bound as pairs in particles called "mesons."

All of the particles have complementary antiparticles, which are exactly alike in all respects, except that they have the opposite electric charges; for example, the electron's positively charged complement is called the "positron." In all interactions, charge, spin, and mass-energy must be conserved.

GRAVITATIONAL FORCE

Physical bodies attract with a force proportional to their mass. It is the force that causes things to fall to the ground on Earth and all other planets and bodies. Galileo's work during the scientific revolution (as P. Machamer notes), Sir Isaac Newton *Principia Mathematica* (written during the Enlightenment and appearing in 1687), and Albert Einstein's modern work on general relativity (1916) led to our current understanding of the force of gravity. Gravity is mediated by a hypothetical gauge boson called the "graviton." This boson works on the mass "charge" of all massive particles. The gravitational force is the weakest of all of the four fundamental forces; however, over cosmological distances the gravitational force is highly important as the only mediator. This is true because the range of the force is infinite, as is the electromagnetic, and most macroscopic objects are electrically neutral, meaning that the electromagnetic force is irrelevant over these large distances.

ELECTROMAGNETIC FORCE

Electric charges and currents act as sources for electric and magnetic fields that permeate nature. James Clerk Maxwell's equations formally describe electromagnetism and show that electricity, magnetism, and optics all fall under a consistent theoretical framework. The electromagnetic force is mediated by photons and acts on the electric charge of a particle. Since particles can have either positive or negative charge, the force can be either attractive, between particles of opposite charge, or repulsive, between particles of similar charge. This attractive and repulsive nature means that particles will tend to form larger electrically neutral groups by attracting oppositely charged and repelling similarly charged. For this reason, atoms, consisting of negatively charged electrons orbiting a central positively charged nucleus of protons and neutrons, tend to be electrically neutral in themselves, or combine to form electrically neutral molecules with other atoms of opposite charge. Finally, moving magnetic fields can induce electric currents and vice versa, as Michael Faraday realized. The phenomenon of induction is key to converting between mechanical energy and electromagnetic energy.

WEAK NUCLEAR FORCE
The weak nuclear force, so called because its relative strength and effective distance are much less than those of the strong nuclear force, acts to change the flavor of fermions within their pair groupings. For example, a down quark may be changed to an up quark, thereby emitting a W boson (to conserve electric charge, the boson must be negatively charged). The W boson will then normally form an electron and antielectron neutrino pair (again, the negative charge is conserved). The process described is the decay of a neutron (udd) to a proton (uud) with the emission of beta radiation (an electron and antielectron neutrino pair) and occurs spontaneously with very high probability.

STRONG NUCLEAR FORCE
The strong nuclear force governs interactions between the quarks within a bound pair or triplet. It is mediated by gluons and acts on the "color charge" of quarks—red, blue, or green—which must be conserved in any interaction. Bound triplets or pairs must be color-neutral, by combining either red, blue, and green in a triplet, or forming color-anticolor pairs. The force has a range of 1 femtometer, or 10-15 meters (within the scale of an atomic nucleus) and has the unusual property of becoming stronger with distance, which will cause any force attempting to separate quarks, if strong enough, to produce more quarks from the vacuum.

BALANCING FORCES IN THE NUCLEUS
Because of the existence of electrically-positive charged protons within the nucleus, there exists an incredibly strong repulsive electromagnetic force. This is balanced within the nucleus by the existence of neutrons, which have no electrical charge but add to the attractive strong nuclear force. The elemental type of nucleus (whether that of hydrogen, helium, or another element) is governed by the number of protons; for example, carbon has six protons. Isotopes of elements have the same number of protons but different numbers of neutrons. Some isotopes are more stable than others; for instance, carbon-14 (with six protons and eight neutrons) is radioactive, whereas carbon-12 is stable. When the number of protons and neutrons within the nucleus allows balance between the strong nuclear and electromagnetic forces, the nucleus is stable.

For large nuclei, since the strong nuclear force acts over only a very short range, more neutrons are required to counteract the electromagnetic force. This means that the balance between protons and neutrons is in favor of neutrons. However, neutrons have a high probability of decaying into protons. If this occurs, the nucleus is unstable and balance is restored by the emission of an alpha particle (a helium nucleus of two protons and two neutrons). In smaller nuclei, the balance between protons and neutrons must be equal. If there are more protons (as is the case in radioactive carbon-14), the electromagnetic force is too strong and the balance is restored by the decay of a proton (via the weak nuclear force) into a neutron and the emission of a beta particle (in this case, in the form of a positron—an antielectron). If the balance is in favor of neutrons, a neutron will decay into a proton, given the high probability of this occurrence with the emission of a beta particle, in this case in the form of an electron (as above). When a nucleus is in a high-energy state, it may emit gamma radiation (a very high-energy, short-wavelength photon), which will not alter the composition of protons or neutrons.

Within stars, the presence of many protons (hydrogen nuclei) at huge pressures and temperatures results in many collisions between particles. If these collisions can overcome the electromagnetic repulsion between protons, helium nuclei (two protons and two neutrons) may be formed (in a process called "fusion") via the decay of two protons into neutrons (the weak nuclear pathway) and the emission of beta particles and neutrinos. The helium nuclei represent a lower energy state than the four protons individually, and hence energy is released.

RELATIONSHIP BETWEEN SPATIAL SCALE AND DOMINATING FORCE

Both of the nuclear forces have a limited spatial range, which limits their influence to interactions on the subatomic scale. The electromagnetic and gravitational forces both have an infinite range and so have effects on macro scales. At large spatial scales, objects tend to form electrically neutral combinations, such that the gravitational force dominates. The rest of this article shall deal primarily with the electromagnetic and gravitational forces.

ENERGY FIELDS AND POTENTIALS

Although all the forces are described in terms of the interaction of particles, the electromagnetic and magnetic forces may be more usefully thought of in terms of fields whereby the presence of charges (positive and negative electric charges in the case of the electromagnetic and mass in the case of gravitational force) give rise to a field with a potential gradient, which then influences the behavior of other objects within that field. The classic example is of massive objects deforming the gravitational field, likened to a bowling ball placed on a rubber sheet. The sheet bends such that other objects will be pulled toward the bowling ball. The magnitude of the force between the objects is a function of both their mass and their separation. Despite its smaller radius, the surface gravity of the moon is about one-sixth that of the Earth because of its smaller mass. In the case of the electromagnetic force, there are both attractive and repulsive aspects, such that the presence of a negative charge will pull in (or be pulled toward) other positive charges and push away (or be pushed away by) other negative charges.

LINKING FORCES TO ENERGY

All of the energy sources currently exploited (and some potential sources for the future) are manifestations of the four forces. The U.S. Energy Information Administration lists energy sources, in order of use from greatest to least, as oil (both conventional and unconventional sources), natural gas (again coming in both conventional and unconventional sources), coal, nuclear energy, hydroelectric power, biomass, wind, geothermal power, and solar power (both solar thermal and photovoltaic, or PV). Energy sources under research and development include tidal energy, wave energy, and ocean thermal energy conversion (OTEC).

The electromagnetic force governs not only electric and magnetic but also chemical interactions. The combustion of a fuel (both fossil fuels and biofuels) is entirely describable in terms of the creation and destruction of electronic (in the sense of electrons) bonds between molecules. As such, this force governs any of the energy sources that are described in terms of thermodynamics, including all of the fossil fuels, biomass, solar thermal power, and geothermal power (although geothermal power can be said to be a result of gravitational energy, both because it is heat energy left over from the cooling of the planet and the interaction of radioactivity—hence the weak nuclear force—and gravity as the heating and convection of material through Earth's mantle). The electromagnetic force also governs photovoltaic interactions, whereby light is converted directly into electricity.

Within the hydrologic cycle, incoming energy from the sun—in the form of infrared (IR), visible, ultraviolet (UV), and X-ray photons, all manifestations of the electromagnetic force—evaporate water, which then rises through the atmosphere. After the water vapor condenses and collects (again, a manifestation of the electromagnetic force), it falls out of the sky as precipitation, which then forms streams and rivers through which it flows to the ocean. This downward leg is governed entirely by the gravitational force. Hydropower makes use of the work done by the sun in lifting vast amounts of water in order to power turbines. The amount of energy available is a function of the distance the water falls (called the "head"), the gravitational pull of Earth, and the amount of water that flows.

Geothermal energy is the slow release by Earth of heat generated by radioactive decay of elements within Earth's mantle. The emission of these high-speed alpha and beta particles and gamma radiation thermally excites the surrounding material. If these particles are also unstable, this may cause further radioactive decay.

Nuclear energy from fission makes use of elements with large nuclei for fuel. The fuel nuclei are bombarded with neutrons, which make the nucleus unstable. The nucleus undergoes fission by splitting into smaller nuclei and emitting up to three neutrons. These neutrons are utilized to generate heat and to enable the fission of more fuel nuclei. The heat generated is used to power a turbine and generate electricity, much like in a fossil fuel power station. The fuel most often used for nuclear power generation is uranium-235. Since this element makes up less than 1 percent of naturally occurring uranium (the bulk being made of the more stable isotope uranium-238), the stable isotope is usually enriched to increase the concentration of U-235 to enable the fission process.

Research is under way to replicate fusion. Different fuel pathways are being investigated, including deuterium (a hydrogen-2 nucleus with one proton and one neutron) and tritium (a hydrogen-3 nucleus with one proton and two neutrons) collisions. The positively charged deuterium or tritium fuels are squeezed by large magnetic fields to induce fusion to occur, and the heat thereby generated is used to power a turbine and generate electricity. If the electrical energy produced is greater than the energy required to produce the magnetic fields (via electromagnets), then the process offers positive net energy.

Tidal forces are also a manifestation of the gravitational force. Water on Earth closest to the Moon is acted upon more strongly by the pull of the Moon than the bulk of Earth. Water on the far side from the Moon feels the least pull, such that there are two bulges in the water, one toward and one away from the moon. This gives rise to two tides per day as Earth rotates through these two bulges. The Sun also has an effect on the tide (about one-third that of the Moon), meaning that tides are increased when the three bodies (Sun-Earth-Moon) are aligned, such as at the full moon (called a "spring tide"), and decreased when a right angle is formed, such as at new moon (called a "neap tide"). In certain places on Earth, the topology of the land acts to amplify the effects of the tide, leading to large tidal ranges—as much as 10 meters (32.8 feet) in some locations.

Wind energy exploits differential (electromagnetic) heating of Earth's surface by incoming solar radiation. As air over a "hot" region rises, it pulls in air from the surrounding regions. This is particularly noticeable in coastal regions because of the different thermal mass of land and water. During the day, solar radiation increases the temperature of the land more than the water, creating an onshore breeze. After sunset, the land loses heat more quickly, creating an onshore breeze. Differential heating between the equator and the poles also creates global weather systems. These are added to by the Coriolis effect, caused by the rotation of Earth.

Solar energy generation comes in two types: that using the direct insolation from the sun to generate heat (and sometimes thereby to generate electricity) and photovoltaic (PV) panels, which generate electricity directly. Thermal solar comes in a number of configurations, from small, flat panels for generating low-grade heat (generally less than 100°C for residential or commercial settings) to large-scale arrays of concentrating troughs, dishes, or mirrors focused onto a central tower. PV cells are made from a number of semiconductive materials (most often high-purity silicon), which conduct electrons when exposed to photons of a particular wavelength.

Wave energy exploits the energy carried in waves to operate a mechanical system to generate electricity. There are a wide variety of designs in development. Surface waves are generated when the wind moves across the surface of a body of water. Once small waves have been created, the effect may be amplified by the differential Bernoulli effect between the moving air and the peaks and troughs of the waves. Due to the greater density of water, the power density of waves is often greater than that of wind. Ocean thermal energy conversion (OTEC) exploits the difference in temperature between the surface water (often around 20°C in the tropics) and the cold, deep water (typically that greater than 1 kilometer, or 0.62 mile deep, which may be near 0°C). This temperature range may be used to drive a heat engine and generate electricity.

EXPLOITING ENERGY GRADIENTS

In all cases, extracting energy from the environment requires exploitation of a naturally occurring energy gradient. These occur because of (sometimes minute) differential conditions between regions in the universe. The Sun is obviously a great source of high-quality, low-entropy energy, which floods Earth. This has given rise to plants, which exploit this resource, and animals, which feed on plants. Plants store solar energy within their structure as highly ordered carbon-oxygen bonds that are far from chemical equilibrium with the surrounding atmosphere. Combustion (of fossil fuels and biofuels) breaks these bonds to release the energy stored. Differential heating of the land and oceans can be exploited to drive turbines. Hydropower and tidal power exploit gravitational gradients. Geothermal power exploits naturally occurring thermal gradients, and nuclear power exploits instability in the balance between electromagnetic and strong nuclear forces within the nuclei of radioactive fuels.

RELATIVE COMPARABILITY OF ENERGY SOURCES

Energy sources are far from equivalent in the service they provide. Obviously, some sources are replenished on the timescale at which they are being used by humanity, and as such they are renewable. Sources that are exploited more quickly than the timescale at which they are replenished are said to be nonrenewable. These include the fossil fuels (coal, oil, and gas) and fissile nuclear fuels. Some energy sources represent stores of energy to be used as and when required; others are intermittent and must be supplemented by means of storage, if so desired.

NET ENERGY ANALYSIS

Energy is also required for the extraction and conversion of energy. Energy requirements may include direct inputs of energy (the fuel needed to power an oil rig) but also the energy embodied in any materials used (the energy required to manufacture the oil rig). If an energy source produces more energy than is used in its delivery, the delivery process is said to have a positive net energy. The ratio of the energy delivered by a process to the energy required in its delivery is called variously the net energy ratio, or NER, the energy-return-on-(energy)-investment, or ERO(E)I, and the energy yield ratio, or EYR, which may vary greatly among energy sources.

Energy sources differ also in their energy density, either by unit mass (megajoules divided by kilograms, or MJ/kg) or by unit volume (megajoules divided by cubic meters, or MJ/m3). Nuclear fuels offer the greatest energy densities, followed by oil, gas, and coal, and then by biomass. Renewable energy sources may be ranked in terms of their power density (megajoules divided by square meters, or MJ/m2).

EXERGY AND ENERGY QUALITY

Not all forms of energy are of equal use to humankind. An equal amount of energy enters Earth's atmosphere in the form of insolation as leaves it in the form of "waste" heat; however, the incoming solar radiation is of much higher quality (is more usable) than the outgoing infrared radiation. A measure of energy quality, exergy is the amount of work that may be extracted as a system comes into equilibrium with its surroundings.

The exergy of 100 joules of low-temperature steam is much lower than the exergy of 100 joules worth of oil. Since most produce electricity directly, renewable energy sources are often rated in terms of the electricity that they produce. A joule of electricity has an exergy of 1 joule, since conversion of electricity to work is presumed to be completely efficient. Fossil fuels are often rated in terms of their higher (or lower) heating values (HHV or LHV), which have a lower exergy due to inefficiencies in conversion to work. The exergy perspective is important, since many of the economic roles of energy are in its ability to produce work, as opposed to heat, directly.

CONCLUSION

Whenever nature undergoes a change, energy transforms from one form to another and is transferred from one location to another or exchanged between material objects. These physical interactions are mediated by one or more of the four fundamental forces: electromagnetism, gravitation, weak-nuclear, and strong-nuclear. Currently, society obtains more than 80 percent of its primary energy from fossil fuel sources: petroleum, natural gas, and coal. It follows that our current energy system is dominated by the electromagnetic force, because the chemical energy present in these hydrocarbons are a manifestation of the electromagnetic force. When these chemical bonds are broken to form stronger bonds with oxygen, infrared radiation is released to heat engine cycles that drive turbines and dynamos, which generate electricity—again, thermodynamic phenomena driven by electromagnetic forces. Society also utilizes the gravitational force to harness the flow of water, whose kinetic energy was obtained from gravitational potential energy. Fundamentally, research in clean energy technologies—turbines, solar PV, energy storage, fission, and fusion—focuses on harnessing the four forces in a manner that permits safe, environmentally responsible, economically viable, and thermodynamically efficient conversion of various forms of energy into electrical energy for society's use.

—*Charles Barnhart*

Further Reading

Einstein, Albert. "Die Grundlage der allgemeinen Relativitatstheorie." *Annalen der Physik*, vol. 354, no. 7, 1916.

Faraday, Michael. *The Correspondence of Michael Faraday: 1811-December 1831*. Edited by Frank A.J.L. James. Institution of Electrical Engineers, 1991.

Glashow, Sheldon L. "Partial-Symmetries of Weak Interactions." *Nuclear Physics*, vol. 22, no. 4, 1961. (1961).

Griffiths, D.J. *Introduction to Elementary Particles*. John Wiley, 1987.

Higgs, Peter W. "Broken Symmetries and the Masses of Gauge Bosons." *Physical Review Letters*, vol. 13, no. 16, 19 Oct. 1964.

International Atomic Energy Agency. IAEA Nuclear Energy Series Establishment of Uranium Mining and Processing Operations in the Context of Sustainable Development. IAEA Nuclear Energy Series NF-T-1. International Atomic Energy Agency, 2009.

Machamer, P. "Galileo Galilei." *The Stanford Encyclopedia of Philosophy*, edited by E.N. Zalta. Metaphysics Research Lab, Center for the Study of Language and Information, Stanford U, 2010. http://plato.stanford.edu/entries/galileo/.

MacKay, David J.C. *Sustainable Energy—Without the Hot Air*. UIT, 2009.

Maxwell, James Clerk. "A Dynamical Theory of the Electromagnetic Field." *Philosophical Transactions of the Royal Society of London*, vol. 155, 1865. Reprint. Wipf and Stock, 1996.

Newton, Isaac. *Philosophiae Naturalis Principia Mathematica*. 1687. Reprint.

Newton's Principia: The Central Argument, edited by Dana Densmore and William H. Donahue. Green Lion Press, 2010.

Salam, Abdus. "Weak and Electromagnetic Interactions." *Il Nuovo Cimento*, vol. 11, no. 4, Feb. 1959.

Salam, Abdus, and N. Svartholm, eds. Elementary Particle Physics: Relativistic Groups and Analyticity. Eighth Nobel Symposium. Almqvist and Wiksell, 1968.

U.S. Energy Information Administration. *Annual Energy Review 2010*. Technical Report DOE/EIA-0387(97). U.S. Energy Information Administration, 2011.

Weinberg, Steven. "A Model of Leptons." *Physical Review Letters*, vol. 19, no. 21, 1967.

G

Gas Energy Transmission

FIELDS OF STUDY
Geology; Petroleum Engineering; Pipe Fitting Trades

ABSTRACT
Pipelines are the principal means for transporting large quantities of gas over long distances. Pipelines have been a proven method for gas transmission for more than a century. Currently, the most important alternatives to pipelines involve conversions to liquefied natural gas (LNG) and from gas to liquids (GTL).

KEY CONCEPTS
clathrate: a solid material in which a small molecule of a gas is enclosed within the crystal structure of a solid
methane hydrate: a solid material formed when molecules of methane gas become encased within a structure of water ice; sometimes called "ice that burns." Methane hydrate occurs naturally on cold ocean floors and often clogs pipes when water vapor is present in natural gas lines

THAT'S A LOT OF PIPE!

Pipelines have been a proven method for gas transmission for more than a century. The first significant long-distance gas pipeline was built in 1891. It was 192 kilometers (120 miles) long and connected gas fields in Indiana to customers in Chicago. Technological improvements in metallurgy, welding, construction methods, compressors, control systems, corrosion management, and other operations have made pipelines over longer distances feasible.

Today, the United States has more than 488,000 kilometers (305,000 miles) of transmission pipelines. Europe also has an extensive gas infrastructure that delivers gas from Russia, the North Sea, and North Africa. Gazprom, the Russian state-owned company, operates more than 442,416 kilometers (276,510 miles) of pipeline in Europe and Asia.

In the 1970s, it became possible to install pipelines on the seabed over long distances using specialized vessels. This allowed the development of offshore oil and gas reservoirs (such as in the North Sea and the Gulf of Mexico), as well as increased international trade in gas. A recent example is the 1,192 kilometer (745 mile) Langeled Pipeline, which connects the Ormen Lange gas field in the Norwegian sector of the North Sea to an onshore terminal in the eastern coast of England.

LIQUEFIED NATURAL GAS (LNG)

Significant gas reserves can be stranded if they are too far from the market and it is not economical to construct a pipeline. An alternative method for transporting gas over long distances is to convert the gas to liquid using pressure and refrigeration, ship the liquid using specialized vessels, convert the liquid back to gas, and finally deliver the gas to end users through conventional pipelines.

In 1959, the *Methane Pioneer* delivered a shipment of LNG across the Atlantic from Louisiana to the United Kingdom. By the mid-1970s, LNG projects were operational in Brunei (exporting to Japan), Algeria, and Libya (both exporting to Europe). Today, there are 15 LNG-exporting countries, including Indonesia, Malaysia, Brunei, Australia, Qatar, the United Arab Emirates, Algeria, Libya, and Russia. LNG projects are capital intensive and are economical only for large gas reservoirs.

GAS TO LIQUIDS (GTL)

The conversion of gas to liquid fuels and specialty chemicals (naphtha, kerosene, gas oil, detergent feedstock, and waxes) requires complex technology and expensive infrastructure. Although the basic technology for the GTL process was developed in the 1920s, it has taken many decades to apply it at a commercial scale. Shell's Bintulu GTL plant (supplied from gas fields in offshore Sarawak) has been operating in Malaysia since 1993 with the capacity to produce 14,700 barrels per day.

The Pearl GTL project in Qatar, operated by Shell, is an integrated gas and GTL project that became fully operational in 2012. Gas is supplied from two offshore platforms 37 miles from Qatar. The GTL plant is the world's largest and has the capacity to produce 140,000 barrels of GTL products per day. Construction commenced in 2006, and the total project cost is estimated at $18 billion.

GAS TO POWER

A significant portion of gas consumption is used for power generation. For example, 31 percent of the gas consumed in the United States in 2008 was for power generation. An alternative to constructing a long-distance pipeline is to construct a power-generation plant close to the source of a gas supply and deliver electricity through high-voltage power transmission lines. This option may be feasible if the cost of constructing the power transmission lines is much less than the cost of constructing a pipeline. An issue that needs to be managed is the energy loss associated with long-distance power transmission.

COMPRESSED NATURAL GAS (CNG)

The bulk transport of CNG by truck, rail, or ship can be an alternative to long-distance pipelines or the use of LNG ships. CNG refueling stations that cannot be supplied by pipeline can be supplied from a CNG "mother station" by trucks with specialized CNG tanks. The technology for bulk transport of CNG by ship (marine CNG) is available but has not been widely adopted.

OTHER OPTIONS

If gas can be converted into liquid, it can also be converted into a solid (gas hydrates). However, gas-to-solid technology is still at the experimental stage and is not yet used in significant commercial scale. It is believed that very large reserves of the clathrate material 'methane hydrate' exist on cold ocean floors, but the technology to harvest this material economically has yet to be developed. Another option is to use the gas to produce commodities (for example steel, copper, aluminum, glass, and cement) that have an energy-intensive production process. The commodities can then be exported using conventional land or marine transport.

—*Kiril Caral*

Further Reading

Elbashir, Nimir O., Mahmoud M. El-Halwagi, Ioannis G. Economou, and Kenneth R. Hall, eds. *Natural Gas Processing from Midstream to Downstream.* Wiley, 2019.

Farmer, Edward J. *Detecting Leaks in Pipelines.* ISA, 2017.

Grigas, Agnia. *The New Geopolitics of Natural Gas.* Harvard UP, 2017.

Miesner, Thomas, and William Leffler. *Oil and Gas Pipelines in Nontechnical Language.* PennWell, 2006.

Mokhatab, Saeid, John Y. Mak, Jaleel V. Valappil, and David A. Wood. *Handbook of Liquefied Natural Gas Transmission.* Gulf Professional, 2014.

Shively, Bob, and John Ferrare. *Understanding Today's Natural Gas Business.* Enerdynamics, 2009.

Sluyterman, Keetie, and Joost Dankers. *Keeping Competitive in Turbulent Markets, 1973-2007: A History of Royal Dutch Shell.* Oxford UP, 2007.

Speight, James G. *Natural Gas: A Basic Handbook.* 2nd ed. Gulf Professional, 2019.

Wang, Xiuli, and Michael Economides. *Advanced Natural Gas Engineering.* Gulf Professional, 2009.

Gasoline and Other Petroleum Fuels

FIELDS OF STUDY

Organic Chemistry; Petroleum Engineering; Refinery Operations; Thermodynamics

ABSTRACT

Gasoline is the most important product from petroleum and is the dominant transportation fuel in the world. Other petroleum products with important fuel uses include kerosene (usually refined to jet fuel), diesel oil, and heating oils.

KEY CONCEPTS

alkanes: organic molecules composed solely of carbon and hydrogen; in normal alkanes the carbon atoms are joined together linearly, in branched alkanes the carbon atoms are joined together as groups attached to a simple linear structure

aromatics: organic molecules composed solely of carbon and hydrogen atoms, in which some or all of the carbon atoms are joined together to form ring structures with more energetic bonds; they are not termed "aromatic" for their odor, but for the special nature of the interatomic bonds in the ring structures

cetane rating: a comparative rating assigned to diesel fuel blends on the basis of their combustion properties relative to pure *n*-hexadecane, a normal alkane consisting of 16 carbon atoms and 34 hydrogen atoms

jet fuel: a particular fraction of petroleum distillates containing a limited variety of alkanes and aromatics of a particular molecular weight and boiling point range

octane rating: a comparative rating assigned to gasoline blends on the basis of their combustion properties relative to pure iso-octane (2,2,4-trimethylpentane)

oxygenates: organic molecules like alkanes but containing one or more oxygen atoms as part of their basic molecular structure

BACKGROUND

Petroleum is the source of nearly all the world's transportation fuels: gasoline for automobiles, light trucks, and light aircraft; jet fuel for airplanes; and diesel fuel primarily for locomotives, heavy trucks and agricultural vehicles, but increasingly for cars as well. Heating oils (also called fuel oils or furnace oils) are used for domestic heating and industrial process heat; they are also used in oil-fired electric generating plants. Petroleum fuels are a vital component of the energy economies of industrialized nations.

The first step in making all petroleum fuels is distillation of the petroleum or crude oil. Kerosene, diesel oil, and heating oils require comparatively little refining thereafter to be ready for market. Considerable effort is put into gasoline production both to ensure adequate engine performance and to guarantee that sufficient quantities will be available to meet market requirements.

GASOLINE

The most important characteristic of gasoline is its combustion performance. When gasoline is ignited in the cylinder, the pressure rises as combustion proceeds. The pressure can, potentially, get so high that the remaining unburned gasoline-air mixture detonates rather than continuing to burn smoothly. The explosion, which can readily be heard, is usually called "engine knock." Engine knock puts undue mechanical stresses on the engine components, is wasteful of fuel (which the driver will experience as reduced mileage), and reduces engine performance, such as acceleration. Several factors contribute to engine knock. One is the compression ratio of the engine—the ratio of volumes of the cylinder when the piston is at the upward and downward limits of its stroke. Generally, the higher the compression ratio, the more powerful the engine and the greater the acceleration and top speed of the car. A higher compression ratio results in higher pressures inside the cylinder at the start of combustion. If the cylinder pressure is higher to begin with, the engine is more likely to knock.

A second characteristic affecting knocking tendency is the nature of the fuel. The dominant family of chemical components of most gasolines is the alkanes. These compounds contain carbon atoms arranged in chains, either straight (the normal alkanes) or with branched structures (iso-alkanes). Normal alkanes have a great tendency to knock, whereas branched alkanes do not. An "octane rating scale" was established by assigning the normal octane *n*-heptane the value 0 and the iso-alkane "iso-octane" (2,2,4-trimethylpentane) the value 100. The octane rating of a gasoline is found by comparing its knocking characteristics (in a carefully calibrated and standardized test engine) to the behavior of a heptane/iso-octane blend. The percentage of iso-octane in a blend having the same knocking behavior of the gasoline being tested is the octane number of the gasoline. Gasoline is sold in three grades, a regular gasoline with octane number 87, a premium gasoline of about 93 octane, and a medium grade of about 89 octane.

Another important property of gasoline is its ability to vaporize in the engine, measured by the vapor pressure of the gasoline. Gasoline with high vapor pressure contains a large number of components that vaporize easily. This is desirable for wintertime driving in cold climates, since easy vaporization helps starting when the engine is cold. It is not desirable for driving in hot weather, because the gasoline could vaporize in the fuel system before it gets to the engine, leading to the problem of vapor lock, which temporarily shuts down the engine. Oil companies adjust the vapor pressure of their gasolines depending on the region of the country, the local climate, and the season of the year.

Many process streams within a refinery are blended to produce the gasolines that actually appear on the market. Gaseous molecules that would be by-products of refining can be recombined to produce gasoline in processes called alkylation or polymerization. Some gasoline, called straight-run gasoline, comes directly from distillation of the petroleum. Refinery streams of little value can be converted into high-octane gasoline by catalytic cracking. The octane numbers of straight-run gasoline, or a related product called "straight-run naphtha," can be enhanced by catalytic reforming. Other refinery operations can also yield small amounts of material boiling in the gasoline range. Various of these streams are blended to make products of desired octane, vapor pressure, and other characteristics.

Environmental concerns about gasoline have centered on the emission of unburned hydrocarbons (including evaporation from fuel tanks), carbon monoxide and nitrogen oxide emissions from combustion, and the presence of aromatic compounds, some of which are suspected carcinogens and contribute to smoke or soot formation. These concerns have led to the development of reformulated gasolines. One aspect of production of reformulated gasoline is increased vapor pressure, which retards evaporation. A second is removal of aromatic compounds; removal actually complicates formulation because aromatics have desirably high octane numbers. A third step is the addition of oxygen-containing compounds, oxygenates, which serve several purposes: They reduce the flame temperature, for example, and change the combustion chemistry to reduce formation of carbon monoxide and nitrogen oxides. Oxygenates also have high octane numbers, so they can make up for the loss of aromatics. An example of an oxygenate useful in reformulated gasoline is methyl tertiary-butyl ether.

JET FUEL

Jet fuel is produced by refining and purifying kerosene. Kerosene is a useful fuel, particularly for some agricultural vehicles, but the most important fuel use of kerosene today is for jet aircraft engines. Because many jet planes fly at high altitudes, where the outside air temperature is well below zero, the flow characteristics of the fuel at very low temperature are critical. When the fuel is cooled, large molecules of alkanes settle out from the fuel as a waxy deposit. The temperature at which the formation of this wax first begins, noticeable as a cloudy appearance, is called the cloud point. Eventually a fuel can be cooled to an extent where it cannot even flow, not even to pour from an open container. This characteristic temperature is the pour point.

Smoke emissions from jet engines are an environmental concern. The "smoke point" measures an important property of jet fuel combustion. Aromatics are the most likely compounds to produce smoke, while alkanes have the least tendency. A jet fuel with a low smoke point will have a high proportion of alkanes relative to aromatics. The sulfur content of jet fuel can be important, both to limit emissions of sulfur oxides to the atmosphere and because some sulfur compounds are corrosive. Both sulfur and aromatics contents of jet fuel can be reduced by treating with hydrogen in the presence of catalysts containing cobalt, or nickel, and molybdenum.

DIESEL FUEL

A familiar automobile engine operates by igniting the gasoline-air mixture with a spark plug. Diesel engines operate differently: They have no spark plugs, but rely on compression heating of the air in the cylinder to ignite the fuel. A diesel engine has a much higher compression ratio than a comparable spark-ignition engine. In a crude sense, a diesel engine actually operates by knocking. The desirable composition for diesel fuel is essentially the inverse of that for gasoline: Normal alkanes are ideal components, while iso-alkanes and aromatics are not. The combustion behavior of a diesel fuel is measured by the "cetane number," based on a blend of cetane (hexadecane), assigned a value of 100, and alpha-methylnaphthalene, assigned 0, as the test components. A typical diesel fuel for automobile and light truck engines would have a cetane rating of about 50.

Many of the physical property characteristics of jet fuel are also important for diesel fuel, including the cloud and pour points and the flow characteristics (viscosity) at low temperature. Sulfur and aromatic compounds are a concern. Aromatics are particularly undesirable because they are the precursors to the formation of soot. As environmental regulations continue to become more stringent, refiners will face additional challenges to reduce the levels of these components in diesel fuels.

HEATING OILS

Heating oils, also called furnace oils or fuel oils, are often graded and sold on the basis of viscosity. The grades are based on a numerical classification from number 1 to number 6 (though there is no number 3 oil). As the number increases, so do the pour point, the sulfur content, and the viscosity. Number 1 oil is comparable to kerosene. Number 2 is an oil commonly used for domestic and industrial heating. Both have low pour points and sulfur contents and are produced from the distillation of petroleum. The other oils (numbers 4-6) are obtained by treating the residuum from the distillation process. They are sometimes called bunker oils because they have such high viscosities that they may have to be heated to have them flow up, from the storage tank, or bunker, and into the burners in the combustion equipment.

—*Harold H. Schobert*

Further Reading

Black, Edwin. *Internal Combustion: How Corporations and Governments Addicted the World to Oil and Derailed the Alternatives.* St. Martin's Press, 2006.

Chaudhuri, Uttam Ray. *Fundamentals of Petroleum and Petroleum Engineering.* CRC Press, 2011.

Conaway, Charles F. *The Petroleum Industry: A Nontechnical Guide.* PennWell Books, 1999.

Kunstler, James Howard. *The Long Emergency: Surviving the Converging Catastrophes of the Twenty-first Century.* Atlantic Monthly Press, 2005.

Middleton, Paul. *A Brief Guide to the End of Oil.* Constable and Robinson, 2007.

Speight, James G. *The Chemistry and Technology of Petroleum.* 4th ed. CRC Press, 2007.

Srivastava, S.P., and Jeno Hancsók. *Fuels and Fuel Additives.* Wiley, 2014.

Witten, Mark L., Errol Zeiger, and Glenn D. Ritchie, eds. *Jet Fuel Toxicology.* CRC Press, 2011.

Yergin, Daniel. *The Prize: The Epic Quest for Oil, Money, and Power.* New ed. The Free Press, 2008.

GEOTHERMAL AND HYDROTHERMAL ENERGY

FIELDS OF STUDY
Civil Engineering; Environmental Engineering; Geology; Volcanology

ABSTRACT
Geothermal energy is the energy associated with the heat in the interior of Earth. The common usage of the term refers to the thermal energy relatively near the surface of Earth that can be utilized by humans. Hydrothermal energy is the energy associated with hot water, whereas geothermal is a more general term. Geothermal energy has been exploited since early history.

KEY CONCEPTS

dry steam: steam that consists only of water in the gas phase, unaccompanied by any liquid water

geothermal gradient: the gradual and consistent increase in ambient temperature with depth below the ground surface

geyser: an eruption of geothermally heated water from some depth below the surface in geothermally active areas

magmatic intrusion: a location at which molten rock—magma—from Earth's mantle layer has moved into a location above it in Earth's crust

wet steam: steam at a lower temperature than dry steam that is accompanied by liquid water

EARTH AS A HEAT SOURCE

A geothermal system is made up of three elements: a heat source, a reservoir, and a fluid that transfers the heat. The heat source can be a magmatic intrusion or Earth's normal temperature, which increases with depth. The reservoir is a volume of hot permeable rock from which circulating fluids extract heat. Fluid convection transports the heat from the higher-temperature low regions to the upper regions, where it can be accessed and used. It is a source of energy with a low pollution potential that can be used for producing electricity as well as for heating and cooling and helping with a number of other needs.

Turbogenerator. (Siemens via Wikimedia Commons)

Geothermal energy is the energy associated with the heat in the interior of Earth. The common usage of the term refers to the thermal energy relatively near the surface of Earth that can be utilized by humans. Hydrothermal energy is the energy associated with hot water, whereas geothermal is a more general term. Geothermal energy has been exploited since early history. It is a source of energy with a low pollution potential that can be used for producing electricity as well as for heating and cooling and helping with a number of other needs.

CAUSES OF GEOTHERMAL PHENOMENA

While individuals in early mining operations may have noted the general increase in temperature with depth, not until the eighteenth century were subsurface temperature measurements performed. The results often showed an increase in temperature with depth. The rate of increase varied from site to site. An average value that is often used today is a 2.5°C to 3°C increase per 100 meters increase in depth from the surface. The geothermal gradient suggested that the source of Earth's heat was below the surface, but the exact cause of the heat was open to discussion for many years. It was not until the early part of the twentieth century that the decay of radioactive materials was identified as the primary cause of this heat. The thermal energy of Earth is very large; however, only a small portion is available for capture and utilization. The available thermal energy is primarily limited to areas where water or steam carries heat from the deep hot regions to, or near, the surface. The water or steam is then available for capture and may be put to such uses as electricity generation and heating.

The interior of Earth is often considered to be divided into three major sections, called the crust, mantle, and core. The crust extends from the surface down to about 35 kilometers beneath the land surface and about 6 kilometers beneath the ocean. Below the crust, the mantle extends to a depth of roughly 2,900 kilometers. Below, or inside, the mantle is Earth's core. The crust is rich in radioactive materials, with a much lower density in the mantle and essentially none in the core. The radioactive decay of these materials produces heat. Earth is also cooling down, however. The volume of the mantle is roughly forty times that of the crust. The combination of the heat generated from the decay of radioactive materials and the cooling of Earth results in the flow of heat to Earth's surface. The origin of the total heat flowing to the surface is roughly 20 percent from the crust and 80 percent from the mantle and core.

The outermost shell of the planet, made up of the crust and upper mantle, is known as the lithosphere. According to the concept of plate tectonics, the surface of Earth is composed of six large and several smaller lithospheric regions or plates. On some of the edges of these plates, hot molten material extends to the surface and causes the plates to spread apart. On other edges, one plate is driven beneath another. There are densely fractured zones in the crust around the plate edges. A great amount of seismic activity occurs in these regions, and they are where large numbers of volcanoes, geysers, and hot springs are located. High terrestrial heat flows occur near the edges of the plates, so Earth's most important geothermal regions are found around the plate margins. A concentration of geothermal resources is often found in regions with a normal or elevated geothermal gradient as well as around the plate margins.

HISTORY OF DEVELOPMENT
The ancient Romans used the water from hot springs for baths and for heating homes. China and Japan also used geothermal waters for bathing and washing. Similar uses are still found in various geothermal regions of the world. Other uses of thermal waters were not developed until the early part of the nineteenth century. An early example occurred in the Larderello area of Italy. In 1827, Francesco Larderel developed an evaporation process that used the heat from geothermal waters to evaporate the thermal waters found in the area, leaving boric acid. Heating the water by burning wood had been required in the past.

Also in the early nineteenth century, inventors began attempting to utilize the energy associated with geothermal steam for driving pumps and winches. Beginning in the early twentieth century, geothermal steam was used to generate electricity in the Larderello region. Several other countries tried to utilize their own geothermal resources. Geothermal wells were drilled in Beppu, Japan, in 1919, and at The Geysers, California, in 1921. In the late 1920s, Iceland began using geothermal waters for heating. Various locations in the western United States have used geothermal waters for heating homes and buildings in the twentieth century. Among these are Klamath Falls, Oregon, and Boise, Idaho.

After World War II, many countries became interested in geothermal energy; geothermal resources of some type exist in most countries. Geothermal energy was viewed as an energy source that did not have to be imported and that could be competitive with other sources of electricity generation. In 1958, New Zealand began using geothermal energy for electric power production. One of the power plants in the United States began operation at The Geysers, California, in 1960. Mexico began operating its first geothermal power plant at Cerro Prieto, near the California border, in 1973.

By 2015, the United States was a leading country in electric power production from geothermal resources, with 3,548 megawatts of installed electrical capacity (28 percent of the world's total). As of that year, Costa Rica, El Salvador, Iceland, Kenya, and the Philippines had significant geothermal energy outputs that accounted for at least 15 percent of each country's energy production—as much as 51 percent in the case of Kenya, where it has become the main power source. Nonelectric uses of geothermal energy occur in most countries. In 2000, the leading nonelectric users of geothermal energy in terms of total usage were, in descending order, China, Japan, the United States, Iceland, Turkey, New Zealand, the Republic of Georgia, and Russia.

CLASSIFICATION OF GEOTHERMAL RESOURCES

Geothermal resources are classified by the temperature of the water or steam that carries the heat from the depths to, or near, the surface. Geothermal resources are often divided into low temperature (less than 90°C), moderate temperature (90°C to 150°C), and high temperature (greater than 150°C). There are still various worldwide opinions on how best to divide and describe geothermal resources. The class or grouping characterizing the geothermal resource often dictates the use or uses that can be made of the resource.

A distinction that is often made in describing geothermal resources is whether there is wet or dry steam present. Wet steam has liquid water associated with it. Steam turbine electric generators can often use steam directly from dry steam wells, but separation is necessary for the use of steam from wet steam wells. In various applications the water needs to be removed from wet steam. This is achieved through the use of a separator, which separates the steam gas from liquid hot water. The hot water is then re-injected into the reservoir; used as input to other systems to recover some of its heat; or, if there are not appreciable levels of environmentally threatening chemicals present, discharged into the environment after suitable cooling.

EXPLORATION

The search for geothermal resources has become easier in the twenty-first century than it was in the past because of the considerable amount of information and maps that have been assembled for many locations around the world and because of the availability of new instrumentation, techniques, and systems. The primary objectives in geothermal exploration are to identify geothermal phenomena, determine the size and type of the field, and identify the location of the productive zone. Further, researchers need to determine the heat content of the fluids that are to be discharged from the wells, the potential lifetime of the site, problems that may occur during operation of the site, and the environmental consequences of developing and operating the site.

Geological and hydrological studies help to define the geothermal resource. Geochemical surveys help to determine if the resource is vapor- or water-dominated as well as to estimate the minimum temperature expected at the resource's depth. Potential problems later in pipe scaling, corrosion, and environmental impact are also determined by this type of survey. Geophysical surveys help to define the shape, size, and depth of the resource. The drilling of exploration wells is the true test of the nature of the resource. Because drilling can be costly, use of previous surveys in selecting or siting each drill site is important.

ELECTRICITY GENERATION

The generation of electrical energy from geothermal energy primarily occurs through the use of conventional steam turbines and through the use of binary plants. Conventional steam turbines operate on fluid temperatures of at least 150°C. An atmospheric exhaust turbine is one from which the steam, after passing through the turbine, is exhausted to the atmosphere. Another form of turbine is one in which the exhaust steam is condensed. The steam consumption per kilowatt-hour produced for an atmospheric exhaust unit is about twice that for a condensing unit, but atmospheric exhaust units are simpler and cheaper.

The Geysers has one of the largest dry-steam geothermal fields in the world. Steam rises from more than forty wells. Pipes feed steam to the turbogenerators at a temperature of 175°C. Some of the wells are drilled to depths as great as 2,700 meters. The geothermal field at Wairakei on North Island of New Zealand has been a source of electric power for several decades. The hot water (near 300°C) rises from more than sixty deep wells. As the pressure falls, the hot water converts to steam. The flashing of hot water to steam is the major source of geothermal energy for electric power production.

Binary plants allow electricity to be generated from low- to medium-temperature geothermal resources as well as from the waste hot water coming from steam/water separators. Binary plants use a secondary working fluid. The geothermal fluid heats the secondary fluid, which is in a closed system. The working fluid is heated, vaporizes, drives a turbine, is cooled, condenses, and is ready to repeat the cycle. Binary plant technology is becoming the most cost-effective means to generate electricity from geothermal resources below 175°C.

In cascaded systems, the output water from one system is used as the input heat source to another system. Such systems allow some of the heat in waste water from higher temperature systems to be recovered and used. They are often used in conjunction with electric generation facilities to help recover some of the heat in the wastewater or steam from a turbine.

SPACE HEATING

Space heating by geothermal waters is one of the most common uses of geothermal resources. In some countries, such as Iceland, entire districts are heated using the resource. The nature of the geothermal water dictates whether that water is circulated directly in pipes to homes and other structures or (if the water is too corrosive) a heat exchanger is used to transfer the heat to a better fluid for circulation. Hot water in the range from 60°C to 125°C has been used for space heating with hot-water radiators. Water with as low a temperature as 35°C to 40°C has been used effectively for heating by means of radiant heating, in which pipes are embedded in the floor or ceiling. Another way of using geothermal energy for heating is through the circulation of heated air from water-to-air heat exchangers. Heat pumps are also used with geothermal waters for both heating and cooling.

In district heating, the water to the customer is often in the 60°C to 90°C range and is returned at 35°C to 50°C. The distance of the customers from the geothermal resource is important. Transmission lines of up to 60 kilometers have been used, but shorter distances are more common and desirable. When designing a district heating system, the selection of the area to be supplied, building density, characteristics of the heat source, the transmission system, heat loss in transmission, and heat consumption by customers are all important factors.

There are more than 600 geothermal wells serving a variety of uses in Klamath Falls, Oregon. Utilization includes heating homes, schools, businesses, and swimming pools as well as snow-melting systems for sidewalks and a section of highway pavement. Most of the eastern side of the city is heated by geothermal energy. The principal heat extraction system is the closed-loop downhole heat exchanger utilizing city water in the heat exchangers. Hot water is delivered at approximately 82°C and returns at 60°C.

Hot water from springs is delivered through pipes to heat homes in Reykjavík, Iceland, and several outlying communities. This is the source of heating for 95 percent of the buildings in Reykjavík. Hot water is delivered to homes at 88°C. The geothermal water is also used for heating schools, swimming pools, and greenhouses and is used for aquaculture.

GREENHOUSE HEATING

Using geothermal resources to heat greenhouses is similar to using it to heat homes and other buildings. The objective in this case is to provide a thermal environment in the greenhouse so that vegetables, flowers, and fruits can be grown out of season. The greenhouse is supplied with heated water, and through the use of radiators, embedded pipes, aerial pipes, or surface pipes, the heat is transferred to the greenhouse environment. Forced air through heat exchangers is also used. The United States, Hungary, Italy, and France all have considerable numbers of geothermal greenhouses.

AQUACULTURE

One of the major areas for the direct use of geothermal resources is in aquaculture. The main idea is to adjust the temperature of the water environment in a production pond so that freshwater or marine fish, shrimp, and plants have greater growth rates and thus reach harvest size more quickly. There are many schemes to regulate the temperature of the pond water. For supply wells where the geothermal water is near the required temperature, the water is introduced directly into the pond. For locations having a well-water temperature too high, the water is spread in a holding pool where evaporative cooling, radiation, and conductive heat loss to the ground can all be used to reduce the temperature to a level at which it can be added to the main production pond.

INDUSTRIAL APPLICATIONS

The Tasman Pulp and Paper Company, located in Kawerau, New Zealand, is one of the largest industrial developments to utilize geothermal energy. Geothermal exploration started there in 1952; it was directed toward locating and developing a geothermal resource for a pulp and paper mill. In 1985, the company was using four wells to supply steam to the operations. The steam is used to operate log kickers directly, to dry timber, to generate clean steam, and to drive an electricity generator. Geothermal energy supplies about 30 percent of the total process steam and 4 percent of the electricity for the plant. Geothermal energy in the form of steam is used to dry diatomaceous earth in Námafjall, Iceland. The diatomaceous earth is dredged from the bottom of a lake and pumped 3 kilometers by pipeline to a plant where it is dried.

Numerous other industrial applications of geothermal resources exist in the world. These range from timber drying in Japan to salt production from evaporating seawater in the Philippines, vegetable drying in Nevada, alfalfa drying in New Zealand, and mushroom growing in Oregon.

ENVIRONMENTAL IMPACT
The environmental impacts associated with the use or conversion of geothermal resources are typically much less than those associated with the use or conversion of other energy sources. The resource is often promoted as a clean technology without the potential radiation problems associated with nuclear energy facilities or the atmospheric emissions problems often associated with oil and coal electric plants. Nonetheless, although associated environmental problems are low, there are some present. In the exploration and development phases of large-scale geothermal developments, access roads and platforms for drill rigs must be built. The drilling of a well can result in possible mixing of drilling fluids with the aquifers intersected by the well if the well is not well-cased. Blowouts can also pollute the groundwater. The drilling fluids need to be stored and handled as wastes.

Geothermal fluids often contain dissolved gases such as carbon dioxide, hydrogen sulfide, and methane. Other chemicals, such as sodium chloride, boron, arsenic, and mercury, may also be associated with the geothermal water. The presence of these gases and chemicals must be determined, and appropriate means must be selected to prevent their release into the environment. In some cases this problem is reduced by the re-injection of wastewater into the geothermal reservoir.

The release of thermal water into a surface water body such as a stream, pond, or lake can cause severe ecosystem damage by changing the ambient water temperature, even if only by a very few degrees. Any discharge of hot water from the geothermal site needs to involve a means of cooling the water to an acceptable level—one that will not cause environmental damage. This result is often achieved through the use of holding ponds or evaporative cooling. The removal of large volumes of geothermal fluid from the subsurface can also cause land subsidence. This is irreversible and can cause major structural damage. Subsidence can be prevented by the re-injection of a volume of fluid equal to that removed.

Noise pollution is one of the potential problems with geothermal sites where electricity generation is conducted. Noise reduction can require costly measures. Because many geothermal electric generation sites are rural, however, this is often not a problem. The noise generated in direct heat applications is typically low.

ECONOMICS
The initial cost of a geothermal plant is usually higher than the initial cost of a similar plant that uses a conventional fuel. On the other hand, the cost of the energy for operating a geothermal plant is much lower than the cost of conventional fuels. In order to be economically superior, the geothermal plant needs to operate long enough to at least make up for the difference in initial cost.

Cascaded systems can be used to optimize the recovery of heat from the geothermal water and steam and therefore to decrease the overall costs. Systems can be cascaded such that the wastewater and heat from one is the input heat source to the next. An example is the cascading of systems used for electricity generation, fruit drying, and home heating. Finally, the distance between the geothermal source and the plant or user should be minimized, as there can be significant transmission losses in heat as well as high costs for pipe, pumps, valves, and maintenance.

ELECTRICITY: CURRENT AND FUTURE PROSPECTS
The United States leads the world in electrical generating capacity. The U.S. installed geothermal electrical generating capacity has moved from 2,534 megawatts in 2005 to 3,548 megawatts as of 2015. This U.S. generating capacity is spread over seven states but is concentrated in California, which has

2,760 MW of installed capacity. The Geysers, the United States' first geothermal energy site, was still home to the largest concentration of geothermal power plants in the world as of 2015. The other states with geothermal electrical generating capacity are Alaska, with 1 MW of installed capacity; Idaho, with 18; Hawaii, with 47 (supplying about 20 percent of the island's electrical needs); New Mexico, with 4; Nevada, with 580; Oregon, with 35; and Utah, with 77.

As of 2015 there were also a number of projects at least at stage one of development; that is, they had secured rights to the resource and had begun initial exploratory drilling. Many of the projects were farther along than that, with some in the facility construction and production drilling stage. In addition to expansions of geothermal capacity in Idaho, California, Alaska, Utah, Nevada, Oregon, and New Mexico, geothermal energy projects have begun or have been planned in Arizona, Colorado, Washington, Texas, North Dakota, Louisiana, Montana, Mississippi, and Wyoming.

Total worldwide geothermal power generation (based on installed capacity) rose from 8,933 megawatts in 2005 to 12.8 gigawatts in 2015, according to the Geothermal Energy Association. As of early 2015, the United States was the world's top generator, followed by the Philippines (1,870 megawatts), Indonesia (1,340 megawatts), Mexico (1,017 megawatts), New Zealand (1,005 megawatts), Italy (916 megawatts), Iceland (665 megawatts), Kenya (594 megawatts), Japan (519 megawatts), Turkey (397 megawatts), and fourteen more nations (producing fewer than 300 megawatts each).

DIRECT USE: CURRENT AND FUTURE PROSPECTS

More geothermal energy is directly used as thermal energy than is used to generate electricity, both in the United States and worldwide. Direct use of geothermal energy includes space heating (both district heating and individual space heating), cooling, greenhouse heating, fish farming, agricultural drying, industrial process heat, snow melting, and swimming pool and spa heating.

In the United States, installed capacity for direct use of geothermal energy increased from 7,817 megawatts in 2005 to 17,415 megawatts in 2015. The U.S. yearly direct use increased from 8,678 gigawatt-hours in 2005 to 21,074 in 2015. The greatest direct use for geothermal energy in the United States, by a wide margin, is geothermal heat pumps. Of the 2015 direct-use figures, 16,800 megawatts are for geothermal heat pumps. The 2015 U.S. capacities and yearly use rates for the other direct-use categories were as follows: individual space heating (140 megawatts, 1,360 terajoules per year); district heating (82 megawatts, 840 terajoules per year); cooling (2 megawatts, 48 terajoules per year); greenhouse heating (97 megawatts, 800 terajoules per year); fish farming (142 megawatts, 3,074 terajoules per year); agricultural drying (22 megawatts, 292 terajoules per year); industrial process heat (15 megawatts, 201 terajoules per year); snow melting (2 megawatts, 20 terajoules per year); and swimming pool and spa heating (113 megawatts, 2,557 terajoules per year).

The countries with the largest capacity for direct use of geothermal energy, as of 2015, included China, the United States, Sweden, Turkey, and Germany, in that order. Together, these five countries accounted for 65.8 percent of the global capacity. The countries with the largest annual use were China, the United States, Sweden, Turkey, and Japan, accounting for 63.6 percent of world use. However, some smaller countries, such as Iceland, have higher nonelectric geothermal energy capacity and use relative to the size of the population. Thailand, Egypt, India, South Korea, and Mongolia had the largest increases in nonelectric geothermal capacity between 2010 and 2015. A total of thirty-six countries have a direct-use geothermal energy capacity of over 100 megawatts.

Geothermal heat pumps are economical, energy efficient, and available in most places. They provide space heating and cooling and water heating. They have been shown to reduce energy consumption by 20 to 40 percent. In 2015, geothermal heat pumps accounted for more than two-thirds of worldwide direct-use geothermal capacity, and more than half of worldwide use.

Enhanced geothermal systems constitute an emerging technology. Most current geothermal systems use steam or hot water that is extracted from a well drilled into a geothermal reservoir.

Geothermal resources available for use can be expanded greatly, however, by using geothermal resources that do not produce hot water or steam directly but can be used to heat water to a sufficient temperature by injecting water into the hot underground region using injection wells and extracting it through production wells. The term "engineered geothermal system" is also used for this type of system. For this system, increasing the natural permeability of the rock may be necessary, so that adequate water flow in and out of the hot rock can be obtained. Estimates indicate that use of geothermal resources requiring enhanced geothermal systems would make more that 100,000 megawatts of economically usable generating capacity available in the United States. This is more than thirty times the 2009 U.S. geothermal generating capacity.

—*William O. Rasmussen and Harlan H. Bengtson*

Further Reading
Batchelor, Tony, and Robin Curtis. "Geothermal Energy." *Energy: Beyond Oil.* Ed. Fraser Armstrong and Katherine Blundell. Oxford UP, 2007.
Bundschuh, Jochen, and Barbara Tomaszewska, eds. *Geothermal Water Management.* CRC Press, 2018.
Dickson, Mary H., and Mario Fanelli, eds. *Geothermal Energy: Utilization and Technology.* 1995. Reprint. Earthscan, 2005.
DiPippo, Ronald. *Geothermal Power Generation: Developments and Innovation.* Woodhead Publishing/Elsevier, 2016.
Glassley, William E. *Geothermal Energy: Renewable Energy and the Environment.* 2nd ed. CRC, 2015.
Gupta, Harsh K., and Sukanta Roy. *Geothermal Energy: An Alternative Resource for the Twenty-first Century.* Elsevier, 2007.
Ismail, Basel I., ed. *Advances in Geothermal Energy.* IntechOpen, 2016.
Lund, John W. "Characteristics, Development and Utilization of Geothermal Resources." *Geo-Heat Center Quarterly Bulletin*, vol. 28, no. 2, 2007.
Lund, John W., and Tonya L. Boyd. "Direct Utilization of Geothermal Energy 2015 Worldwide Review." *Proceedings of the World Geothermal Conference 2015.* Melbourne, Australia. International Geothermal Association, 2015. PDF file.
Matek, Benjamin. *2015 Annual US & Global Geothermal Power Production Report.* Geothermal Energy Association, 2015. PDF file.
McCaffrey, Paul, ed. *U.S. National Debate Topic, 2008-2009: Alternative Energy.* Wilson, 2008.
Simon, Christopher A. "Geothermal Energy." *Alternative Energy: Political, Economic, and Social Feasibility.* Rowman, 2007.
Slack, Kara. *U.S. Geothermal Power Production and Development Update.* Geothermal Energy Association, 2008.

Geothermal Energy

FIELDS OF STUDY
Geothermal Engineering; HVAC Trades Seismology; Thermodynamics

ABSTRACT
Geothermal energy is the heat energy that exists deep inside the Earth. Geothermal energy is observed in many places around the world including active volcanoes, naturally occurring hot springs, and oil wells. Geothermal technologies aim to harness this emerging thermal heat to prevent it from dissipating into the atmosphere so that it can be used to generate electricity and for other applications.

KEY CONCEPTS
feed-in tariff: a rated amount paid back to energy consumers who also produce energy through the installation of geothermal, solar or other type of renewable energy and deliver their excess output into the general supply
heat pump: the central component of a heating/cooling system that uses the constant heat content of a large mass such as Earth's soil below the depth affected by surface temperature variations. Heat pumps use a cooled liquid to accept heat from the depths, then recapture that heat above ground by passing the heated liquid through a heat exchanger that functions like an air conditioner to recover the heat and cool the liquid. Reversing the process allows excess heat to be taken from above ground to be dissipated below ground.
joule: defined electrically as the work done by a current of one ampere through a potential difference of one volt
magma: molten rock within Earth's mantle layer; magma that has extruded onto the surface through fissures or volcanic eruption is called "lava"

radioactive decay: the spontaneous fission, or splitting apart, of the nuclei of certain unstable atoms, accompanied by the release of the corresponding quantity of the nuclear-binding energy of those atoms

watt: defined as one joule per second

EARTH'S HOT HEART

Contained deep beneath the crust, Earth's thermal energy is directed toward a heat sink medium, such as magma, that holds the energy within its mass. Subsequently, pressure created from the heat and hot gases becomes great enough that magma rises to the surface through cracks and fissures in the crust, in volcanic eruptions. Similarly, groundwater can act as a heat sink, giving rise to hot springs that emerge at the surface.

Geothermal technologies aim to harness this emerging thermal heat to prevent it from dissipating into the atmosphere so that it can be used to generate electricity and for other applications. The direct application of geothermal energy as heat, predominantly for space and water heating, worldwide represents the greater percentage of geothermal use, greater than electricity generation. This may be partly because lower temperatures from shallower ground can be exploited for space heating, and there is minimal energy loss because of direct application of the harnessed thermal heat without converting it into electricity or any other form of energy. Geothermal heat is also considered to be a renewable resource, because it occurs continuously as a result of the natural radioactive decay of minerals in Earth's interior. The thermal flow through porous rock to Earth's surface is therefore sustainable and does not involve artificial enhancement or modification for its exploitation.

The extent of geothermal energy resources in Earth is summarized by researcher Ladislaus Rybach, in his 2003 article "Geothermal Sustainability."

The ultimate source of geothermal energy is the immense heat stored within Earth; 99 percent of Earth's volume has temperatures greater than 1000°C (1830°F), with only 0.1 percent at temperatures less than 100°C (212°F) The total heat content of Earth is estimated to be about 1013 EJ (Exajoule=10^{18} joules) and it would take over 109 years to exhaust it through today's global terrestrial heat flow of 40 million megawatts (Megawatt thermal). The internal heat of Earth is mainly provided by the decay of naturally radioactive isotopes, at the rate of 860 EJ/yr—about twice the world's primary energy consumption (443 EJ in 2003). Thus, the geothermal resource base is sufficiently large and basically ubiquitous.

TAKING ADVANTAGE OF THE ENERGY

Utility-scale electricity is generated from the hot water and steam that emerge from Earth. Electricity generation requires high geothermal temperature resources, and since temperatures are higher at deeper depths within Earth, suitable natural hot water conduits for this purpose can only come from very deep underground. Of the total geothermal energy resources, only a fraction reach Earth's surface through natural fluid conduits. Apart from this, a vast amount of geothermal energy is still trapped under Earth's dry and nonporous crust of dense rock. In order to stimulate geothermal resources in hot, dry rock, artificial wells are dug several miles deep to permit electricity generation through enhanced geothermal techniques. This involves pumping cold water at high pressure down an injection well into the hot, dry rock. This induced fluid pressure creates fractures in the rock, allowing the water to absorb the heat as it passes through these fractures before being forced out as very hot water and steam through an outlet borehole. A steam turbine or other power plant system is used to convert the steam into electricity. Such enhanced geothermal systems are under development in many countries to produce utility-scale electricity. However, an enhanced geothermal electric project in Basel, Switzerland, was curtailed in December 2009 after a seismic hazard evaluation was conducted; a 7.9 magnitude earthquake occurred at a similar project in Baja California, reportedly the result of induced seismic activity caused by building the project.

Even so, geothermal electricity-generating plants are becoming a popular choice for generating electricity in many countries. The increase in geothermal electric capacity in developing countries has been from 75 to 1,495 megawatts electrical, showing a rate of increase of 500 percent between 1975 and 1979 and 223 percent between 1980 and 1984, respectively, with a further increase of 150 percent between 1984 and 2000, as noted by M. H. Dickson and M. Fanelli of the International Institute for Geothermal Research. In 2001 the electric energy produced from geothermal resources represented 27% of the total electricity generated in the Philippines, 12.4 percent in Kenya, 11.4 percent in Costa Rica, and 4.3 percent in El Salvador.

Nonelectric uses of geothermal energy are mainly for space and district heating, and some other minor applications, such as in greenhouses, for aquaculture, and for some industrial processes. Geothermal heat pumps are used for space heating and use a heat exchanger in the form of pipes buried in the ground or submerged into a water body, through which a heat exchange fluid with antifreeze properties is circulated. The fluid absorbs heat from the ground and through the heat pump, transfers the heat to a water reservoir that supplies underfloor heating systems in buildings. The cooled fluid then returns to the ground through the closed-loop system to reabsorb heat and bring it to the surface. In summer, the direction of flow can be reversed, with the cooled water flowing through the building, absorbing heat and returning it to the cooler ground outside. Since the belowground temperatures are warmer in winter and cooler in summer than the outdoor ambient air temperatures, higher heating and cooling efficiencies can be achieved with lower energy use. This heat pump technology makes geothermal heating economically viable in any geographic location.

Larger geothermal district heating systems are capital intensive, requiring a substantial initial investment for production and injection wells, down-hole and transmission pumps, pipelines and distribution networks, monitoring and control equipment, peaking stations, and storage tanks. Operating expenses, however, may be comparatively lower than those of conventional systems and consist of pumping power, system maintenance, control, and management. A critical factor in assessing the feasibility of large-scale geothermal heating applications is the thermal load intensity or heating demand per unit area of the district. Higher heating demand indicates greater economic feasibility because of a potentially greater number of users and subscribers for the service. Systems that combine both heating and cooling are more economically feasible, as the load factor in a system with combined heating and cooling is higher than the factor for heating alone, and the cost to consumers per unit heating or cooling would be cheaper than for a single application.

Greenhouses and aquaculture (fish farming) are the two primary uses of geothermal energy in the agribusiness industry. Geothermal water is useful in any industry that requires steam or hot water. Some uses include timber processing, pulp and paper processing, washing wool, dyeing cloth, drying diatomaceous earth (a light, abrasive soil used as a filtering material and insecticide), drying fish meal and stock fish, canning food, drying cement, drying organic materials (such as vegetables, seaweed, and grass), and refrigeration. Using geothermal water and steam saves the economic cost of heating water and reduces the environmental pollution that would be caused by burning the fuel for heating the water. Most greenhouse operators estimate that using geothermal resources instead of traditional energy sources saves about 80 percent of fuel costs—about 5 to 8 percent of total operating costs.

GEOTHERMAL AS A RENEWABLE ENERGY SOURCE

In recent years, governments around the world have been promoting renewable energy technologies to address some of the future energy demand. Renewable energy also enables democratization of energy resources among small and large producers in every country and is a preferred policy in the light of recurring energy security issues for governments. Hence, to incentivize small communities, individuals, businesses, and others to install and produce electricity and heat from renewable resources, feed-in tariffs are offered for each kilowatt of energy produced from geothermal, solar, wind, and other renewable resources. Schemes differ from technology to technology and country to country, but many producers are finding it worthwhile to invest in such projects with a view to using free energy in the long term and having the payback period reduced through such government incentives.

—*Swati Ogale*

Further Reading

Boden, David R. *Geologic Fundamentals of Geothermal Energy*. CRC Press, 2017.

Dickson, Mary H., and Mario Fanelli. *Geothermal Energy: Utilization and Technology*. United Nations Educational, Scientific and Cultural Organization, 2003.

Glanz, James. "Quake Threat Leads Swiss to Close Geothermal Project." *New York Times*, 10 Dec. 2009.

http://www.nytimes.com/2009/12/11/science/earth/11basel.html?_r=2.

Glassley, William E. *Geothermal Energy. Renewable Energy and the Environment.* 2nd ed. CRC Press, 2015.

International Geothermal Association. "What Is Geothermal Energy?" http://www.geothermal-energy.org/314,what_is_geothermal_energy.html.

Manzella, Adele, Agnes Allansdottir, and Anna Pellizzone, eds. *Geothermal Energy and Society.* Springer, 2019.

National Geothermal Collaborative. "Geothermal Direct Use." http://www.geocollaborative.org/publications/Geothermal_Direct_Use.pdf.

Rosen, Mark A., and Seama Koohi-Fayegh. *Geothermal Energy: Sustainable Heating and Cooling Using the Ground.* Wiley, 2017.

Rybach, Ladislaus. "Geothermal Sustainability." *Geothermics*, vol. 32, nos. 4-6, 2003, pp. 463-70.

Gibbs, Josiah Willard

FIELDS OF STUDY
Physics; Thermodynamics

ABSTRACT
The "father of vector analysis," Josiah Gibbs made thermodynamics mathematically operational. His study of energy relationships in physical systems, especially the relationship of enthalpy and entropy, resulted in the "Gibbs free energy" concept, a difference between enthalpy and temperature-dependent entropy.

KEY CONCEPTS
degrees of freedom: the maximum number of states of a system according to the number of independent variables (e.g., temperature, pressure, volume, etc.) operating on the system

enthalpy: the energy of a system expressed as heat

entropy: essentially, the degree of randomness within a system

heterogenous: a system consisting of two or more phases, such as solid and liquid, etc.

matrix: an array of numerical factors that can be manipulated mathematically to solve a number of equations simultaneously

thermodynamics: the science of the movement of heat energy

BACKGROUND
Josiah Willard Gibbs was born in New Haven, Connecticut, on February 11, 1839, the only son in a family of four girls. His mother, Mary Anna Van Cleve, was an amateur ornithologist and contributor to his scientific interests. His father, the Reverend Doctor Josiah Willard Gibbs, was an exemplary professor of biblical or sacred literature at Yale College (1824-61). As a child, Josiah survived scarlet fever, but it left him vulnerable to illness for the remainder of his life. He lost both of his parents and two of his sisters by the end of his undergraduate days.

Gibbs's elementary education was in local schools in New Haven. He was a decorated Latin student prior to entering Yale in 1854, from which he re-

Josiah Willard Gibbs. (Wikimedia Commons)

ceived a BA at the age of 19. In 1863, Gibbs graduated with the first doctorate in engineering, which was one of the first doctorates of any kind awarded by an institution of higher learning in America. His dissertation was on spur gear design. Between 1863 and 1866, Gibbs tutored in Latin. In 1866, he received a patent for his redesign of railroad car brakes. That same year, he went to Europe for postdoctoral work. He heard lectures at institutions in Paris, Berlin, and at the University of Heidelberg, which was one of the most advanced centers for the study of physics and chemistry.

THE SCIENTIST
The royalties Gibbs received from his patent paid for his travels. It also paid his living at Yale College, to which he returned in 1869. In 1871, he was appointed professor of mathematical physics without pay because he was still unpublished. In 1873, Gibbs published "Graphical Methods in the Thermodynamics of Fluids" in the *Transactions of the Connecticut Academy*. In 1877, Gibbs founded the Yale Math Club and was its leader for the next 10 years. His studies in multiple algebra and matrix studies may have been presented there. These studies of multiple variants and how to solve for them were preparing the way for his development of vector analysis.

By 1878, Gibbs had published seven papers that were collectively entitled *On the Equilibrium of Heterogeneous Substances*. In his papers, Gibbs applied the principles of thermodynamics to chemicals. Thermodynamics, a branch of physical science, studies heat and its relation to other forms of energy and work as they affect temperature, pressure, and volume in physical systems. Gibbs's study of the thermodynamics of chemical reactions produced dynamic new knowledge.

Gibbs created graphs to plot entropy, temperature, and pressure in relation to volume. His work put the first and second law of thermodynamics on entropy and mechanical energy into mathematically comprehensible rules that were scientifically useful. Gibbs had laid the foundations for physical chemistry with his phase rule. The phases refer to changes of an object from solid to liquid or gas. The rule stated as a formula is $f = n + 2 - r$.

In the formula, "f" is the degrees of freedom in temperature and pressure, "n" is the total number of chemical elements involved, and "r" is the number of phases. Others would take Gibbs's work with a single system method and turn it into a Gibbsian ensemble that is foundational to statistical mechanics. Using his ensembles, researchers can avoid a lot of trial and error experimentation when synthesizing new compounds or alloys.

From the 1880s, Gibbs taught vector analysis (vector calculus) to his students. His vector analysis went beyond representation of a property, such as heat, with a scale representing the flow of heat or representing physical quantities such as displacement, velocity, acceleration, force, impulse, momentum, torque, and angular momentum as vectors. Gibbs also applied his vector analysis to astronomy in order to improve the descriptions of the orbits of the planets and comets. He eventually put together an 85-page outline that he sent to Oliver Heaviside in 1888. Heaviside used the work, credited Gibbs when he did, but noted that it was too condensed to be used.

By the 1890s, Gibbs's publications were being translated into French and German and published in European scientific journals. In 1901, he was awarded the Royal Society's Copley Medal. In 1901, Gibbs was heavily involved in writing *Elementary Principles in Statistical Mechanics*, so he was unable to write on vector analysis for the Yale University Bicentennial.

Dean A. W. Phillips at Yale had a graduate student expand Gibbs's vector analysis class notes into a book. Edwin Bidwell Wilson took Gibbs's course on vectors and wrote *Vector Analysis: A Text-Book for the Use of Students of Mathematics and Physics, Founded Upon the Lectures of J. Willard Gibbs Ph.D. LL.D*. Gibbs spent his latter adult years in the home of his sister, Julia, who married Addison Van Name, Yale's librarian. A member of the Connecticut Academy, Addison aided Gibbs in getting his articles published by the Connecticut Academy. Gibbs died at home on April 28, 1903.

—*Andrew J. Waskey*

Further Reading
Bumstead, H.A., and R.G. Van Name, eds. *The Scientific Papers of J. Willard Gibbs*. 2 vols. Longmans, Green and Co., 1906.
Crowther, J.G. *Famous American Men of Science*. Books for Libraries Press, 1969.
Gibbs, J. Willard. *The Collected Works of J. Willard Gibbs*. [1906]. Yale UP, 1948.

Gibbs, J. Willard. *Elementary Principles in Statistical Mechanics Developed with Special Reference to the Rational Foundation of Thermodynamics.* Reprint. Creative Media LLC, 2019.

Rukeyser, Muriel. *Willard Gibbs.* Ox Bow Press, 1988.

Wheeler, Lynde Phelps. *Josiah Willard Gibbs: The History of a Great Mind.* Ox Bow Press, 1998.

GREEN BUILDINGS

FIELDS OF STUDY
Architecture; Environmental Studies

ABSTRACT
Residential and commercial buildings generate more than 30 percent of the world's emissions of carbon dioxide, a greenhouse gas that has been linked to global warming. Green buildings are structures designed and constructed to increase resource efficiency and reduce negative impacts on human health and the environment, reduce carbon emissions and provide environmental benefits by reducing solid waste, efficiently using energy and other resources, reducing air and water pollution, and conserving natural resources.

KEY CONCEPTS
roof garden: a roofing design that uses a significant layer of soil as its outermost layer, or to at least include raised garden beds, providing an insulating and water absorbing layer while also allowing the roof to be used as a garden space

sick building syndrome: a characteristic of buildings that are constructed with poor ventilation, insulation, lighting, etc., that causes them to be environments that are not conducive to the proper health of the people who live and work in them

TYPICAL HOMES AND BUILDINGS
Residential and commercial buildings generate 38 percent of the total carbon dioxide emission in the United States. Green buildings reduce carbon emissions substantially and provide significant environmental benefits by reducing solid waste, efficiently using energy and other resources, reducing air and water pollution, and conserving natural resources.

Although energy efficiency and sustainability were not major concerns at the time, early green buildings originated during the mid-nineteenth century. The Galleria Vittorio Emanuele II in Milan, Italy, designed in 1861, and the Crystal Palace in London, England, built in 1851, both used underground air cooling and roof ventilators to control the interior temperature.

From the 1930s to the 1960s technological advances such as the inventions of reflective glass, structural steel, and air-conditioning resulted in the proliferation of high-rise buildings that consumed huge amounts of cheap fossil fuels. During the 1960s, however, environmental consciousness grew, and visionaries began defining the "green building." During this period scientist James Lovelock formulated the Gaia hypothesis, a holistic concept of Earth as a single, complex organism. In 1969 landscape architect Ian L. McHarg published *Design with Nature,* which helped define green architecture.

BEGINNINGS OF THE MOVEMENT
On the first Earth Day, in April 1970, millions of Americans showed their concern about the environment. The 1973 and 1979 oil crises demonstrated the need for the nation to seek energy from diversified sources and become less dependent on fossil fuels. The U.S. government and many corporations began investing in research into methods of energy conservation and alternative energy sources.

During the 1980s architect Malcolm Wells designed green underground and earth-sheltered buildings. In

Earth Day activists. (Volodymyr Kryshtal via iStock)

1982, physicist Amory Lovins and his wife, environmentalist Hunter Lovins, emphasized the basic green principle of using regional resources in founding their Rocky Mountain Institute, a nonprofit resource policy center that promotes resource efficiency and global security. Beginning during the mid-1980s, popular environmental organizations— such as the Sierra Club, Greenpeace, the Nature Conservancy, and Friends of the Earth—became increasingly active. Growing awareness of the problem of sick building syndrome raised concerns regarding the indoor environments of some workplaces. In 1984 architect William McDonough designed a headquarters building for the Environmental Defense Fund in New York City using a high-performance building approach (the building was completed in 1985). During the late 1980s Pliny Fisk III designed Blueprint Farm—a green agricultural community—in Laredo, Texas, using recycled materials, wind power, and photovoltaic panels.

MILESTONES DURING THE 1990S

In 1992, the first local green building program began in Austin, Texas, and the U.S. Environmental Protection Agency (EPA) launched the Energy Star program, a voluntary energy-efficiency labeling program for consumer products. By 2012, Energy Star labels were appearing on 65 product categories and Energy Star ratings had become the standard for major appliances, homes, commercial buildings, and heating systems. By 2012, 1.4 million Energy Star-qualified homes had been built throughout the United States. Many other countries adopted the Energy Star idea, including Japan, Taiwan, China, New Zealand, South Africa, and the nations of the European Union.

In 1993, U.S. Green Building Council (USGBC), Bill Clinton's presidential administration began the successful "Greening of the White House" initiative, and the nonprofit U.S. Green Building Council (USGBC) was created to promote the construction of environ-

Crowd at Earth Day celebrations in Montreal (2012). (MmeEmil via iStock)

Example of a Green building complex in Japan. (ahei via iStock)

mentally responsible, healthy, and profitable buildings. USGBC is a national, voluntary consensus coalition with members from all sectors of the building industry. In 1995, USGBC began developing its green building certification program, known as LEED (for Leadership in Energy and Environmental Design), which became available for public use in 2000. This voluntary system provides third-party certification that certain standards have been met in the construction of high-performance, sustainable buildings, with an emphasis on reducing carbon dioxide emissions and increasing energy efficiency. LEED certification covers a wide range of existing and new commercial and residential buildings, including offices, schools, medical facilities, private homes, and stores.

ENVIRONMENTAL BENEFITS

The key areas measured in the LEED certification process reflect the environmental benefits of green building. Sustainable site development involves preserving natural resources for future generations and can include reusing existing buildings, planting around buildings, roof gardens, and underground or earth shelters. Building for water savings and efficiency involves monitoring water supplies and usage, recycling gray or previously used water, and constructing rainwater catchment systems. To improve energy and atmosphere efficiency, buildings can use geographically and climatically appropriate energy resources, including renewable energy. Efforts to conserve materials and resources include using renewable, recycled, local, chemical-free, nonpolluting, and durable materials. Indoor environmental quality can be improved through the use of nontoxic materials, adequate ventilation and insulation, energy-efficient temperature controls, and materials that emit few or no volatile organic compounds.

In the twenty-first century, as environmental knowledge and building technologies continue to improve, the green building movement is gaining worldwide momentum. Given that commercial and residential buildings generate more than 38 percent of carbon dioxide emissions and represent 68 percent of total electricity consumption in the United States, the benefits of green building have become increasingly obvious. By 2015 more than forty countries had developed their own LEED initiatives, including Australia, Brazil, Canada, France, India, Israel, Mexico, the United Arab Emirates, and the United Kingdom. Green buildings offer a number of economic benefits in the long term, including reduced costs for heating, cooling, and electricity.

—Alice Myers

Further Reading

Ching, Francis D.K., and Ian M. Shapiro. *Green Building Illustrated.* Wiley, 2014.

Fisanick, Christina, ed. *Eco-architecture.* Greenhaven Press, 2008.

GreenSource Magazine. *Emerald Architecture: Case Studies in Green Building.* McGraw-Hill, 2008.

Johnston, David, and Scott Gibson. *Green from the Ground Up: A Builder's Guide—Sustainable, Healthy, and Energy-Efficient Home Construction.* Taunton, 2008.

Kao, Jimmy C.M., Wen-Pei Sung, and Ran Chen, eds. *Green Building, Materials and Civil Engineering.* CRC Press, 2015.

Kruger, Abe, and Carl Seville. *Green Building: Principles and Practices in Residential Construction.* Delmar, 2013.

U.S. Environmental Protection Agency. "Sources of Greenhouse Gas Emissions: Commercial and Residential Sector Emissions." *EPA.gov.* Environmental Protection Agency, 17 Apr. 2014. Web. 30 Jan. 2015.

"Why Build Green?" *EPA.gov.* Environmental Protection Agency, 9 Oct. 2014. Web. 30 Jan. 2015.

Yudelson, Jerry. *The Green Building Revolution.* Island, 2008.

Green Energy Certification

FIELDS OF STUDY
Environmental Management; Marketing; Policy Studies

ABSTRACT
The terms "green energy certification" and "renewable energy certification" can refer to various certification schemes that confirm or reward the efficient or renewable use of energy in a given application.

KEY CONCEPTS

biodiesel: fuel for use in diesel engines, produced by processing vegetable oils and animal fats through a transesterification process and refining the product for quality

commodify: to use the certifications provided as part of a program as a tradable commodity

green energy: energy resources and practices that function to reduce energy consumption and the environmental impact of energy-consuming products

renewable energy: energy resources that replenish themselves naturally and so are not considered finite resources that will be depleted through consumption

renewable identification number (RIN): an identification and tracking number assigned to a specific identifiable batch or production run of biofuel, typically ethanol or biodiesel; the RIN is provided by and registered with the organization or branch of government that oversees the biofuel industry, and must be reported to that body when the product is used in preparing a commercial fuel blend

REWARDING ENERGY CONSERVATION

Renewable energy certificates, also known as green certificates in Europe, are tradable intangible commodities representing the generation of a megawatt-hour of electricity from a renewable energy source. They are used to meet renewable energy goals and requirements, and by commodifying renewable energy, they allow electric companies that need to meet those goals to avoid the need to build their own renewable energy infrastructures. Other green certification schemes verify that a project, building, or appliance meets certain energy-use requirements.

WHITE CERTIFICATES

White certificates certify a particular reduction of energy consumption and, like green certificates, are tradable commodities used to meet reduction goals and requirements. Energy producers that exceed their reduction goals, therefore, have a surplus of white certificates, which they may sell to other producers to put toward their own goals. White certification is a newer scheme than green certification, and it is not tied specifically to the use of renewable energy but rather to energy efficiency. So far, it has been less widely adopted; several European countries have adopted white certificate programs, and in the United States, Connecticut, Nevada, and Pennsylvania have included white certificates in their energy efficiency portfolio standards.

Essentially, the requirement on energy utilities is to meet a particular percentage of their energy needs through reducing consumption—either by somehow convincing their customers to do so (through energy efficiency awareness programs and the like) or by purchasing white certificates. Part of the innovation here is in putting the responsibility for energy efficiency on the utility's shoulders instead of simply relying on consumers to be well-informed and act in their self-interest to keep their bills down. The requirement also encourages utilities to cooperate with customers installing renewable energy generation systems at home.

BIOFUEL TRACKING

As per the Energy Policy Act of 2005, the Environmental Protection Agency (EPA) oversees the production and use of biofuels—whether ethanol or biodiesel—and assigns each batch of produced biofuel a renewable identification number (RIN), which is used for tracking purposes. The EPA is also responsible for setting each year's quota for the percentage of American motor fuel consumption that must be made up of biofuel blends. Companies demonstrate their compliance with those quotas by submitting to the EPA a list of the RINs that they used in blending their gasoline products, which is then checked against the EPA's database for verification.

THE ENERGY STAR PROGRAM

An increasingly well-known green energy certification program is the Energy Star program, developed in the 1990s by the EPA but now used by other countries as well, including the European Union and Japan. The Energy Star logo may be displayed by products that meet a specific efficiency standard, determined by the product type; the logo is not available for all products but has expanded to a vast array of them

Energy Star Logo. (Wikimedia Commons)

since its introduction. Generally the appearance of the logo on a product indicates that the product uses between 15 and 30 percent less energy than the average amount used for that product type. The first products eligible for Energy Star logos were computers and printers, which in the mid-1990s were experiencing a surge in usage and sales and were thus becoming a more significant contributor to energy consumption. Since 2009, the Energy Star 5.0 specifications have been in effect for computers, requiring more efficient power supplies. That same year, the first Energy Star specifications were published for stand-alone computer servers.

Energy Star appliances generally use 20 percent less energy than the standard, but specifics vary wildly. Dishwashers use 41 percent less, for instance, whereas room air conditioners are only 10 percent more efficient and cordless telephones and battery chargers are 90 percent more efficient.

Energy Star-certified homes are designed to use 15 percent less energy than a standard home built according to the 2004 International Residential Code. Given the gains of green building and sustainable design and the inefficiency found in so much home construction, this is an easy target to shoot for; it can be attained with tighter and better-insulated construction, Energy Star-certified lighting and appliances (including the water heater, heating system, and air-conditioning system), and high-performance windows.

The Energy Star program has numerous shortcomings, however. The most obvious is that it considers only the energy consumed by the product's usage—not the efficiency of its manufacture, transport, marketing, recycling, or other elements of the product's life cycle. The Energy Star program is essentially geared to appeal to consumers by promising products that will cost them less to use, but this is not coequal with products that are less energy-intensive over the course of their lifetimes. Certainly, there is overlap, but for products that remain in consumer possession for a short period of time, such as iPods, cellular phones, many computers, and especially computer peripherals, the energy consumption of operations is a smaller percentage of the product's total life-cycle energy cost. Furthermore, achieving efficiency gains may sacrifice other aspects of the product: It may shorten the product's life span, which over the long term can negate the efficiency gains by requiring that a greater number of units of the product be manufactured. It may also offer reduced functionality in some aspect, such as limiting the depths of cold that a freezer can achieve. Such issues can pose serious long-term problems if they lead consumers to associate the Energy Star logo—or, more disastrously, energy efficiency in general—with poor quality.

In some cases, the Energy Star standard simply does not seem high enough to merit special attention. The standard set for Energy Star homes is easily achieved by most new construction without special painstaking; arguably, the standard should represent not a special achievement but rather a requirement

The U.S. Dept of Energy: appliance and equipment standards are saving consumers and businesses billions of dollars. (U.S. Dept of Energy)

imposed on all new construction, particularly since it puts no limit on the activity of the residents or their consumption. (It puts no limit on the use of air conditioning, for instance, as long as the air-conditioning unit itself uses electricity efficiently.)

In other cases, the circumstances under which the product applying for certification is tested are problematic. A 2008 audit of the program found that the reported energy savings of many certified products were unreliable or had been measured improperly. For instance, manufacturers of refrigerators test their appliances with the ice makers—which cause a major drain on power—turned off, which does not reflect how the product is intended to be used.

The practice of basing a product's certification on its performance relative to other products in its category amounts to grading on a curve. Certain product categories are simply inherently inefficient: side-by-side refrigerators, nearly all air conditioners, room air conditioners that are not expertly installed in well-insulated homes, laundry dryers, and so forth. Other products, like dishwashers or washing machines, may be designed for better energy efficiency but still consume inefficient amounts of water.

OUTSIDE THE UNITED STATES

The EKOenergy label is managed by the Finnish Association for Nature Conservation, which has operated it since 1998, and applies to electricity generated in Finland, Denmark, Norway, and Sweden. The label is granted when a company meets EKOenergy criteria for the promotion of energy conservation and the use of sustainable, renewable energy sources such as solar, wind, biofuels, and hydropower. Hydropower sources must already exist and must have been constructed before 1996 (the Finnish are concerned with the ecological impact of further hydropower development), and wind farms must not be located in nature reserves or on culturally significant sites. Because residential and business customers can choose from numerous electricity providers in Finland, the EKOenergy label helps to make energy efficiency and renewable energy competitive practices.

Future Energy was a green electricity accreditation program in the United Kingdom that ran from 1999 to 2002, but it was allowed to expire for funding reasons. Green advocates in the United Kingdom continue to press for a resumption or replacement of the program.

—Bill Kte'pi

Further Reading

Cottrell, Michelle. *Guidebook for the LEED Certification Process.* Wiley, 2011.

Enescu, Diana. *Green Energy Advances.* IntechOpen, 2019.

Mulder, Machiel. *Regulation of Energy Markets: Economic Mechanisms and Policy Evaluation.* Springer, 2021.

Mulvaney, Dustin, and Paul Robbins, eds. *Green Energy: An A-to-Z Guide.* SAGE, 2011.

Reeder, Linda. *Guide to Green Building Rating Systems: Understanding LEED, Green Globes, Energy Star, the National Green Building Standard, and More.* Wiley, 2010.

U.S. Congress. Senate. Committee on Homeland Security and Governmental Affairs. *Energy Star Program: Covert Testing Shows the Energy Star Program Certification Process Is Vulnerable to Fraud and Abuse.* U.S. Government Accountability Office, 2010.

Wimberly, Jamie. *Energy Star Shining Bright? National Consumer Survey of the Energy Star Brand.* Ecoalign, 2010.

Greenhouse Gases and Human Industry

FIELDS OF STUDY
Atmospheric Science; Environmental Science

ABSTRACT
Carbon dioxide, methane, nitrogen oxides, sulfur hexafluoride, chlorofluorocarbons, and water vapor are the most important greenhouse gases. Over the past two centuries they have been emitted in increasing amounts as a result of human (industrial) activities.

KEY CONCEPTS

anthropogenic: refers to conditions and consequences that are caused by human activities

greenhouse effect: the environmental condition in which atmospheric gases absorb infrared radiation, or heat energy, emanating from Earth's surface and then emit the absorbed energy back into the atmosphere as heat, thus maintaining an elevated atmospheric temperature

greenhouse gas: any atmospheric gas capable of absorbing heat energy emanating from Earth's surface and subsequently acting to return that heat energy into the atmosphere

infrared radiation: electromagnetic waves having wavelengths longer than those of the visible red color, associated with and detectable as heat

tonne: a mass of 1,000 kilograms, also called a "metric ton," equivalent to 2,200 pounds or 1.1 standard tons

TAKING EARTH'S TEMPERATURE

Over the past century, ongoing measurements demonstrate that the temperature on Earth has increased by about 0.75°C. As generally agreed among scientists, this increase tied to greater atmospheric concentration of key greenhouse gases (GHGs)—including carbon dioxide, methane, and nitrous oxide, as well as sulfur hexafluoride and chlorofluorocarbons—is due to human activities, and is considered to be a direct cause of the observed climate changes.

Levels of several important GHGs have risen by about 25 percent since large-scale industrialization began around 150 years ago. Since the early 1990s, about three-quarters of anthropogenic (human-induced) emissions have come from the burning of fossil fuels. Concentrations of carbon dioxide in the atmosphere are naturally regulated by numerous processes, which together are designated the carbon cycle. Human impact on the global climate since the Industrial Revolution has been difficult to interpret, given our incomplete understanding of how some factors operate and interact with changes in surface temperature. Also, while emissions of GHGs such as carbon dioxide and methane have had a net warming effect, emissions of sulfate aerosols have had a net cooling effect.

The movement of carbon between the atmosphere, land, and oceans is dominated by natural processes such as plant photosynthesis and weather. Although these processes can absorb some of the net 6.2 billion tonnes of anthropogenic carbon dioxide emissions produced each year—that is, 7.2 billion tonnes less approximately 1 billion tonnes reabsorbed by what are termed carbon sinks (areas such as forests that absorb carbon dioxide), as measured in carbon equivalent terms—an estimated 4.1 billion tonnes are added to the atmosphere annually. This positive imbalance between GHG emissions and absorption results in the continuous net increase in atmospheric concentrations of GHGs.

EARTH'S NATURAL ENERGY BUDGET

Earth, understood as a physical system, has an energy budget that includes incoming and outgoing energy. Thermal solar radiation is absorbed by Earth's surface, causing it to warm. Part of the absorbed energy is then radiated back into the atmosphere as long-wave infrared radiation. Some of this radiated energy escapes into space, but some is absorbed by atmospheric GHGs. Then, the GHGs re-radiate the thermal waves in all directions. Part of this re-radiation goes back toward Earth's surface, transferred to the lower atmosphere again, resulting in higher temperatures. This mechanism differs from that of an actual greenhouse, in that a greenhouse isolates warm air inside the structure so that the heat is not lost by convection. Human activities since the Industrial Revolution—especially fossil fuel combustion, land-use change, increasingly intensive agriculture, and an expanding global human population—are primarily responsible for the recent steady increases in atmospheric concentrations of various GHGs, causing the atmosphere to warm.

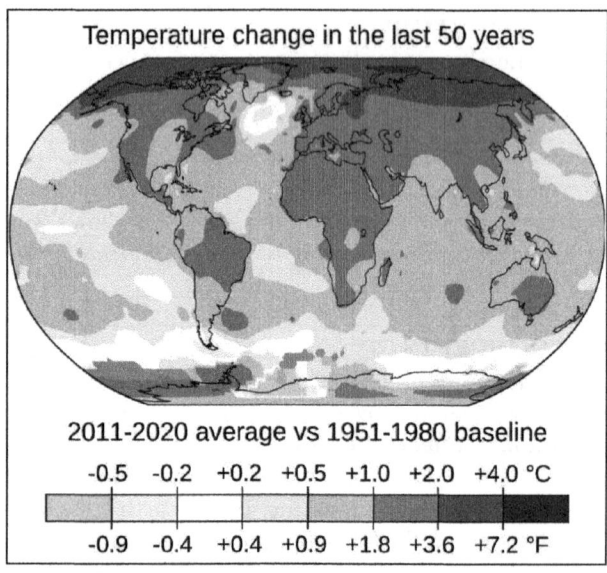

Average global temperatures from 2010 to 2019 compared to a baseline average from 1951 to 1978, according to NASA's Goddard Institute for Space Studies. (Wikimedia Commons)

HEATING POTENTIAL OF GHGS

The global warming potential (GWP) of the known GHGs is a factor given in relation to that of carbon dioxide. Methane, for example, has a GWP of 25. That is, every kilogram of methane in the atmosphere has, on a timescale of 100 years, the equivalent global warming potential of 25 kilograms of carbon dioxide. Methane is much more effective than carbon dioxide at absorbing infrared radiation, but its lifetime in the atmosphere is shorter, about 12 years, compared to anywhere from 30 to 1,000 years for a molecule of carbon dioxide.

The Intergovernmental Panel on Climate Change (IPCC), an organization sponsored by the United Nations and made up of 2,500 scientists from around the world, projected in its 2007 report that global warming would have severe impacts on human health, natural ecosystems, agriculture, and coastal communities. However, an opposite view, taking longer periods into account, states that the global warming of the last century is part of the planet's natural cycle driven by emissions by tectonic activities, wetlands, oceanic sources and sinks, and other factors. Currently, only one thing is certain: Human-produced GHGs have been emitted at extremely high rates. Fossil fuels including petroleum, coal, and natural gas, made up of hydrogen and carbon, release carbon dioxide and other GHGs upon combustion.

One scientific forecast projects that during our children's lifetimes, global warming will raise the average temperature of the planet by 1°C to 3.5°C (2°F to 6°F). In contrast, Earth is only about 3°C to 6°C (5°F to 9°F) warmer today than it was 10,000 years ago, during the last ice age. Human-driven global warming is thus occurring far more quickly than warming at any other time in at least the last 10,000 years. Higher temperatures, changes in precipitation patterns, and acidification of the oceans may lead to reduction of the land and ocean carbon sinks, thereby unleashing a feedback-induced acceleration in the concentration of GHGs in the atmosphere.

Another phenomenon to be taken into account is the so-called Little Ice Age, which occurred between 1550 and 1850. Conditions around the world were cooler than usual; many bodies of water froze over. The average global temperature since then has risen by just 1°F, which some consider an argument against global warming.

CARBON DIOXIDE

Carbon dioxide (CO_2) is a colorless, odorless, non-flammable gas. It is one of the many trace gases that naturally occur in the atmosphere, making up 0.03 percent of the gaseous composition. It is considered to be the most important GHG because of its predominance in emissions yielded from combustion of fossil fuels.

CO_2 is produced naturally: as an exhalation product of all respiring organisms, during the decay of organisms, in the weathering of carbonate-containing rock strata, and as volcanic emissions. Forests and oceans are natural carbon sinks. CO_2 is recycled by photosynthesis. Besides the day-and-night change in plant respiration, plants absorb CO_2 as they grow. Terrestrial ecosystems emit approximately 119 billion tonnes of carbon each year via the process of respiration and absorb approximately 120 billion tonnes of carbon each year via photosynthesis—a net sink of 1 billion tonnes of carbon. However, with the decline of forests, especially alarming in the tropics, less CO_2 can be absorbed. Approximately 88 billion tonnes of carbon are emitted annually by the oceans, and about 90 billion tonnes of it are absorbed, a net sink of about 2 billion tonnes of carbon. Over the last 150 years, these sinks have absorbed about 40 percent of the CO_2 emissions released by human activities.

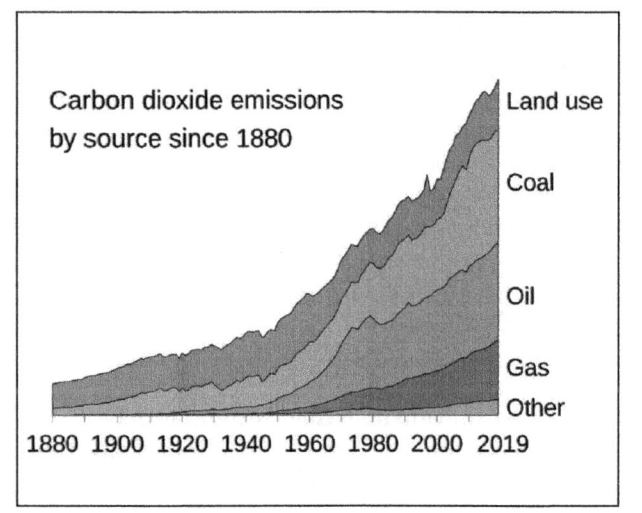

Carbon dioxide emissions by source since 1880 as calculated for the 2019 Global Carbon Budget. Carbon dioxide generated by land use changes (deforestation) has been added to as coal, oil, and natural gas consumption have each ramped up in turn. Data source is the Global Carbon Budget 2019. (Efbrazil via Wikimedia Commons)

For as long as 600,000 years before 1750, the generally recognized date of the beginning of the Industrial Revolution, atmospheric levels of carbon dioxide were about 0.028 percent (280 parts per million). Today, measurements show concentrations of CO_2 in the atmosphere of about 0.0425 percent (425 ppm), an approximately 50 percent increment or more than 30 billion tonnes per year, chiefly due to industrial development. Besides the combustion of fossil fuels, yielding about 65 percent of the CO_2 emissions, deforestation and land-use change have caused the remaining 35 percent. This is an estimated number that takes tree logging, charcoal production, slash-and-burn agricultural practices, pulpwood and fuelwood production, and forest degradation into account. Another important contributor is the clearing of new farmland and rangeland. Natural ecosystems can store 20 to 100 times more CO_2 per unit area than agricultural systems. Current land use activities in Africa, Asia, and South America contribute the greatest CO_2 emissions due to deforestation.

CARBON DIOXIDE AND OTHER GREENHOUSE GASES

The fossil fuels, such as coal, oil, and natural gas, were created chiefly by the decay of bacteria and plants from a great many millions of years ago. Today they are used to generate electricity and produce heating energy to power factories, as well as provide many comforts of civilization. With energy stored in their hydrocarbon molecules, upon burning the energy is released—as is CO_2. The World Energy Council reported that global CO_2 emissions from burning fossil fuels rose 12 percent between 1990 and 1995. The United States was responsible for 25 percent of all emissions worldwide, the leading share at that time, with China on the cusp of overtaking that rank, which it did in 2006.

Carbon footprint estimates and analyses can be useful in revealing the consequences of consumption activities to individuals, businesses, and societies. Still, it is emissions that can be most directly measured and around which national policy effects can be structured. The Organisation for Economic Co-operation and Development (OECD) in 2012 projected that world GHG emissions would rise by 37 percent above year 2005 levels by 2030, and by 52 percent above 2005 levels by the year 2050—if no new policies were adopted. The OECD projected that a CO_2 atmospheric concentration of 0.045 percent (450 ppm)—well above historical maximums and dangerously close to the current measured level—could be achieved as a GHG stabilization target if countries reduced year 2050 GHG emissions by 39 percent below year 2000 levels.

METHANE

Methane (CH_4) is a colorless, odorless, flammable gas. Its colloquial name, swamp gas, indicates that it is formed when plants and other organic matter decays under anaerobic (oxygen-free) conditions. Since 1750, methane emissions have doubled, and they could double again by 2050. Each year, 600 million tonnes) of methane are emitted into the air by livestock (especially cattle), coal mining, drilling for oil and natural gas, crop production in paddy fields, and direct emissions from organic matter breaking down in landfills. The changes in agriculture and land use in response to the world's growing population have caused additional methane emissions that are currently under research in the effort to find ways to reduce them.

Frozen methane clathrate deposits found on the sea floor and in deep permafrost are additional natural sources of GHGs and have the potential for future exploitation and energy use, as well as the potential to boost rates of global warming. The potentially uncontrolled release of methane into the atmosphere could occur as permafrost areas warm; also, releases related to exploitation of these reservoirs could occur, and

Methane bubbles can be burned on a wet hand without injury. (Scohen2017 via Wikimedia Commons)

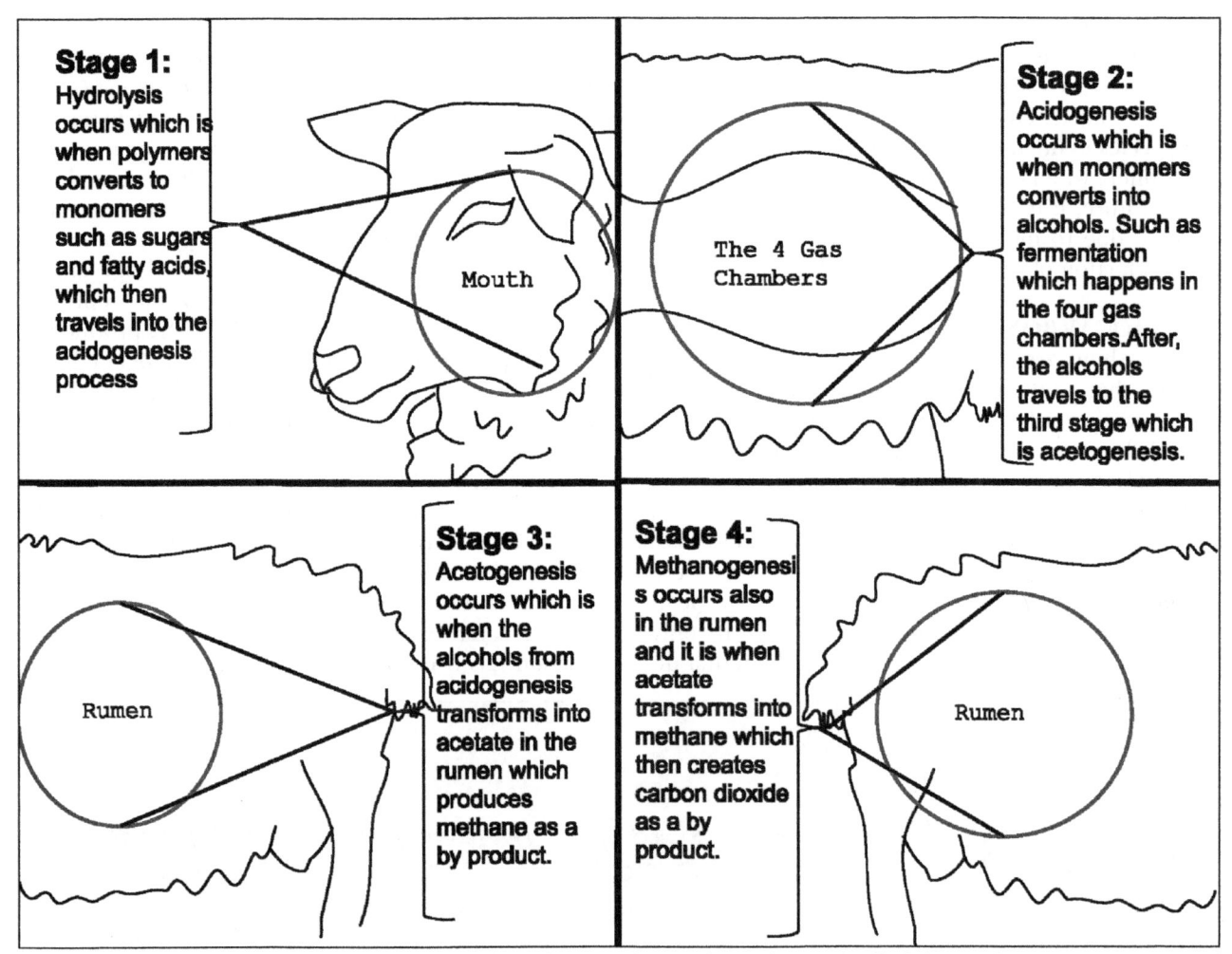

Stage 1: Hydrolysis occurs which is when polymers converts to monomers such as sugars and fatty acids, which then travels into the acidogenesis process

Stage 2: Acidogenesis occurs which is when monomers converts into alcohols. Such as fermentation which happens in the four gas chambers. After, the alcohols travels to the third stage which is acetogenesis.

Stage 3: Acetogenesis occurs which is when the alcohols from acidogenesis transforms into acetate in the rumen which produces methane as a by product.

Stage 4: Methanogenesis occurs also in the rumen and it is when acetate transforms into methane which then creates carbon dioxide as a by product.

This image represents a ruminant, more specifically a sheep producing methane within the four stages of hydrolysis, acidogenesis, acetogenesis, and methanogenesis. (Wazzupitsalice via Wikimedia Commons)

from slippage of the continental shelf once the supporting reservoirs are reduced. Recent scientific research has reported significant amounts of methane released into the atmosphere already from methane clathrate deposits found in the Arctic, due to global warming.

NITROUS OXIDE

Nitrous oxide (N2O) is a colorless gas with a slightly sweet odor. Its colloquial name is laughing gas; it is used as an anesthetic and has other uses as well. Occurring naturally, it is emitted from oceans and by bacteria in soils; industrially, it is mostly an exhaust product. The current 13 million tonnes of nitrous oxide gas emitted annually seems to be low compared to CO_2 emissions, but N_2O emissions have increased more than 15 percent since 1750, and they endure in the atmosphere for 100 years. This long atmospheric lifetime leads to a GWP of 298. The reduction of these emissions is a big challenge because since hydrochlorofluorocarbons (HCFCs) have been largely controlled, nitrous oxide is the most hazardous ozone-depleting substance released by human activities.

Most of the N_2O added to the atmosphere each year comes from deforestation and the conversion of forest, savannah, and grassland ecosystems into agricultural fields and rangeland; and from the use of nitrate and ammonium fertilizers. In the past 15 years, the use of nitrogen-based fertilizers has doubled, but plants absorb only 30 percent of the added nitrogen. In extreme cases, artificial fertilization can lead to the death of forests, eutrophication of aquatic biotopes, and species extinctions. N_2O is also released into the atmosphere when fossil fuels and biomass are burned. In the future, it is estimated that N_2O emissions will increase due to more agriculture activity related to the production of supposedly "green" biofuels.

FLUOROCARBONS AND HALONS
Fluorocarbons are a group of synthetic organic compounds that contain fluorine and carbon. Many of these compounds, including chlorofluorocarbons (CFCs), have properties that are favorable for many technical applications. They are relatively nontoxic, nonflammable, odorless, and colorless, and they can be easily converted from gas to liquid or liquid to gas.

In the 1970s, studies showed that when CFCs were emitted into the atmosphere, they destroyed ozone in the ozone layer of the stratosphere, which is the only barrier protecting Earth's surface from the lethal ultraviolet radiation coming from the Sun. These compounds are highly resistant in the atmosphere, which makes them thousands of times more potent as GHGs than carbon dioxide. Currently, all halogenated hydrocarbons together add 0.337 watt per square meter to the greenhouse effect. Chlorodifluoromethane (HCFC-22), is, after carbon dioxide and methane, the third most important anthropogenic greenhouse gas. The Montreal Protocol, an international agreement adopted in 1989, which phases out ozone-depleting substances, requires the end of HCFC-22 production by 2020 in developed countries and by 2030 in developing countries. CFCs and HCFCs are being replaced by less-damaging hydrofluorocarbons (HFCs) for such widespread uses as the coolant in air conditioners and refrigerators.

Halons are organic compounds derived from methane or ethane, including bromine. They were used as extinguishing agents. Like CFCs, they are ozone-depleting substances, and their destructive potential is 10 times greater than that of CFCs. Since 1994, their production has been banned globally, with an extended

Surface temperature rise is greatest in the Arctic, where it has contributed to melting permafrost which releases more greenhouse gases and the retreat of glaciers and sea ice. (ea-4 via iStock)

deadline for developing countries. The Montreal Protocol described a special regulation for methyl bromide, a highly reactive compound that readily damages deoxyribonucleic acid (DNA), enacting a global successive phaseout of its production by 2005.

FUTURE EFFECTS AND STRATEGIES

The long-term environmental problems resulting from the increased atmospheric concentration of GHGs are intertwined, and pose a huge ecological challenge. One GHG, water vapor, is increasing in the atmosphere. Water vapor has an immense greenhouse effect due to its great capacity to absorb thermal energy and lock it within the atmosphere. That causes further warming, leading to still more water vapor rising from ocean surfaces; this positive feedback loop leads to a more intense greenhouse effect. Consequences are likely to include more and stronger tropical cyclones, as well as changing ocean current patterns, the potential for more tidal waves or tsunamis, and faster erosion along the coastlines.

Besides such looping effects, there are one-way developments that escalate with each step and directly cause secondary effects. Melting glaciers and polar ice caps, more severe floods and droughts, and rising sea levels (on average between 10 and 25 centimeters, or 4 and 10 inches, since 1990) are considered in this manner. Rises in sea level, for example, can increase the salinity of wetlands and upstream freshwater, and endanger coastal lands and communities. Global warming is also regarded as introducing health concerns, such as the spread of tropical diseases into temperate zones.

The Kyoto Protocol, adopted in 1997 as part of the United Nations Framework Convention on Climate Change, was the first international treaty to establish carbon emissions reduction programs. However, it did little to enforce the parties' meeting their emissions targets. The United States, until recently the leading emitter of GHGs, has not ratified the Kyoto Protocol, while China, now the leading emitter, retains its status as a developing country and is therefore not required to reach the strict targets imposed on developed nations.

One core strategy for reducing greenhouse gas emissions is to enhance energy-saving technologies while generating more energy from renewable sources, thereby reducing the demand for fossil fuels. By the year 2050, renewable sources could provide 40 percent of the energy needed in the world. Use of renewable energy can help to slow global warming even while reducing air pollution. The Greenhouse Gas Protocol, a joint initiative of the World Resources Institute and the World Business Council for Sustainable Development, has become the most commonly used international accounting framework for government and business leaders to understand, quantify, and manage greenhouse gas emissions.

—*Manja Leyk*

Further Reading

"Greenhouse Gas." *The Encyclopedia of Earth*, edited by Cutler J. Cleveland. National Council for Science and the Environment. http://www.eoearth.org/article/Greenhouse_gas?toxic=49554.

Hansen, James. *Storms of My Grandchildren: The Truth About the Coming Climate Catastrophe and Our Last Chance to Save Humanity*. Bloomsbury Press, 2009.

Lankford, Ronald D. *Greenhouse Gases*. Greenhaven Press, 2009.

McCarthy, James E., Cianni Marino, Nico Costa, Larry Parker, and Gina McCarthy. *EPA Regulation of Greenhouse Gases: Considerations and Options*. Nova Science Publishers, 2011.

Solomon, S., D. Qin, M. Manning, Z. Chen, M. Marquis, K.B. Avery, M. Tignor, and H.L. Miller, eds. *Contribution of Working Group I to the Fourth Assessment Report of the Intergovernmental Panel on Climate Change, 2007*. Cambridge UP, 2007.

Moya, Bernardo Llamas, and Juan Pous, eds. *Greenhouse Gases*. IntechOpen, 2016.

U.S. Department of Commerce. "Emissions of Potent Greenhouse Gas Increase Despite Reduction Efforts." U.S. Department of Commerce, 2010. http://www.noaanews.noaa.gov/stories2010/20100127_greenhousegas.html.

U.S. Department of State. "The Kyoto Protocol on Climate Change." http://www.state.gov/www/global/oes/fs_kyoto_climate_980115.html.

H

HEAT TRANSFER

FIELDS OF STUDY
Climatology; Oceanography; Physics; Thermodynamics

ABSTRACT
Heat transfer is the movement of thermal energy between physical systems, moving from the warmer to cooler matter and leading to equilibrium. Heated matter cannot remain at a higher temperature than its surroundings; it constantly transfers heat by one or more processes: radiation, conduction, and convection.

KEY CONCEPTS
conduction: the transfer of energy by physical contact between substances of different temperatures

convection: the transfer of energy in fluids as a result of changes in density according to changes in temperature

global conveyor belt: a globe-spanning system of surface and deep ocean currents driven by changes in salinity and density according to changing temperatures; also known as thermohaline circulation

Hadley cell convection: atmospheric convection currents caused by the influx of solar energy along the equator; the warmed air rises up from the cooler sea-level air, moves to the north and south away from the equator, then cools at high altitudes and sinks back toward the surface as its density increases, ultimately to flow back toward the equatorial region

radiation: the transfer of energy through space, without contact, generally as electromagnetic waves, but energetic particles such as electrons and alpha particles emitted from an appropriate source can also carry out the same function

RUNNING HOT TO COLD
The flow of heat from a hot solid, liquid, or gas to cooler matter in its vicinity is termed heat transfer. Heated matter constantly transfers heat either by direct contact of materials—conduction and convection—or by the wave action we call radiation. Heat can be transferred by these three processes between any types of matter, and the process continues until local equilibrium is achieved.

THERMAL RADIATION
Matter is made up of atoms consisting of electrons that are constantly in motion around a nucleus. Electric and magnetic fields exist between the positively charged protons in the nucleus and the negatively charged electrons, and energy is created by their activity. This type of energy—along with the full spectrum of related electromagnetic energy types—is emitted in the form of self-propagating waves that transmit heat to cooler bodies completely removed from the source of the heat. The waves excite the particles on the surfaces and within objects they strike, causing a rise in the temperatures of the receiving objects.

Electromagnetic radiation is classified according to wave frequency; it ranges from short-wavelength types (such as gamma, ultraviolet, and X-radiation) to long-wavelength (such as microwaves and radio waves). Infrared radiation and visible light have medium wavelengths. All of these can transmit both heat and light from the source, which is partly converted to thermal energy in the irradiated objects. An example is the sun's infrared and visible light reaching Earth's surface; that is, the infrared and visible spectra from the sun travel through the vacuum of space and through Earth's atmosphere; they generate heat in the objects receiving the radiation on Earth's surface. Heat is also radiated into the surrounding air, because radiation can take place in a vacuum or in the presence of atmospheric gases, water vapor, and other liquids, as well as solids.

THERMAL CONDUCTION
Conduction involves the transfer of heat between substances in direct contact. The orbiting electrons in the

heated materials interact with electrons of adjacent atoms, and in this process the hotter atoms gradually transfer heat to the cooler atoms and molecules in succession, until the whole object is at the same temperature on all sides or surfaces. The rate at which heat flows depends in part on the dimensions of the object and the temperature differential between its different surfaces.

For example, consider a square solid block where l = length, d = thickness, and h = height:

area of heated surface = l × h
temperature at heated surface = th
temperature at cooler surface = tc.

The rate of heat transfer, denoted by r, is directly proportional to:

(th - tc) × (area of heated surface [l × h]) / thickness of block (d).

Therefore, if (th - tc) increases, the rate of heat transfer also increases proportionately. If, however, (th - tc) is a constant and the thickness of the block were to be doubled, then the rate of heat transfer would be halved. Similarly, for the same condition, if the length or height of the block were doubled, the rate of heat transfer would also double. Any increase in the area of the heated surface results in a corre-

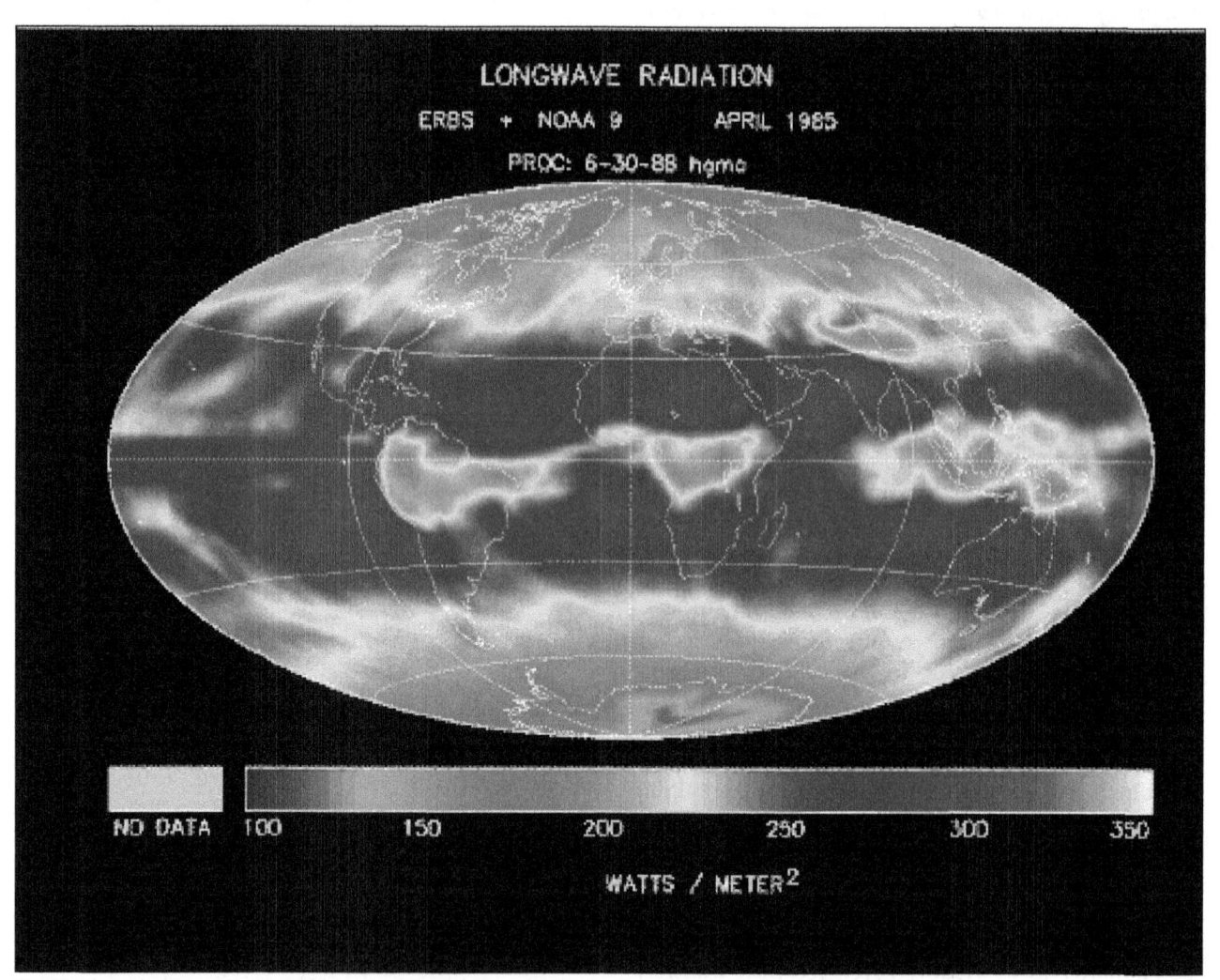

Earth's long wave thermal radiation intensity, from clouds, atmosphere and surface. (NASA via Wikimedia Commons)

spondingly larger area of the cooler surface getting heated quickly, whereas an increase in the thickness of the object—that is, an increase in the distance between the heated surface and the cool surface—results in a slower rate of heat transfer, because it takes longer for the heat to reach the opposite side. In a three-dimensional object having tc that is surrounded by a warm medium having th, heat conducts to the center of the object rather than to the opposite surface, and it does so irregularly according to the three-dimensional shape of the object until thermal equilibrium is attained.

In addition to these dynamics, the rate of heat transfer depends upon the material properties of the object. Some materials, such as air and wood, are poor conductors of heat; others, like the metals copper and silver, are good conductors. The property describing how well a material transfers heat is expressed as the thermal conductivity (k) of the material. Another way to express this is by defining the thermal resistance of the material or object, called the R value. It is expressed as a reciprocal of the thermal conductivity. Thermal resistance of a composite structure or assembly can be measured as a cumulative value of the individual resistances of its component materials.

THERMAL CONVECTION

Convection is the typical means of heat transfer in liquids and gases. It involves two processes, diffusion and advection. Diffusion is the motion of individual particles in a liquid or gaseous medium, and advection is the large-scale cumulative flow of such particles. Such flows can also be classified as natural or forced; natural convection occurs in nature as atmospheric currents, oceanic currents, and volcanic movements within the Earth's surface.

Generally, convective flows occur due to the difference in the densities of the heated matter and cooler matter in the liquid or gas. Heated matter is more buoyant because it is less dense, and therefore easily rises to the upper layers of the medium, whereas cooler matter sinks to lower levels in the medium because of its greater density. It follows that natural convection can occur only in the presence of gravity. The upward-flowing heated medium directly exchanges heat with the cooler medium and as a result loses some of its heat. In this process, the cooler particles progressively sink to the lower levels in a finite mass until the whole mass is at an equilibrium temperature. When the mass is not locally finite, such as in nature, convective currents give rise to atmospheric phenomena on a global scale. The greatest examples on Earth of the interplay of differences in temperature and fluid density are the Hadley cell circulation of air rising above the equator to produce the westerly "trade" winds, and the thermohaline circulation of the "global conveyor belt" in the oceans.

The formation of land and sea breezes is due to such large-scale convective forces (in combination with conduction). Water has a larger heat capacity than land; it holds heat better. It therefore takes the sea longer to change temperature, either upward or downward. During the day, the air above the water will thus be cooler than that over the land, the temperature of the air being determined by the conduction of heat from either the land surface or water surface. As the air over land is heated by conduction and rises due to convection, breezes blow in toward land from the cooler air above the water. On the other hand, during the night water cools off more slowly than the land, leaving air above the water warmer than that over land. Again, the warmer air rises, creating the gap into which the cooler air—this time from above the faster-cooling land—flows in. The greater the temperature differences, the stronger the convective air movements in this transfer of heat from the warmer to the cooler areas over land and water.

These principles of heat transfer by liquid or gas flows also apply to forced convection, but in this case, the convective flows occur by artificial means such as a fan or a pump—that is, without depending on gravitational forces. Air-conditioning systems in buildings are examples of forced convection systems; they must routinely be balanced to accommodate the natural convection forces that result in lower floors becoming cooler than upper floors. Similarly, wet radiator heating systems work on the forced convective flows of hot water for space heating in cold climates. The mechanisms of diffusion and advection are effectively utilized by controlling the heat transfer into a system for heating, or heat flowing out of a system for cooling.

—*Swati Ogale*

Further Reading

Annaratone, Donatello. *Transient Heat Transfer*. Springer, 2011.

Bergman, T.L. *Fundamentals of Heat and Mass Transfer.* Wiley, 2011.

Biswas, Gautam, Amaresh Dalal, and Vijay K. Dhir. *Fundamentals of Convective Heat Transfer.* CRC Press, 2019.

Kaviany, Massoud. *Essentials of Heat Transfer.* Cambridge UP, 2011.

Lienhard, John H. *A Heat Transfer Textbook.* Dover, 2011.

Rathakrishnan, Ethirajan. *Elements of Heat Transfer.* CRC Press, 2012.

Rathore, M.M., and Raul Raymond Kapuno. *Engineering Heat Transfer.* Jones and Bartlett, 2011.

Watts, Naomi. *Heat Transfer: Fundamentals and Applications.* NY Science Press, 2020.

HELMHOLTZ, HERMANN VON

FIELDS OF STUDY
Optics; Physics; Physiology

ABSTRACT
Hermann von Helmholtz was trained as a physiologist and began work as a surgeon, but approached all his studies from a mathematical perspective, leading him to important discoveries in physics.

KEY CONCEPTS
conservation of energy: a physical law that states energy is neither created nor destroyed, but merely changes from one form to another; also stated as the total energy of any system before a change to the system must be equal to the total energy of the system after that change

enthalpy: the heat energy associated with a system

entropy: the energy associated with the degree of randomness or disorder within a system

free energy: also Gibbs free energy, the useful work obtainable from a thermodynamic system at constant temperature and volume

metabolism: the biochemical processes by which energy is extracted from food and oxygen in living biological systems

vitalism: an old philosophical concept that claimed a special "vital" force within living bodies was responsible for muscle movement, it was discredited when it was demonstrated that muscle movement was produced by electrical stimulation of muscle tissue

HELMHOLTZ'S LIFE...
Hermann von Helmholtz (August 31, 1821-September 8, 1894) was a Prussian scientist whose work was significant to physics, physiology, and the philosophy of science. The son of Ferdinand Helmholtz, headmaster of the Potsdam Gymnasium and friend of Immanuel Hermann Fichte, Helmholtz was encouraged by his father to study medicine because of the funding available to future doctors. For a time, he worked as a surgeon in the Prussian army. He was trained in physiology and made several important contributions to the field. However, even while employed as a physiology professor, he continued his studies in physics, and 22 years into his teaching career he accepted a position as professor of physics at the University of Berlin. He was influenced both by the Immanuel Kant and his father's friend Fichte, and extended their work. He published major scientific papers nearly every year, from his years as a student until the year he died.

...AND SCIENCE
Two years before his first teaching position, Helmholtz published his first significant treatise—not

Hermann von Helmholtz. (Wikimedia Commons)

in physiology, but in physics, albeit in an area inspired by his medical studies. *On the Conservation of Force* dealt with the conservation of energy as he had observed it in his studies of muscle metabolism, which he found to disprove claims made by *Naturphilosophie*, a school of thought dominant in the philosophical tradition of German idealism at the time. Specifically, *Naturphilosophie* advocated "vitalism" the notion that in addition to mechanical and physical forces, a vital force animates the body and causes its parts to move. Helmholtz inherited some of his hostility toward vitalism from Johannes Muller, the professor who had overseen his work in 1841-42. (Muller also published Helmholtz's first essay in 1843, an antivitalist work called "On The Nature of Fermentation and Putrefaction.") Helmholtz's work straddled the line between his personal commitment to physics and furthering his mentor's quest to rid physiology of vitalism and describe physiology purely in terms of physical forces. One of the ways he demonstrated that vital forces were not necessary to explain physiology was by applying electrical currents to frogs' muscles, and demonstrating that the resulting heat could be accounted for by metabolism and muscle movement.

The work was not well-received, not because it was inaccurate, but because his colleagues believed he had plagiarized the work of Julius Robert von Mayer, who had discovered the law of conservation of energy a few years earlier. While Helmholtz explicitly refers to the work of others, James Joule and Sadi Carnot, he makes no mention of Mayer; it is very likely that, though he was a member of the Berlin Physical Society, as a student whose attention was actually engaged by another field, he simply was unaware of Mayer and came to the same conclusions by a different path. Thomas Kuhn, the historian and philosopher of science, has argued that credit for discovering the conservation of energy belongs to Helmholtz, whose formulation was more complete than Mayer's or others'.

Helmholtz later collaborated with Scottish engineer and mathematician William Rankine, one of the fathers of thermodynamics, on the "heat death" of the universe, a natural conclusion drawn from thermodynamics. William Thomson (Lord Kelvin), another thermodynamicist, was the first to outline the idea, which said that the ultimate fate of the universe was to reach a point where thermodynamic free energy has diminished beyond the capacity to sustain motion, life, or work: a state of maximum entropy. Helmholtz met with Rankine, Lord Kelvin, and major figures of the field (including Michael Faraday and James Joule) on a trip to England in 1864, having delivered a lecture on electrical distribution that included the same conclusions Lord Kelvin had come to independently. Though physics has changed much since the nineteenth century, heat death is still one of the most popular and well-supported theories for the final state of the universe.

Helmholtz's most significant work on electromagnetism began shortly before he took the physics position at Berlin. After some minor work on electrical oscillations, he published a series of articles on theories of electrodynamics, and the Helmholtz equation is named for him, as is the Gibbs-Helmholtz equation:

$$\Delta G = \Delta H - T\Delta S$$

for determining changes in a system's Gibbs free energy as a function of temperature. Gibbs free energy is the useful work obtainable from a thermodynamic system at constant pressure and temperature. Helmholtz free energy, sometimes just called "free energy," is the useful work obtainable from a thermodynamic system at constant temperature and volume. He had proposed it in order to extract a value independent of heat or entropy, which provided a foundation for theoretical chemistry in the coming decades. One of his physics students, Heinrich Hertz, went on to surpass his mentor in the field of electromagnetism. Some of Hertz's most significant experiments, which demonstrated the existence of radio waves, were prompted by the prize problem Helmholtz had set for the Prussian Academy of Science, on the relationship between electromagnetic forces and the dielectric polarization of insulators. Hertz's work also ended a years-long dispute between Helmholtz and William Weber, which depended on a view of electrodynamics that Hertz displaced.

Helmholtz's other major works included explanations of the physiology of perception and stereoscopic vision, and the role of inference in the mind's construction of perceptual space; the horopter problem, in which the eye perceives points on the horopter (a set of points perceived as equidistant from the point on which the eye is focused) as a straight line, when they are actually a curve; Riemannian geometry; and pioneering work on acoustics, including his 1863 *On the Sensations of Tone as a Physiological Basis for the Theory of Music*, which provided an exact mathematical model of the vibrations of sound in an open cylindrical tube. He approached all his work with mathematic rigor in mind, but was clearly informed by his early encounters with philosophy.

—*Bill Kte'pi*

Further Reading

Cahan, David, ed. *Hermann von Helmholtz and the Foundations of Nineteenth-Century Science.* U of California P, 1994.

Cahan, David. *Helmholtz: A Life in Science.* U of Chicago P, 2018.

Schiemann, Gregor. *Hermann von Helmholtz's Mechanism: The Loss of Certainty: A Study on the Transition from Classical to Modern Philosophy of Nature.* Springer, 2009.

Wise, M. Norton. *Aesthetics, Industry and Science: Hermann von Helmholtz and the Berlin Physical Society.* U of Chicago P, 2018.

HERTZ, HEINRICH

FIELDS OF STUDY

Communication; Computer Technology; Electronics Engineering; Physics

ABSTRACT

Heinrich Hertz was a German-born physicist who made many discoveries that relate to the efficient use of energy. Hertz expanded the electromagnetic theory of light first mathematically predicted by English scientist James Clerk Maxwell in 1873. In addition to his role in clarifying the theory, Hertz discovered that electromagnetic waves could transmit electricity using an electrical apparatus.

KEY CONCEPTS

electromagnetic spectrum: the continuous range of frequencies of electromagnetic waves

Hertz (Hz): the standard unit of frequency for cyclically repeating phenomena; one complete cycle of the phenomenon over a period of one second defines a frequency of one Hertz

induction coil: a type of transformer in which a secondary coil of a large number of turns of thin wire is mated to a primary coil of a few turns of thick wire about an iron core. A low-voltage direct electrical current flowing through the primary coil is turned on and off at a regular frequency, inducing a high-voltage current of the same frequency in the secondary coil.

photoelectric effect: the physical phenomenon in which light striking the surface of susceptible materials causes the surface atoms of the material to emit electrons

polarization: the phenomenon in which electromagnetic waves, which oscillate radially about the axis of their direction of motion from a source, are made to oscillate in a single plane along the axis of their direction of motion

ultraviolet light: electromagnetic radiation having frequencies higher than the highest visible light frequency in the electromagnetic spectrum

BACKGROUND

Heinrich Hertz was a German-born physicist who made many discoveries that relate to the efficient use of energy. Hertz expanded the electromagnetic theory of light first mathematically predicted by English scientist James Clerk Maxwell in 1873. In addition to his role in clarifying the theory, Hertz discovered that electromagnetic waves could transmit electricity using an electrical apparatus. These developments later led to the creation of the wireless telegraph and the radio. Hertz also was a founder of the field of contact mechanics, which resulted in important developments related to the energy-efficient design of technical systems. Hertz remains a significant foundational source for research related to energy and has been honored through a variety of prizes and awards bearing his name that are given to researchers in the field.

Born in 1857, Hertz demonstrated an aptitude for sciences and languages at an early age. He first decided to study engineering at the University of Munich but later transferred to the University of Berlin to study physics, where he ultimately received his doctorate in philosophy, *magna cum laude*, in 1880. After graduating from the University of Berlin, Hertz continued to study in the physical lab under Hermann von Helmholtz, one of the first German scientists to work in the field of electrodynamics. In 1883 Hertz lectured in theoretical physics at the University of Kiel, where he began his studies in James Clerk Maxwell's electromagnetic theory. From 1885 to 1889, Hertz served as a professor of physics at Karlsruhe Polytechnic, and he made his first successful attempt at detecting electromagnetic radiation in 1886, encouraged by Helmholtz to attempt to solve a prize problem put forth by the Berlin Academy of Sciences in 1879.

ELECTROMAGNETIC RESEARCH

After Maxwell predicted that electromagnetic waves moved at the speed of light, light itself was seen as just one type of such waves. Maxwell's successors worked to generate and detect electromagnetic radiation using electrical apparatus. In order to generate and detect the electromagnetic radiation, Hertz created a radio-wave transmitter built from a high-voltage induction coil, a condenser, and a spark gap. An oscillator made of polished brass knobs was connected to the induction coil and separated to determine if the sparks could leap over the spark gap. In order to test this theory produced by Maxwell, Hertz created a receiver of looped wire. If the theory was correct, the current in the looped wire would send sparks across the gap. In Hertz's experiment, the induction coil produced a high voltage, which produced electromagnetic waves. An electric current was created, producing a small spark in the spark gap. Hertz was the first to confirm the electromagnetic theory proposed by Maxwell experimentally. He was the first to broadcast and receive radio waves through his experiments. He later performed more complicated experiments to measure the velocity of electromagnetic radiation, which was found to be the same speed as the velocity of light. Hertz also studied the reflection and refraction of radio waves, finding them to be the same as light.

Hertz published his findings in the scientific *Annalen der Physik* and then in his first book, *Untersuchungen über die Ausbreitung der elektrischen Kraft* (Investigations on the Propagation of Electrical Energy) in 1892. His scientific papers were translated into English and published in three volumes: *Electric Waves* (1893), *Miscellaneous Papers* (1896), and *Principles of Mechanics* (1899). Considered one of the most important contributions to science, his experiments led to the invention of the Hertz antenna receiver ra-

Heinrich Hertz. (Wikimedia Commons)

dio and other "wireless" technologies. In 1886 he developed the Hertz antenna receiver, a set of terminals that is not electrically grounded for its operation. Hertz helped to establish the photoelectric effect, spurred by his observation that a charged object loses its charge more rapidly when illuminated by an ultraviolet light. Albert Einstein later was able to explain the photoelectric effect fully. Hertz's work helped to explain such concepts as reflection, refraction; polarization, interference, and velocity of electric waves, steps that would help usher in the wireless age of the next century.

CONTACT MECHANICS

Hertz's work with contact mechanics also had lasting effects on the study of, and conceptions about, energy. Regarded by many as the founder of the field, Hertz published two groundbreaking papers in 1886 and 1889 that continue to be influential. Hertz summarized how axisymmetric objects placed in contact behave under loading. In initiating these studies, Hertz was examining how the optical properties of multiple, stacked lenses might be affected if the force holding them together was altered. Hertz found that as two curved surfaces come in contact with each other, they deform slightly, as a result of the localized stress caused by the imposed loads. Today, this is known as Hertzian contact stress. Hertzian contact stress provided the foundation for studies related to the load-bearing capabilities and fatigue life for bodies where two surfaces are in contact, such as bearings, gears, and aircraft fuselages. One area that Hertz neglected in his contact mechanics work was the effect of adhesion between the two solids. As there existed at the time no experimental method of testing for adhesion, which has ultimately proven to be significant, Hertz failed to take this into account in developing his theory.

At the age of 34, Hertz was diagnosed with an infection, and despite treatment he died two years later of Wegener's granulomatosis in Bonn. In 1930, the International Electrotechnical Commission (IEC) established the hertz (Hz) as the unit of frequency used by the International System of Units (SI), a measurement of the number of times a repeated event occurs per second. The hertz was also adopted by General Conference on Weights and Measurements in 1964, replacing the prior name for the unit, cycles per second (cps). The term *gigahertz*, used commonly to measure computer processor clock rates and radio frequency applications, is derived from this term. Beginning in 1987, the Institute of Electrical and Electronics Engineers (IEEE) has annually awarded the Heinrich Hertz Medal for outstanding theoretical or experimental achievements in Hertzian waves.

—Danielle J. Daly

Further Reading
Baird, Davis, R.I.G. Hughes, and Alfred Nordmann. *Heinrich Hertz: Classical Physicist, Modern Philosopher.* Kluwer Academic, 1998.
Buchwald, Jed Z. *The Creation of Scientific Effects: Heinrich Hertz and Electric Waves.* U of Chicago P, 1994.
Lützen, J. *Mechanistic Images in Geometric Form: Heinrich Hertz's Principles of Mechanics.* Oxford UP, 2005.
Susskind, Charles. *Heinrich Hertz: A Short Life.* San Francisco Press, 1995.

High-Intensity Light-Emitting Diodes (LEDs) and Energy Conservation

FIELDS OF STUDY
Electrical Engineering; Electronics; Materials Science

ABSTRACT
LED lightbulbs consume much less electricity than traditional incandescent bulbs and are often associated with efforts to conserve electricity for environmental reasons. They are often used as a symbol of sustainability programs across the United States.

KEY CONCEPTS
CFL: compact fluorescent light
diode: an electronic component made of semiconductive material that allows current flow only in one direction when activated; assemblies of interconnected diodes are etched onto a silicon substrate to produce transistors and complex transistor combinations
filament: in incandescent lightbulbs, a thin tungsten wire that resists the flow of electricity to the point that it becomes hot enough to emit white light
fluorescence: the property of a material that it will absorb ultraviolet light, as from electrically excited mercury vapor in a CFL, and re-emit energy as visible light, or fluoresce
semiconductor: a material that does not normally conduct electricity, but can be made electrically conductive by the application of a biasing voltage

LIGHT-EMITTING DIODES (LEDS)
Diodes are solid-state electronic components that have the property of emitting light when activated in an electrical circuit. The particular properties of the semiconductor materials from which they are made currently allow light-emitting diodes (LEDs) to emit a variety of different wavelengths of light, from infrared to ultraviolet, and even as laser sources. Of special application are high-intensity LEDs that emit pure white light equivalent to sunlight. These have been applied to use in lights of all kinds, from typical household lighting to street lighting.

LEDs generate the same amount of light as traditional incandescent bulbs; however, they accomplish this goal while using very much less electricity and producing much less wasted energy that is lost as heat. LEDs function normally at direct current (DC) voltages of 3.5V to 5V with power ratings of as little as 3.5 W to produce the same amount of light as an incandescent lightbulb requiring 120V with a power rating of 60W - 100W. The tradeoff for this application, however, is that LED lights have to work at standard 120 V household voltages, and therefore must have a stepped-down power management circuit built in. This tends to mitigate their efficiency somewhat relative to incandescent and CFLs. A more positive trade-off is the long life of LED lightbulbs, which will easily last as long as 150,000 hours, as compared to the approximately 10,000 hours available with the best incandescent and CFL lightbulbs. Additionally, LED lightbulbs are far more resistant to physical failure than other types of lightbulbs, since they do depend on the presence of a filament that can be easily destroyed by a simple jolt or burning out. Thus, although LED lightbulbs cost more to purchase than incandescent bulbs and CFLs, their much lower electricity consumption and very long lives make them a lot more efficient and much less costly in terms of energy consumption.

INCANDESCENTS AND CFLS

Even though CFLs appear regularly as components of energy conservation and environmental protection campaigns, critics of CFLs point out that the bulbs contain mercury, a very hazardous substance, and thus require special collection and containment programs when they break or eventually burn out.

Incandescent bulbs generate light when an electric current is passed through a filament. The filament resists the flow of electrons and begins to glow very brightly; this is the source of the light. Along with the light comes a significant amount of heat, an effect most have noticed by sitting near a very bright incandescent light, or by touching a bulb that has just failed and needs replacement. Unfortunately, this heat is a secondary byproduct of the bulb's main task, to produce light, and thus can be classified as waste. In fact, incandescent bulbs waste most of the electricity that they consume. CFLs, on the other hand, use electricity directly to stimulate gases inside the bulb to produce ultraviolet (UV) radiation. This UV radiation passes through a filter coating the glass part of the bulb that converts the UV radiation into light in the visible spectrum—that is, radiation we can see. Because they do not rely on resistance to produce light, CFLs do not generate as much heat as incandescent bulbs, nor do they consume as much electricity.

Although such alternatives to incandescent bulbs have existed for much of the twentieth century, technological challenges (a long "warm-up time" in particular) limited the use of CFLs until the 1990s. Since then, however, advances in technology have improved the quality of CFLs while making them easier to manufacture, lowering costs. Modern CFLs contain an internal circuit board and other electronic or magnetic equipment (called a "ballast") that regulates the flow of electricity to the glass tube containing the gas.

The reality of CFLs, however, is that the use of mercury in their construction has made their disposal so problematic as an environmental and health hazard that their production and use in North America has essentially disappeared. It is now impossible to find replacement CFLs in stores, as their sale has been made illegal.

Comparison of compact fluorescent light bulbs with 105W, 36W, 11W. (Tobias Maier via Wikimedia Commons)

LED LIGHTBULBS

LEDs are quickly becoming a focal point in many energy conservation programs. This is in part due to their low cost when compared to other home energy efficiency measures, such as improved insulation, windows with high R-values (which measure the resistance to the flow of heat through a given material), or the installation of small-scale solar, wind, or geothermal energy systems. LEDs are also popular because of their immediate impact on energy bills, lowering users' power bills from the first day they are installed in lighting fixtures. Furthermore, LED bulbs last approximately 15 to 20 times longer than incandescents, reducing the money spent on replacement bulbs while also removing potentially hazardous glass and metal objects from the solid waste stream.

Most CFLs have a rather distinctive shape: a thin glass tube bent into a spiral shape, with the ballast component at the base. Several other interesting shapes have been produced for CFLs, but the main impact is that CFLs have also become a highly visible and memorable piece of energy conservation programs. By contrast, the small size of LEDs means they can be structured into essentially any suitable shape. They are commonly found in devices ranging from miniature flashlights with high-intensity beams, to multi-LED "light bars," to lightbulbs that maintain the standard lightbulb shape and lightbulbs that are a simple tube shape. Since they have become routinely available, their use has been aggressively pursued by governments around the world, some of which have gone as far as to ban traditional incandescent bulbs from home lighting purposes.

LEDs differ from incandescent lightbulbs in that the internal components have circuitry and other "high-tech" pieces to them to control the voltage at which they operate. Variability in manufacturing procedures may produce LEDs with shorter-than-usual

A house lit up with Christmas decoration. (Oleg Albinsky via iStock)

life spans right out of the box. However, the production of the diode material itself is a very well controlled and consistent operation, meaning that the problem in any less-than-perfect LED lightbulbs is the result of inconsistencies in assembling the supporting circuitry.

Another complaint that plagued CFLs concerns the quality of their light. Incandescent bulbs generate white light from the heat produced by the filament when electricity passes through it. This light is already part of the visible spectrum that humans can see, and it can be modified with dimmer switches and various types of tinted glass. CFLs, on the other hand, had to process their radiation with special UV filters to make it visible, which some complain lends an especially artificial quality to the light, in some instances triggering headaches after extended use. LED lightbulbs do not have any of these particular drawbacks, since the light they produce is perhaps even whiter than that produced by incandescent lightbulb filaments and they do not require filtering of any kind to produce such clear, white visible light. If anything, people complain about how bright they actually are, especially when driving at night and an oncoming car has high-intensity LED headlights.

—*Jordan P. Howell and Richard M. Renneboog, MSc*

Further Reading

Cangeloso, Sal. *LED Lighting: A Primer to Lighting the Future.* O'Reilly, 2012.

Energy Star. "Light Bulbs (CFLs)." http://www.energystar.gov/index.cfm?fuseaction=find_a_product.showProductGroup&pgw_code=LB.

General Electric Consumer and Industrial Lighting. "Compact Fluorescent Light Bulbs." http://www.gelighting.com/na/home_lighting/ask_us/faq_compact.htm#how_work.

Johnson, Alex. "Shining a Light on Hazards of Fluorescent Bulbs." http://www.msnbc.msn.com/id/23694819/ns/us_news-environment/t/shining-light-hazards-fluorescent-bulbs.

Lenk, Ron, and Carol Lenk. *Practical Lighting Design with LEDs.* John Wiley & Sons, 2017.

Schubert, E. Fred. *Light-Emitting Diodes.* 2nd ed. Cambridge UP, 2005.

Hybrid Vehicles and Energy Security

FIELDS OF STUDY
Automotive Engineering; Electrical Engineering; Mechanical Engineering

ABSTRACT
Concerns about fuel dependency and cost, energy security, and greenhouse gas emissions have renewed public- and private-sector interest in hybrid vehicles.

KEY CONCEPTS
barrel: a benchmark volume used as a *de facto* standard measurement for petroleum production, with 1 barrel being standardized as a volume of 159 liters

hybrid: a light-duty vehicle that uses a small internal combustion engine in conjunction with an electric motor as its power unit

regenerative braking: a system in which part of the kinetic energy lost in braking is captured during braking and used to regenerate electrical charge in the battery

LIGHT-DUTY VEHICLES AND PETROLEUM USE

Since the 1970s, the dependence of vehicles on petroleum has contributed to economic and environmental concerns, particularly in industrialized countries. The National Renewable Energy Laboratory (NREL) of the U.S. Department of Energy reports that 69 percent of total petroleum consumption went to the transportation sector in the United States in 2007. Light-duty vehicles (LDVs) such as cars, sport utility vehicles, and light trucks, have the highest fuel use within this sector, which makes them particularly relevant in discussions of energy consumption. In 2006, LDVs used 8.86 million barrels of petroleum per day, followed by medium to heavy vehicles, which used 2.47 million barrels per day. An increasing reliance on oil imports also impacts national energy security. In 2007, 58 percent of petroleum consumed in the United States was imported oil, compared to 43 percent in 1979.

With steep increases in fuel prices and use, there are clear implications for the individual and national economy and carbon footprint. Developing a long-term solution to energy needs requires reducing the petroleum consumption of vehicles, which could come from changes in consumer behavior, increases in vehicle efficiency, or greater diversification of energy sources.

HYBRID VEHICLE VARIATIONS AND VALUE

Hybrid vehicles, which use a combination of technologies to perform, are increasingly seen as a plausible means to address these issues. Some hybrid variations include plug-in hybrid electric vehicles (PHEVs), which have the ability to charge the battery by plugging a vehicle cable into the electricity grid; battery electric vehicles (BEVs), which use rechargeable lithium ion batteries; and hybrid electric vehicles (HEVs), the most common, which use both gasoline and electricity for power sources.

HEVs allow the use of a smaller engine that can run in the most efficient part of its operating range, and some models can be driven in an all-electric mode. Their continuously variable transmissions also prevent the engine from operating when it is unnecessary, such as when idling or when the vehicle is stopped. Finally, hybrids can store some of the kinetic energy normally lost while braking, where the batteries store some of this kinetic energy and then release it to assist the gasoline engine as needed; this process is known as "regenerative braking." These features contribute to reducing both petroleum usage and greenhouse gas emissions.

Various hybrid technologies may attract consumers by providing different options for reduced fuel use. A 2009 NREL report quantifying fuel savings from hy-

Vehicle being charged. (nrqemi via iStock)

brid vehicles estimates that between 1999 and 2007, hybrid vehicles saved nearly 385 million gallons of fuel in the United States, or more than 9.2 million barrels of petroleum. Although these savings were small in comparison to the total amount of fuel consumed by LDVs, the savings are projected to grow as additional hybrid models penetrate the market. The International Energy Agency (IEA) reports that the percentage of hybrid vehicles purchased in the United States between 2005 and 2009 increased from 1.2 to 2.8 percent, even in the midst of an economic recession; however, the automotive website Edmunds.com reports that hybrids peaked at 3.1 percent of US new-car sales in 2013, falling to 2 percent market share by 2016 while plug-in hybrids reached 0.4 percent penetration that same year. Although these growing numbers still represent a small percentage of the overall market, the 2017 annual energy report of the U.S. Energy Information Administration (EIA) predicts that hybrid vehicles will comprise 4 percent of the market share of new-car sales by 2040.

Government groups and businesses alike, keen to take on a "greener" image and reduce costs, are considering hybrid buses, trucks, and commercial vehicles as possible solutions. In pickup and delivery service, truck fuel economy can be improved from 30 to 50 percent with hybrid technologies. According to the U.S. Environmental Protection Agency (EPA), a typical

Hybrid vehicle. (nrqemi via iStock)

step van could save as much as $1,200 in fuel costs and reduce greenhouse emissions by more than 7 tonnes per year, with even greater benefits for a typical enclosed delivery truck. If public and private entities do continue to adopt hybrid vehicles according to research projections, there is the potential to reduce imported fuel for U.S. transportation needs significantly and to improve the economic and ecological outlook.

OBSTACLES TO WIDESPREAD ADOPTION
Public interest in rising fuel prices, concerns over global warming, and recovery from an economic recession helped to promote hybrid vehicles, but consumers

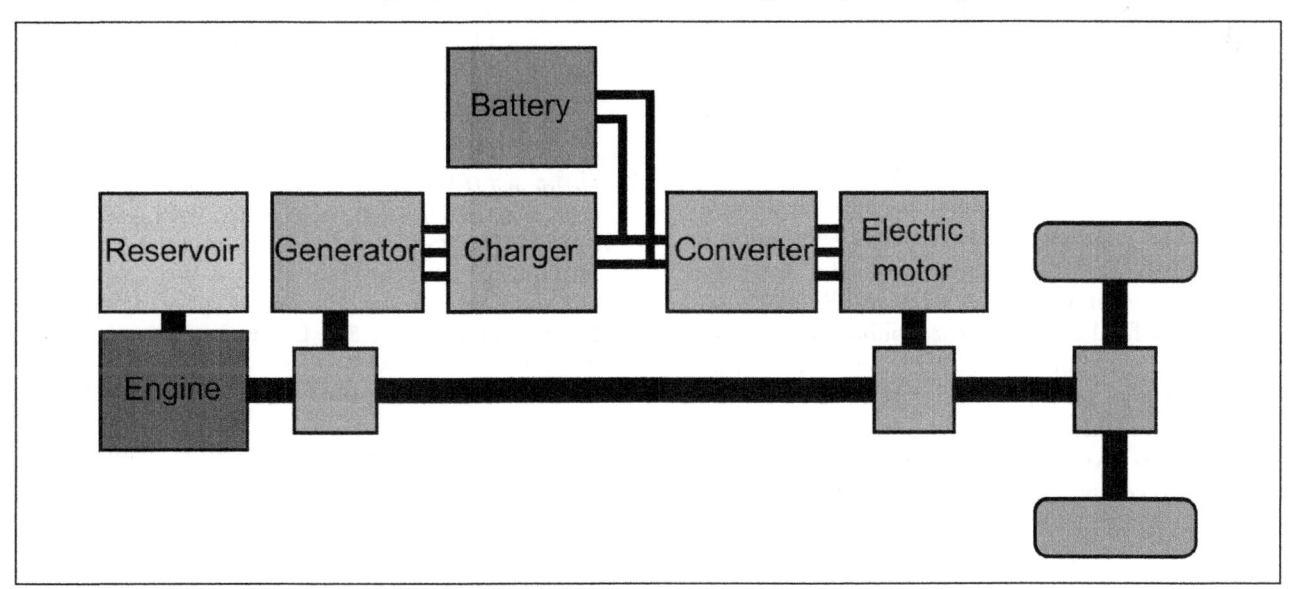

Structure of a combined hybrid vehicle. (Fred the Oyster via Wikimedia Commons)

still face significant barriers of cost, fuel infrastructure, and a viable market. Cost remains an impediment, as hybrid vehicle production requires a large amount of intricate electronic and engineering construction, and models can have expensive parts, such as lithium ion batteries. Moreover, domestic petroleum production was reinvigorated with the shale-oil boom of the 2010s thanks to hydraulic fracturing, thus lowering the price of gasoline again and reducing interest in high-efficiency vehicles for many. At the same time, improvements in battery technology and the slow, but increasing, installation of charging infrastructure have made all-electric vehicles more appealing to environmentally conscious consumers.

GOVERNMENT MANDATES AND INCENTIVES

The U.S. government has been pursuing research in hybrid vehicles for decades to address the nation's economic and environmental concerns. In 1966, Congress introduced the earliest bills recommending the use of electric vehicles as a means of reducing air pollution. In 1976, Congress passed the Electric and Hybrid Vehicle Research, Development, and Demonstration Act to spur the development of new technologies, including improved batteries, motors, and other hybrid-electric components. In 1990, California passed the Zero Emission Vehicle (ZEV) Mandate, which required 2 percent of the state's vehicles to have no emissions by 1998 and 10 percent by 2003, although the mandate was incrementally weakened by the early 2000s. These early legislative efforts attempted to use new technologies to reduce gas use and emissions, although both industrial and public adoption of changes was limited.

In 2008, the U.S. Department of Energy awarded $8 billion in automaker loans to support the development of fuel-efficient vehicles and assist the automotive industry after the impacts of the economic recession. The American Recovery and Reinvestment Act of 2009 further allocated $2 billion for development of electric vehicle batteries and related technologies, and another $400 million was slated to fund building the infrastructure necessary to support plug-in electric vehicles. These initiatives, along with tax credit incentives for consumers, focused on building self-sufficiency and sustainability in energy resources.

Technological advances have produced several models of hybrid vehicles with viable options for energy and economic savings. Continued progress toward energy sustainability requires changes in consumers' attitudes and behavior, support from government entities, and profitable incentives for automobile manufacturers. Government and industry researchers forecast increased public acceptance of and myriad benefits from hybrid vehicles that reveal a hopeful alternative to current energy dependence, economic strain, and environmental pollution.

—*Nicole Menard*

Further Reading

Annual Energy Outlook 2017, with Projections to 2050. U.S. Energy Information Administration, 5 Jan. 2017, www.eia.gov/outlooks/aeo/pdf/0383(2017).pdf. Accessed 10 Oct. 2018.

Bennion, K., and M. Thornton. "Fuel Savings from Hybrid Electric Vehicles." National Renewable Energy Council Technical Report. Mar. 2009. http://www.nrel.gov/docs/fy09osti/42681.pdf.

Donateo, Teresa, ed. *Hybrid Electric Vehicles.* IntechOpen, 2017.

Josephs, Leslie. "Long before the Combustion Engine, the Hybrid Car Is Facing Obsolescence." *Quartz*, 14 July 2017, qz.com/1029464/what-percent-of-us-car-sales-are-hybrids. Accessed 10 Oct. 2018.

International Energy Agency. *Hybrid and Electric Vehicles: The Electric Drive Advances.* Paris, International Energy Agency, 2010. http://www.ieahev.org/pdfs/2009_annual_report.pdf.

Khajepour, Amir, Saber Fallah, and Avesta Goodarzi. *Electric and Hybrid Vehicles. Technology, Modeling and Control: A Mechatronic Approach.* Wiley, 2014.

MDPI. *Emerging Technologies for Electric and Hybrid Vehicles.* MDPI, 2018.

"Oil: Crude and Petroleum Products Explained—Oil Imports and Exports." *EIA*, U.S. Energy Information Administration, 1 May 2018, /www.eia.gov/energyexplained/index.php?page=oil_imports. Accessed 10 Oct. 2018.

Van Mierlo, Joeri, ed. *Plug-in Hybrid Electric Vehicle (PHEV).* MDPI, 2019.

Industrial Revolution and Machine Power

FIELDS OF STUDY
History; Materials Science; Mechanical Engineering

ABSTRACT
The Industrial Revolution in the eighteenth and nineteenth centuries in Europe and the United States led to a shift from traditional power sources to machine power, fueled mainly by coal-fired steam engines. It also saw increases in the scale of manufacturing production.

KEY CONCEPTS
American system: a production manufacturing system that relied on specialized or single-purpose machines rather than general or multipurpose machines

charcoal: an essentially pure carbon source produced by burning wood at low temperatures with insufficient air to promote combustion

coke: an essentially pure carbon source produced by heating coal, without promoting its combustion, to drive out contained tars and oils

craft industry: nonmechanized industries that produce traditional handmade goods such as textiles and ceramic ware

Newcomen engine: an external combustion engine that functioned by injecting high-pressure steam into a closed cylinder to raise a piston within the cylinder, then relied on the reduced pressure inside the cylinder, produced by the condensation of the steam, to draw the piston back down inside the cylinder before the next introduction of high-pressure steam

INDUSTRIAL REVOLUTIONS?
There is no universal agreement about the dates or extent of the Industrial Revolution, although clearly the word *revolution* is a misnomer, because the Industrial Revolution took place over many decades and occurred at different times in different places. The most common use of the term is to refer specifically to the first phase of European industrialization, starting in Britain in the 1730s and lasting until about 1820. Others limit it to a narrower period from the 1740s to the 1780s and characterize the following decades (until about 1820) as a period of geographical expansion of manufacturing and a consolidation and standardization of technological processes.

Still others recognize a "Second Industrial Revolution" in Germany and the United States from the 1870s until the outbreak of World War I in 1914, when specialization in steel production, along with universal education and new management techniques and systems, propelled these countries into global leadership in industrial production. For former European colonies in Latin America, Africa, and Southeast Asia, industrialization occurred much later, well into the twentieth century; thus, in a sense, these countries are still experiencing industrial revolutions of their own.

WHY BRITAIN?
Historians have pointed to several explanations for why the Industrial Revolution began in Britain: Britain's place in the global economy in the early 1700s; the availability of natural resources in the British Isles, such as coal and iron ore; an intellectual climate that fostered practical engineering and innovation in machine technology; relatively high wage levels that made it desirable for factory owners to substitute machines for human labor; existing transportation and financial infrastructures suitable for adaptation to the economies of scale enabled by machine-based manufacturing; and a relatively well-educated surplus pool of labor that was available to work in the factory system.

The first blast furnace of Germany as depicted in a miniature in the Deutsches Museum. (Wikimedia Commons)

First, in the early 1700s, Britain was already a major trading power in the world. Raw materials flowed into British ports from its far-flung colonies, including fur, lumber, and cotton from North America; sugar (in both crystalline and molasses forms) from the West Indies; and cotton from the Indian subcontinent. This global presence helped to create a new merchant class that was looking to invest its profits to generate additional capital. Moreover, Britain's island location meant that it was relatively isolated from wars on the European continent and the disruptions to industry and trade that such conflicts caused.

Second, Britain had ample raw materials for industrializing and expanding textile manufacture, in the form of wool from sheep and cotton from its colonies overseas. It also produced wrought-iron tools for agriculture and industry, using readily available iron ore, charcoal (a reducing agent made from burning wood in insufficient oxygen), and the power of swiftly running streams to run the hammers and bellows on forges. These resources were located close to densely populated areas, making it efficient to transport them to sites where industrial production could take place.

However, timber began to grow scarce: British forests had been cleared for shipbuilding and agriculture, burned for heating and cooking fuel, harvested for the production of charcoal, and cut for lumber for building construction. Thus, iron makers turned to coal, which could be burned to produce coke, another reducing agent in the iron-making process. Fortunately, British coal ore was low in phosphorus, which allowed higher-strength iron (and later, steel) to be developed.

Third, practical inventions, such as the spinning jenny, cotton gin, and above all the steam engine, enabled raw materials to be processed more efficiently, and in larger quantities. Continued refinements and improvements of these inventions by a range of British and European scientists, tinkerers, and engineers allowed development of even safer, more fuel-efficient machines.

Fourth, wage levels for workers were high, relative to other countries in Europe. Workers were more likely to have the means to purchase manufactured goods in Britain than elsewhere. This increased the demand for manufactured goods but also provided the impetus for owners of craft industries to seek ways in which machine power could substitute for human labor, thereby reducing production costs and increasing profits.

Fifth, Britain had an existing craft industry system in which linkages between producers of raw materials, intermediate manufactured products (such as yarn in textiles), and finished goods were well developed. These producers were also connected through a comprehensive barge/canal transportation network, making it convenient and cost-effective to move goods to local markets and to port cities for export. Thus, the British were well positioned to seek greater efficiency in using those new machines and types of production to make greater-value goods that could be sold both domestically and on international markets.

Last, improvements in agricultural technology and knowledge of farming led to greater efficiency in pro-

A Roberts loom in a weaving shed in 1835. Textiles were the leading industry of the Industrial Revolution, and mechanized factories, powered by a central water wheel or steam engine, were the new workplace. (Wikimedia Commons)

duction of food and fiber crops, freeing rural workers for other kinds of work. This led to mass migrations to the new industrial cities, where wage labor (payment by the piece or by the hour) was the norm. Cities like Manchester and Birmingham experienced enormous population growth as rural inhabitants streamed into these new manufacturing centers seeking work. This population was relatively well educated and receptive to the disciplines of the factory system, where production was governed by the clock rather than the rhythms of the day and seasons.

GEOGRAPHICAL DIMENSIONS

From their beginnings in Britain, the innovations of the Industrial Revolution, particularly steam-powered engines and machine-based manufacture, spread first to northwestern Europe: along the Rhine- Ruhr Valley in northwestern Germany, and to Belgium, the Netherlands, and France. Subsequently, these innovations were taken up in the northeastern area of the United States, central Germany, northern Italy, and the industrial heartland of southern Poland and the Czech Republic. Somewhat later, the Russian Empire began to industrialize in the Moscow region, in the Volga River area, and later in the Ural Mountains and Siberia (where natural resources, especially energy sources such as oil, coal, and natural gas, were plentiful). In all these areas, as in Britain, densely populated areas, access to raw materials, and existing communication and transportation networks made larger-scale manufacturing not only possible but also profitable.

The Industrial Revolution was accompanied by social, economic, and political changes as well. For example, after the overthrow of the monarchy during the French Revolution and the subsequent Napoleonic Wars, the French government emphasized universal primary education, the establishment of a national language, the teaching of mechanical skills in schools, and the professionalization of science. These factors enabled the rise of a workforce well suited to work in factories and an engineering profession that continued to innovate in mechanical engineering and the emerging field of electrical technology.

DEVELOPMENT OF TECHNOLOGY

Thomas Newcomen, a British engineer and businessman, invented the first commercially used steam engine around 1710. This device heated a cylinder of water that had a piston at the top. As the water in the cylinder turned to steam, it expanded, pushing the piston upward. The piston was connected to a rocker arm (like a seesaw) that could move up and down, allowing materials to be lifted mechanically. Such a device was highly desirable in mines, where periodic flooding limited access to new, deeper beds of ore. The rocker arm governed a pump that moved water out of the mine shafts to the surface.

Although hundreds of Newcomen steam engines were built, they were inefficient, because fuel was needed to reheat the water in the cylinder after it had cooled and condensed. By the 1760s, the Scottish inventor James Watt had refined the basic steam engine principle by condensing steam in a separate cylinder, allowing the main cylinder to maintain its high temperature. This machine could produce more power with the same input of heating. Watt patented his device in 1769.

The next step was to develop a more useful mechanical device using a camshaft to produce rotary motion rather than piston (seesaw) motion, and Watt patented a machine to do that in 1781. In the decades that followed, many other inventors and tinkerers contributed useful refinements to the original steam engine concept: indicator gauges, safety valves, platforms allowing the engines to be portable, and ultimately more compact and lighter-weight steam engines that could be used in ships and railroad engines.

Parallel with the development and refinement of the steam engine came machines that would automate the process of producing textiles, one of Britain's largest craft industries. The production of cloth involved multiple steps, including washing, dying, spinning yarn, and weaving the yarn into fabric. An Englishman, John Kay, patented the flying shuttle in 1733; his device had wheels, allowing the shuttle to pass easily across the loom and speeding up the weaving process. This put pressure on yarn makers to keep up, and around 1750, James Hargreaves invented the spinning jenny, which allowed spinners to produce multiple spindles of yarn simultaneously.

Using these and other technologies, by 1789 the first steam-driven cotton textile factory had been established, in Manchester, England. Cotton fibers were strong and particularly amenable to being worked with machines. This was a good fit with the resources

An iron worker dwarfed by an industrial blast furnace (sdlgzps via iStock)

of Britain, whose cotton came from North America and India. In 1790, a British textile mechanic named Samuel Slater smuggled the British technology out of England by memorizing the design of the machines, and he helped to establish the first American textile mills, in Pawtucket, Rhode Island. By 1800, the steam engine was being used in textile manufacture in Germany, and by about 1810 in France.

In the United States, Eli Whitney invented the cotton gin to separate cotton seeds from the boll fibers by mechanical means. It is generally believed that others, working in Europe, probably invented similar machines at about the same time. Whitney's true innovation, however, was in what is called the "American system" of manufacture, in which specialized machines that did only one task were produced, rather than more generalized machines that could be configured to do multiple kinds of tasks. Moreover, these specialized machines were designed to have standardized parts, allowing adequate parts inventories and efficient repair and replacement of parts. This approach eventually led to the development of the assembly-line system introduced by Henry Ford in the period known as the Second Industrial Revolution.

The development of steam-powered, machine-based technology in Europe and the United States thus can be viewed as a network in which incremental improvements responded to bottlenecks of workflow within a geographically dispersed production system, each bottleneck spurring new innovations and improvements in machines for the manufacturing process.

CHANGES IN TRANSPORTATION

Innovations in manufacturing were accompanied by changes in transportation technologies and networks.

In 1800, Richard Trevithick developed a small, portable steam engine that was suited for use in boats and railroads. Separately, in 1802, William Symington adapted the steam engine for use in a tugboat. In the next decade, steam engines were placed in locomotives to pull freight on iron tracks; this system was used at first to haul coal out of mines before the overland system was developed. Iron wheels running on iron tracks produced less friction than iron or wooden wheels running on roadbeds, and as a result less energy was needed to move goods from place to place.

Despite refinements to the steam engine, it remained a relatively inefficient and rigid means for producing power, and its use began to decline after the 1880s. From the mid-nineteenth century on (in the period known as the Second Industrial Revolution), inventors and engineers sought more efficient ways to power machines. In 1859, the first prototype of the internal combustion engine was developed in Belgium. By the 1870s, several modern forms of the internal combustion engine had been developed, using gasoline or diesel. The internal combustion engine offered several benefits for manufacturing: It required less labor to operate; it could run at different speeds; and it could easily stop and start.

The German inventor Karl Benz adapted the technology of the internal combustion engine for automobiles and patented it in 1889. Ford adopted this technology (earlier automobiles had run on coal gas or electric power or a combination), and by 1910 the United States was the largest producer of gasoline-powered automobiles in the world, using the mass-production system to build automobiles quickly and inexpensively.

SOURCES OF ENERGY

Before the Industrial Revolution, mills (for example, textile mills and sawmills) could be run with hydropower, using the energy of water running downstream in order to power belts, shafts, and other mechanical devices. However, these systems were vulnerable to extreme changes in water levels such as droughts and floods. As discussed above, wood as a fuel was in short supply by the early eighteenth century, and interest turned to coal as a fuel source.

One of the benefits of coal was that it had several byproducts that were useful in industry, including coal gas and coal tar. Both were byproducts of the process of making coke from coal. Coal gas was widely used in the eighteenth and nineteenth centuries for lighting, cooking, and heating. The development and availability of coal gas permitted widespread outdoor lighting; by 1820, most major European city streets were lit with coal gas. This form of technology lasted about 50 years, giving rise to the term the "Gaslight Era." Coal gas was also widely used in interior lighting, both in homes and in factories. The use of artificial lighting in addition to daylight in factories allowed factories to be larger and their hours of production to be extended.

Coal tar, another byproduct, was a useful raw material that spurred the advance of the industrial chemical sector, which developed organic chemicals such as paints, synthetic dyes, photographic materials, and medicines.

Although petroleum was known (it seeped naturally out of the ground), there was little commercial use of it until 1853, when Samuel Kier used it to make kerosene, a cheap substitute for whale oil, which was used, along with coal gas, for interior lighting. In 1859, Edwin Drake drilled the first oil well in Pennsylvania, and in 1863 the first oil pipeline was constructed, also in Pennsylvania.

Electrical power, a secondary energy source produced from water flows or from the burning of fossil fuels, came into wide commercial and industrial use in the Second Industrial Revolution, although the properties of electricity had been known for more than a century.

SECOND INDUSTRIAL REVOLUTION

While the first phase of the Industrial Revolution had centered on greater automation of textile production, automating steel production was the basis of the Second Industrial Revolution. Steel is made from iron by blasting a stream of air through molten iron in a "blast furnace" to pull off the impurities of the iron. Steel had previously been made only in small quantities because it was such an energy-intensive process. The English inventor Henry Bessemer invented a furnace that would allow steel to be produced much more inexpensively; he patented his furnace (called the Bessemer converter) in 1855. In the 1850s, the German engineer Karl Siemens developed the open-hearth furnace, which recovered the waste heat of the furnace to preheat the gases used for the next round of iron. In this process, even larger quantities of steel could be produced, allowing steel to become competitive in price with iron. By 1914, Germany was the European leader in steel, producing as much in that year as England, France, Italy, and Russia combined.

The availability of inexpensive steel fundamentally changed the urban landscape, making possible large-span bridges and skyscrapers. Steel was also used to increase the scale of other manufacturing processes, since larger generators, turbines, and other machines could be constructed that were much lighter in weight than their cast-iron counterparts.

Beginning about 1870, the United States began to dominate global industrial production through the development of the so-called American system, in which specialized, single-purpose machines made products with standardized, interchangeable parts. (The American government supported this system, in particular in its requirements and specifications for weapons.) Ford pioneered the assembly-line system, in which the product itself, not the workers or the machines, moved along a conveyor belt, and this technique was subsequently used widely in American manufacturing. (European countries did not adopt assembly-line production techniques until World War I.) The American system allowed for efficiency in machinery, efficiency in use of labor, and levels productivity that ultimately led to the worldwide domination of American manufacturing in the early twentieth century.

—*Judith Otto*

Further Reading
Allen, Robert C. *The British Industrial Revolution in Global Perspective*. Cambridge UP, 2009.
Bunch, Bryan. *The History of Science and Technology*. Houghton Mifflin, 2004.
Farr, James R., ed. *The Industrial Revolution in Europe, 1750-1914*. Gale Group, 2002.
Hahn, Barbara. *Technology in the Industrial Revolution*. Cambridge UP, 2020.
Mokyr, Joel. "The Second Industrial Revolution, 1870-1914." August 1998. http://faculty.wcas.northwestern.edu/~jmokyr/castronovo.pdf.
Schwab, Klaus. *The Fourth Industrial Revolution*. Crown Business, 2016.
Stearns, Peter N. *The Industrial Revolution in World History*. 4th ed. Routledge, 2018.

Internal Combustion Engine

FIELDS OF STUDY
Automotive Engineering; Mechanical Engineering; Power Engineering

ABSTRACT
The technology of the internal combustion engine internal combustion engine applies the reaction of an oxidizer and a fuel within the engine cylinders in a combustion that produces the working fluids (in the form of expanding gases) to supply kinetic energy.

KEY CONCEPTS
four-stroke engine: an engine requiring four piston strokes for a single fuel combustion; one stroke draws in air, a second compresses the air, fuel is injected and ignited to drive the piston through the third stroke, and the fourth stroke pushes out the combustion gases. All four strokes are timed to the opening and closing of intake and exhaust valves.
oxidizer: a compound that supports the combustion (oxidation) of the fuel; most commonly atmospheric oxygen, although other materials such as nitrous oxide are used in specialty applications
rotary engine: an internal combustion engine design that uses a central rotor driven by successive fuel ignitions rather than individual pistons connected to an eccentric crankshaft and driven in sequence by fuel ignitions
two-stroke engine: an internal combustion engine in which piston movement within the cylinder opens and closes intake and exhaust ports within the cylinder during operation, rather than using a camshaft timed to the rotation of the crankshaft to coordinate the opening and closing of separate intake and exhaust valves as in the four-stroke engine

ENGINES DRIVE THE WORLD
For all practical purposes, the internal combustion engine is the sole type in use in 100 percent of automobiles and trucks today—and that dominance of the transportation sector is likely to be little changed

through 2021. At the same time, as of 2017, commentators were raising more questions about whether the incorporation of electric motors would supersede the use of internal combustion engines, at least in automobiles, in the not-so-distant future. Advancements in battery technology, specifically regarding efficient lithium-ion batteries, combined with global efforts to fight climate change led to some arguments that it would not be long before automakers would be shifting more to electric technology—or at least more of a hybrid between the two. As electric cars are simpler to make and maintain and drastically reduce carbon emissions as opposed to cars that use internal combustion engines, the call for electric cars has further increased, to the point that all three major North American automakers are investing heavily in electric vehicle research and production facilities in 2021. Yet others claim that with some companies working on methods for making the internal combustion engine more efficient, it will continue to remain in use for some time.

An internal combustion engine is one in which the working fluids are comprised of reactants of combustion—oxidizer and fuel—along with the products of combustion. The heat released by combustion of the oxidizer-fuel mixture propels the gaseous products of combustion against the moving surfaces of the engine, whether a piston, turbine blade, or other device. The chief types of internal combustion engines are reciprocating and rotary (Wankel). Gas turbine types are also in use.

RECIPROCATING ENGINES

Reciprocating engines come in three general categories:

- *Continuous- and intermittent-combustion engines:* A steady flow of fuel and oxidizer into the engine is the signature of the continuous-combustion engine. A stable flame is maintained; jet engines are typical. The intermittent-combustion engine applies periodic ignition of air and fuel; this is most often called a "reciprocating engine"; gasoline and diesel piston engines are the most common examples.
- *Spark ignition/compression ignition engines:* Fuel and air are mixed prior to the intake stroke, or just after inlet valve closure, in the electric spark ignition, or Otto, engines. In compression ignition versions, the fuel—diesel—is injected after the compression process; rather than a spark, the fuel is ignited by the high temperature of the compressed gas.
- *Four-stroke (4s) and two-stroke (2s) engines:* The work cycle in a four-stroke engine takes two crankshaft revolutions, divided into intake, compression, expansion, and exhaust stages. This is sometimes called the "Otto cycle," in honor of inventor Nikolaus Otto. One crankshaft revolution is sufficient for a complete two-stroke cycle, which has no intake or exhaust strokes. Gas exchange in these engines occurs when the piston is near the bottom center position, between the expansion and compression strokes.

ROTARY ENGINES

The rotary-piston engine, invented by Felix Wankel, generates power in the familiar four-stroke cycle of compression, ignition, and expansion of a gasoline-air mixture. However, the moving parts work in a continuous rotary motion, instead of a reciprocating movement.

The Wankel rotary engine delivers one power stroke for each full crankshaft rotation. Thus its displacement volume is used twice as often as in four-stroke engines. This allows the average Wankel to be engineered to roughly half the size and weight of a conventional engine. There are also far fewer components, usually about 40 percent of the number of moving parts a V-8 engine would have. There is a similar kind of advantage, although not as dramatic, over a four-cylinder engine. By reducing the number of components and the complexity of their interactions, engine manufacture costs are also reduced.

Ironically, the Wankel engine also embodies some manufacturing drawbacks, mainly the need for expensive materials and the requirement for higher-precision manufacturing techniques. For all that, the Wankel has tended toward low fuel economy and high emissions of incompletely combusted hydrocarbons. The high hydrocarbon emissions result from poor sealing between the rotor and housing. However, design and engineering improvements since 2003 have brought about production models of the Wankel type

Reciprocating engine of a car. (Christopher Ziemnowicz via Wikimedia Commons)

that meet contemporary fuel economy and emissions standards.

GAS TURBINE ENGINES

The gas turbine, or combustion turbine, engine is a rotary, continuous internal combustion engine wherein the fuel flows to a burner supplied with an excess of compressed air. The expanding combustion gases apply pressure to turbine blades; the turbine power is transferred via gearing to an output shaft.

The gas turbine engine embodies four main operations, as described in Mehrdad Ehsani, et al:

- *Compression:* Air enters the gas turbine and is compressed.
- *Heat exchange:* Heat is drawn from the exhaust gases and communicated to the compressed air.
- *Combustion:* Fuel is mixed with hot air and ignited. The pressure increases.
- *Expansion:* The hot exhaust gases drive the turbine, thus releasing their energy. The turbine turns the compressor and the output shaft.

The advantages of gas turbines include high rotational speed, yielding a very compact engine; rotating movement yields vibration-free operation; ability to operate on a wide variety of fuels; and continuous combustion yields reduced hydrocarbon and carbon monoxide emissions compared to internal combustion engines.

However, these advantages must be balanced against the following drawbacks when gas turbines are considered for automotive applications: quenching of the gases by the turbine and compressor yields high noise levels; low efficiency of the dynamic compressor and turbine at smaller scales yields relatively higher fuel consumption; to obtain the high rotating speeds needed to maximize efficiency, sophisticated and costly materials must be incorporated; and high cost materials must also be used to tolerate the higher temperature levels compared to other engine types.

COMPARISONS WITH EXTERNAL COMBUSTION ENGINES

An internal combustion engine is a heat engine; its thermal energy is derived by a chemical reaction within the working fluid. The working fluid itself is then exhausted to the environment, aiding excess heat rejection. In external combustion engines, heat is transferred to the working fluid through a solid wall and also expelled to the environment via another solid wall. Steam engines are in this class.

Internal combustion engines have two intrinsic advantages: Except for auxiliary cooling, they require no heat exchangers, thereby reducing weight, volume, cost, and complexity. With no requirement for high-temperature heat transfer through walls, the design of internal combustion engines permits the maximum temperature of the working fluid to exceed maximum allowable wall material temperature.

Intrinsic disadvantages include the fact that for all practical purposes the working fluids are limited to air and products of combustion. Nonfuel heat sources, such as waste heat, cannot be used to generate motive energy. Additionally, there is little flexibility in combustion conditions—they are largely set by engine requirements. This factor makes achieving low-emissions combustion more difficult.

—Reza Fazeli

Further Reading

"The Death of the Internal Combustion Engine." *The Economist*, 12 Aug. 2017, www.economist.com/news/leaders/21726071-it-had-good-run-end-sight-machine-changed-world-death. Accessed 23 Aug. 2017.

Ehsani, Mehrdad, et al. *Modern Electric, Hybrid Electric, and Fuel Cell Vehicles, Fundamentals, Theory, and Design*. CRC Press, 2005.

Fijalkowski, B.T. *Automotive Mechatronics: Operational and Practical Issues*. Vol. II. Springer, 2011.

Hiereth, Hermann, and Peter Prenninger. *Charging the Internal Combustion Engine (Powertrain)*. Springer, 2010.

Josephson, Paul R. *Motorized Obsessions: Life, Liberty, and the Small-Bore Engine*. Johns Hopkins UP, 2007.

Kornhauser, Alan A. *Internal Combustion Engines: The Engineering Handbook*. 2nd ed., CRC Press, 2004.

Mayersohn, Norman. "The Internal Combustion Engine Is Not Dead Yet." *New York Times*, 17 Aug. 2017, www.nytimes.com/2017/08/17/automobiles/wheels/internal-combustion-engine.html. Accessed 23 Oct. 2017.

Smil, Vaclav. *Creating the Twentieth Century: Technical Innovations of 1867-1914 and Their Lasting Impact*. Oxford UP, 2005.

ISOTOPES, RADIOACTIVE

FIELDS OF STUDY

Archaeology; Geology; Nuclear Engineering; Nuclear Medicine

ABSTRACT

Radioactive isotopes are unstable nuclides that decay ultimately to stable nuclides by emission of alpha, beta, gamma, or proton radiation, by K capture, or by nuclear fission.

KEY CONCEPTS

chromatographic separation: a preparatory technique in which compounds in a mixture are separated from each other by differential adhesion to a stationary solid phase within a moving fluid phase

nuclide: a particular radioactive atomic nucleus

pharmoacokinetics: the time-related behavior of pharmacologically active compounds within living systems

radiation: the emission of subatomic particles and electromagnetic energy due to nuclear fission or fusion processes

radiopharmaceuticals: drugs and other biological compounds whose molecules have been constructed with a radioactive component

RADIOACTIVE ISOTOPES

All the known elements have at least one radioactive isotope, either natural or artificially produced. Therefore, the radionuclides are found in the Earth's crust, in its surface waters, and in the atmosphere. Radioisotopes are used in many areas of science and industry as tracers or as radiation sources. They provide fuel for the nuclear generation of electricity and have found both diagnostic and therapeutic uses in medicine.

DESCRIPTION, DISTRIBUTION, AND FORMS OF RADIATION

Alpha, beta, and gamma radiation are the three types of naturally occurring radioactivity; they result in the transmutation of one chemical nucleus to another. Alpha decay is the ejection from the nucleus of a particle equivalent in size to a helium nucleus. The daughter nucleus has an atomic number (Z) two less than that of the parent and a mass number (A) four less than the parent. The equation below represents the emission of an alpha particle from a polonium nucleus to produce an isotope of lead (a gamma ray is also emitted in rare cases):

$$^{210}_{84}Po \rightarrow {^{206}_{82}Pb} + {^{4}_{2}He} + \gamma$$

Beta decay results from the change within the nucleus of a neutron into a proton. Z increases by one, while A is unchanged. The equation below illustrates beta emission by phosphorus to become sulfur:

$$^{32}_{15}P \rightarrow \beta^- + {^{32}_{16}S}$$

In gamma decay, electromagnetic radiation is emitted as a nucleus drops to lower states from excited states. It is the nuclear equivalent of atomic line spectra that show wavelengths of visible light emitted by atoms when electrons drop from higher to lower energy levels. Nuclear fission is an extremely important process by which isotopes of the heavy elements such as uranium 235 capture a neutron and then split into fragments:

$$^{235}_{92}U + {^{1}_{n}} \rightarrow {^{140}_{56}Ba} + {^{94}_{36}Kr} + 2 \text{ neutrons}$$

The neutrons produced are captured by other nuclei, which in turn undergo fission, producing a chain reaction. This is the process that resulted in the first atomic bomb and is now used in nuclear plants to produce electric power.

HISTORY

The story of radioactivity begins with Wilhelm Conrad Röentgen's work with cathode-ray tubes. Roentgen allowed cathode rays to impinge on various metal surfaces and observed that highly penetrating radiations, which he called X-rays, were produced. He noted similarities between the X-rays and sunlight in that both could expose a photographic plate and could cause certain metals and salts to fluoresce.

This fluorescence was of interest to Antoine-Henri Becquerel, who discovered by accident that crystals of uranium salt left on a photographic plate in a drawer produced an intense silhouette of the crystals. Although his understanding of the phenomenon was limited at the time, what Becquerel had observed was the effect of uranium radioactivity.

Marie and Pierre Curie pursued the study of this phenomenon with other minerals. They worked to isolate and characterize the substances responsible and were able to isolate and purify samples of polonium and radium. Other scientists worked at the same time to characterize the radiations emitted. In 1903, Ernest Rutherford and Frederick Soddy proposed that the radiations were associated with the chemical changes that radiation produced, and they characterized three types of radiation: alpha (α), beta (ß), and gamma (γ) rays.

OBTAINING RADIOISOTOPES

The use of nuclear fission to produce energy is based on a principle formulated by Albert Einstein, $E=mc^2$. E is energy, m refers to mass, and c is a constant equal to 3.0×10^8 m/c. The complete conversion of one gram of matter per second would produce energy at the rate of nine trillion watts.

The main particles contained in the nucleus of an atom are protons and neutrons. The mass of a given nucleus is less than the sum of the masses of the constituent protons and neutrons. This mass defect has been converted, according to the equation above, to energy (binding energy) in the process of forming the nucleus. The separation of the nucleus into its constituent particles would require replacement of this energy. The binding energy per nucleon is a measure of the stability of a particular nucleus. Those nuclei having mass numbers between 60 and 80 have the highest binding energy per nucleon and are therefore the most stable. A large nucleus such as uranium can split into fragments with sizes in the 60 to 80 mass range. When this happens, the excess binding energy is released.

USES OF RADIOISOTOPES

Radioisotopes are used in a number of ways in the fields of chemistry and biology. Radioimmunoassay (RIA) is a type of isotopic dilution study in which labeled and unlabeled analytes compete for limited amounts of a molecule that binds the analyte very specifically. RIA is used worldwide in the determination of hormones, drugs, and viruses. The technique is so specific that concentrations in the picomolar region can be measured. Another major use of radioisotopes is as tracers that determine metabolic pathways, transport processes, and reaction mechanisms. A compound labeled with a radioactive isotope is introduced into the process, and the radioactivity allows the compound to be followed through the mechanism.

Pharmacokinetics is the study of the rates of movement and biotransformation of a drug and its metabolites in the body. Many kinetic parameters, such as a drug's half-life in the body, can be determined by using radiolabeled drugs and measuring radioactivity after some type of chromatographic separation of the parent drug from its metabolites.

Radiopharmaceuticals are substances labeled with radionuclides that are used in the visualization of organs, the location of tumors, and the imaging of biochemical processes. This usage is based on the fact that a substance that is found in a healthy cell at a certain concentration has a different concentration in damaged cells. The particular isotope used depends on the organ or biochemical process under study.

Radioisotopes are used in many ways in industry. Gamma rays from cobalt-60 are used to examine objects for cracks and other defects. Radioisotopes can be used to measure thickness of all types of rolled materials and as tracers in locating leaks in pipes carrying liquids or gases. The fill level of closed containers is monitored by absorption or scattering of radiation.

In the chemical industry radioisotopes are used to indicate the completeness of a precipitation reaction. A radioisotope of the element to be precipitated is added to the solution to be precipitated. When the filtrate is free of radioactivity, precipitation is complete.

Radioisotopes are used in dating ancient rocks and fossils. Carbon is used in dating recent fossils. All living organisms are assumed to be in equilibrium with their environment, taking in carbon in food and expelling it through respiration and other processes. A living organism is assumed, when it dies, to have a certain percentage of carbon-14, the radioactive isotope of carbon. As the fossil ages the carbon-14 decays by beta emission, and its percentage is reduced. Since the decay rate is known, a reasonable age estimate can be obtained by measuring the rate of radioactive emission (proportional to percentage carbon-14) from the fossil. Uranium is used in a similar way to date rock samples that contain a mixture of uranium and lead, which is at the end of its decay chain.

—*Grace A. Banks*

Further Reading

Abdel Rahman, Rehab O., and Hosam El-Din Em Saleh, eds. *Principles and Applications in Nuclear Engineering-Radiation Effects, Thermal Hydraulics, Radionuclide Migration in the Environment*. IntechOpen, 2018.

Faure, Gunter, and Teresa M. Mensing. *Isotopes: Principles and Applications*. 3rd ed. Wiley, 2005.

Plekhanov, Vladimir G. *Isotopes in Condensed Matter*. Springer, 2013.

Serway, Raymond A., Chris Vuille, and Jerry S. Faughn. *College Physics*. 8th ed. Brooks/Cole Cengage Learning, 2009.

World Nuclear Association. "Radioisotopes in Industry." http://www.world-nuclear.org/info/default.aspx?id=548&terms=radioisotopes.

World Nuclear Association. "Radioisotopes in Medicine." http://www.world-nuclear.org/info/default.aspx?id=546&terms=radioisotopes.

ISOTOPES, STABLE

FIELDS OF STUDY

Chemistry; Geology; Nuclear Science; Physics

ABSTRACT

Stable isotopes comprise the bulk of the material universe. Some elements are found in only a single form, while others have several isotopes. For study and application, it is necessary to separate the various isotopes from one another. A number of methods have been developed to accomplish isotope separation.

KEY CONCEPTS

atomic weight: generally, the weighted average of the atomic weights of the isotopes of an element according to their natural abundances; specifically the weight of an atom as determined by the number of protons and neutrons in its nucleus

element: a substance whose atoms contain a specific number of protons

isotopes: atoms of an element in which all atoms contain the same number of protons but different numbers of neutrons

mass spectrograph: or mass spectrometer; an analytical device that uses the deflection of a charged particle in a magnetic field according to its mass-to-charge ratio to identify and isolate particles according to their mass

natural abundance: the relative proportions in which the isotopes of an element are found in nature

ELEMENTS AND ISOTOPES

Stable isotopes comprise the bulk of the material universe. Some elements are found in only a single form, while others have several isotopes. For study and application, it is necessary to separate the various isotopes from one another. A number of methods have been developed to accomplish isotope separation. Analysis of stable isotopes and isotopic composition is used extensively in a wide variety of fields. These include soil and water analysis, plant tissue analysis, determination of metabolic pathways in plants and animals (including humans), archaeology, forensics, the geosciences, and medicine.

An isotope is one of two or more species of atom that have the same atomic number (number of protons) but different mass numbers (number of protons plus neutrons). Stable isotopes are those which are not radioactive. Because the chemical properties of an element are almost exclusively determined by atomic number, different isotopes of the same element will exhibit nearly identical behavior in chemical reactions. Subtle differences in the physical properties of isotopes are attributable to their differing masses.

There are approximately 260 stable isotopes. While most of the 81 stable elements that occur in nature consist of a mixture of two or more isotopes, 20 occur in only a single form. Among these are sodium, aluminum, phosphorus, and gold. At the other extreme, the element tin exhibits ten isotopic forms. Two elements with atomic numbers less than 84, technetium and promethium, have no stable isotopes.

The atomic weight of an element is the weighted average of its isotope masses as found in their natural distribution. For example, boron (atomic number 5) has two stable isotopes: boron-10 (an isotope with 5 protons and 5 neutrons, with mass number 10), which accounts for 20 percent of naturally occurring boron, and boron-11 (an isotope with 5 protons, 6 neutrons, and mass number 11), which accounts for 80 percent. The average atomic weight of naturally-occurring boron is therefore $(0.2) \times (10) + (0.8) \times (11) = 10.8$. In those elements that have naturally occurring isotopes, the relative abundance of the various isotopes is found to be remarkably constant, independent of the source of the material. There are cases in which the abundances are found to vary, and these are of practical interest.

HISTORY

In the early part of the twentieth century, the discovery of radioactivity, radioactive elements, and the many distinctly different products of radioactive decays showed that there were far more atomic species than could be fit into the periodic table. Although possessing different physical properties, many of these species were chemically indistinguishable.

In 1912, Joseph John Thomson, discoverer of the electron, found that when a beam of ionized neon gas was passed through a properly configured electromagnetic field and allowed to fall on a photographic plate, two spots of unequal size were exposed. The size and location of the spots were those that would be expected if the original neon consisted of two components—about 90 percent neon-20 and 10 percent neon-22. Later Francis William Aston improved the experimental apparatus so that each isotope was focused to a point rather than smeared out. The device he developed, known as a mass spectrograph, allows much greater precision in the determination of isotope mass and abundance.

OBTAINING ISOTOPES

All methods for separating stable isotopes are based on mass difference or on some isotopic property that derives from it. The difficulty of isotope separation depends inversely upon the relative mass difference between the isotopes. For example, the two most abundant isotopes of hydrogen are ordinary hydrogen (hydrogen-1) and deuterium (hydrogen-2).

These isotopes have a relative mass difference of (2-1)/1 = 1, or 100 percent. The mass difference between chlorine-35 and chlorine-37, by contrast, is only (37-35)/35 = 0.057, or 5.7 percent.

There are two types of separation methods. The only single-step method is electromagnetic separation, which operates on the principle that the curvature of the path of a charged particle in a magnetic field is dependent on the particle mass. This is the same principle on which the mass spectrograph is based. Though it is a single-step technique, the amount of material that can be separated in this way is extremely small. All other processes result in a separation of the original material into two fractions, one slightly enriched in the heavier isotope. To obtain significant enrichment the process must be repeated a number of times by cascading identical stages. Such multistage methods include gaseous centrifugation, aerodynamic separation nozzles, fractional distillation, thermal diffusion, gaseous diffusion, electrolysis, and laser photochemical separation. For example, in centrifugation a vapor of the material to be separated flows downward in the outer part of a rotating cylinder and upward in the center. Because of the mass difference, the heavier isotope will be concentrated in the outer region and can be removed to be enriched again in the next stage.

USES OF STABLE ISOTOPES

Most stable isotope applications are based on two facts. First, isotopes of a given element behave nearly identically in chemical reactions. Second, the relative abundances of isotopes for a given element are nearly constant. The three principal types of applications are those in which deviations from the standard abundances are used to infer something about the environment and/or history of the sample, those in which the isotopic ratio of a substance is altered so that the substance may be traced through a system or process, and those in which small differences in the physical properties of isotopes are used to understand process dynamics.

As an example of the first type of application, consider that the precise isotopic composition of water varies with place and time as it makes its way through Earth's complex hydrologic cycle. Knowledge of this variation allows for the study of storm behavior, identification of changes in global climatic patterns, and investigation of past climatic conditions through the study of water locked in glaciers, tree rings, and pack ice. The cycling of nitrogen in crop plants provides an example of stable isotope tracer methods. Fertilizer tagged by enriching (or depleting) with nitrogen-15 is applied to a crop planting. Subsequent analysis makes it possible to trace the quantities of fertilizer taken up by the plants, remaining in the soil, lost to the atmosphere by denitrification, and leached into runoff water.

—*Michael K. Rulison*

Further Reading

Carter, James F., and Lesley A. Chesson, eds. *Food Forensics: Stable Isotopes as a Guide to Authenticity and Origin.* CRC Press, 2017.

Dawson, Todd E., and Rolf T.W. Siegwolf, eds. *Stable Isotopes as Indicators of Ecological Change.* Elsevier, 2007.

Fry, Brian. *Stable Isotope Ecology.* Springer, 2006.

Hobson, Keith A., and Leonard I. Wassenaar, eds. *Tracking Animal Migration with Stable Isotopes.* Academic Press, 2008.

Meier-Augenstein, Wolfram. *Stable Isotope Forensics: Methods and Forensic Applications of Stable Isotope Analysis.* 2nd ed., Wiley, 2018.

Northern Arizona University, Colorado Plateau Stable Isotope Laboratory. "What Are Stable Isotopes?" http://www.mpcer.nau.edu/isotopelab/isotope.html.

JOULE, JAMES

FIELDS OF STUDY
Electrical Engineering; Physics; Thermodynamics

ABSTRACT
James Joule made major contributions to the fields of heat, electricity, and thermodynamics, and was the first scientist to grasp the connection between heat and other forms of energy. He is best known for his articulation of the second law of thermodynamics, Joule's law, and the Joule-Thomson effect.

KEY CONCEPTS
caloric theory: an early theory that heat existed as a type of fluid in all materials and flowed from objects with a high caloric content to objects with lower caloric content, and held that the caloric content of the universe was constant

magnetostriction: the phenomenon in which metal objects deform somewhat when they become magnetized

mechanical theory of heat: the theory that heat is produced by mechanical movement of atoms and molecules within materials (NB: the concept of atoms and molecules was very rudimentary in Joule's time; a workable theory of atomic structure had not yet been devised and the existence of subatomic particles would not be demonstrated until several years after Joule's death)

perpetual motion machine: the fallacious concept of a machine that would produce at least as much work or energy as that required for its operation

BACKGROUND
One of the most important physicists of the nineteenth century, James Prescott Joule was born on Christmas Eve, 1818, in Salford, Lancashire, near Manchester, England. Joule's father was a wealthy brewer. Because he was ill as a young child, Joule was tutored at home until the age of 16 when he began studying mathematics and natural philosophy with eminent chemist John Dalton (1766-1844), who taught at the scientific academy of the Manchester Literary and Philosophical Society (later Manchester University). Under Dalton's tutelage, Joule examined current hot topics in science such as the caloric theory of heat, mechanics of the steam engine, and complexity of electric motors. He had a personal interest in such issues because the brewery was considering replacing its steam engine with the newer electric motor. That interest led Joule to conduct experiments in electrolysis, magnetic attraction, and electric motors.

James Joule. (Wikimedia Commons)

While still a teenager, Joule attempted to build a perpetual motion machine, but eventually gave it up.

Because Joule continued to work at the brewery, most of his experiments were done in the early morning before the brewery opened, or late at night after it closed. Joule and his elder brother Benjamin took over the brewery when their father's health began to fail, and his passion for science was often viewed by family and friends as only an interesting hobby. In 1847, Joule married Amelia Grimes of Liverpool. She died in 1854, leaving him with a son and a daughter. His experimentation with heat led him to reject the popular caloric theory and replace it with a new mechanical theory of heat. Eventually, the success of his experiments and his conviction that he had viable contributions to make to the field won him respect from his peers.

Major changes brought on by the Industrial Revolution were occurring in the understanding of scientific phenomena during Joule's lifetime. He was the first scientist to recognize the relationship between heat and mechanical energy. Joule's early work was influenced by that of German physician and physicist Julius Robert von Mayer (1814-78), who is considered the founder of the principle of the conservation of energy. By 1840, Joule had articulated Joule's law, which states that heat produced within a wire by a constant direct current is always equal to the resistance to the current times the square of the current:

$$P = I^2 \times R$$

This is now the general expression for the power (P) expressed in watts (W) expended by an electrical component while operating, with resistance R expressed in ohms (O) and current I expressed in amperes (i).

THE JOULE EFFECT

Joule also described the Joule effect, which states that when gas is allowed to freely expand into a vacuum, there is no apparent change in temperature. Joule's findings on the mechanical equivalent of heat became an essential element in teaching basic physics classes. Joule developed this theory as a result of his "paddle experiment" in which he used an insulated container holding liquid and paddle wheels. The system was controlled by a system of pulleys and weights. As the weights fell, they caused the paddles to agitate the liquid inside the container, resulting in a rise in temperature. The experiment allowed Joule to estimate the mechanical equivalent of heat, which became known as the Joule, referred to as "J" or "kJ."

Joule met Scottish mathematician and physicist William Thomson, Lord Kelvin (1824-1907), at the British Association for the Advancement of Science in 1847. Thomson's interest in the result of Joule's paddle-wheel experiments also generated interest among other scientists at the meeting. The two men subsequently developed a prolonged and collaborative professional relationship. In 1854, they identified the cooling effect that became known as the Joule-Thomson effect as a result of the "porous-plug experiments," which allowed them to ascertain that temperatures of gases decrease when they are expanded through a porous plug under external pressure.

Joule's law of conservation of energy became accepted as the first law of thermodynamics and has major implications for chemistry. The law states that changes in energy that occur within a system are equivalent to the heat in the system, plus the system itself, minus the labor produced by the system. Joule also conducted experiments on magnetostriction, determining that the shape of materials such as nickel, iron, and cobalt changes slightly in response to magnetization. That process is still referred to as Joule magnetostriction because he was the first to identify it.

Joule's contributions were officially recognized by his peers in 1850 when he was elected as a fellow of the Royal Society. In 1870, he was awarded the Copley Medal by the society. Joule lost much of his fortune in 1875, and his health began to decline in subsequent years. He died in Sale, England, in 1889. His legacy continues to have major implications for modern electrical and refrigeration systems. Designers and manufacturers of contemporary refrigerators, freezers, air conditioners, heat pumps, and dehumidifiers all regularly employ his findings when creating their products. Utility companies also use his theories to determine supplies of voltage to their customers.

—*Elizabeth Rholetter Purdy*

Further Reading

Cardwell, Donald S.L. *James Joule: A Biography.* Manchester UP, 1989.

Cropper, William H. *Great Physicists: The Life and Times of Leading Physicists from Galileo to Hawking.* Oxford UP, 2001.

North, John, ed. *Mid-Nineteenth Century Scientists.* Pergamon Press, 1969.

Segrè, Emilio. *From Falling Bodies to Radio Waves: Classical Physicists and Their Discoveries.* W.H. Freeman, 1984.

Steffens, Henry John. *James Prescott Joule and the Concept of Energy.* Science History Publications, 1979.

K

KINETIC ENERGY

FIELDS OF STUDY
Chemistry; Engineering; Mechanics

ABSTRACT
Kinetic energy occurs from motion and the object in motion's ability to do work. Kinetic energy can take three forms: rotational, translational, and vibrational. Rotational kinetic energy comes from rotational motion, translational kinetic energy exists when an object moves from one position to another, and vibrational kinetic energy comes from vibration.

KEY CONCEPTS
rotation: angular motion of a mass about a fixed axis
scalar quantity: a physical property that has magnitude that is not associated with a specific direction
translation: movement from one locus to another through space in a straight line
vector quantity: a physical property that has both magnitude and an associated direction
vibration: back-and-forth movement, or oscillation, of a physical mass about a central fixed point; vibrational motion may be longitudinal along a central axis, lateral across that axis, or a combination of both

KINETIC ENERGIES
The word *kinetic* is derived from the Greek word for motion (*kinesis,*) and the word *energy* is the ability to do something (*energia*). Kinetic energy, then, occurs from motion and the object in motion's ability to do work. Kinetic energy can take three forms: rotational, translational, and vibrational. Rotational kinetic energy comes from its rotational motion, such as that produced from the spinning of Earth on its axis. Translational kinetic energy exists when an object moves from one position to another, such as water flowing in a river or a train traveling. Vibrational kinetic energy comes from vibration, such as the oscillation of a loudspeaker's cone.

HISTORY
From a historical perspective, Gottfried Leibniz (1646-1716) and Johann Bernoulli (1667-1748) depicted kinetic energy as a living force that was equal to mass times velocity squared. Gaspard-Gustave de Coriolis (1792-1843) first coined the term *work* and described the concept of kinetic energy in its contem-

The word energy *comes from the ancient Greek word energeia, meaning "activity," as it appeared in the work of Aristotle in 4th century b.c.e. (Justus van Gent, oil on panel, c. 1476). (Wikimedia Commons)*

233

The cars of a roller coaster reach their maximum kinetic energy when at the bottom of the path. When they start rising, the kinetic energy begins to be converted to gravitational potential energy. The sum of kinetic and potential energy in the system remains constant, ignoring losses to friction. (Wikimedia Commons)

porary form, equal to the expression $1/2\ mv^2$. William Thomson (Lord Kelvin) (1824-1917) coined the term *kinetic energy* in 1856.

The total kinetic energy of an object is the sum of its translational and rotational forms. In classical physics, the amount of translational kinetic energy can generally be determined by the equation $KE = 1/2\ mv^2$, with m as the mass of the object and v its speed at the center of the object. (This equation is for an object that is not also rotating, and when v is less than the speed of light.) As with work and potential energy, the quantity of kinetic energy is expressed in joules, with a joule expressed as kgm^2s^2, where kg is a kilogram, m is a meter, and s is a second. Kinetic energy is a scalar quantity, having magnitude but not direction. Rotational kinetic energy is equal to one-half the product of the moment of inertia around the axis of rotation times the square of the angular velocity, expressed as $E_{rot} = 1/2\omega^2$. In relativistic physics, kinetic energy is equal to the increase in mass caused by motion multiplied by the square of the speed of light, as $E_{rel} = mc^2$. The kinetic energy of a system is the total kinetic energy of the objects within the system, and is also the capacity to do work on other systems or objects.

From a point of zero kinetic energy when an object is at rest, the level increases as work is done on it and the object accelerates, while the kinetic energy level decreases with deceleration. According to the Work-Energy Principle, the amount of work can be determined by subtracting the initial quantity of kinetic energy from its current quantity. The amount of kinetic energy is relational, subject to the frame of reference. For example, a car being observed by a person moving at the same speed would have zero kinetic energy relative to the person.

Kinetic energy can be transferred to another object(s) or forms. For example, soccer players can supply kinetic energy from their swinging legs to move a ball a distance on the field or in the air, while the brakes of automobiles dissipate kinetic energy into heat as they slow the vehicle. In the case of the soccer players' kicks, the energy is conserved in what is known as an elastic collision. In inelastic collisions, the total kinetic energy is not conserved as some is turned into other forms, such as the vibrating atoms in heat.

—*Kirk S. Lawrence*

Further Reading
Firk, Frank W.K. "Essential Physics I."
　http://www.physicsforfree.com/essential.html.
Georgia State University. "Kinetic Energy."
　http://hyperphysics.phy-astr.gsu.edu/hbase/ke.html.
Halliday, David, Robert Resnick, and Jearl Walker. *Fundamentals of Physics*. John Wiley & Sons, 2009.
Ling, Samuel J., Jeff Sanny, and William Moebs. *University Physics Vol. 1*. Samurai Media Limited, 2017.
McCall, Martin W. *Classical Mechanics from Newton to Einstein: A Modern Introduction*. 2nd ed. Wiley, 2011.
Young, Hugh D. *College Physics*. Addison-Wesley, 2012.

Liquid Fluid Energy Transmission

FIELDS OF STUDY
Hydraulic Engineering; Mechanical Engineering

ABSTRACT
The term "fluids" refers to both liquids and gases. With regard to energy transmission the most common referents are water- and oil-based transmission media. Liquids are incompressible, and this allows smooth transmission of the input energy. Water- and oil-based fluids in closed systems or in motion are used to transmit energy from source to point of application.

KEY CONCEPTS
actuator: a device that when activated performs a motion that causes a function such as extending a robotic arm or extending an aircraft's landing gear to be carried out
end effector: a device that functions as the "end user" of a fluid power system to carry out a function
incompressible fluid: a fluid that does not change its volume when an external pressure is applied
momentum: the propensity of an object in motion to remain in motion or if at rest to remain at rest, in accord with Newton's First Law
Pascal's Law: the pressure exerted on a fluid is exerted in all directions within the fluid

FLUID POWER
The term "fluids" refers to both liquids and gases. With regard to energy transmission, however, the most common referents are water- and oil-based transmission media. Liquids, unlike gases, are incompressible, and this allows smooth transmission of the input energy.

Along with electrical and mechanical power transmission, fluid power is one of the most popular and practically effective means of transmitting energy in industries. Force and power are transmitted through fluids when pressure is applied to a liquid in a closed system or by increasing the kinetic energy of the liquid. Fluid energy transmission is flexible and widely preferred where power or forces have to be transferred in different directions through different angles from one place to another. Such transmission has a source where energy originates and an actuator or

Daniel Bernoulli, whose Hydrodynamica *(1738) initiated the kinetic theory. (Wikimedia Commons)*

235

end effector where it is applied. Machines that utilize energy transmission through fluids are called "hydraulic machines." Commonly used actuators are cylinders and rotating devices.

Fluid transmission is typically used for transmitting large power over short distances, for example, in aircraft, ground vehicles, cranes, construction equipment, industrial machines, and within buildings or industrial setups. Hydraulic transmission has the ability to convert small input forces into large output force or power, and this property is used in controlling machines of all sizes, automobiles, and aircraft. For example, fluids containing glycol ethers are used to transmit force in braking systems in order to stop the motion of wheels. These brakes are called "hydraulic brakes," and they are very effective in converting the small forces applied to a brake pedal into the large forces needed to stop spinning wheels and degrade the momentum of a heavy vehicle. Lockheed Martin, which started as the Lockheed Hydraulic Brake Company, first made these brakes. Since hydraulic power can be stored, auxiliary or backup power in aircraft, to rotate turbines and start engines, is provided through hydraulics. For long distances, electrical transmission is convenient and more economical as compared to fluid transmission.

Fluid power transmission has been in use since ancient times for tasks such as lifting heavy loads, rotating objects, and facilitating agriculture. Force and power transmission was achieved almost entirely using fluids in the early part of the 1900s. The typical application was lifting loads. The basis for fluid energy transmission comes from seventeenth-century developments in the fields of hydrostatics and hydrodynamics. Blaise Pascal, in the year 1647, discovered an important property of fluids that pressure applied to a fluid at rest is transmitted in all directions. This law, known as Pascal's law, was the basis for development of hydrostatic power transmission using fluids. One of the first hydraulic machines, called the "Bramah press," was based on Pascal's law. An English inventor Joseph Bramah invented the press. It can convert small forces acting on a small area into large forces acting on larger areas by transmitting pressure through a fluid in a closed system. This feature is an integral part of most hydraulic applications even today.

HYDRAULIC SCIENCE

The scientific basis for hydrodynamic power transmission, or the use of fluids in motion for transmitting energy, can be explained by the Bernoulli principle, first stated in 1738 by mathematician Daniel Bernoulli, although hydrodynamic transmission was in use long before. This type of transmission imparts kinetic energy to the liquid, and the momentum of the liquid is used to transmit power. The kinetic energy, or flow of liquid, is obtained with the help of a pump. The choice of pump used depends on the operating conditions. Such systems are used primarily for obtaining rotational motion. The long history of fluid power transmission has resulted in the establishment of well-known engineering standards in hydraulics today. Numerous flow-control and pressure-control valves are available and routinely used to control fluid transmission.

The first hydraulic fluid used was water. The increasing use of metallic piping and mechanical actuators led to the replacement of this fluid with oil-based or synthetic liquids, as water causes corrosion and freezes at lower temperatures. Mineral-oil-based liquids that provide lubrication and can be used over a larger temperature range were superior. Hydraulic transmission liquids today can contain a wide range of chemical compounds, including water-soluble oils, silicones, esters, water-glycol mixtures, and complex hydrocarbons. The recent introduction of ceramics and water-resistant materials for transmission equipment has restored the feasibility of water as a medium.

The advantages of fluid-based energy transmission are its safety in hazardous applications such as areas with harmful radiation, the consistency of the force and power output, and easy control. Such systems can produce a wide range of rotation speeds and have high accuracy. The advent of modern electronics and its combination with fluid transmission have led to the development of smart actuators such as robotic hands, arms, and precision machines with sensors.

One of the major disadvantages of fluid energy transmission is the leakage of fluids. Oil-based fluids are dangerous to use in applications where there is a possibility of leakage in high-temperature areas or around electrical equipment. Hydraulic transmission is also accompanied by a consistent and loud hissing sound.

—*Dinseh Kapur*

Further Reading

Cundiff, John S., and Michael F. Kocher. *Fluid Power Circuits and Controls: Fundamentals and Applications.* 2nd ed. CRC Press, 2020.

Daines, James R. *Fluid Power: Hydraulics and Pneumatics.* Goodheart-Wilcox Company, 2012.

Johnson, James. *Introduction to Fluid Power.* Delmar Thomson Learning, 2002.

Klette, Patrick J. *Fluid Power Systems.* American Technical Publishers, 2014.

National Fluid Power Association. "Our Industry Overview." http://www.nfpa.com/OurIndustry/OurInd_Overview.asp.

Rabie, M. Galal. *Fluid Power Engineering.* McGraw-Hill, 2009.

Skinner, Steve. *Hydraulic Fluid Power: A Historical Timeline.* Steve Skinner Productions, 2014.

M

Mechanical Energy Transmission

FIELDS OF STUDY
Material Science; Mechanical Engineering; Physics

ABSTRACT
Lifting devices such as levers are among the simplest devices for mechanical energy transmission and have been in use since prehistoric times. During the Industrial Revolution many more types of mechanical energy transmission devices were developed.

KEY CONCEPTS
coupling: a connection in which the axis and the direction of rotation of the two separate parts are identical

fulcrum: the supporting point upon which a lever operates

helical gear: a gear in which the teeth follow an angled, or helical, direction across the thickness of the gear; prone to thrust forces due to the angle of the teeth and the force applied against them

herringbone gear: a combination gear in which helical gears with teeth angled in opposite directions are interfaced and used as a single gear; not prone to thrust forces as helical gears are

lever: a rigid bar or bar-shaped structure that operates upon a fulcrum

mechanical advantage: the force needed to carry out a task using a leverage function relative to the force required to carry out the task without the leverage function

pulley: essentially a gear without teeth used to facilitate the application of mechanical advantage

sprocket: a type of gear with pointed teeth designed to fit into the spaces of a flat drive chain; in an inverted design for use with link chains the chain fits into matching spaces cut into the edge of a pulley

worm gear: a type of gear in which a single tooth spirals around its central axis; typically used where the axis of rotation of the worm gear and the axis of rotation of the regular gear it meshes with are at right angles to each other

LEVERS AND LEVERAGE
Probably the simplest form of mechanical energy transmission is the lever. Although the lever has existed since the first unknown human used a tree branch as a prybar, Archimedes has been identified as the first-known scientist to describe the lever mathematically sometime before his death in 212 BCE. Three main types of levers exist, and all have four components. The first component is the fulcrum, the pivot point about which the lever rotates. The second component is the lever itself, which is usually a stiff rod or linear structure. The third component is the load the lever is to work against, and the fourth component is the applied force to move the load.

Class one levers work by pivoting on the fulcrum, with the load on one side of the fulcrum and the applied force on the other. If the lever is longer on the side of applied force, the load can be lifted with an applied force less than the force needed to raise the load directly. The distance the applied force moves will be greater than the distance the load moves. Conversely, if the lever is shorter on the side of applied force, then the applied force to move the load must be greater than the weight of the load, and the load will move a greater distance than the applied force. The simple crowbar is an example of a class one lever.

A class two lever has the load between the fulcrum and the applied force. A standard wheelbarrow is an example of a class two lever. A class three lever is similar to a class two lever; it has the applied force between the load and the fulcrum. Most mobile cranes are class three levers. A deep-sea fisherman using a fishing rod attached to a belt on his waist turns the fishing rod into a class three lever when he is pulling up on the middle of the fishing rod to lift a fish as the load.

GEARS AND GEARING

Gearing and gears constitute another common and often used means of transmitting mechanical energy transmission. Early gears existed in simple machines, such as potter's wheels, at the dawn of civilization at least 4,000 years ago in the Middle East. Both early Greeks and Romans used gears made of metal, wood, and stone in primitive machines. Systems of chain-driven gears were depicted by Leonardo da Vinci in the fifteenth century, and chain-driven energy transmission is common today in bicycles and many relatively low-speed machines.

Many types of gearing exist, and most gears are used to change torque and speed from a primary drive to a secondary drive. When two gears mesh, the smaller of the two is commonly referred to as the pinion. The simplest type of gear is the spur gear. Gears use "teeth" to transmit power, and in spur gears the teeth are perpendicular to the face of the gear. This is the type of gear often depicted in period-piece movies with old windmills or other constructs, with the mechanisms made of wood and the spur gear "teeth" being wooden pegs protruding from a circular base. In such depictions, it is fairly easy to see that two gears operating in concert with each other can be thought of as sets of levers working with each other, and in fact the directional forces that apply to tooth-on-tooth contact are identical to those of levers. This motion, however, tended to wear out wooden teeth rather frequently due to the repetitive friction of the teeth rubbing against each other as the machinery operated. In modern gear technology the teeth are designed and shaped in such a way that the surfaces of teeth that are meshed actually roll against each other rather than slide off of each other.

Helical gears represent a slight improvement on spur gears; the main difference is that the gear teeth are not perpendicular to the face of the gear but at an angle. The angle makes the tooth have greater surface area for transmitting force, which is the advantage that helical gears have over spur gears. Gears of this type can have a hunting tooth design, which means that one gear has an odd number of teeth and its meshing gear has an even number of teeth. In this arrangement, the teeth do not contact the same corresponding tooth with every revolution, and thus wear on the gear teeth is reduced.

Herringbone gears are an improvement over helical gears; the herringbone design comprises two helical gear teeth with opposite angles on the same gear. This type of gear can resist axial forces, an advantage over helical gears. When forces need to be transmitted by gears over right angles, bevel gears are used. The simplest form of bevel gearing is a cog-style gear, and early examples used wooden pegs. Worm gears are 90-degree transmission gears, like bevel gears, but instead of a bevel, a fluted worm is used for one of the gears. These gears are often used for fine-motion applications.

Many other types of gearing exist, including rack-and-pinion gears, but all gears transmit energy through teeth and rotational motion. Sprocket gearing can allow gears to be belt- or chain-driven instead of requiring direct contact between teeth. The belt or chain thus can be the weak point and can break if the machine driving, or driven by, the belt locks up or malfunctions. Belts exist in many forms and can be smooth or have teeth or ribs to assist in the transmission of force. In the early days of belt transmission, most belts were made of leather or cotton fibers but most modern belts are constructed of synthetic elastomers such as Viton. Most twenty-first century energy transmission chains are link chains. High-strength links are made and can be added in series to form a chain of any length desired. Chains can be self-lubricating or can require external lubricators, but all chains need some form of lubrication to keep friction low enough to make the chains practical for energy transmission and extend their functional lifetime by reducing wear.

Couplings are also used to transmit energy. Couplings exist as two main types, flexible and rigid. Rigid couplings are usually solid metal and joined together by friction fit or by being secured with bolts. Flexible couplings are usually more tolerant of misalignment and perform better than rigid couplings at higher shaft speeds. Flexible couplings take many forms; some of the main types are gear couplings, fluid couplings, wire grid couplings, disk pack couplings, and magnetic couplings. Gear couplings use gear teeth to transmit the force, typically as a splined shaft and a cylindrical fitting with matching splines on the inside, and slight misalignment between the gear teeth of the two machines is tolerated but unusual. Wire grid couplings use wire springs to transmit the force between the driver machine and the driven machine. The flexibility of the wire grid also allows for some misalignment of the coupling and still lets the force and energy be transmitted. Disk pack couplings are very similar to wire grid couplings but use large metal disks or washers instead of wire as the springs to transmit the force. Even magnetic force can be used to transfer power between two machines; the magnetic coupling uses magnets that do not touch each other to transmit the force. Hydraulic couplings work the same way as magnetic couplings, but fluid friction transmits power instead of magnetic force.

—*Damon Eric Woodson*

Further Reading

Abersek, B., and J. Flašker. *How Gears Break*. WIT, 2004.

Childs, Peter R.N. *Mechanical Design Engineering Handbook*. Elsevier, 2014.

Gears and Stuff. "Different Types of Gears." http://www.gearsandstuff.com/types_of_gears.htm.

Gears Manufacturers. "Gears History." http://www.gears-manufacturers.com/gears-history.html.

Goldfarb, Veniamin, Evgenii Trubachev, and Natalya Barmina, eds. *New Approaches to Gear Design and Production*. Springer Nature, 2020.

Howse, Jennifer. *Levers*. Weigl, 2010.

Isogawa, Yoshihito. *Gears*. No Starch Press, 2010.

Kapelevich, Alexander L. *Direct Gear Design*. CRC Press, 2013.

Radzevich, Stephen P. *Advances in Gear Design and Manufacture*. CRC Press, 2019.

Ryan, V. "Levers." http://www.technologystudent.com/forcmom/lever1.htm.

N

Natural Energy Flows

FIELDS OF STUDY
Environmental Engineering; Environmental Studies; Thermodynamics

ABSTRACT
Natural energy flows are sources of energy that do not need to be processed with additional energy but, rather, naturally flow through the world and need only be harvested or otherwise put to work.

KEY CONCEPTS
anthropogenic emissions: byproduct emissions of materials like carbon dioxide from the burning of fossil fuels caused by human activities

photosynthesis: the process in which chlorophyll-containing cells produce glucose from atmospheric carbon dioxide and water in the presence of sunlight

EARTH'S ABUNDANT ENERGIES
The term *natural energy flows* was popularized in environmentalism by the Hannover Principles, a document drafted by William McDonough and Michael Braungart in 1992, in anticipation of the Expo 2000 at Hannover, Germany. Natural energy flows are those sources of energy that are naturally flowing through the world already and need only be harvested or otherwise put to work; they include the flow of running water, the movement of the seas, the kinetic energy of wind, the solar radiation of sunshine, and the geothermal energies beneath the surface of Earth. They are not coequal with renewable energy sources, which would include biofuel.

CONSTANT MOVEMENT
Massive amounts of energy flow along the surface of Earth constantly. The sun radiates some 28×10^{32} calories of energy every year, about 13×10^{23} calories of which are intercepted by Earth. Of those intercepted calories, approximately one-third are reflected back into space. (This quantity, called the "albedo," varies locally according to weather conditions and type of surface coverage; snow-covered land reflects the most energy, as much as 90 percent, while ocean surfaces reflect less than 10 percent, and cloud cover also influences albedo.) The rest of the solar energy is absorbed by Earth or its atmosphere. Almost half is converted into heat, some of which powers the water (hydrologic) cycle of evaporation and precipitation. A small amount of sunlight, about 0.2 percent, is converted into the energy of wind, water currents, and waves. Even less than that, 0.1 percent, is used by photosynthesis. A tiny amount of the Earth's energy (tiny, that is, relative to this solar energy) comes from other sources, the geothermal heat emanating from Earth's interior and the tidal energy resulting from the gravitational interactions between Earth and the Moon. By contrast, sunlight accounts for tremendous amounts of Earth's energy and leads to most other forms. For example, photosynthesis by plants turns solar energy into chemical energy and stores it as carbohydrates in the plant, which over a geologic timescale when buried under great pressures eventually becomes gas, coal and petroleum deposits. Thus, even when humans are not harnessing natural energy flows, these flows are nevertheless the source of most of our energy.

McDonough was a New York City architect; Braungart, a German chemist. Together they collaborated on a sustainable design they called "cradle to cradle." The Hannover Principles articulated guidelines for the design of objects and buildings. The principles focused on interdependence with the natural world, the consequences of design decisions, waste and life-cycle assessments, the interrelationship between sustainability and human rights, and the responsible use of energy. With regard to natural energy flows, the document says, "Rely on natural energy flows. Human designs should, like the living world, derive their creative forces from perpetual solar income. Incorporate this energy efficiently and safely for responsible use."

McDonough has elaborated on this in later writing—indeed, he and Braungart have spent much of their subsequent careers refining and elaborating on the ideas in the Hannover Principles. In particular, McDonough has downplayed the idea of interpreting the directive to "rely on natural energy flows" in a quantitative way, and he does not appear to support renewable energy quotas or quantified energy efficiency goals. Instead, he and Braungart call for a fundamentally new approach to design, not simply a more efficient version of current designs. He has proposed buildings that sequester carbon, fix nitrogen, and make oxygen and distilled water while powered by solar energy; in general, the Hannover Principles seem primarily focused on solar power among the renewable energy sources.

ENERGY MANAGEMENT
The call for reliance on natural energy flows has been referred to as energy management, analogous to waste or water management. It is concerned with more than just the sustainability of its energy sources, in other words. Using natural energy flows reinforces and explicates humankind's connection to the planet by thriving from its already extant mechanisms, such as wind and water currents. The usual view of advocates is that natural energy flows alone—indeed, even solar power alone—are sufficient to meet humans' energy needs and that not attempting to use them to meet those needs is letting them go to waste.

There has been some suggestion that the use of natural energy flows may have environmental consequences that have not yet been made clear because their use is not yet widespread enough to understand the consequences of their use. John Holdren, one of President Bill Clinton's science advisers and Director of the Office of Science and Technology Policy under President Barack Obama, while publishing extensively on climate change and the need for proactive policies to reduce anthropogenic emissions, has also insisted that there is no known energy technology with negligible environmental impact. Windmills may harm avian life; hydropower frequently disrupts ecosystems; reservoirs and solar collectors need land to occupy; and some collection of geothermal energy is nonrenewable.

—*Bill Kte'pi*

Further Reading
Bunn, James H. *The Natural Law of Cycles: Governing the Mobile Symmetries of Animals and Machines.* Transaction Publishers, 2014.
McDonough, William, and Michael Braungart. *Cradle to Cradle: Remaking the Way We Make Things.* North Point Press, 2002.
McDonough, William, and Michael Braungart. *The Hannover Principles: Design for Sustainability.* W. McDonough Architects, 2003.
Roaf, Sue, and Fergus Nicol, eds. *Running Buildings on Natural Energy: Design Thinking for a Different Future.* Routledge, 2018.
Rogers, Elizabeth, and Thomas Kostigen. *The Green Book: The Everyday Guide to Saving the Planet One Simple Step at a Time.* Foreword by Cameron Diaz and William McDonough. Three Rivers Press, 2007.

Natural Gas

FIELDS OF STUDY
Environmental Sciences; Geology; Petroleum Engineering

ABSTRACT
Natural gas is largely composed of methane and is one of the three major fossil fuels used to generate energy. Natural gas may lead to the development of economically viable and efficient alternatives to fossil fuels.

KEY CONCEPTS
biogenic: produced by the decomposition of organic matter under anaerobic conditions
clathrate: a class of compounds in which a small molecule such as methane becomes trapped within the crystal lattice structure of water ice crystals as they form at cold temperatures in a wet environment
odorant: a compound having a strong, easily detectable and unpleasant odor, such as the sulfur-containing mercaptans, an odor characteristic of skunks

reserve-to-production ratio (RP): the amount of gas in a reserve relative to the annual rate at which gas is being removed from the reserve, stated in years

thermogenic: produced by the thermal decomposition of carbonaceous matter

AN IMPORTANT FUEL

Natural gas is one of the three major fossil fuels currently used to generate energy. In 2015, it was the fastest-growing fossil fuel in use, providing approximately 25 percent of the world's total primary energy, according to the International Energy Agency (IEA) supply. However, uses for natural gas vary between less developed and more developed nations; the former using natural gas largely for household heating and cooking and the latter for industry, electricity production, residential uses, and a small percentage of transportation fuels.

Natural gas consists primarily of methane and other hydrocarbon compounds. It is odorless and colorless and is often found associated with other fossil fuels, such as coal or petroleum beds. However, there are nonassociated stores of natural gas in isolated beds. Like other fossil fuels, natural gas is a highly combustible substance that, with recent infrastructure developments, is relatively easy to transport and store. Its combustibility makes it a highly effective resource for generating energy and heat.

THE CREATION PROCESS

It is generally believed that two processes create two types of natural gas. Biogenic gas is created when bacteria decompose organic material at shallow depths below Earth's surface. For example, these gases can be observed when wetlands are disturbed and "swamp gas" is released. Thermogenic gas develops deep underground, resulting from compression and heat. Holes must be drilled to access thermogenic gas deposits. Once accessed, in most cases, the natural gas must then be pumped to the surface. In a few cases, however, the gas will flow freely because of natural pressure. Both thermogenic and biogenic gases are created from biomass that has been buried in anaerobic environments. There is also a third theory that nonassociated gas stores exist far below Earth's surface, which were created during the formation of the earth.

The naturally occurring clathrate solid known as methane hydrate represents an intermediate formation of natural gas. Methane hydrate typically forms in cold, wet environments such as cold ocean floors and in terrestrial permafrost. It has also been observed to form as liquefied natural gas (LNG) is vented from storage tanks on humid days. Methane molecules become encased in the pores of a cage structure formed as water molecules freeze about them. Naturally occurring methane hydrate is believed to exist in large quantities on cold ocean floors, forming there with methane from either or both of biogenic and thermogenic sources. Also known as "ice that burns," the technology for harvesting the material from the oceans has not yet been developed.

One of the earliest known extractions and uses of natural gas by human society occurred during the Han dynasty (200 BCE). Chinese laborers drilled for natural gas with bamboo and used it to boil sea water for salt extraction. In the late 1700s, cities and towns in Britain used natural gas for lighting. In the United States, the first known intentional drilling for natural gas occurred in New York State in the 1820s. However, it was not until the invention of the Bunsen burner in 1885—allowing for more conventional uses such as cooking and heating—that natural gas became a viable option for generating energy around the world.

NONRENEWABLE RESOURCE

Natural gas is considered a nonrenewable resource because recoverable reserves are being exhausted at a rate that is a tiny fraction of the amount of time needed to create them. The Russian Federation, the United States, Iran, and Canada produced about 48 percent of the world's natural gas, as of 2014, and the United States and Russia accounted for more than one-third of the world's total natural gas consumption in 2013, based on data from the IEA and the U.S. Energy Information Administration. If extracted with contemporary technologies and known costs, current reserve-to-production ratio estimates (R/P ratio, measured in years) suggest there are enough conventional reserves to satisfy society's consumption needs for the next several decades. The vast majority of these proven reserves are located in Russia and the Middle East. The largest deposits are in Iran, Russia, and Qatar. The largest conventional gas field is the Urengoy

Field in the Western Siberian Basin, east of the Gulf of Ob, within the Arctic Circle. The Urengoy Gas Field was discovered in the 1960s and held an estimated initial total of 280 trillion cubic feet (8 trillion cubic meters).

Historically, natural gas was considered a low-value byproduct found interspersed with oil and coal deposits. Until the early twentieth century, natural gas was too inefficient for use as a large-scale energy resource. Because of a lack of technological development and insufficient infrastructure, producers could not get natural gas to markets in a feasible manner, and it was often burned off or allowed to vent into the air on site. Natural gas production remained slow during the industrial era, until post-World War II, when improvements in pipeline construction (and hence fuel transport) and safer infrastructure made natural gas technologically possible and economical. In 1937, after an undetected leak caused an explosion at the New London School in Texas, killing at least 300 people, minute amounts of odorants (such as mercaptan) were added to retail natural gas. This allows consumers to detect leaks in order to prevent fires or explosions.

IMPACT OF WORLD WAR II
During World War II, large pipelines were built in the United States from Texas to the northeast states to ensure energy security for the country during wartime. These pipelines were known as "Big Inch" and "Little Big Inch," and were responsible for transporting more than 350 million barrels of crude oil and refined products to the northeast before the war's end in the summer of 1945. After the end of the war, there were debates to determine if the pipelines should continue to transport oil or be converted to transport natural gas. The issue was settled in 1946, when an influential coal miner strike motivated the Senate and War Assets Administration to award the Tennessee Gas and Transmission company a lease to supply natural gas commercially. This transition made natural gas readily available to the northeastern United States, where there was a high-demand home heating market.

A COSTLY ENDEAVOR
Although current techniques (such as seismology) have reduced the costs of finding, extracting, and processing the fuel, natural gas production is an uncertain, complex, and costly endeavor. Natural gas straight out of the well is often accompanied by water and liquid hydrocarbons, including benzene, toluene, ethylbenzene, and xylene (BTEX), hydrogen sulfide (H_2S), and other organic compounds that must be removed. To make "pipeline quality" natural gas, it must be passed through units (called heater treaters) with chemical substances that absorb the byproduct water from the gas. Once the chemical extraction solution is saturated with water, the heaters raise temperatures to boil off the water. When cool, these large volumes of byproduct water are pumped to a "produced water" tank. The chemical separating fluid, which has a higher boiling point than water, cools and is recycled into reuse. Byproduct oily substances that were produced with the gas and water become volatile and recondense in a separate holding tank. This "condensate water" is commonly re-injected underground or hauled offsite to waste evaporation pits. In some cases, temporary pits are constructed, which hold waste materials and drilling mud so they can be reused through the drilling process. In order to reclaim drilling pads and sites, reserve pits must be drained and covered with topsoil or a capping material within a month of drilling completion.

After natural gas is processed and impurities are removed, it is often liquefied and compressed for storage and transport. Storage is an important issue, as most gas is used for heating during winter. A refrigeration process is used to condense the gas into liquified natural gas (LNG) by cooling it to -160°C (-260°F). LNG is then stored in insulated tanks, specially engineered to hold a cold-temperature liquid. LNG storage tanks are typically double-walled, composed of an outer wall of thick concrete and an inner wall made of steel. Between the walls is a thick layer of insulation. Often such storage facilities are underground to increase insulation. LNG will boil off and evaporate as natural gas, no matter how efficient the storage or refrigeration. This gas is then removed from the tank and used as a fuel on site, or refrigerated again to return it to the liquid state and placed back into the storage tank. If no pipeline is available to immediately transport gas, LNG also makes natural gas easier to store and transport. When LNG is transported (by train, truck, or ocean tanker) to its destination, or when it is removed from storage, it must be regasified.

Regasification is accomplished by heating LNG and allowing it to evaporate back into its gas state at typical temperature and pressure conditions. Regasification is usually done at a facility where the gas can be placed into storage or directly into a pipeline for transport.

REGASIFICATION TERMINALS

The two types of regasification terminals are called liquefaction terminals and regasification terminals. Liquefaction terminals, which turn natural gas into LNG, are on the export end of transactions. Regasification terminals, which turn LNG back into natural gas, are on the import side of operations. Currently, most natural gas is transported domestically within enormous infrastructure-intensive pipeline networks. Natural gas pipelines are often constructed of carbon steel to withstand the extremely high pressure of transporting compressed gas over large distances. Gas pipelines typically have small gathering systems that are composed of small diameter pipelines (5—20 centimeter, or 2—8 inches). These "gathering lines" tap into gas fields and gather to larger "trunk lines," which typically have an 20-120 centimeter, or 8- 48 inch, diameter and are large transnational or international transmission pipelines. Natural gas as a final product is delivered through another set of local pipelines directly to consumers. For example, the United States has natural gas gathering lines over a length of 32,000 kilometers (20,000-plus miles), and additionally, more than 432,000 kilometers (270,000 miles) of natural gas transmission lines.

The Federal Energy Regulatory Commission (FERC) regulates natural gas in the United States. Their primary responsibilities are to regulate transmission and sale, approve facility siting, render penalties for FERC energy market violations, oversee environmental matters, and administer reports. Safety rules fall under the purview of the Department of Transportation's Pipeline and Hazardous Materials Safety Administration (PHMSA). In early 2016, PHMSA required older pipelines to be pressured tested and previously unregulated gathering lines to be marked and undergo repairs and preventative maintenance in response to several accidents in the 2010s and concerns about leaks along pipeline routes contributing to greenhouse gas emissions.

DEREGULATION

Since the mid-1980s, the demand for, and production of, natural gas has risen significantly. This is largely due to deregulation, geopolitical dynamics, rising energy demands, and new technologies. In this context, natural gas is often characterized as an energy source that can potentially bridge society's current fossil fuel dependence to the development of economically viable alternatives to fossil fuels. Natural gas burns much cleaner than other fossil fuels and the development of infrastructure and technology has made it much easier to extract and transport. In the early part of the twenty-first century, T. Boone Pickens (an oil tycoon and prominent alternative energy promoter) advocated the development of natural gas-powered automobiles in the United States. Pickens claimed that by

Natural gas drilling rig in Texas. (David R. Tribble via Wikimedia Commons)

transitioning the transportation fleet to natural gas, U.S. society could fight the rising costs of oil with a smaller environmental impact, particularly in terms of reducing air polluting auto emissions.

Deregulation is a particularly important aspect of natural gas use. Natural gas has been regulated since the mid-1800s. The U.S. Congress passed national-scale regulations in the form of Natural Gas Act (1938) in order to manage interstate natural gas transmission and control monopolies. By the 1970s, however, many consumer states in the Midwest were experiencing shortages, despite adequate levels of supply in producer states. To remedy this, congress passed the Natural Gas Policy Act (1978), which created a single natural gas market and equalized supply and demand. This market-based approach was further embraced by FERC Orders (436 and 636), which unbundled transport, storage, and marketing, resulting in more consumer choice. As a result, prices decreased for large commercial and industrial customers but declined only slightly for residential consumers.

TRADED COMMODITIES

In the United States, a large percentage of natural gas is traded on the New York Mercantile Exchange (NYMEX) located in New York City. NYMEX is the exchange for energy products, metals, and other commodities, and transactions reflect the prices of traded commodities. Thus, U.S. gas prices are closely correlated with trading on NYMEX. Currently, the price of natural gas in the United States is signifi-

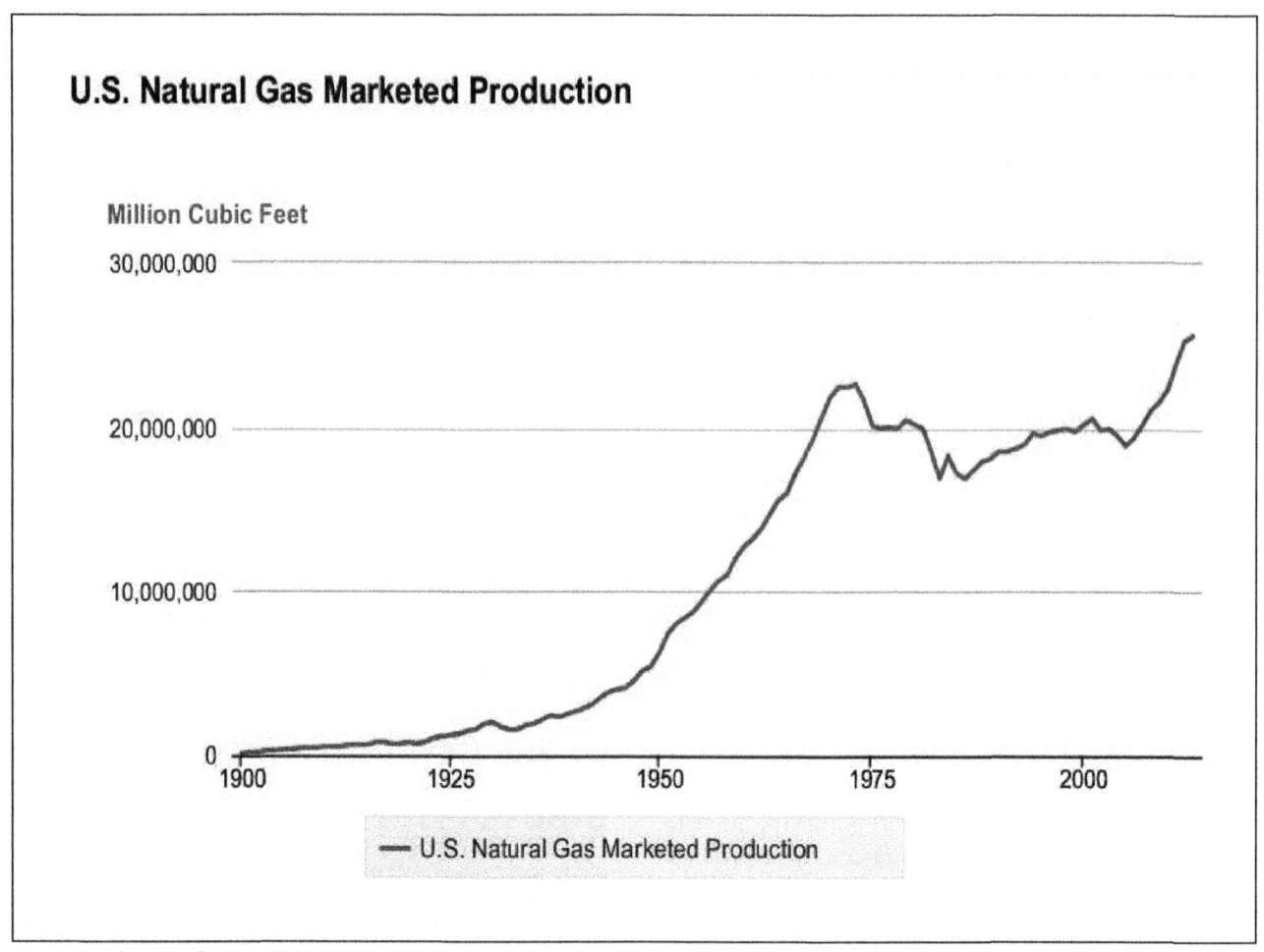

U.S. natural gas production, 1900–2013. (U.S. Energy Information Administration)

cantly lower than in the rest of the world. Therefore the United States is becoming a global exporter of natural gas, particularly for firms with access to LNG export terminals.

GLOBAL BENEFITS
Globally, the popularity of natural gas as a fuel around the world has initiated numerous large-scale and high-profile international natural gas pipeline projects in Asia, Europe, and North America. Today, LNG is exported from many regions where natural gas production exceeds consumption, such the Middle East and northern Africa, Russia, Trinidad and Tobago, Australia, and several southeast Asian countries. The low price of natural gas as a commodity often offsets the costs incurred when building liquefaction plants, converting the gas into LNG, and transporting the product to distant markets. Currently, the United States is the largest gas producer in the world but has little capacity for exporting gas. The U.S. Department of Energy is therefore promoting the construction of liquefied natural gas export terminals, and many new LNG export terminals are expected to be built in ports across the country. The larger markets for LNG include South Korea, Japan, and Taiwan. These are countries with densely-populated areas and little to no available domestic fossil fuel sources. Having access to this global LNG market has provided a relatively clean-burning fuel that can be easily distributed in pipelines.

It should be noted, however, that there are a relatively limited amount of remaining conventional deposits of natural gas and that, since it is a nonrenewable resource, it is not a suitable long-term substitute for oil and coal. While several large deposits of natural gas have been located throughout the world, experts believe that the majority of proven conventional reserves have been accessed. Estimates of dry natural gas reserves vary, but in 2014, the EIA approximated the world total at around 6,973 trillion cubic feet. Additionally, although natural gas is considered the cleanest of the three major fossil fuels, producing it often releases methane (the main constituent of natural gas and a potent greenhouse gas) directly into the atmosphere. Further, burning natural gas releases carbon dioxide (CO_2) on a molecule per molecule basis, the most prolific greenhouse gas implicated in global climate change. Natural gas production and combustion, therefore, impact water quality, air quality, and contribute to global warming.

There are, however, several other "unconventional" deposits of natural gas located around the world that have traditionally been considered cost-prohibitive to extract. Growing demand and improved technology have increased the potential of extracting and producing unconventional natural gas deposits. The majority of these unconventional deposits are located in interbedded geologic layers such as in coalbeds and shales, mixed with heavy crude oil, and as methane hydrates in cold regions or near the ocean floor. Additionally, recent research suggests that extraction of unconventional natural gas deposits inadvertently emits large amounts of methane, one of the most potent greenhouse gases. Despite concerns, unconventional natural gas resources are growing rapidly in importance. In 2000, unconventional natural gas resources contributed to only 1 percent of the U.S. natural gas supply. By 2016, however, they represented 50 percent and could grow to 69 percent by 2040, according to EIA estimates in 2016. Therefore, much of the current speculation, investment, and technological development concerning natural gas focuses on the greater extraction, production, and consumption of what is considered, at present, unconventional natural gas resources. R/P ratio estimates for recoverable unconventional reserves could be up to one hundred times the current R/P ratio.

In the last several decades, more developed nations have taken the lead in natural gas resource development and operations. Technological advances have enabled drilling firms to drill deeper and expand the industry, tapping into formerly unrecoverable reserves with greater efficiency and reduced costs. Firms are currently expanding their pursuit of unconventional resources in several African and European countries.

—*Laura J. Stroup*

Further Reading
Al-Megren, Hamid A., and Rashid H. Altamimi, eds. *Advances in Natural Gas Emerging Technologies.* IntechOpen, 2017.
Colborn, T., C. Kwiatkowski, K. Schultz, and M. Bachran. "Natural Gas Production from a Public Health Perspective." *International Journal of Human and Ecological Risk Assessment*, vol. 17 no. 5, 2011.

Cusick, Marie, and Susan Phillips. "U.S. Proposes New Safety Rules for Natural Gas Pipelines." *StateImpact: Pennsylvania*. WITF, 18 Mar. 2016. Web. 23 May 2016.

Geology.com. "Marcellus Shale-Appalachian Basin Natural Gas Play." http://geology.com/articles/marcellus-shale.shtml.

International Energy Agency. *Key World Energy Statistics 2015*. OECD/IEA, 2015. PDF file.

National Research Council. *Hidden Costs of Energy: Unpriced Consequences of Energy Production and Use*. National Academy of Sciences Press, 2009.

Palmer, Jerrell Dean, and John G. Johnson. "Big Inch and Little Big Inch." http://www.tshaonline.org/handbook/online/articles/dob08.

"A Primer." http://www.netl.doe.gov/technologies/oil gas/publications/EPreports/Shale_Gas_Primer_2009.pdf.

Smil, Vaclav. *Energy in Nature and Society: General Energetics of Complex Systems*. MIT Press, 2008.

Speight, James G. *Natural Gas: A Basic Handbook*. 2nd ed. Gulf Professional Publishing, 2019.

NUCLEAR POWER PLANTS

FIELDS OF STUDY
Electrical Engineering; Nuclear Engineering; Nuclear Science

ABSTRACT
Nuclear power plants are promoted as a solution to energy problems and as a way to reduce greenhouse gas emissions. However, in the United States, the nuclear industry has not upgraded technologies or constructed new plants since the 1990s.

KEY CONCEPTS
graphite: one of the many forms of pure carbon, the same material that is used as pencil lead

half-life: for any first-order reaction such a spontaneous nuclear fission, the length of time for one-half of any quantity of the material to undergo the reaction process

heavy water: water molecules in which the normal hydrogen atoms of H_2O have been replaced by their heavier isotope deuterium as D_2O

light-water: normal H_2O; the designation is used to indicate which type of moderator is used in a particular nuclear reactor; heavy-water indicates that D2O is used

neutron: one of the three main subatomic particles that compose atoms; protons and neutrons are located in the positively-charged nucleus, and electrons surround the nucleus as an ordered "cloud" of negative charge

nuclear fission: the process in which the nucleus of an atom such as the 235U isotope splits into two smaller nuclei with the release of energy, some subatomic particles and electromagnetic energy

nucleus: the central core of an atom, containing a specific number of protons for each element, but variable numbers of neutrons

A BRIEF HISTORY OF NUCLEAR POWER GENERATION
Soon after the discovery of nuclear fission in 1938, scientists started working on producing electricity using nuclear fission. The first nuclear power station, known as APS-1 Obninsk (Atomic Power Station 1 Obninsk), was built in 1954 about 110 kilometers (65 miles) southwest of Moscow, Russia. The power plant had a capacity of about 6 megawatts (MW) and functioned until 2002. In the United States, the Shipping Port Reactor in Pennsylvania was the first commercial nuclear power plant to become operational, in 1957. This reactor provided about 60 MW of energy and was located about 40 kilometers (25 miles) from Pittsburgh. It remained operational until 1982.

There are about 440 nuclear power plants located in 30 countries that are producing reliable electricity and providing 14 percent of all electricity. France and Lithuania produce over three-quarters of their electricity from nuclear power. Belgium, Bulgaria, Czech Republic, Hungary, Slovakia, South Korea, Sweden, Switzerland, Slovenia, and Ukraine generate one-third or more of their electricity from nuclear power.

NUCLEAR POWER PLANTS IN THE UNITED STATES
As of January 2016, there were sixty-one operational nuclear power plants with ninety-nine reactors among them in the United States, providing 19 percent of the total electricity, which amounts to approximately 100,000 MW. These reactors are located in thirty states. Illinois has the largest number of nuclear power plants with 11, followed by Pennsylvania with

nine, and South Carolina with seven. In contrast, in the mid-1970s, new nuclear power plants were being constructed in the United States and experts predicted that the country would have approximately 1,000 nuclear power plants in operation by the year 2000. The United States has not constructed any nuclear power plants in the past 30 years.

Because of the population growth in America, many old power plants are currently located in densely populated areas. The nuclear industry is at a critical juncture. To maintain about a 19 percent share of electricity production, America would need to construct hundreds of nuclear power plants by 2050 or earlier.

The U.S. Nuclear Regulatory Commission (NRC) approved licenses to construct two nuclear reactors about 170 mi (270 km) east of Atlanta in 2012, the first such issuance of licenses since 1979. The Obama administration allocated $8.3 billion in federal loan guarantees to build these two reactors. This is the first loan guarantee to a nuclear power plant under the Energy Policy Act of 2005. The effort was made to promote the nuclear technologies in use in the Westinghouse AP1000 reactor, which make reactors more efficient and safer.

NUCLEAR POWER GENERATION

In traditional power plants, coal, natural gas, or petroleum are commonly used to heat water. Similarly, in the case of a nuclear power plant, the nuclear fission process is used to produce heat to boil water, and using a turbine, electricity is produced. Nuclear power is derived by splitting uranium-235 with neutrons. This reaction produces two or three neutrons that can further be used to split other uranium nuclei and produce a "self-sustaining" chain reaction. This chain reaction produces an enormous amount of energy with each fission—about 200 milli-electron volts (meV) per fission.

The first self-sustaining chain reaction was successfully carried out at the University of Chicago's Amos Alonso Stagg Field on December 2, 1942. Neutrons with a 0.4 electron volt (eV) of energy or less are most effective in the fission of U-235. Neutrons ejected during fission have energies that are comparable to millions of electron volts. These neutrons simply escape and do not produce another reaction. A medium, known as a *moderator*, which dampens the energies of neutrons without absorbing them, is required to slow down the neutrons.

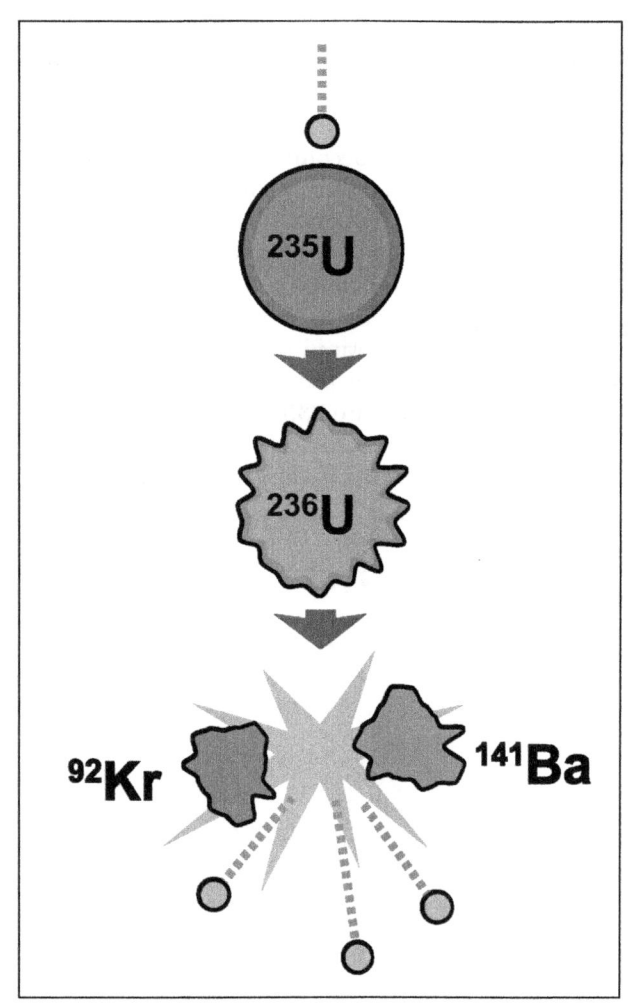

Induced fission reaction. A neutron is absorbed by a uranium-235 nucleus, turning it briefly into an excited uranium-236 nucleus, with the excitation energy provided by the kinetic energy of the neutron plus the forces that bind the neutron. The uranium-236, in turn, splits into fast-moving lighter elements (fission products) and releases a small amount of free neutrons. At the same time, one or more "prompt gamma rays" (not shown) are produced, as well. (Wikimedia Commons)

Graphite and water are two common moderators. The hydrogen nuclei in water play a key role in the slowing of neutrons. In a collision process, for example, when a moving tennis ball strikes another stationary tennis ball it will lose more energy than a tennis ball striking a bowling ball. It takes about 10-5 seconds for a 2-MeV neutron to slow down to 0.025 eV in a mere 18 collisions with hydrogen nuclei.

251

To maintain a steady state, it is essential that the flux of neutrons that produce fission remain constant. Since more neutrons are produced than consumed in U-235 fission reaction, it is imperative to remove some neutrons. This is accomplished by using control rods that are made of either boron or cadmium. Both elements are quite efficient in absorbing neutrons. The rate of absorption is controlled by the movement of these rods in and out of the reactor core. If the control rods do not absorb neutrons, the neutron density increases and causes a rapid release of energy. However, since nuclear fuel used in a reactor contains about 3 to 4 percent pure 235U versus about 95 percent purity in an atom bomb, a nuclear reactor cannot explode like a bomb even in the event of a meltdown.

TYPES OF REACTORS

There is no single design to produce the fission process: approximately 82 percent of reactors in the world are light-water reactors, implying that the moderation and cooling is done by regular water; about 10 percent are moderated and cooled by heavy water; 4 percent are gas-cooled reactors; while the remaining 4 percent are moderated by graphite and water-cooled. The term *light-water* is used to distinguish it from heavy-water reactors. Water is used as a moderator, as a coolant to remove heat from the reactor, and as a source to produce steam for the turbine.

Some 35 percent of light-water reactors are boiling-water reactors (BWRs), while 65 percent are pressurized water reactors (PWRs). In PWRs, the water in

Nuclear power plant in Cattenom, France. (Stefan Kühn via Wikimedia Commons)

the reactor core is kept at a high pressure of about 2,200 lbs/in² (psi) and the reactor core water temperature is kept at about 315°C (600°F).

Even at this high temperature, the water remains in liquid form due to high pressure. This hot water gives off its heat to the water in the steam generator through a heat exchanger. In BWRs, water in the primary core turns into steam and hits the turbine blades to produce electricity. Afterward, it is condensed and pumped back to the reactor core. Radioactivity is better contained in PWRs than in BWRs.

Both types of reactors have more or less the same efficiency rate of about 33% and consume the same amount of water. The remaining 67 percent of the energy is released into the environment. Most nuclear power plants use somewhere between 400 and 720 gallons of water per 1MW.hr production of electricity. Thus, the water consumption of nuclear power plants is relatively high. Ocean water is usually avoided for use as a coolant due to the corrosive nature of salt water.

Heavy-water reactors use deuterium oxide rather than hydrogen oxide as a moderator. The method has several benefits: it does not absorb neutrons and it effectively slows down fast neutrons. It also can achieve a sustainable chain reaction with naturally occurring uranium. This type of reactor, called a Candu reactor, is common in Canada.

A new reactor design involves uranium fuel being placed inside a 2.5-inch-deep bed of pyrolytic graphite pebbles. These pebbles are covered with silicon carbide ceramic. Both materials are highly resistant to high heat and the chance of radioactive release is minimal. With this design, if a reactor becomes hotter than desired, the uranium-238 in the pebbles absorbs high-energy neutrons in a nonchain reaction and slows the fission process. Thus, this type of reactor protects the public more effectively from accidental meltdown or fires. Instead, any damage generally affects one pebble at a time. Another advantage to pebble-bed reactors is higher efficiency. Traditional nuclear power plants produce electricity at about a 35 percent efficiency rate, while pebble-bed reactors range from 40 to 50 percent efficiency.

ENVIRONMENTAL IMPACT

Nuclear power plants do not generate greenhouse gases like fossil fuel plants. They do release trace amounts of radioactive gases; namely krypton, xenon, and iodine vapor. However, these power plants do indirectly cause greenhouse gas emissions. In a typical reactor, about 1.6 million tonnes of steel and about 14 million tonnes of concrete are manufactured, transported to the site, and used in construction. This process consumes millions of gallons of fossil fuel. For example, for every ton of Portland cement manufactured, about a ton of carbon dioxide is released into the atmosphere. This indirect emission is prominent only during the construction phase.

Nuclear reactors generate both high-level and low-level radioactive waste. High-level waste is waste generated in the nuclear reactor fuel cycle. Low-level waste includes gloves, clothing, tools, machine parts, and other items that may be contaminated with radioactivity. Extreme care needs to be taken in handling and storing these wastes.

A typical 1,000-MW nuclear power plant produces about 30 tonnes of high-level waste. This waste includes uranium, plutonium, cesium, strontium, and neptunium. The plutonium in this spent fuel is a major concern as it can be separated into dangerous and valuable materials using chemical techniques. Smuggling of radioactive waste is a major concern for nuclear power plants in several developing countries and in eastern Europe, where security measures may be inadequate.

In the United States, consumers end up paying about 1 cent per kilowatt-hour of electricity used toward nuclear waste management. This fee has raised billions of dollars, and yet a proper waste repository has yet to be built. Additionally, the half-life of uranium-238, uranium-235, and plutonium-239 is quite high; 24,000 years for plutonium, 713,000,000 years for uranium-235, and 4,500,000,000 for uranium-238. The half-life of strontium and cesium are 29 and 30 years, respectively. These wastes will need to be secured for extremely long periods of time.

As of 2012, the world population had faced two major nuclear power accidents: the first at the Chernobyl reactor in Ukraine (then a part of the Soviet Union) on April 26, 1986, and the second on March 11, 2011, at the Fukushima Daiichi Nuclear Power Station in Japan after a major undersea earthquake produced an extremely large tsunami. In both disasters, radioactive debris spread to the soil that produces food, groundwater used for drinking, and the air.

In the United States, the most severe nuclear accident occurred at the Three Mile Island power plant in Pennsylvania on March 28, 1979. Due to a valve malfunction, a large amount of radioactive coolant escaped. In an emergency measure, some radioactive wastewater was dumped into the Susquehanna River. The cleanup cost related to the disaster cost around $1 billion and was completed in 1993, over 14 years later.

MAJOR ISSUES
In the United States, nuclear power has encountered much public opposition, in part due to the coincidental timing of the Three Mile Island incident and the release of a major motion picture called *The China Syndrome*, and turning many politicians against its proliferation. The International Atomic Energy Agency (IAEA) acknowledges the task of "achieving and retaining public confidence in nuclear power" as a challenge in many parts of the world. If nuclear power is to continue to develop, public acceptance of its importance as an energy resource needs to improve. This is true for the United States as well as for many other nations.

The IAEA acknowledges that there is a shortage of nuclear reactor design, architecture, engineering, and project management organizations internationally. Nuclear proliferation in developing countries is another major concern, where the safety of used fuel for short- or long-term storage is an issue. Since most nuclear power plants are large, producing more than 1,000 MW, proper grids for power transmission are also essential. Such grids are not widely available in many countries.

The life span of a typical nuclear reactor is about 50 to 60 years. Reactors need to be decommissioned and sealed for thousands of years after their use period. The decommissioning alone can cost $500 million or more for each reactor.

In public perception, the nuclear industry functions under a veil of secrecy. For example, the IAEA prohibited the World Health Organization (WHO) from releasing information about the adverse health impacts of Chernobyl without getting prior clearance from the IAEA. For this reason, some information has never been released to the public. In Japan, the Japanese government and the Tokyo Electric Power Company (Tepco), the owner of the nuclear reactors, have admitted there was a lack of communication to the general public about the magnitude of the problem in the early stages of the disaster. As a result of the Chernobyl and Fukushima Daiichi fallouts, German officials abandoned plans to build new nuclear power plants, and became a nuclear-free country. However, China is still proceeding with the construction of about 25 new nuclear power plants.

—*Alok Kumar*

Further Reading

Alonso, Gustavo, Edmundo del Valle, and Jose Ramon Ramirez. *Desalination in Nuclear Power Plants*. Woodhead, 2020.

Baker, Keith, and Gerry Stoker. *Nuclear Power and Energy Policy: The Limits to Governance*. Palgrave, 2015.

Funabashi, Yoichi, and Kay Kitazawa. "Fukushima in Review: A Complex Disaster, a Disastrous Response." *Bulletin of the Atomic Scientists*, vol. 68, no. 2, 2012.

Malin, Stephanie A. *The Price of Nuclear Power: Uranium Communities and Environmental Justice*. Rutgers UP, 2015.

Mittica, P. *Chernobyl: The Hidden Legacy*. Trolley, 2007.

Muller, Richard A. *Physics for Future Presidents*. Norton, 2008.

Riznic, Jovica, ed. *Steam Generators for Nuclear Power Plants*. Woodhead, 2017.

Stockton, Nick. "Nuclear Power Is Too Safe to Save the World from Climate Change." *Wired*. Conde Nast, 3 Apr. 2016. Web. 16 May 2016.

Walker, J.S. *Three Mile Island: A Nuclear Crisis in Historical Perspective*. U of California P, 2004.

Xie, Wei-Chau, Shun-Hao Ni, Wei Liu, and Wei Jiang. *Seismic Risk Analysis of Nuclear Power Plants*. Cambridge UP, 2019.

Ocean Current Energy

FIELDS OF STUDY
Electrical Engineering; Marine Engineering; Oceanography

ABSTRACT
The use of ocean currents as an energy source carries great potential, but development has proceeded slowly because the cost is not competitive with that of other energy sources.

KEY CONCEPTS
gyre: a broadly circular flow of surface waters in the oceans of the world
salinity: the concentration of dissolved mineral salts in ocean waters
thermohaline circulation: global oceanic water circulation driven by changing temperatures and salinities as the water moves

BACKGROUND
Just as winds flow through Earth's atmosphere, currents flow throughout the world's oceans. These currents are a potential power source as great as wind, although winds harnessed for power have greater speed than the currents. The energy available in a fluid flow varies both with velocity (as the square) and with density:

$$\text{Kinetic Energy} = (\text{Density}) \times (\text{Velocity})^2$$

Because water is nearly eight hundred times denser than air (1,000 and 1.27 kilograms per cubic meter, respectively), a current of 1.6 kilometers per hour has as much energy as a wind of 45 kilometers per hour, which is considered an excellent average speed for wind energy. Furthermore, currents are more dependable than winds and flow in a constant direction.

OCEAN TEMPERATURE AND SALINITY
Ocean currents are driven by surface winds and differences in temperature and salinity. For example, as water flowing toward the poles is cooled, its density increases, and much of this cooler water sinks toward the ocean floor. From there it flows toward the equator, displacing warmer water as it goes. Meanwhile, water near the equator is warmed, becoming less dense. It tends to flow along the surface toward the higher latitudes to replace the sinking denser water. The result is a continuous current flow like a continuous loop.

The Gulf Stream is such a current. It starts from an area of warm water in the equatorial Atlantic Ocean off the west coast of Africa and flows into and out of the Gulf of Mexico. This warm water then flows generally northward parallel to the coast of North America and bends gradually to the right due to the rotation of the Earth. This tendency to curve (right in the Northern Hemisphere, left in the Southern Hemisphere) is called the Coriolis effect, and it bends the flow northeast as the West Wind Drift, bringing warm, moist air to Western Europe. It continues south as the Canaries Current (carrying cooler water) past western North Africa. Finally, the bending turns back west toward North America as the North Equatorial Current.

Similar circular patterns, or gyres, occur in all the world's oceans, with many locations having great potential for electrical power generation. For instance, the Gulf Stream has more energy than all the world's rivers combined. The area off Florida might yield 10,000 megawatts (10 billion watts) without observable change in the heat flow to Europe.

Salinity differences also drive major flows in what is called the thermohaline circulation. The most easily tapped salinity currents are those between a sea with high evaporation and the open ocean. High-salinity water flows along the bottom from the Mediterranean Sea, for instance, while less saline Atlantic water flows in to replace it. (German submarines used these currents during World War II for drifting silently past the major British base at Gibraltar.) Two lesser potential sources of current power are tidal currents and the currents at the mouths of rivers.

METHODS FOR HARNESSING OCEAN CURRENTS

Electrical power generation from currents requires three things: mooring power stations to the ocean floor, generating power, and transmitting power to customers on shore.

Mooring and transmitting power are related economic constraints on ocean current power. Although an underwater cable from a mid-Atlantic power station could technically supply power, deeper mooring lines and longer cables eventually cost more than the power delivered. Thus, ocean current stations, if built, will tend to be near shore on the continental shelf and slope before investors attempt to moor a plant to the depths of the ocean floor.

Using currents in deeper and more distant waters will require some means of energy storage. This issue has been considered in design studies for ocean thermal energy conversion (OTEC) power stations, which would harness the temperature difference between warm tropical waters and the colder deep waters. Electricity could be used for some energy-intensive process (such as refining aluminum) or for electrolyzing hydrogen from water. Hydrogen could be used to synthesize chemical products, such as ammonia or methanol. Once the potential of current power is proven, investors may consider the second set of risks inherent in such mid-ocean ventures.

Among various proposals, two methods have been studied in detail: turbines and sets of parachutes on cables. Turbines were first proposed by William Mouton, who was part of a study team led by Peter Lissaman of Aerovironment, Inc. Their design is called Coriolis. In the study design, one 83-megawatt Coriolis station has two huge counter-rotating fan blades (so it does not pull to one side), roughly 100 meters in diameter. The blades move slowly enough for fish to swim through them.

With blades so large, neither rigid blades nor the central hub could be made strong enough without being too heavy and expensive. However, a catenary (free-hanging, like the cables of the Golden Gate Bridge), flexible blade can be held in the proper shape by the current while the generators are in a rim around the blades. The rim also acts as a funnel to increase current speed past the blades and as an air reservoir for raising the station when necessary.

Another concept is parachutes on cables, which was proposed by Gary Steelman. His water low-velocity energy converter (WLVEC) design is an endless loop cable between two pulleys, much like a ski-lift cable. Parachutes along the cable are opened by the current when going downstream and closed when coming back upstream. The WLVEC is cheaper than Coriolis, but there is a question of how well any fabric could withstand sustained underwater use.

Ocean currents are sufficiently powerful and predictable to supply electricity effectively. However, costs of competing fossil fuels must rise significantly before investors will overcome their timidity about constructing offshore power plants. Test projects in the United States, China, Japan, and the European Union, in particular Britain, Ireland, and Portugal, continue with high expectations. Currently, however, the attention of those who would harvest energy from the oceans is focused on the far more accessible potential of power generated by ocean waves.

—*Roger V. Carlson*

Further Reading
Minerals Management Service, U.S. Department of the Interior. "Ocean Current Energy." http://ocsenergy.anl.gov/guide/current/index.cfm.
Neill, Simon P., and M. Reza Hashemi. *Fundamentals of Ocean Renewable Energy: Generating Electricity from the Sea.* Academic Press, 2018.
Yang, Zhaoqing, and Andrea Copping, eds. *Marine Renewable Energy: Resource Characterization and Physical Effects.* Springer, 2017.

Ocean Thermal Energy Conversion

FIELDS OF STUDY
Electrical Engineering; Marine Engineering; Oceanography; Thermodynamics

ABSTRACT
In some tropical regions of the Earth, there is virtually limitless energy in the ocean for possible conversion to electric power. The efficiency of the conversion is very low, however, and the engineering problems are challenging. Development of ocean thermal energy conversion (OTEC) has been slow.

KEY CONCEPTS
closed cycle: an ocean thermal difference technology that uses transfer heat energy from a warm water stream to vaporize a low-boiling working fluid that is kept in a closed system to drive a generator, and is then condensed by heat exchange with the cold water stream

heat engine: a system that converts heat energy into another form of energy such as mechanical or electrical by taking advantage of the difference in heat energy between two bodies, usually fluids

noncondensable gases: in an ocean thermal energy conversion system, dissolved gases such as nitrogen and carbon dioxide that do not condense to liquid form in the cooling phase of the operating system since they have boiling points far below that of the working fluid

open cycle: an ocean thermal difference technology that uses flash vaporization of cold seawater under reduced pressure to produce steam directly to be used to drive a generator; subsequent condensation of the used steam produces desalinated or freshwater

solar energy: the electromagnetic radiation emanating from the Sun that strikes Earth's surface

BACKGROUND
In tropical oceans, the temperatures of warm and cold layers of water may differ significantly even though the layers are less than 1,000 meters apart. This phenomenon results from global circulation currents caused by the Sun. Solar energy warms water near the surface, and colder, more dense water moves to lower depths. At the same time, the rotation of Earth causes the cold water to flow from the poles toward the tropics. As it is warmed, this cool water then rises toward the surface as its density decreases, causing the warm surface water to flow toward the polar regions, where it is cooled.

Differences of 20°C to 25°C over a distance of 500 to 1,000 meters are found in the Caribbean Sea and the Pacific Ocean near the Hawaiian Islands. In accordance with the second law of thermodynamics, thermal energy from the warm layer can be used as a "fuel" for a heat engine that exhausts energy to the cool layer. Typically, the warm layer has a temperature between 27°C and 29°C, and the cool layer is between 4°C and 7°C. The second law of thermodynamics indicates that the maximum efficiency of the conversion from thermal energy to mechanical energy will be very low. For example, if the warm layer is at 25°C and the cold layer is at 5°C, the maximum efficiency will be less than 7 percent; even this figure is between two and three times the actual efficiency that can be achieved in an energy conversion plant.

HISTORY
The concept of OTEC was first suggested in 1882 by the French physicist Jacques Arsène d'Arsonval, but it was not until 1926 that the French scientist Georges Claude made an attempt to implement the idea at Matanzas Bay, Cuba. The facility in Cuba was a small, land-based plant which was so inefficient that it required more power to operate than it produced. It ran for only a few weeks. Beginning in the 1960s, improvements in design and materials led to considerable research. Feasibility as a practical method of power generation was first demonstrated in the 1980s.

Advances in OTEC have depended on governmental support. In the mid-1970s, only the U.S. and Japanese governments were supporting research and development. The French government later became interested, and sponsorship followed in the Netherlands, the United Kingdom, and Sweden.

BASIC DESIGNS
Broadly speaking, designs are either open cycle (OC) or closed cycle (CC). In the OC method, the incoming warm seawater is continuously sent into an evaporator operating at low pressure, where a small portion of the water "flashes" into steam. The steam in turn passes through a turbine connected to an electric power generator. The low-pressure steam leaving the turbine is then cooled and condensed in a heat exchanger by the cold seawater stream. The condensed water is freshwater, the salt of the ocean having been left behind in the evaporator. Hence, this

water can be used for drinking and other household uses.

In the CC process, heat from the warm stream is transferred in a heat exchanger to a "working fluid" such as propane or ammonia. This fluid is vaporized and passed through a turbine generator in the same fashion as in the OC process. The vapor leaving the turbine is then condensed in a second heat exchanger. The condensate is recycled to the first exchanger, where it is again vaporized. Thus, the working fluid is never in direct contact with the seawater. Some hybrid plants have been designed that are combinations of OC and CC technology.

Though the first plant was a land-based unit, some plant designs involve plants located offshore, possibly floating or submerged. One of the key elements in the process is the water pipe which carries the cold water to the plant. This pipe is typically between 1 and 2 kilometers long. Originally, Claude used a corrugated steel pipe, 1.6 meters in diameter, which was fragile and not corrosion-resistant. Steel has been replaced by fiberglass-reinforced plastic or high-density polyethylene. Diameters larger than this have been considered in some studies but are not feasible owing to a lack of flexibility.

ENGINEERING PROBLEMS

Designs for OTEC plants with power capacities on the order of 10 megawatts or more have been made, but actual plants have been much smaller, with outputs on the order of tens of kilowatts. In spite of these relatively small outputs, the equipment and the engineering problems are challenging. Both cold and warm water flow rates are large because the efficiency of the conversion process is so low. The seawater carries considerable dissolved gases, notably nitrogen and oxygen, and these gases must be vented if flash evaporation is used. The presence of noncondensable gases poses difficult problems both in the evaporator which precedes the power turbine and in the condenser which follows it. These gases not only increase the sizes of the units but also, because they are below atmospheric pressure, must be pumped out to maintain the vacuum levels in the process. The CC method can avoid some of these problems. The operating pressures in the cycle using propane are relatively high, so a turbine of reasonable size can be used. Moreover, because the pressures are greater than atmospheric, vacuum and deaeration problems are eliminated. The CC process introduces additional problems, however, owing to the heat-transfer steps between the working medium and the hot and cold water.

ADVANTAGES

In view of the very low efficiency of OTEC, it may seem hard to imagine how the process can be profitable. However, the "fuel" is free and virtually unlimited. In addition, the OC process can produce sizable quantities of freshwater, which is often valuable in places where OTEC plants are located. Some OC plants may even be profitable on the basis of their freshwater production alone. Nevertheless, OTEC, even in the best of circumstances, poses both engineering and economic challenges that will continue to hamper its development for many years.

—*Thomas W. Weber*

Further Reading

Coiro, Domenico P., and Tonio Sant. *Renewable Energy from the Ocean: From Wave, Tidal and Gradient Systems to Offshore Wind and Solar,* Institution of Engineering and Technology, 2019.

Kim, Albert S., and Hyeon-Ju Kim. *Ocean Thermal Energy Conversion (OTEC): Past, Present and Progress.* IntechOpen, 2020.

Khaligh, Alireza, and Omer C. Onar. *Energy Harvesting: Solar, Wind and Ocean Energy Conversion Systems.* CRC Press, 2010.

Neill, Simon P., and M. Reza Hashemi. *Fundamentals of Ocean Renewable Energy: Generating Electricity from the Sea.* Academic Press, 2018.

State of Hawaii, Department of Business, Economic Development, and Tourism. "Ocean Thermal Energy." http://hawaii.gov/dbedt/info/energy/renewable/otec.

U.S. Department of Energy. "Ocean Thermal Energy Conversion." http://www.energysavers.gov/renewable_energy/ocean/index.cfm/mytopic=50010.

OCEAN WAVE ENERGY

FIELDS OF STUDY
Electrical Engineering; Marine Engineering; Mechanical Engineering

ABSTRACT

A number of designs for harnessing wave energy have been proposed, and some are in use on various scales, but the vast potential of this power source has not been tapped because of the uncertainties and expense involved.

KEY CONCEPTS

in phase: the condition when the peaks and troughs of a wave tend to occur in the same place at the same time, thus adding to the amplitude of the resultant wave

out of phase: the condition when the peaks of one wave and the troughs of another tend to occur in the same place at the same time, producing a resultant wave with less amplitude than either of the incident waves

penstock: a tubular structure that directs water flow to the turbine to drive a generator in a power station

pneumatic pressure: the force of air being compressed in a confined space

rectifier: a device that restricts the flow of water to one direction, or electrons if the rectifier is an electronic device in an electrical circuit

BACKGROUND

Waves crashing against a beach are a vast, almost mystical, display of mechanical power. For centuries people have sought ways of tapping it. In 1799, a father and son named Girard applied to the French government for a patent on a wave-power device. They noted that waves easily lifted even mighty ships. Hence, a lever from a ship to shore could power all manner of mills. (There are records of Girard mills on rivers, but the wave machine was probably never built.) Hundreds of patents later, wave power is still largely a dream, although a dream that is approaching reality.

THE NATURE OF WAVES

Waves are the product of wind blowing on the ocean surface. The energy available comes from the wind speed and the distance (or "fetch") that the wind blows: A breeze blowing on a small bay produces ripples, whereas a hurricane blowing across several hundred kilometers builds hill-sized waves. Waves hitting a beach can be the result of a storm on the opposite side of an ocean. From that standpoint, waves are a collecting and concentrating mechanism for wind power. However, there is some loss of wave energy over great distances, so the best places to take advantage of wave potential are along high-wind coasts of the temperate and subpolar latitudes. Specific regions of the world with strong wave actions include the western coasts of Scotland, northern Canada, southern Africa, Australia, and the northwestern coasts of the United States.

Water waves mostly consist of a circular motion of the water molecules as the wave energy continues until it meets a barrier, such as a shoreline. Then the energy hurls water and pieces of the shore until gravity pulls them back. Ultimately, the energy is transformed into heat, hardly noticed in the water. Along the way, the energy is vast. The North Pacific is estimated to have a flux of 5 to 50 megawatts of mechanical energy per kilometer.

One limitation of wave energy is that timing and power are variable (although not as much as with winds). The crests and troughs of one storm may be out of phase with another, in which case they largely cancel each other out. Winds may be low, or they may be directly against waves approaching the power plant. Any of these factors can limit power production at unpredictable times. Conversely, waves from two or more storms may be in phase and stack, creating monster waves that have been observed as high as 34 meters and more in the open ocean. Extraordinary waves have been the destruction of countless ships and of more than one wave power station. They are probably the greatest obstacle to widespread use of wave energy.

METHODS FOR HARNESSING OCEAN WAVES

Electrical power generation from waves requires three things: mooring the power stations to the ocean floor or building along the coast, generating power, and transmitting the power to customers inland. As with wind energy, a useful fourth item would be storage to deal with low-wave days.

Building and power transmission are straightforward operations, because most wave-harvesting designs are on or near shore. Even though these installations must be reinforced against especially strong waves, they do not have the cost and complexity of deepwater structures.

Proposed energy-harvesting techniques have great variation because many researchers have been attempting to harness wave potential. The researchers face three major problems. First, generators face the previously mentioned fluctuations in awesome power. Second, wave power is large but moves at a slow pace, and the machinery to obtain high speed (needed for an electric generator) is expensive. Third, complex hinges, pistons, and other moving parts need frequent replacement in the salty ocean environment.

The simplest approach is a ramp and dam facility that traps water splashing above sea level. Draining water goes down a pipe (penstock) to turbines, just as in a hydroelectric dam. The "Russel rectifier" is sort of a dam with chambers and flaps so that both rising and falling waves cause water in a turbine to flow continuously in the same direction. The various dam schemes are familiar and can be built on land. The disadvantage is that power dams must be large and capable of surviving the surf; thus they are expensive.

The "dam atoll" is an open-ocean variant of the ramp. A half-submerged dome in the ocean bends waves around it so that waves come in from all sides, just as with coral atolls. The water sloshes to a central drain at the top and drains back through a penstock. The central collection increases efficiency, and being a floating structure allows submerging below the waves during major storms. However, increased distance from shore increases power transmission costs.

Air pressure can translate slow wave motion into a fast spin. In many schemes, waves rise and fall either in a series of open rooms at the bottom of a floating structure or in cylinders at the end of funnel-shaped passages facing the waves. In both cases, the waves alternately push air out and suck it in. Both processes run turbines at high speed. Extensive work on air-pressure designs has been done by the British, the Norwegians, and the Japanese, who tested the *Kaimei*, an 80-meter ship with a number of chambers for testing various turbine designs, in the 1970s. In the 1980s and 1990s, the Japanese worked on the Mighty Whale project, which consists of near-shore floating structures with three large air chambers that convert wave energy into pneumatic energy.

Directly harnessing wave motion has some advantages to offset the slow motion and exposure of moving parts. The necessary equipment can be much smaller (thus cheaper) per unit of electricity generated than the other schemes. For many years, Japanese buoys have used pendulums and pulling units to power lights and horns. Scaling these units to larger sizes is difficult and expensive. Experimental units have used hinges between rafts (Cockerell's design), "nodding duck" cam-shaped floats to activate rotary hydraulic pumps to turn a generator (Salter's design), paddles on rollers, and many other techniques.

In 2008, the world's first commercial-scale wave-power station went live off the coast of Portugal. This British-designed and Portugese-financed station is about 5 kilometers off the northern coast of Portugal and consists of several semisubmerged 142-meter-long, 3.5-meter-diameter "snakes" of carbon steel, each with four articulated sections. The wave action drives hydraulic rams in the snake's hinges, creating energy that generators convert to electricity, which is relayed to a substation in Portugal via seabed cables. At peak output, three machines can generate 2.25 megawatts, which is enough to serve fifteen hundred family homes for a year.

Once produced in quantity, ocean wave power may have economics similar to hydroelectric plants—expensive to build but inexpensive overall because of low operating costs. Development continues steadily.

—*Roger V. Carlson*

Further Reading

Babarit, Aurélien. *Ocean Wave Energy Conversion: Resource, Technologies and Performance.* Elsevier, 2017.

Craddock, David. "Catching a Wave: The Potential of Wave Power." *Renewable Energy Made Easy: Free Energy from Solar, Wind, Hydropower, and Other Alternative Energy Sources.* Atlantic, 2008.

Cruz, João, ed. *Ocean Wave Energy: Current Status and Future Perspectives.* Springer, 2008.

Greaves, Deborah, and Gregorio Iglesias, eds. *Wave and Tidal Energy.* Wiley, 2018.

Minerals Management Service, U.S. Department of the Interior. "Ocean Wave Energy." http://ocsenergy.anl.gov/guide/wave/index.cfm.

Neill, Simon P., and M. Reza Hashemi. *Fundamentals of Ocean Renewable Energy: Generating Electricity from the Sea.* Academic Press, 2018.

Pecher, Arthur, and Jens Peter Kofoed, eds. *Handbook of Ocean Wave Energy.* SpringerOpen, 2017.

Rusu, Eugen, and Vengatesan Venugopal, eds. *Offshore Renewable Energy: Ocean Waves, Tides and Offshore Wind.* MDPI, 2019.

Wright, Glen, Sandy Kerr, and Kate Johnson, eds. *Ocean Energy: Governance Challenges for Wave and Tidal Stream Technologies.* Earthscan, 2018.

Oil and Petroleum

FIELDS OF STUDY
Chemical Engineering; Geology; Petroleum Engineering; Refinery Operations

ABSTRACT
Crude oil, a naturally occurring, flammable fossil fuel, is a mixture of various hydrocarbons and other chemical compounds. Today, crude oil, also called "petroleum," is one of the world's most important energy sources, producing gasoline, kerosene, and heating fuel, among other products. However, there is increasing debate on how long the production of crude oil can continue and whether it will be able to meet market demand.

KEY CONCEPTS
cracking: a thermal process in which hydrocarbons are made to fracture and decompose into a number of different, smaller compounds, according to specific chemical reactions

oil sands: dense deposits of bitumen combined with sand

sour crude: crude oil having a high sulfur content

sweet crude: crude oil with a very low sulfur content

train oil: oil recovered from the harvesting and rendering of whales, and especially from the head cavity of sperm whales

UNREFINED PETROLEUM
Crude oil, a naturally occurring, flammable fossil fuel, is a mixture of various hydrocarbons and other chemical compounds. Today, crude oil, also called "petroleum," is one of the world's most important energy sources, producing gasoline, kerosene, and heating fuel, among other products. However, there is increasing debate on how long the production of crude oil can continue and whether it will be able to meet market demand.

Crude oil—that is, unrefined petroleum—is a flammable fossil fuel in a mixture of up to 17,000 or more various substances, including diverse hydrocarbons, such as alkanes (paraffins), cycloalkanes (naphthenes), aromatic hydrocarbons, and asphaltenes, which bind nitrogen, oxygen, and sulfur and form compounds such as thioethers, alcohols, and resins. Also, metals, such as iron, copper, vanadium, and nickel, may be found in crude oil. Not only does the chemical composition of crude oil vary, so do its physical properties, including color and viscosity, from transparent, light yellow and thin fluid to deep black and viscous, depending on its source. Its density ranges from 0.82 to 0.94 gram per cubic centimeter. Crude oil containing only little sulfur is referred to as "sweet," while crude oil relatively rich in sulfur is called "sour." The latter has an unpleasant odor. Low density refers to light oil, and high density refers to heavy oil. Crude oil is not soluble in water or ethanol, but it is soluble in ether, benzene, and tetrachloromethane.

Currently, crude oil is the most important fuel of modern industrialized societies. More than 80 percent of global oil production goes into fuels. The remaining oil is used in industrial chemical processes to produce pharmaceuticals, fertilizers, foods, plastics, paints, textiles, and construction materials. Industrially developed countries are highly dependent on petroleum. Most chemical products can be synthesized from about 300 bulk chemicals based on natural gas and petroleum, including ethene (more commonly known as ethylene), propene, benzene, toluol, and xylol. About 6 to 7 percent of the global petroleum production is used to synthesize bulk chemicals, avoiding expensive alternative methods. The biggest part is burned as fuel in power plants and engines. In 2019, worldwide oil consumption was approximately 98.3 million barrels per day. (One barrel equals about 159 liters.)

Global processes of exploration, extraction, refining, and transporting (by means of both supertankers and pipelines), and marketing of petroleum products are major areas of the oil industry. Until 1985, global oil prices followed a pricing system administered by the Organization of Petroleum Exporting Countries (OPEC). After this system collapsed, oil-exporting countries adopted a market-linked pricing mechanism for international crude oil trade. Currently, the benchmarks for pricing are the Brent, West Texas Intermediate (WTI), and Dubai/Oman markers.

Like all fossil fuels, petroleum fuels, when combusted, pose environmental risks, mainly with respect to exhaust products. Too little oxygen during combustion yields carbon monoxide. Exhaust gases from car engines include nitrogen oxides, which cause photochemical smog.

HISTORY

In a technical sense, petroleum refers only to crude oil, but in a broader sense it describes all liquid, gaseous, and solid hydrocarbons. *Petroleum* means literally "rock oil" (from the Latin *oleum petrae*). Its density is lower than that of water, so it wells up to Earth's surface through vugs (cavities) in sediments of shale, sand, or carbonate. Crude oil at Earth's surface, mostly in the form of bitumen due to a reaction with oxygen, was known 12,000 years ago in Mesopotamia.

The ancient Babylonians used bitumen to pave roads, seal ship planks, and impregnate textiles. Naphtha can be traced back to the Babylonian word *naptu*, which means "to glow." The first known public regulation of oil respective of bitumen was written by Hammurabi, first king of the Babylonian Empire, in 1875 BCE. The Roman army probably used crude oil as a lubricant for axles and wheels. In early medieval Byzantium, oil from bitumen was used for flamethrowers during battles—perhaps called "Greek fire"—and in the Renaissance so-called St. Quirin oil was sold by German monks as a medicine. The Pechelbronn oil field in Alsace was the site of the first European oil well, productive from 1498 until 1970.

In 1854, the Canadian physician and geologist Abraham P. Gesner obtained his first U.S. patent for the synthesis of kerosene from coal and petroleum, providing the initial impetus for the oil industry. The development of kerosene was motivated by the search for a reasonable alternative to train oil (from whales), at that time the common lamp fuel. Only a few years later, in 1859, Colonel Edwin L. Drake drilled for oil in Pennsylvania and found a big deposit only 70 feet (21.2 meters) underground. In the same year, the first oil refinery was established, soon followed by many others. After the introduction of electric lighting, oil lost its attraction for a while, only to return in the form of benzine, or petroleum ether. Henry Ford considered ethanol as fuel for his automobiles, but John D. Rockefeller, founder of Standard Oil Company, used benzine as motor fuel. It was Standard Oil of California that discovered the first oil deposit in Saudi Arabia close to Dammam in 1938. Since then, giant amounts of crude oil have been used as motor fuels.

John D. Rockefeller. (Wikimedia Commons)

Welcome sign for Titusville, Pennsylvania, where in 1859, the United States' first oil well was drilled, 69 feet (21 meters) below the surface. (Wikimedia Commons)

GEOPHYSICAL PROSPECTION

Precise maps provide fundamental information prior to geophysical prospection (or geophysical surveys), a systematic search for oil in place. In certain areas, such as Iran, deposits can be identified by means of aerial photographs. In cases where there are rocks at the surface that are typical for oil in place, samples are taken.

During prospection, physical properties, such as magnetism, density, sonic speed, electrical resistance, and radioactivity, are measured. The most often applied method is seismic reflection. To receive computable data, a controlled seismic source, such as an artificially induced dynamite explosion, an air gun, or a seismic vibrator, is required; these methods produce acoustic waves that propagate within Earth's crust. The waves need different times to pass through different rock formations, and reflect at the border of two formations. Reflected waves are received by sensitive geophones or hydrophones and are logged in a seismogram. The result is an image of the underground, providing information about existing trap structures.

Based on seismic data, sample drills are carried out. In a next step, three-dimensional measurements are taken in selected areas. In combination with geophysical measures at the borehole, a quantitative model of oil or natural gas deposits, as well as plans for further boreholes and production, can be prepared.

PRODUCTION

When oil occurs subaerially (at or near Earth's surface), it can be recovered by means of opencast mining. For deeper deposits, boreholes are drilled to recover the oil. Deposits under the sea are recovered by means of offshore drilling. At a water depth of up to 100 meters (328 feet), the oil platforms stand on stilts. Another type of drilling platform is the semisubmersible floating platform, carrying heavy ballast weights to stabilize the platform. The deepest successful offshore wells have been drilled at depths of several thousand feet below sea level. Drillships can even drill in depths of more than 3,600 meters (up to 12,000 feet). The installation of a permanent drilling platform costs several billion dollars. However, offshore drilling has made enormous additional oil reservoirs available.

Traditional oil drilling is done using the rotary drilling technique. Steel drilling rods hang at the oil derrick and are attached at the rotary table at the bottom of the derrick. Scavenging pumps flush water into the borehole, forcing the drilling cuttings through the hollow spaces between rods and borehole to the surface. The mixture of water and cuttings covers the borehole wall, preventing it from collapsing and thus making extremely deep boreholes possible. Once the drill reaches the oil reservoir, pressure can be so high that the oil rockets upwards like a fountain, producing a "gusher."

Commonly the recovery of conventional crude oil takes place in three stages. The natural reservoir pressure is exploited for the primary oil recovery. Natural gas in the reservoir enables an eruptive recovery of about 10 to 30 percent of the oil available in that reservoir. For the second oil recovery phase, two primary methods are used to boost production: the water flood method, which is the injection of water into the reservoir to increase pressure, and the thermal flood method, which is injection of steam (primarily used to give very viscous oil a lower viscosity and thus force the oil to the borehole). Because water is denser than oil, it sinks beneath it, filling up the reservoir from the bottom without mixing with the oil. More specialized methods include the injection of solvents via the chemical flood technique. These methods enable the recovery of 10 to 30 percent more oil, so that altogether 20 to 60 percent of the available oil can be recovered.

In the final, third stage of recovery, complex substances such as polymers, carbon dioxide, and microorganisms are injected to increase the capacity of the well again. High prices and global market dynamics can intensify efforts at this tertiary stage of oil recovery, not only at currently producing reservoirs but also at old reservoirs. Usually, the cost of recovering oil increases as the amount available at the well decreases until production becomes too expensive and the borehole is shut down. However, as world reserves decline and the price of oil rises, the economics of oil recovery may change to encourage more recovery at later stages.

When crude oil is found in semisolid form mixed with sand and water, it is called "crude bitumen," or "oil sand." There are abundant reservoirs known in Canada, the Athabasca oil sands, and in Venezuela, the Orinoco oil sands. Together these deposits are estimated to be able to yield 3.6 trillion barrels of crude heavy oil, which is much more than prospected global

Offshore drilling rig. (James Juntree via iStock)

reserves of conventional oil. These sands are considered unconventional oil sources, because the oil cannot be recovered by means of traditional drilling methods. In addition, oil shales represent an indirect source of oil. These rocks contain the waxy, high-carbon substance called "kerogen." Using heat and high pressure, the process that leads to the geological formation of crude oil can be simulated, converting the trapped kerogen into oil. This method it not a modern discovery but has been known for centuries. It was even patented in 1694 under British Crown Patent No. 330. An advanced version known as hydraulic fracturing, or fracking, became increasingly prevalent in the twenty-first century.

Factors for the classification of crude oil, conducted by the oil industry, are the geographic location of its source, the API gravity (a measure, devised by the American Petroleum Institute, of the density of a petroleum liquid compared to water), and the sulfur content. The geographic location determines transportation costs to the refinery. Lighter grades of crude oil are more eligible than heavier grades, because upon refining they produce more gasoline. Also, sweet oil is traded at higher prices than sour oil because it causes less environmental problems. Refineries have to meet strict sulfur standards, just as fuels for consumption must. The specific molecular characteristics of the different crude oil grades are listed in crude oil assay analysis, used in petroleum laboratories.

PROCESSING
First, the recovered crude oil is separated into oil, natural gas, and saltwater by means of a gas separator. When the gas has completely escaped, oil and saltwater can be separated by density separation.

Petroleum refineries then decompose crude oil into light and heavy components, such as heating oil, kerosene, jet fuel, and gasoline, by means of distillation columns, called "fractional distillation" or "rectification." Because of these distillates' different boiling points, they can be fractionated into boiling ranges. During this first distillation, fractioned gases, such as methane, ethane, propane, and butane, are important heating fuels. Light and heavy petrols (with boiling points of 30°C-180°C) are used as gasoline for cars. The medium distillate (180°C-250°C) is used as lamp oil or is processed into jet fuel. In a subsequent vacuum distillation of the residue, additional important petroleum products are produced, such as lubricants, used to grease engines, and bitumen, used to pave roads. The heavy fuel oil fuels power plants and ship's engines. Special refineries recycle waste oil.

The amounts of naphtha produced directly from crude oil per fractional distillation are not sufficient to cover the market. However, a process that takes place at high temperatures, called catalytic cracking, cracks long-chain alkanes into short-chain alkanes. With this method, petrol can be produced from paraffin oil. For example, decane cracks into propene and heptane. A trap inside the reactor separates the cracking products from the spent catalyst. Again, the cracked hydrocarbons become fractionated in a subsequent distillation. At the surface, the catalyst segregates carbon, making the catalyst ineffective. Upon mixing with hot air, the carbons burn up, regenerating the catalyst. This process is also used to crack heavy crude oil, because it often contains too much carbon and too little hydrogen.

GASOLINE (PETROL) PRODUCTION

The compression and heat in the cylinders of a gasoline engine may cause untimely spontaneous combustion of the gasoline-air mixture, resulting in engine knocking. Unramified hydrocarbons tend toward untimely ignitions, while unsaturated, ramified hydrocarbons are relatively knockproof. The antiknock rating is indicated by an octane number. The higher the octane number is, the more knockproof the gasoline is. Pure isooctane, the upper standard for measurement, is ascribed an octane number of 100, while pure n-heptane, the lower standard for measurement, is ascribed an octane number of 0. Common octane numbers for gasoline are 87, 95, and 98, and the higher the antiknock rating of the gasoline used, the longer the life of an engine. Therefore, naphtha is converted into gasoline with high antiknock ratings by means of a platinum catalyst in a process called platinum reforming. Byproducts of this process are hydrogen and gaseous alkanes. Prior to reforming, the naphtha needs to be desulfurized, releasing hydrogen sulfide, because sulfur would destroy the catalyst.

DESULFURIZATION, HYDROFINING, AND THE CLAUS METHOD

After fractional distillation, heating fuel and lubricants still contain much sulfur, leading to toxic sulfur dioxide exhaust upon combustion; these emissions lead to environmental degradation in the form of acid rain and forest decline. In a process called "hydrofining," the oil is mixed with hydrogen and heated. Then the hydrogen in the hot mixture reacts in a catalytic reaction with the contained sulfur to form hydrogen sulfide. Subsequently, the hydrogen sulfide is burned in the Claus desulfurization process, forming sulfur and water, as follows:

$$6H_2S + 3O_2 \rightarrow 6S + 6 H_2O - 664 \text{ kJ/mol}.$$

SYNTHESIS GAS

Other residues from fractional distillation are used to produce synthesis gas. That is a mixture of carbon monoxide and hydrogen, used to produce many other, organic compounds, such as ammonia and methanol. Synthesis gas production from methane occurs as follows:

$$CH_4 + H_2O \rightarrow CO + 3H_2 \text{ (endothermic)}$$

$$2CH_4 + O_2 + 4N_2 \rightarrow 2CO + 4N_2 + 4H_2$$
$$\text{(exothermic)}$$

The carbon monoxide can be converted with water vapor into hydrogen and carbon dioxide, which can be washed out with water. The carbonated water is sold to the soft-drink industry. The pure gases nitrogen and hydrogen are left over as chemical precursors for other products.

Another refining process is called pyrolysis, used to crack light gasoline at very high temperatures into ethene, acetylene, and propene. A mixture of methane and oxygen is heated to 2,500°C. When the light gasoline is conveyed into this mixture, it cracks. For example, pyrolysis for n-heptane occurs as follows:

$$\text{n-heptane} \rightarrow \text{ethene} + \text{acetylene} + \text{propene} + \text{hydrogen}$$

Ethene and propene, also commonly known as ethylene and propylene, are important key precursors for the production of plastics. Compared to catalytic cracking, pyrolysis takes place at much higher temperatures and without a catalyst.

OIL SPILL DISASTERS

Generally, transport of oil via oil tanker is not only very expensive but also hazardous with regard to environmental and ecological issues. Supertankers carry up to 300,000 tons of oil. About 100,000 tons of crude oil leak every year into the ocean as a result of tanker accidents. An especially drastic accident happened in 1989, when the *Exxon Valdez* lost an estimated 11 million barrels of crude oil off Alaska's coast. After the accident occurred, immediate efforts to use dispersants to clean it up failed because of a storm that rendered the dispersants ineffective, and booms were used to collect oil and burn it. Nevertheless, the oil slick spread and contaminated about 2,100 kilometers (1,300 miles) of coastline, killing wildlife and destroying habitat. Catastrophic ecological consequences led to broad public discussions about risks and dangers of maritime oil transports. Eventually this incident resulted in more regulations for oil tankers and an investigation of how to avoid such disasters.

In 2010, the largest marine disaster in the history of the petroleum industry occurred in the Gulf of Mexico, known as the Deepwater Horizon oil spill or BP oil spill. After the explosion of the offshore platform Deepwater Horizon, which drilled in the BP-operated Macondo Prospect, on April 20, 2010, oil from a leak at a seafloor oil gusher flowed for three months into the sea. Although the gushing wellhead was capped on July 15, it took until September 19 to complete the process; by then, the leaking well was effectively dead. Official estimates state that about 4.9 million barrels, or 780 million liters (205.8 million gallons), of crude oil spilled into the Gulf; some estimates are higher.

GLOBAL RESOURCES, PRODUCTION, AND CONSUMPTION

The question of how long global oil resources will be able to cover the world's consumption is debated. In the late twentieth and early twenty-first centuries many forecasters began to warn that "peak oil" was imminent based on ever-increasing consumption and limited known reserves. Many scientists agreed that the peak of oil discoveries was reached in 1965, and from 1980 oil consumption typically exceeded oil discoveries each year. However, the discovery of some new reserves and especially improvements in extraction methods such as fracking allowed further growth in production into the 2020s, exceeding some estimates. Optimistic observers, including some in the oil industry, asserted that with today's technology, prospected areas, and consumption, Earth's oil resources will cover the world's consumption for at least another century; this prediction assumes that only 10 percent of all liquid oil resources have been recovered.

Still, even the most staunch supporters of the oil industry acknowledge that oil is a finite, nonrenewable resource. Key publications, such as the International Energy Agency (IEA)'s annual *World Energy Outlook*, present different outlooks based on potential scenarios for the global energy landscape. The 2019 edition suggested that if current world policies and consumption continued, demand would grow until about 2025 and then slow considerably. If a sustainable development model was followed, demand would drop much more sharply.

The most important oil producing countries in 2019 were the United States, Russia, Saudi Arabia, and Iraq, followed by Canada, China, and the United Arab Emirates. The United States, China, India, Japan, and Saudi Arabia were the biggest consumers of oil in 2019. The annual consumption in the United States was about 19.4 million barrels per day (mBPD), representing 19.7% of the global total but showing a slight decline from the previous year, while China consumed 14.1 mBPD, or 14.3% of the global share, with the fastest growth of any country.

WORLDWIDE OIL PRICES

Oil prices vary depending on type, but several varieties are used as benchmarks for buyers and sellers. Among the most important reference oils are West Texas Intermediate (WTI, considered a high-quality, sweet, light oil from Cushing, Oklahoma, used as a reference for North American oil; Brent Crude, which integrates 15 fields in the East Shetland Basins of the North Sea and is used to price oil produced in Europe, Africa, and the Middle East flowing west; Dubai/Oman, used as a benchmark for Middle East sour crude oil flowing toward the Pacific; Tapis, from Malaysia, for light Far East oil; Minas, from Indonesia, for heavy Far East oil; Midway Sunset Heavy, for heavy oil in California; and the "reference basket" of the OPEC, a weighted average of oil blends from various OPEC countries.

Prices for all types of crude oils fluctuate according to various market factors. The importance of oil to global energy production means that oil prices have a major impact on the world economy, as well as the economies of individual countries. Periods of significantly elevated or depressed oil prices can have widespread effects, which are often most visible to average citizens in the price of gasoline for automobiles. Gasoline prices themselves have been correlated to consumer habits, notably including a tendency to purchase large and less fuel-efficient vehicles when prices are low and, conversely, favoring smaller, more efficient vehicles when prices are high.

—*Manja Leyk*

Further Reading

Clayton, Blake C. *Market Madness: A Century of Oil Panics, Crises, and Crashes.* Oxford UP, 2015.

Cox, Lydia. "The Surprising Decline in US Petroleum Consumption." *World Economic Forum.* World Economic Forum, 10 July 2015. Web. 16 May 2016.

Deffeyes, Kenneth S. *Hubbert's Peak: The Impending World Oil Shortage.* Princeton UP, 2008.

Glasby, G.P. "Abiogenic Origin of Hydrocarbons: An Historical Overview." *Resource Geology*, vol. 56, no. 1, 2006.

Kolesnikov, Anton, et al. "Methane-Derived Hydrocarbons Produced Under Upper-Mantle Conditions." *Nature Geoscience*, vol. 2, 2009.

Maugeri, Leonardo. *Beyond the Age of Oil.* Praeger, 2010.

"Oil: Crude and Petroleum Products Explained." *US Energy Information Administration.* Department of Energy, 5 Nov. 2015. Web. 16 May 2016.

"Oil Information: Overview." *International Energy Agency*, July 2020, www.iea.org/reports/oil-information-overview Accessed 18 Aug. 2020.

"Petroleum & Other Liquids." *US Energy Information Administration*, www.eia.gov/petroleum/. Accessed 18 Aug. 2020.

Ruppert, Michael C. *Confronting Collapse: The Crisis of Energy and Money in a Post Peak Oil World.* Chelsea Green, 2009.

"World Energy Outlook. "2019: Oil." *International Energy Agency,* Nov. 2019, www.iea.org/reports/world-energy-outlook-2019/oil. Accessed 18 Aug. 2020.

OIL SHALE AND TAR SANDS

FIELDS OF STUDY
Environmental Science; Geology; Petroleum Engineering

ABSTRACT
Oil shale and tar sands are sources of oil and gas fuel, lubricants, and chemical feedstock. Tar sands are rocks with pore spaces filled by solid or semisolid bitumen. Oil shale is any fine-grained sedimentary rock containing kerogen and yielding petroleum when heated in the absence of oxygen.

KEY CONCEPTS
alkaline: mineral material capable of acting as a base, as opposed to an acid

bitumen: high molecular weight hydrocarbon materials, commonly seen as roofing or paving tar

kerogen: an insoluble organic material found in shale and the pores of other types of rock; kerogen must be heated in the absence of oxygen to be made suitable for processing

lacustrine: formed in the sediment of ancient freshwater lakes

petroliferous rock: rock that contains or can be used as a source of petroleum

retorting: heating of a material in the absence of oxygen in order to alter its characteristics

saline: containing salt

tar sand: a heavy, rock-like deposit of sand intermixed with bitumen

BACKGROUND
Tar sands are most abundant in sandstone and limestone. Most large deposits occur near sedimentary basin margins in deltaic, estuarine, or freshwater rocks.

Oil shale occurs in lacustrine (lake) sediments, associated with coal, or in marine shale.

Kerogen is a waxy, insoluble organic compound with a large molecular structure. Almost all sedimentary rocks contain some kerogen; those that both contain kerogen and yield a few liters of oil per metric ton are considered oil shale. About 45 liters per tonne generally is the minimum figure used for calculating reserves. Ninety to 135 liters are required for development. Some shales contain more than 225 liters per tonne. Worldwide shale oil resources have been estimated at 3 trillion barrels, of which the United States is believed to have 2 trillion.

ORIGIN OF OIL SHALE

Oil shale forms in oxygen-deficient environments where organic debris accumulates more rapidly than it is destroyed by oxidation, scavengers, or decay. Deep, confined ocean basins with stagnant water or restricted water circulation may preserve organic debris. Baltic and Manchurian oil shales are of this origin. Swamp lakes, with slow circulation and rapid accumulation of plant debris, also may produce oil shale. Oil shale accompanying coal in Scotland and North America are examples. Lakes with noncirculating water at the bottom also may accumulate oil shale. The gigantic Green River oil shale deposit of Wyoming and Colorado is of this type.

PRODUCING SHALE OIL

Very great amounts of shale are needed for economically significant petroleum production. Open-pit mining is much more economical than underground mining, although there are some large underground mines. Pits 300 meters deep and 3 kilometers across have costs equal to those of underground mining in the Green River deposit. Further expansion reduces expense, making gigantic pits the most economical mining option. Heating shale in the absence of oxygen (retorting) converts kerogen (solid organic material that is insoluble in petroleum solvents) to liquids. The liquid then requires hydrogenation to make petroleum.

Retorted shale is a saline and/or alkaline powder. Open pits are ready disposal sites for waste material, and underground mines may be backfilled. The remaining 10 percent or more of retort waste, however, requires disposal elsewhere. Finally, the waste must be isolated from surface water and groundwater to prevent contamination. In situ processing solves some disposal problems. In this method large blocks of oil shale are undermined and collapsed, creating an underground porous rubble. Gas introduced to the top of the rubble is ignited, after which the shale burns on its own. Gas and oil "cooked" from the rock are withdrawn from the base of the rubble, leaving the spent shale underground.

SHALE OIL HISTORY

Shale oil for medicinal purposes was produced in 1350 at Seefeld, Austria. The manufacture of illuminating oil and lubricants from oil shale began in France around 1830 and quickly spread through Europe and North America. Petroleum almost entirely supplanted shale oil during the late nineteenth century. Afterward, shale oil production was largely limited to periods of oil shortage or of military or economic blockade. Flammable oil shale, rich in kerogen, has been burned to generate steam in Latvia. Scotland, the former Soviet Union, Manchuria, Sweden, France, Germany, South Africa, the United States, Brazil, and Australia all have produced shale oil, but total world production through 1961 was only about 400 million barrels.

TAR SAND OCCURRENCE

Tar sand bitumens are larger, heavier, and more complex hydrocarbon compounds than those in liquid petroleum, and they include substantial nitrogen and sulfur. ("Bitumen" is a term for a very thick, natural semisolid material such as asphalt or tar.) Deposits are most abundant in sandstone or limestone. Most large deposits are in deltaic, estuarine, or freshwater sandstone. The largest occur at depths of less than 1,000 meters on sedimentary basin margins where inclined layers of petroliferous rocks approach the surface. Here, upward migrating petroleum could lose volatiles and, with oxygenation and biodegradation, leave asphalt-impregnated rock. Some solid bitumens may be hydrocarbons not yet sufficiently altered to form liquids rather than residues of once liquid material.

United States reserves, although large, are insignificant compared to those of Canada and Venezuela. Additional deposits are known in Albania, Siberia,

Madagascar, Azerbaijan, the Philippines, and Bulgaria.

TAR SAND HISTORY

Tar sands have been used since ancient times for surfacing roads, laying masonry, and waterproofing. The Athabasca tar sand deposit of northern Alberta, Canada, was discovered in 1778 when Peter Pond, a fur trader, waterproofed his canoes with tar. Geologic exploration began in the 1890s, and by 1915 tar sand was being shipped to Edmonton, Alberta, to pave streets. Pilot plant extraction of oil began in 1927. Afterward, exploitation continued with provincial and federal subsidies and support. By the end of the twentieth century, operations were self-supporting.

TAR SAND EXPLOITATION

Canadian tar sand is mined in large open pits and transported to processing plants where steam treatment produces bitumen froth and sand slurry. Naphtha steam removes the remaining sand, leaving viscous bitumen. Raw bitumen then is "cracked," a chemical process by which the large organic molecules in the bitumen are broken into smaller, more liquid molecules, gas, and coke. Finally, cracked oil is hydrogenated to produce synthetic crude oil. Sulfur is a salable byproduct. Sand ultimately is returned to the pit, overburden is replaced, and the site is reforested.

Open-pit production, however, is feasible only at the shallow periphery of the deposit, so in situ extraction will be required for about 90 percent of the Canadian deposit. In one system, wells drilled into the deposit are injected with steam to liquefy the bitumen. Bitumen then is pumped until flow ceases, after which the well is again steamed. In another system wells sunk into the tar sand are ignited. Heat then cracks the bitumen, producing liquid and gas that flow to production wells.

—*Ralph L. Langenheim, Jr.*

Further Reading

Bartis, James T., et al. *Oil Shale Development in the United States: Prospects and Policy Issues.* Rand Institute, 2005.

Chastko, Paul. *Developing Alberta's Oil Sands: From Karl Clark to Kyoto.* U of Calgary P, 2004.

Clarke, Tony. *Tar Sands Showdown: Canada and the New Politics of Oil in an Age of Climate Change.* J. Lorimer, 2008.

Gonzalez, George A. *American Empire and the Canadian Oil Sands.* Palgrave, 2016.

Laxer, Gordon. *After the Sands: Energy and Ecological Security for Canadians.* Douglas & McIntyre, 2015.

Nikiforuk, Andrew. *Tar Sands: Dirty Oil and the Future of a Continent.* Greystone Books, 2009.

Speight, James G. *Deep Shale Oil and Gas.* Elsevier, 2017.

U.S. Department of the Interior, Bureau of Land Management. "About Oil Shale." http://ostseis.anl.gov/guide/oilshale/index.cfm.

U.S. Department of the Interior, Bureau of Land Management. "About Tar Sands." http://ostseis.anl.gov/guide/tarsands/index.cfm.

U.S. Geological Survey. "Heavy Oil and Natural Bitumen: Strategic Petroleum Reserves." http://pubs.usgs.gov/fs/fs070-03/fs070-03.pdf.

U.S. Geological Survey. "Natural Bitumen Resources of the United States." http://pubs.usgs.gov/fs/2006/3133/pdf/FS2006-3133_508.pdf.

Wang, Xiangzeng. *Lacustrine Shale Gas: Case Study from the Ordos Basin.* Elsevier, 2017.

OTTO, NIKOLAUS

FIELDS OF STUDY
Combustion Science; Electrical Engineering; Mechanical Engineering

ABSTRACT
German engineer Nikolaus Otto designed the Otto-cycle engine, the first practical four-stroke, internal combustion engine, the type which today powers almost all automobiles as well as other machines that rely on a self-contained power source. Diesel engines are a modification that utilizes the heat of compression to ignite the fuel, rather than a high-energy electrical spark.

KEY CONCEPTS
four-stroke engine: an internal combustion engine requiring four-timed piston strokes to complete one cycle of operation per cylinder: an intake stroke to take in fuel and air, a compression stroke in which the air-fuel mixture is compressed, a power stroke in which the air-fuel mixture is ignited to drive the piston down with force, and an exhaust stroke to drive the combustion products out of the cylinder

internal combustion engine: an engine in which fuel is combusted within the engine itself in order to extract energy from the fuel as an integral part of the engine operation

stratified-charge engine: an internal combustion engine in which the fuel required for one cycle is delivered at successive points in the power stroke, rather than all at once at the beginning of the power stroke

two-stroke engine: an internal combustion engine in which piston movement within the cylinder opens and closes intake and exhaust ports within the cylinder during operation, rather than using a camshaft timed to the rotation of the crankshaft to coordinate the opening and closing of separate intake and exhaust valves as in the four-stroke engine

BACKGROUND

German engineer Nikolaus August Otto achieved renown in the field of engine design despite a limited education and lack of formal training. He and partner Eugen Langen founded what became Gasmotoren-Frabrik Deutz AG, known simply as Deutz AG since 1997. Working alongside fellow designers Gottlieb Daimler and Wilhelm Maybach, Otto designed the first practical four-stroke, gasoline-powered internal combustion engine, which became popularly known as the Otto, or Otto-cycle, engine. His design was a revolutionary advance in the field of energy that led to the replacement of steam power and the introduction of the automobile.

Otto was born in Holzhausen, Germany, on June 14, 1832. His father's early death and Germany's political and economic instability led him to drop out of high school and become a clerk in a Frankfurt grocery store. His brother Wilhelm soon found him work as a traveling salesman. Although he lacked formal training, he decided to pursue his interest in the mechanical design of engines. He married Anna Katherina Gossi-Rouply in 1886. Otto's letters to Anna during their courtship would later become a key source of information on his early experiments. The couple had one son, Gustav.

Otto's interest in engine design was sparked by his belief in the potential future of internal combustion engines and his desire to improve upon the two-stroke, gas-powered internal combustion engine designed by Étienne Lenoir. Otto experimented with

Nikolaus Otto. (Wikimedia Commons)

the development of a liquid gasoline-fueled engine through models built by Cologne manufacturer Michael Zons. He introduced the first gasoline-powered engine in 1861, filing for a patent alongside his brother in that same year.

Otto was able to demonstrate that internal combustion engines fueled by liquid gasoline were more efficient than that of Lenoir, which used a system of natural gas and air. Otto's early successes gained him the support of businessman Eugen Langen in 1864, who supplied the capital and business knowledge necessary to form the *N. A. Otto & Cie* company to manufacture internal combustion engines based on Otto's designs. Otto and Langen's fuel-saving atmospheric gas engine received a gold medal at the 1867 Paris World Exhibition, and the partners quickly moved to patent their design.

Otto and Langen were soon overwhelmed with the demand for their new engine. Their company expanded with the addition of German businessman and investor Ludwig August Roosen-Runge in 1869, and the company came to be known as Langen, Otto, and Roosen; in 1872, it reincorporated as Gasmotoren-Frabrik Deutz AG. Otto, who had no financial stake in the company, was not a stockholder. He instead signed a long-term employment contract. New designers were also added, notably Daimler and Maybach. Otto's most significant contribution would come as the direct result of one of the projects he and Maybach undertook, the design of an internal combustion engine capable of powering road vehicles, first known as horseless carriages and later as automobiles.

Otto designed a four-stroke-cycle internal combustion engine in 1876 alongside Daimler and Maybach. He first used the engine to power a motorcycle. Otto created the model engine, while Maybach made the necessary alterations for mass production. It became popularly known as the Otto engine, or Otto-cycle engine. The Otto cycle is a four-stroke piston cycle consisting of intake, compression, power, and exhaust. The Otto-cycle engine was the first practical gasoline-powered internal combustion engine, offering an alternative to steam power and introducing an energy and transportation revolution. It incorporated the main components of the modern four-stroke engine and paved the way for Karl Benz's 1879 two-stroke engine.

Otto continued to experiment and improve on his engine design throughout the remainder of his career, culminating in his 1884 design for a practical low-voltage magnetic ignition system for the four-stroke engine. Otto also continued to pursue his interest in stratified-charge theory, or shockless combustion. He believed that a stratified-charge engine would run more smoothly and cleanly. Although this belief was never widely accepted, the Japanese automobile corporation Honda would introduce such an engine in the 1970s.

Otto would lose his German patent rights in 1886 after the emergence of a French patent granted to French engineer Alphonse Beau de Rochas for his pamphlet detailing the concept of the Otto cycle, even though Rochas had failed to pay the required publication fees and had never built a working model engine based on his design. The Otto engine was still patented in Britain, where Francis and William Crossley held the license for its production. Despite the overturning of Otto's German patents in 1886, the terms *Otto engine* and *Otto cycle* still enjoy widespread usage.

Gasmotoren-Frabrik Deutz AG emerged as a European leader in engine manufacturing through its licensing agreements and subsidiary arrangements with other companies. It has been known as Deutz AG since 1997. Daimler and Maybach left to form their own company in 1882, and Daimler's name lent itself to one of the first automobiles ever produced, based on the four-stroke Otto-cycle engine. Nikolaus Otto died in Cologne, Germany, on January 26, 1891 at the age of 59. His son, Gustav, followed in his footsteps, helping found the Bayerische Motoren Werke AG, commonly known as BMW. The four-stroke engine remains one of the most common of all engine designs and is found in most modern automobiles and trucks, as well as a vast number of self-powered machines and generating equipment.

—*Marcella Bush Trevino*

Further Reading

Eckermann, Erik. *World History of the Automobile.* Society of Automotive Engineers International, 2001.

Pulkrabek, Willard W. *Engineering Fundamentals of the Internal Combustion Engine.* Pearson Prentice Hall, 2004.

"The Steering Column. Nikolaus Otto and His Remarkable Compression Stroke." *Car and Driver*, vol. 39, no. 8, Feb. 1994, p. 7.

Van Basshuysen, Richard, and Fred Schafer. *Internal Combustion Engine Handbook: Basics, Components, Systems, and Perspectives.* Society of Automotive Engineers International, 2004.

P

Photovoltaics (PVs)

FIELDS OF STUDY
Electrical Engineering; Electrical Trades; Materials Science

ABSTRACT
Solar cells, semiconductor devices that convert sunlight into direct-current electricity, are called photovoltaics. Groups of these cells can be used to charge batteries and power electrical loads to operate appliances and even return power to the electrical utility grid.

KEY CONCEPTS
building-integrated photovoltaics (BIPVs): photovoltaic systems that are designed and built into buildings, either at the time of construction or as aftermarket additions

feed-in tariff: a rate paid to individuals who use photovoltaics or other renewable electricity generating technologies for excess electricity that is fed into the grid system

grid system: the regional electricity distribution network

photovoltaic: a term indicating electrical voltage generated by interaction with light

semiconductor: a material that does not normally conduct electricity, but that can be made conductive by stimulation either by applying an electrical biasing voltage or by light

WHAT THEY ARE
Photovoltaics (PVs), or solar cells, are semiconductor devices that convert solar energy into direct-current (DC) electricity. Groups of PV cells are electrically configured into modules and arrays, which can be used to charge batteries or power a variety of electrical loads. With the appropriate power-conversion equipment, PV systems can produce alternating current (AC) compatible with conventional appliances and can operate in parallel to, and interconnected with, the electrical utility grid. This has made PV technology a strongly promoted renewable energy option across the world.

HISTORY
The word "photovoltaic" comes from the Greek word *phos*, meaning "light," and the English word "volt," derived from the name of Italian physicist Alessandro Volta (1745-1827), who was a pioneer in the field of electricity. The photovoltaic effect is the basic physical process through which a solar cell converts sunlight into electricity: briefly, photons of light collide with electrons, knocking them into higher states of energy. French physicist Alexandre-Edmond Becquerel (1820-91) first recognized the photovoltaic effect in 1839.

Credit for building the first solar cell is given to American inventor Charles Fritts (1850-1903). In 1883, Fritts coated the semiconductor selenium with a very thin layer of gold, thus creating a device that was about 1 percent efficient. Five years later, Russian physicist and Moscow University professor Aleksandr Grigorievich Stoletov (1839-96) built the first cell based on the photoelectric effect discovered by Heinrich Hertz (1857-94). The photoelectric effect occurs under the right circumstances when light is used to push electrons, freeing them from the surface of a solid. Albert Einstein (1879-1955) finally explained the photoelectric effect in 1905, for which he received the Nobel Prize in Physics in 1921. In 1946, Russell Ohl (1898-1987) patented the modern solar cell, having discovered it while working on a series of advances that would lead to the transistor.

The modern age of solar PV technology arrived in 1954, when Bell Laboratories discovered that silicon doped with certain impurities is very sensitive to light. This first solar PV module was mostly a curiosity, since it was too expensive to have widespread applications. The first practical use for PVs was in the space industry. The Soviets employed a solar array as early as May 15, 1957, when they launched the *Sputnik 3* satellite. Throughout the 1960s, the technology was employed to power other orbiting satellites and spacecraft.

As PV technology advanced through space programs, its reliability was established, and costs began to decline. During the energy crisis of the 1970s, PV technology gained recognition as a source of power for nonspace applications. Improvements in manufacturing, performance, and quality of PV modules helped open up a number of opportunities for powering remote terrestrial applications, including battery charging for navigational aids, signals, telecommunications equipment, and other critical low-power needs.

THE BASICS

Electricity is essentially a stream of moving electrons. A basic PV cell allows for the direct conversion of solar radiation into electricity at the atomic level. The photons in sunlight hit the semiconductor materials in a panel and are reflected, are absorbed, or pass right through. The absorbed photons energize electrons, allowing them to escape from their atoms, and one or more electric fields act to force the freed electrons to flow in a certain direction. This flow of electrons is an electrical current that can be drawn off for external use, such as a providing light.

PV systems may employ a variety of materials in their construction, displaying different efficiencies. The most prevalent materials are derived from crystalline silicon and include single- or monocrystalline silicon, multi- or polycrystalline silicon, and amorphous silicon. Monocrystalline cells produce the most electricity per unit area, and amorphous cells the least. Amorphous silicon is among a group of other materials (also including cadmium telluride and copper-indium-gallium selenide) that are used in thin-film PV technology, which in 2009 was expected to soon dominate the market, due to its promise of reducing material requirements and manufacturing costs of PV modules and systems. However, by the end

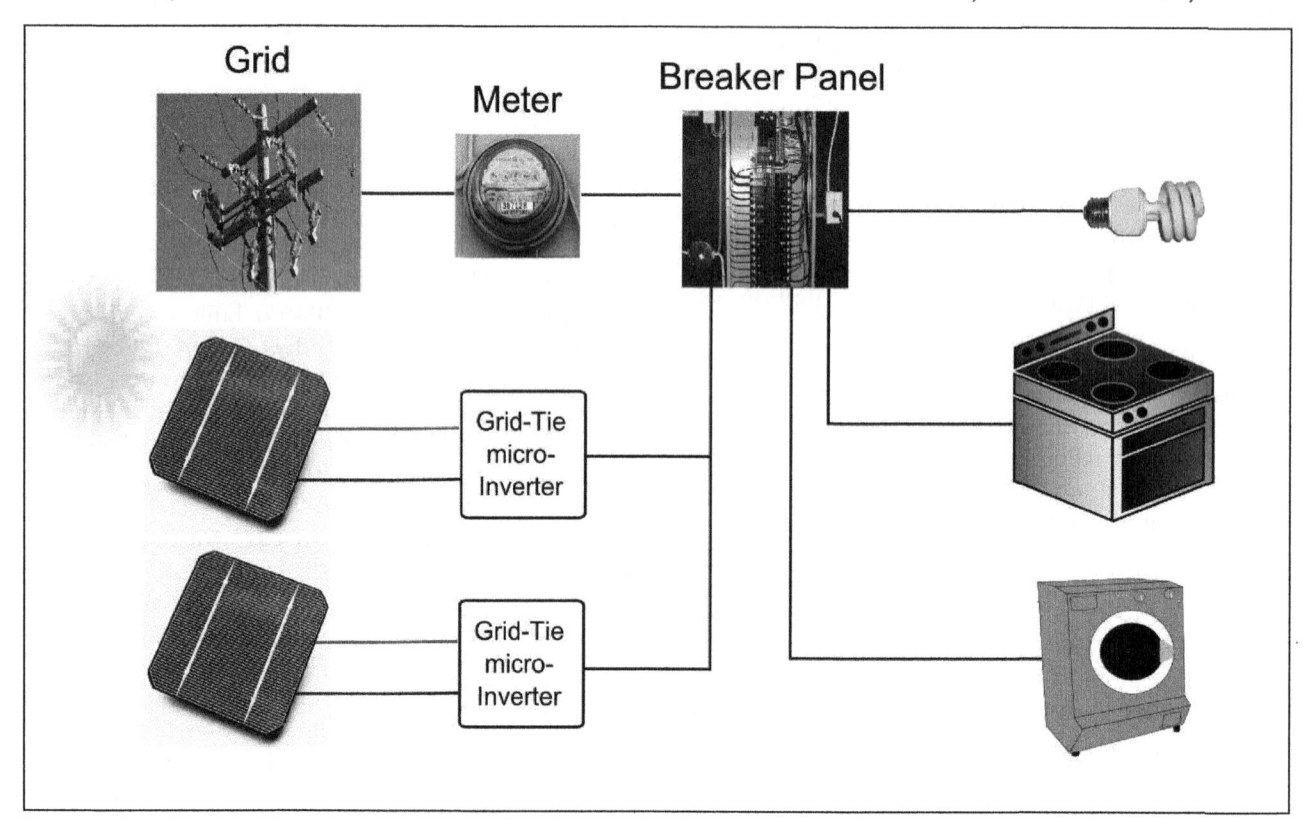

Diagram of a residential grid-connected PV system. (MikieMike via Wikimedia Commons)

Photovoltaic sunshade 'SUDI' is an autonomous and mobile station that replenishes energy for electric vehicles using solar energy. (Tatmouss via Wikimedia Commons)

of 2013, the global market share of thin-film modules had shrunk from 16 percent in 2009 to 10 percent, while other crystalline silicon modules represented about 90% of the market.

A traditional PV cell typically produces a small amount of power. In order to produce more power, PV cells are electrically interconnected and mounted in an environmentally protective support structure to form a panel or module. Modules are designed to supply electricity ranging in output from 10 to 300 watts (W). A number of modules can be wired together to form an array. Another way to increase power output is to move from a fixed-angle array (oriented and inclined according to its latitudinal location) to one that is mounted on a single- or dual-axis tracking device. These types of arrays follow the sun, capturing the most sunlight possible throughout the day.

Today's PV modules are a safe, reliable, low-maintenance source of electricity that produces no on-site pollution or emissions. They boast minimal failure rates and projected service lifetimes of twenty to thirty years. Most major manufacturers offer warranties of twenty-plus years for maintaining a relatively high percentage of the initial rated power output (up to 80 percent).

RENEWABLE ENERGY GENERATION

The 1970s energy crisis prompted the development of PV systems designed for residential and commercial usage. These systems operate interconnected with or independent of the utility grid, and they can be connected with other energy sources and energy storage systems. A grid-tied PV array may be deployed as an electricity generator in a variety of ways. Two of the more prevalent options are rooftop and ground-mounted systems, although building-integrated photovoltaics (BIPV) are increasing in popularity. Rooftop PV systems are particularly ideal in urban settings (though they may also prove useful in some rural areas), whereas ground-mounted systems are favored by rural generators.

Grid-tied systems are primarily connected with one of two different types of metering arrangements, depending on the local utility. Net metering is a program through which a utility charges a consumer for its net consumption of electricity. If the consumer produces a net surplus of electricity over the course of a given billing cycle, the utility will either pay that amount back or credit the consumer's next utility bill.

Consumers contracting into feed-in tariff programs use the second option. In this arrangement, a separate utility meter measures the electricity generated by the PV system. The utility pays the consumer for electricity that is generated at a different rate (a tariff) from that used for what is taken from the grid. For example, in 2009, the Canadian province of Ontario started offering twenty-year fixed-price contracts paying rooftop PV generators CDN$0.802 for every kilowatt-hour (kWh) produced from rooftop systems smaller than 10 kW.

THE PV MARKET

Today, the majority of PV installations supply grid-tied power generation. By the end of 2013,

Solar panels on the International Space Station. (Wikimedia Commons)

twenty-three countries had on-grid PV capacities greater than 100 megawatts (MW), and seventeen had capacities greater than 1 gigawatt (GW); in 2014, three more countries achieved 1 GW capacity. (Off-grid PV generation is a smaller market, accounting for just 1 percent of the market in 2013, and is used for remote homes, boats, recreational vehicles, electric cars, telecommunications, and remote sensing and monitoring.) Between 2003 and 2013, cumulative global PV capacity increased at an annual average rate of 49 percent, leading to over 135 GW of installed capacity worldwide, plus another 40 GW added in 2014.

Driven by advances in technology and increases in manufacturing scale and sophistication, the cost of PVs has declined steadily since the first solar cells were manufactured. The International Energy Agency (IEA) reported in 2014 that the cost of PV modules had decreased to one-fifth of their price in 2008, while full systems had decreased to just over one-third of their 2008 cost. As of 2013, the typical cost of a PV system in the United States was around US$3.30 per watt for utility-scale systems, US$4.50 per watt for commercial systems, and US$4.90 per watt for residential. (Prices were generally lower in China and in non-U.S. Western countries, e.g., US$1.40-1.50/W in China, US$1.70-2.00/W in Australia, US$1.40-2.40/W in Germany). Government- and utility-backed incentive programs, such as net metering and feed-in tariffs targeting solar-generated electricity, have also supported the deployment of solar PV installations in many countries. Global leaders include Germany, Italy, China, Japan, and the United States, particularly California.

In 2013, China and Japan became primary global drivers of PV installations, adding about 11 GW and 7 GW of new capacity, respectively. However, Germany remained the country with the greatest total capacity, with 38.2 GW installed by the end of 2014, compared to 28.2 GW in China, the first runner-up.

PV technology is also experiencing growing demand in the developing countries of Africa, Asia, and Latin America as they purchase very small-scale, off-grid systems. Sales and total capacity of off-grid systems have increased steadily since the early 1980s. In many cases, these systems are already at price parity with fossil fuels.

—*Robert J. Wakulat*

Further Reading

Brabec, Christoph, Ulrich Scherf, and Vladimir Dyakonov, eds. *Organic Photovoltaics: Materials, Device Physics and Manufacturing Technologies*. 2nd ed. John Wiley & Sons, 2014.

International Energy Agency. *Technology Roadmap: Solar Photovoltaic Energy*. 2014 ed. Author, 2014. *International Energy Agency*. Web. 13 May 2016.

Knier, Gil. "How Do Photovoltaics Work?" *NASA Science*. NASA, 6 Apr. 2011. Web. 13 May 2016.

Mertens, Konrad. *Photovoltaics-Fundamentals, Technology and Practice*. Wiley, 2019.

"Photovoltaic (PV) Systems." *Canada Mortgage and Housing Corporation*. Canada Mortgage and Housing, 2010. Web. 13 May 2016.

Reinders, Angèle. *Designing with Photovoltaics*. CRC Press, 2020.

REN21. *Renewables 2015: Global Status Report*. REN21 Secretariat, 2015. *REN21*. Web. 13 May 2016.

Yam, Vivian W.W., ed. *WOLEDs and Organic Photovoltaics: Recent Advances and Applications*. Springer, 2010.

POTENTIAL ENERGY

FIELDS OF STUDY
Engineering; Physical Chemistry; Physics

ABSTRACT
Energy exists in several forms, but most scientists agree that the two basic forms of energy are potential energy and kinetic energy. Potential energy is energy that is stored and can exist in many subforms, including chemical, mechanical, and atomic energy. Kinetic energy is energy associated with some form of motion.

KEY CONCEPTS
activation energy: the amount of energy required to perturb a system sufficiently to render it unstable

flywheel: a relatively large mass that stores energy by virtue of its rotation; a flywheel typically may acquire rotation by converting energy from linear motion or by being directly spun up to speed by an external driver; it can then be used to moderate the energy load being placed on the system

hydraulic system: a system that uses fluid pressure to perform mechanical functions; hydraulic fluids may be compressible gases or incompressible liquids as the working fluid

moderator: in a nuclear reactor core, the moderator is a material such as carbon or heavy water (D2O) that slows the flight of ejected neutrons by absorbing an appropriate portion of their kinetic energy

potential energy surface: a surface defined by its height above a gravitational reference surface

work: the movement of a mass through a distance by the application of a sufficient force

THE ENERGY NOT USED

Potential energy is energy that is stored rather than used and can exist in many subforms, including chemical, mechanical, and atomic energy. Kinetic energy is energy that is being used and is associated with some form of motion. To convert energy to work, it is advantageous to have stored energy in a form that can be converted to both kinetic energy and work in a controlled and metered fashion. Potential energy in sources that are stable and available in abundance are needed so it can be converted for the generation of electricity to power homes, run machinery, provide lighting, and meet myriad other energy needs.

GRAVITATIONAL POTENTIAL ENERGY

The most common or recognizable form of potential energy is the energy stored when an object that has mass is moved from a lower, stable energy state to a second stable position higher in energy and physical elevation than the original state. This is called "gravitational potential energy" and is expressed mathematically as $E = mgh$, where E is the potential energy, m is mass, g is the gravitational constant, and h is height. An easy visualized example of this type of potential energy is demonstrated by a person picking up a round, large rock at the bottom of a hill and carrying it to the top of the hill. The round rock is then set down on a stable ledge, where it can rest without further restraint. The person carrying the round rock provided the kinetic energy to lift it to a higher elevation and transferred this kinetic energy at some efficiency to the rock as a new, higher potential energy state.

The potential energy in the rock could be released and converted to kinetic energy if the person pushed the rock off the ledge on which it is now sitting and set the rock rolling down the hill. The force of gravity would compel the rock to release its potential energy as the kinetic energy of motion, when a small amount of destabilizing energy was added by pushing the rock off its ledge. The push is the energy required to destabilize the potential energy state and cause it to transform to kinetic energy. The formal name of the energy used to destabilize the potential energy state is "activation energy." If the activation energy was not added, the rock would have sat on top of the hill indefinitely with more potential energy than it had before it was carried up the hill.

One very common practical use of stored potential energy is in the construction of dams to store water at an elevated height above the normal water level. Most of the water in rivers and streams originates from rain, which is at an elevated height. If a container can be constructed to capture this water before gravity pulls it to the normal water level, the water will retain potential energy due to its elevated height. This is ex-

Archery is one of humankind's oldest applications of elastic potential energy. (Penny Mayes via Wikimedia Commons)

actly what dams can do; dam reservoirs are really large containers that capture water. The elevated water level in the dam's reservoir represents potential energy; if the water is allowed to fall to normal water level through a controlled opening, the mass and velocity of the falling water can be used to turn a waterwheel or water turbine. The waterwheel or water turbine's rotating energy can be used to run machinery or generate electricity, known in this case as hydroelectricity.

Electricity is very hard to store at large power levels, and one of the few practical ways to do so is to pump water up to an elevated level for storage using electricity-driven pumps. Some hydroelectric generating plants have water turbines that do not just let falling water turn the turbines to generate electricity but have the ability to allow the generator to function as a motor and take power off the electric grid, then pump the water back up into the dam's reservoir. These generating stations are called "pumped-storage units," and they convert electric energy back into potential energy by using electricity to move water to a higher elevation, where it can be held in a stable state.

SPRINGS, ACCUMULATORS, AND FLYWHEELS

Mechanical springs and hydraulic accumulators are good sources of stored potential energy as well. These devices are often used on hydraulic systems to provide stored energy to be turned into kinetic energy on demand. The most common use is to store energy to take a valve or other device to a safe position upon failure of the device's primary drive. The potential energy device, such as a spring, is thus a secondary source of energy used when the primary source of energy is unavailable due to failure.

Springs and accumulators can also store energy from a primary movement to be used to assist in returning to an original position. Large flywheels are also sources of mechanical potential energy, and these devices can be used to store energy and even out load on variable demand energy sources.

NUCLEAR ENERGY

Nuclear energy, another a form of potential energy, is similar to chemical potential energy but with the source of the stored energy in the forces that unite the subatomic particles within the nucleus of the atom,

Uranium. (Wikimedia Commons)

rather than in the bonds between atoms. Splitting the nucleus in a fission process or joining atomic nuclei in a fusion process releases the stored potential energy in the nuclear structure.

Some of the larger atoms have a nucleus that can be broken apart by bombardment with free neutrons in what is called "nuclear fission." For example, when the nucleus of an atom of uranium-235 is struck by neutrons traveling at a sufficiently high velocity, the nucleus will break apart, releasing more neutrons, daughter nuclear products, radiation, and potential energy.

If the uranium is arranged in the right geometry and in the right amount, the neutrons released by one fission reaction will strike the nuclei of other uranium atoms close by in a self-sustaining chain reaction. If the amount and speed of these neutrons can be controlled by a moderator in which the uranium is immersed, the reaction can be controlled and used to heat fluids, and those fluids can then provide the mechanical power to generate electricity.

Potential energy can also be released from nuclear fusion, during which atoms with small nuclei, such as hydrogen (whose nucleus consists of a single proton), are forced to unite to form helium. This form of nuclear energy, called "fusion energy," is not yet practical for electricity generation; the extremely high pressures and temperatures needed to make the reaction occur make it difficult to control, and currently require more energy to maintain than such a reactor would produce. Stars, however, have large reserves of potential fusion energy in hydrogen reserves, enough stored potential energy to continue fusing hydrogen for billions of years.

As fusion in stars consumes the hydrogen and then the helium as well, heavier elements begin to form, with the fusion process being maintained by the intense pressure of gravity within the star's body. Ultimately, the star forms and fuses atoms of iron, at which point the star will explode as a nova. Elements heavier than iron are formed (and continue to be formed) when stars explode using large amounts of stored energy at once to make those heavier elements.

—*Damon Eric Woodson*

Further Reading
Demirel, Yasar. *Energy: Production, Conversion, Storage, Conservation, and Coupling.* Springer, 2012.

Kenward, Michael. *Potential Energy: An Analysis of World Energy Technology.* Cambridge UP, 1976.

PowerStream Technology. "Chemical Changes in the Battery." http://www.powerstream.com/1922/battery_1922_WITTE/batteryfiles/chapter04.htm.

U.S. Energy Information Administration. "Energy Explained: Your Guide to Understanding Energy." http://www.eia.doe.gov/energyexplained/index.cfm.

Viegas, Jennifer. *Kinetic and Potential Energy: Understanding Changes Within Physical Systems.* Rosen, 2005.

World Nuclear Association. "What Is Uranium? How Does It Work?" http://www.world-nuclear.org/education/uran.htm.

Propane

FIELDS OF STUDY
Chemical Engineering; Geology; Process Engineering

ABSTRACT
Propane has been extensively used as a fuel since its isolation in the early twentieth century. It is a "clean" fuel that produces carbon dioxide and water as well as a quantity of heat.

KEY CONCEPTS
alkanes: an infinite series of molecules containing only carbon and hydrogen, beginning from methane, with only single bonds between adjacent atoms

cracking: a refining process in which hydrocarbon molecules are made to break apart into smaller hydrocarbon molecules

ethylene: the simplest member of an infinite series of "unsaturated" hydrocarbon molecules called "alkanes," in which two adjacent carbon atoms are doubly-bonded to each other; more than one such 'site of unsaturation' may exist in alkene molecules

LPG: liquid propane gas, which is effectively pure propane; not to be confused with LNG or liquid natural gas, which contains primarily methane and ethane

propylene: an "unsaturated" form of propane in which two adjacent carbon atoms are doubly bonded to each other which makes such compounds capable of undergoing a wide variety of reactions, formula is C_2H_6 or $H_2C=CH-CH_3$

WHERE PROPANE COMES FROM
Propane occurs in natural gas, to the extent of up to 6 percent, as well as with crude oil. It is commercially available after its separation from natural gas or the gases formed when heavy oil is broken down (cracked) to produce gasoline and other fuels and petroleum products. The principal gas deposits are in the United States, Canada, the former Soviet bloc, and the Middle East.

PRIMARY USES
Propane has been extensively used as a fuel since its isolation in the early twentieth century. It is a "clean" fuel that produces three units of carbon dioxide and four units of water per unit of propane, as well as a quantity of heat.

TECHNICAL DEFINITION
Propane's condensed chemical formula is C3H8 or $CH_3CH_2CH_3$, and it is a colorless, odorless, and nontoxic gas. It has a density of 0.58 gram per cubic meter at -450°C, a melting point of -187°C, and a boiling point of -42.2°C. Its heat of combustion (the energy released when a hydrocarbon is burned in the presence of oxygen to produce carbon dioxide and water) is relatively high and is equal to 531 kilocalories per mole, or 2,220 kilojoules per mole.

DESCRIPTION, DISTRIBUTION, AND FORMS

Propane is the third-lightest alkane of the hydrocarbon family after methane and ethane. Propane is almost insoluble in water, moderately soluble in alcohol, and very soluble in ether.

According to the U.S. Energy Information Administration, the proven world reserves of natural gas (most of which is methane) were 6,973 trillion cubic feet as of 2014. This figure suggests that there are ample supplies of propane gas for the near future. Such sources include ultradeep gas, which is in the form of sedimentary deposits as deep as 15,000 meters, the hydrated gas that exists in the Arctic regions, and the reserves beneath the oceans.

Unlike coal and petroleum products, such as diesel fuel, which create environmental problems when burned, propane and natural gas are clean fuels. Pollution-control equipment is generally unnecessary. Incomplete combustion of propane, however, yields carbon monoxide (instead of carbon dioxide) and water. Carbon monoxide is particularly toxic to humans, since it has an excellent binding ability to the iron of hemoglobin and displaces oxygen during the breathing process. As a result, the tissues die from oxygen starvation. Incomplete propane combustion takes place in areas that do not have sufficient oxygen, such as closed garages and relatively airtight rooms. Because of this danger the use of camping stoves indoors to heat water is highly discouraged. Another area of great concern is the deliberate inhaling of propane by teenagers. "Huffing" propane as an intoxicant for about five minutes may easily lead to death.

Propane and butane have largely replaced the freons as propellants in aerosol cans. Freons, which are chlorinated, brominated, and/or fluorinated hydrocarbons, are easily broken down by light into sta-

Staying warm outdoors in winter with a propane patio heater. (praetorianphoto via iStock)

ble radicals which are prime suspects in causing damage to Earth's protective ozone layer.

HISTORY

In 1910, Walter O. Snelling discovered propane while working for the U.S. Bureau of Mines. Snelling and others started the American Gasol Company to sell propane. The product reached its peak in the post–World War II era, when many houses were equipped with propane stoves. In the 1960s, Chevrolet built and marketed propane trucks. In 1990, propane was designated a clean fuel.

Propane is not a common starting reagent in industrial synthetic processes. Nevertheless, more than 20 percent of the propane obtained from natural gas sources is converted to its unsaturated derivative, propylene. This process involves its decomposition in hot tubes. Propane is also used in the synthesis of the colorless, flammable gas ethylene, which serves as the monomer of the useful polymer polyethylene. Polyethylene is the essential ingredient of many household plastics (such as plastic wraps for food items). Ethylene is formed via cracking, a process that involves high temperatures of about 500°C and several atmospheres of pressure. Isopropyl alcohol, also known as rubbing alcohol, was the first petrochemical synthesized via propane in the 1920s. The high-temperature oxidation of propane to acetaldehyde is also of commercial importance.

USES OF PROPANE

Propane and the next higher alkane, butane, are the main components of liquefied petroleum gas (LPG), which may be carried in tanks or cylinders and used as camping gas and in portable cooking stoves. In certain areas LPG is transported via pipelines, and it may also be used for internal combustion engines. In many agricultural areas propane and butane are more cost-effective tractor fuels than are gasoline and diesel fuel. Commercially, the clean-burning fuel is maintained in the liquid state, which is obtained under conditions of elevated pressures, in a steel container. It converts spontaneously to the gas state upon exposure to the normal atmospheric pressure. Unlike propane, butane condenses to a liquid at 0°C and thus cannot be used for camping under cold conditions. As a result butane tends to be used in the southern United States, while propane (which condenses at -42°C under normal pressure conditions is used in the North.

As noted previously, propane has a high heat of combustion, which, combined with its ease of transport, makes it a convenient fuel. Another way of measuring the fuel capacity of propane is its "heating value," which measures the amount of heat evolved when a gram of propane is burned. Propane's value is about 6,098 kilocalories per liter. Overall the fuel capacity of propane is about 2.5 times that of natural gas.

—*Soraya Ghayourmanesh*

Further Reading

Erjavec, Jack, and Jeff Arias. "Propane/LPG Vehicles." *Hybrid, Electric, and Fuel-Cell Vehicles.* Thomson, 2007.

Gibilisco, Stan. "Propulsion with Methane, Propane, and Biofuels." *Alternative Energy Demystified.* McGraw-Hill, 2007.

Intratec. *Ethylene Production via Cracking of Ethane/Propane: Report Ethylen E21A Basic Cost Analysis.* Intratec Solutions, 2019.

Intratec. *Propylene Production from Propane-Cost Analysis: Propylene E33A.* Intratec, 2016.

Myers, Richard L. "Propane." *The One Hundred Most Important Chemical Compounds: A Reference Guide.* Greenwood, 2011.

Speight, James G. *The Chemistry and Technology of Petroleum.* 5th ed. CRC Press, Taylor, 2014.

U.S. Energy Information Administration. "Proved Reserves of Natural Gas (Trillion Cubic Feet)." *EIA.gov.* U.S. Department of Energy, 2015. Web. 6 Jan. 2016.

U.S. Department of Energy. "Alternative and Advanced Fuels: Propane."
http://www.afdc.energy.gov/afdc/fuels/propane.html.

U.S. Department of Energy. "Alternative and Advanced Vehicles: Propane Vehicles."
http://www.afdc.energy.gov/afdc/vehicles/propane.html.

R

RANKINE, WILLIAM

FIELDS OF STUDY
Thermodynamics

ABSTRACT
One of the originators of the laws of thermodynamics, William Rankine is best known for his complete theory of heat engines.

KEY CONCEPTS
efficiency: the ratio of the actual work output of a machine relative to the ideal, or theoretical, work output of the ideal machine
heat engines: any engines that converts heat energy to mechanical energy
thermodynamics: the study of the movement and characteristic behavior of heat
vapors: typically, molecules of a liquid that have entered the gas phase at temperatures below the boiling point of the liquid

BACKGROUND
William John Macquorn Rankine was a Scottish engineer and physicist who, along with Rudolf Clausius and Lord Kelvin (William Thomson), was one of the fathers of thermodynamics. Rankine's best-known work was his complete theory of heat engines, and his engineering manuals continued to be used for most of the nineteenth century.

Initially educated at home, Rankine later attended the High School of Glasgow, the Military and Naval Academy, and the University of Edinburgh, which he began attending at age 16, working under James David Forbes, who was also mentor to physicist and mathematician James Clerk Maxwell.

HEAT ENGINES
During Rankine's time, there was a continual problem in railway engineering in putting down curved rails, and Rankine's method was significantly more accurate and efficient than preceding methodologies, making use of the recently developed theodolite, a precision surveying instrument used for measuring angles in vertical and horizontal planes. From early in his career, Rankine was interested in representing heat mathematically, but in his university years, he suffered from the dearth of experimental data from which to extrapolate. His interest in heat soon turned into an interest in heat engine mechanics, the me-

William Rankine. (Wikimedia Commons)

chanics of any engine that converts thermal energy into mechanical work, which in Rankine's time included the steam engine and today also includes the internal combustion engine, among many other types. In 1849, Rankine successfully determined the relationship between temperature and saturated vapor pressure. He soon moved on to establish relationships between the temperature, pressure, and density of gases, and he predicted the counterintuitive conclusion that saturated steam has a negative apparent specific heat.

Soon Rankine had succeeded at calculating the efficiency of heat engines, and he extrapolated from his experimental data that the maximum efficiency of such an engine is a function of the temperatures between which it operates (the thermodynamic function). These results were reformulated and rephrased several times and were used to support an energy-centric formulation of dynamics, which described dynamics in terms of energy and energy transformations rather than motion and force: the science of energetics. For a time, Rankine was a vocal critic of James Clerk Maxwell's theories of heat, because they were incompatible with some of his own models, but in 1869 he eventually admitted the validity of Maxwell's equations.

Rankine's best successes were in working toward practical results, analyzing properties of gases, vapors, and steam to achieve more efficient heat engines. He was also among the first to understand the fatigue failure of railway axles; he was able to show how they developed from brittle cracks and described the process now called metal fatigue.

—*Bill Kte'pi*

Further Reading

Karwatka, Dennis. "William Rankine and Engineering Education." *Tech Directions*, vol. 69, no. 3, Oct. 2009.

Lewis, Christopher J.T. *Heat and Thermodynamics*. Greenwood, 2007.

Liszka, John. "Are You Sure, Mr. Carnot? A Re-examination of the Thermodynamic Principles as Formulated by Nicolas Carnot and William Rankine Over One Hundred Years Ago Might Lead to Greater Efficiency in Electrical Power Generating Stations, Together With Reduced Emissions." *Engineering Digest*, vol. 38, no. 3, June 1992.

Muller, Ingo. *A History of Thermodynamics*. Springer, 2010.

Raman, V.V. "William John Macquorn Rankine (1820-1872)." *Journal of Chemical Education*, vol. 50, no. 4, 1973.

Sutherland, H.B. "Professor William John Macquorn Rankine." *Proceedings of the Institution of Civil Engineers Civil Engineering*, vol. 132, no. 4, 1999.

Renewable Energy

FIELDS OF STUDY
Electrical Engineering; Hydraulic Engineering; Mechanical Engineering

ABSTRACT
The environmental movement and the oil crises of the 1970s led to interest in the development of energy sources that would offer alternatives to the use of fossil fuels. Fossil fuels are limited resources, and the burning of fossil fuels to generate energy emits carbon dioxide, toxic chemicals, and other pollutants that harm the environment and human health. Because renewable, or clean, energy systems use natural, local sources that are inexhaustible, such systems have fewer negative impacts on human life and the environment. Governments have provided increasing support for the development of renewable energy technologies.

KEY CONCEPTS

biomass: matter consisting of material components of living organisms, typically plants, animals and their solid wastes

biomass oils: oils that can be obtained from biomass, such as vegetable seed oils and animal fats, for processing into usable fuels

fossil fuels: essentially, ancient biomass such as petroleum from bacteria that existed on Earth billions of years ago and coal from plant matter that existed in great swamps hundreds of millions of years ago

geothermal heat pump: devices that use the consistent temperatures that exist underground as the heat source or heat sink for heating and cooling systems above-ground

solar energy: used interchangeably to mean either the heat of incoming sunlight that reaches Earth's surface, or the electricity that it produced using sunlight

WHY RENEWABLE ENERGY?

The burning of fossil fuels such as coal, natural gas, and petroleum releases emissions that contain greenhouse gases and cause air pollution and acid rain, whereas most forms of renewable energy are nonpolluting. In addition, because Earth has a finite supply of fossil fuels, the development of renewable energy sources is important to the long-term future of humankind.

The environmental movement and the oil crises of the 1970s led to interest in the development of energy sources that would offer alternatives to the use of fossil fuels. Fossil fuels are limited resources, and the burning of fossil fuels to generate energy creates emissions of carbon dioxide, toxic chemicals, and air pollutants that harm the environment and human health. Because renewable, or clean, energy systems use natural, local sources that are inexhaustible and such systems have fewer negative impacts on human life and the environment, governments have provided increasing support for the development of renewable energy technologies.

BIOMASS

The oldest renewable energy source is biomass, which is organic animal and plant material and waste. Biomass resources include grass crops, trees, and agricultural, municipal, and forestry wastes. Since the discovery of fire, humans have burned biomass to release its chemical energy as heat. For example, wood has been burned to cook food and to provide heat. Biomass energy has also been used to make steam and

Aside from large infrastructure projects, hydroelectric generation can also be used on a much smaller scale, such as this micro-generation setup in a village in Northwestern Vietnam. (Shermozle via Wikimedia Commons)

electricity. Biomass oils can be chemically converted into liquid fuels or biodiesel, a transportation fuel. Ethanol, another transportation fuel, comes from fermented corn or sugarcane. Crops such as willow trees and switchgrass are also cultivated for biomass energy generation.

Biomass energy has many environmental benefits when compared with fossil-fuel energy. It contributes little to air pollution, as it releases 90 percent less carbon dioxide than do fossil fuels. But because the carbon in biomass had been removed from the atmosphere as the materials grew, no net increase in atmospheric carbon dioxide is produced when biomass is burned. Fossil fuels, on the other hand, release carbon that has been sequestered within Earth for millions of years, billions in the case of petroleum. Energy crops, such as prairie grasses, require fewer pesticides and fertilizers than do high-yield food crops such as wheat, soy beans, and corn, so they cause less water pollution. Energy crops also add nutrients to the soil. About 4 percent of the energy used in the United States is biomass energy.

SOLAR ENERGY

One of the most promising and popular kinds of renewable energy is solar energy, which uses radiant energy produced by the sun. Solar energy was used as early as the seventh century BCE, when a magnifying glass was used to concentrate sunlight to light fires. In 1767, Swiss scientist Horace-Bénédict de Saussure invented the first solar collector, a device for capturing the Sun's radiation and converting it into a usable form, such as by heating water to create steam. In 1891, American inventor Clarence Kemp patented the first commercial solar water heater.

Sunlight can be converted directly into electricity at the atomic level by photovoltaic (PV) cells, also called solar cells. The photovoltaic phenomenon was first noted in the eighteenth century and became more practical with the use of silicon for the cells in the twentieth century. The cells are joined together in panels, often connected together in an array. They can be placed on rooftops and connected to a grid. In the twenty-first century, solar cells are used worldwide in home and commercial electrical systems, satellites, and various consumer products.

Solar energy has numerous environmental benefits. Photovoltaics produce electricity without gaseous

An illustration of furnace bellows operated by waterwheels, from the Nong Shu, by Wang Zhen, 1313, during the Yuan Dynasty of China. (Wikimedia Commons)

or liquid fuel combustion or hazardous waste by-products. Decentralized PV systems can be used to provide electricity for rural populations, saving more expensive conventional energy for industrial, commercial, and urban needs. By providing electricity for remote and rural areas, solar energy also reduces the use of disposable lead-acid cell batteries, which can contaminate water and soil if they are not disposed of properly. The use of solar energy in rural areas also reduces air pollution by decreasing the use of diesel generators and kerosene lamps.

Since solar energy depends on sunlight, the efficiency and performance of solar energy systems are affected by weather conditions and location. Nevertheless, the amount of inexhaustible solar energy that could be generated around the globe exceeds the amount needed to meet the world's energy requirements. It has been estimated that if PV systems were installed in only 4 percent of the world's deserts, they could supply enough electricity for the entire world. As solar technologies improve and costs decrease, solar energy has the potential to be the leading alternative energy source of the future.

WIND ENERGY

One of the fastest-growing types of renewable energy during the 1990s, wind energy has been used by humans for centuries. Windmills appear in Persian drawings from about 500 CE, and they are known to have been used throughout the Middle East and China. The English and the French built windmills during the twelfth century, and windmills were indispensable for pumping underground water in the western and Great Plains regions of the United States during the nineteenth and early twentieth centuries. These windmills converted wind into mechanical power.

The modern windmills used to convert wind energy into electricity are called wind turbines or wind generators. In 1890 Poul la Cour, a Danish inventor, built the first wind turbine to generate electricity. Another Dane, Johannes Juul, built the world's first alternating current (AC) wind turbine in 1957. In the twenty-first century, large wind plants are connected to local electric utility transmission networks to relieve congestion in existing systems and to increase reliability for consumers. Wind energy is also used on a smaller scale by home owners in what is known as distributed energy; home-based wind turbines, with batteries as backup, can lower electricity bills by up to 90 percent.

The use of wind energy has long-term environmental benefits. Unlike nuclear and fossil-fuel electricity generation plants, wind generation of electricity does not consume fuel, cause acid rain and greenhouses gases, or require waste cleanups. For example, it has been estimated that when Cape Wind, America's first offshore wind farm under development in Nantucket Sound, becomes fully operational, its 130 wind turbines will reduce greenhouse gas emissions by 734,000 tons annually.

In 2010 the annual wind energy generating capacity of the United States was more than 35,000 megawatts, enough electricity to power 9.7 million homes. This amount of electricity generated by fossil-fuel-burning plants would have released some 62 million tonnes of carbon dioxide; avoiding the release of so much carbon dioxide is the equivalent of keeping 10.5 million cars off the roads.

Wind energy technology has advanced to the point that wind power is affordable and can compete successfully with fossil fuels and other conventional energy generation. According to a report by the U.S. Department of Energy, wind energy could provide 20 percent of the U.S. electricity supply by 2030. Wind power as a commercial enterprise has been established in more than eighty countries, and between 2013 and 2014 wind energy capacity increased by 51,477 MW, with China, Germany, and the United States seeing the biggest increases.

The main disadvantage of wind energy is that it is intermittent, because wind velocities are inconsistent even in areas of strong winds. In addition, some environmentalists have objected to the establishment of wind farms because of their potential to harm wildlife and the aesthetic damage they do to natural landscapes.

HYDROPOWER
Long before electricity was harnessed, in about 4000 BCE, ancient civilizations used hydropower, or energy from moving or flowing water, in the waterwheel, the first device employed by humans to produce mechanical energy as a substitute for animal and human labor. Running water in a stream or river moves the wooden paddles mounted around a waterwheel, and the resulting rotation in the shaft drives machinery. The earliest waterwheels were used to raise water for irrigation of crops and to grind grain, and the technology went on to be used worldwide for those purposes, as well as to supply drinking water, drive pumps, and power sawmills and textile mills.

In the nineteenth century, the water turbine replaced the waterwheel in mills, but then the steam engine replaced the turbine in mills. The hydraulic turbine reemerged, however, to power electric generators in the world's first hydroelectric power stations during the 1880s. By the early twentieth century, 40 percent of the U.S. electricity supply was hydroelectric power. Modern large hydropower plants are attached to dams or reservoirs that store the water for turning the turbines and are connected to electrical grids or substations that transmit the electricity to consumers.

Hydropower is the leading renewable energy source for generating electricity. It has both negative and beneficial effects on the environment. Building dams and reservoirs changes the environment and can harm native habitats and their fish, animal, and plant life. In addition, reservoirs sometimes emit methane, a greenhouse gas. Water is a natural and inexpensive energy source, however. No fuel combustion takes place in the generation of hydropower, so the process does not pollute the air, and energy storage is clean. Hydroelectric power accounts for about

16 percent of electricity used worldwide, and 7 percent of electricity in the United States. Hydroelectric energy capacity has been growing by approximately 3.3 percent per year throughout the twenty-first century, although widespread drought in 2015 slowed growth in this area.

GEOTHERMAL ENERGY

Geothermal energy comes from heat produced deep inside Earth. Deep wells and pumps bring underground hot water and steam to the surface to heat buildings and generate electricity. Some geothermal energy sources come to the surface naturally, including hot springs, geysers, and volcanoes. The ancient Chinese, Native Americans, and Romans used hot mineral-rich springs for bathing, heating, and cooking. Food dehydration became the major industrial use of this form of energy. In 1904, the first electricity from geothermal energy was generated in Larderello, Italy.

Although not as popular as a renewable energy source as wind or solar energy, geothermal energy has significant advantages and benefits for the environment. Because Earth's heat and temperatures are essentially constant, geothermal energy is reliable and inexhaustible; it is also not affected by changes in climate or weather. It is very cost-efficient as well; heat pumps can be operated at relatively low cost. The steam and water used in geothermal systems are recycled back into the environment.

Geothermal plants are environmentally friendly. Because they do not burn fuel to generate electricity, they release little or no carbon dioxide and other harmful compounds. Geothermal plants produce no noise pollution and have minimal visual impacts on the surrounding environment, because they do not occupy large surface areas.

The U.S. Environmental Protection Agency and the Department of Energy support the use of geothermal heat pumps. The American Recovery and Reinvestment Act of 2009 (ARRA, also known generally as the Stimulus) provided for grants and tax incentives worth $400 million to the industry, which added 144 geothermal energy plants in fourteen states at the beginning of 2010. ARRA measures benefited the development of all renewable energy sources; it included a Treasury Department grant program for renewable energy developers, increased funding for research and development, and a three-year extension of the production tax credit for many renewable energy facilities. By the end of 2014, the United States had an installed geothermal energy capacity of about 3.5 GW.

It has been predicted that by 2070, 60 percent of all global energy will come from renewable energy sources. The World Bank, the World Solar Decade, and the World Solar Summit have designated $2 billion for projects focused on renewable energy resources and the environment.

—*Alice Myers*

Further Reading

Craddock, David. *Renewable Energy Made Easy: Free Energy from Solar, Wind, Hydropower, and Other Alternative Energy Sources*. Atlantic, 2008.

Da Rosa, Aldo Vieira. *Fundamentals of Renewable Energy Processes*. 3rd ed. Elsevier, 2012.

Goswami, D. Yogi, and Frank Kreith, eds. *Energy Efficiency and Renewable Energy Handbook*. 2nd ed. CRC Press, 2016.

Jacobson, Mark Z. "How Renewable Energy Could Make Climate Treaties Moot." *Scientific American*. Nature America, 23 Nov. 2015. Web. 6 Jan. 2016.

Langwith, Jacqueline, ed. *Renewable Energy*. Greenhaven, 2009.

MacKay, David J.C. *Sustainable Energy—Without the Hot Air*. UIT Cambridge, 2009.

Meyer, John Erik. *The Renewable Energy Transition: Realities for Canada and the World*. Springer, 2020.

Neill, Simon P., and M. Reza Hashemi. *Fundamentals of Ocean Renewable Energy: Generating Electricity from the Sea*. Academic Press, 2018.

Nelson, Vaughn. *Wind Energy: Renewable Energy and the Environment*. CRC Press, 2009.

Pimentel, David, ed. *Biofuels, Solar, and Wind as Renewable Energy Systems: Benefits and Risks*. Springer, 2008.

Raikar, Santosh, and Seabron Adamson. *Renewable Energy Finance: Theory and Practice*. Academic Press, 2020.

Ruin, Sven, and Gören Sidén. *Small-scale Renewable Energy Systems: Independent Electricity Systems for Community, Business and Home*. CRC Press, 2020.

Twidell, John, and Tony Weir. *Renewable Energy Resources*. 3rd ed. Routledge, 2015.

Renewable Energy Resources

FIELDS OF STUDY
Biotechnology; Environmental Engineering; Process Engineering

ABSTRACT
Renewable energy is energy derived from regenerative resources—that is, resources that can be fully replenished in a short time period. These renewable energy resources, if well explored and developed, can be inexhaustible.

KEY CONCEPTS
biogas: a mixture of gases produced by the decomposition of animal and vegetable wastes
biomass: matter of vegetable origin
gasification: the process of heating organic matter with a limited amount of oxygen so that it is driven to decompose to gaseous materials
pyrolysis: the process of heating organic matter in such a way that it breaks down and decomposes to a mixture of liquids, oils, and gases

POTENTIALLY ENDLESS ENERGY SOURCES
Renewable energy is energy derived from regenerative resources—that is, resources that can be fully replenished in a short time period. Renewable energy is generated from elements found in nature and includes the use of biomass in fuels such as ethanol, biodiesel, charcoal, and biogas; the use of water or hydropower by means of waterwheels, water mills, run-of-the-river hydroelectricity facilities, hydroelectric dams, tidal power, and wave power; the use of solar energy in solar heating, photovoltaic (PV) systems, and concentrating solar power; the use of wind energy to drive turbines via windmills; and the use of geothermal energy, the heat energy that emanates from deep inside Earth. These renewable energy resources, if well explored and developed, can be inexhaustible.

By contrast, nonrenewable energy resources, such as fossil fuels and nuclear power, are limited, and their use leads to the depletion of those reserves. For much of human history, the energy used to power societies came from renewable resources; however, with the coming of the Industrial Revolution and its use of coal, and later oil, nonrenewable fossil fuels became predominant.

HISTORICAL ASPECTS
Until the nineteenth century, civilization survived using essentially renewable energy resources, based on burning firewood and biomass for cooking, heating, and building materials. In modern times, the least developed countries still use essentially these kinds of energy resources.

Pont du Gard is the most famous part of the Roman aqueduct which carried water from Uzès to Nîmes until roughly the 9th century when maintenance was abandoned. The monument is 49m high and now 275m long (it was 360m when intact) at its top. It's the highest Roman aqueduct, but also one of the best preserved (with the aqueduct of Segovia). The Pont du Gard is a UNESCO world heritage site since 1985. (Benh Lieu Song via Wikimedia Commons)

The use of fossil fuels became dominant in the twentieth century and continued to increase at the beginning of the twenty-first century. During the Industrial Revolution, the demand for energy increased drastically, and the high use of coal, petroleum, and natural gas led to the large-scale generation of electricity, heat, and fuel. Today, fossil fuels account for more than three-fourths of the energy used in the world. Fossil fuels have higher energy density when compared to raw biomass. They are easy to extract because huge amounts of these materials are found in a single place, which make the raw materials relatively low cost. The use of fossil fuels has had a large impact on civilization's industrial development. However, the reduction of fossil-fuel usage has become an important issue for two main. reasons. The first is related to the depletion of reserves, cost instability, and irregular distribution of these resources around the world, conditions that have led to many conflicts and wars; the second reason is related to climate change due to the greenhouse gases emitted by the burning of these fuels for energy.

WORLD ENERGY SCENARIO

The *2014 Renewable Energy Data Book*, issued by the U.S. Department of Energy, notes that 23.6 percent of energy consumed worldwide comes from renewable sources, with 16.3 percent coming from hydropower, 1.0 percent from solar power, 1.9 percent from biomass, 4.2 percent from wind energy, and 0.3 percent from geothermal energy. In 2009, during the worldwide economic recession, there was a boom in renewable energy; in mid-2010, more than a hundred countries focused their energy production on renewable sources, an increase of almost 100 percent from 2005.

There has thus been a major shift in the global energy scenario with regard to renewable fuels. Between 2004 and 2014, new installations for wind power and solar power, in the form of both photovoltaic (PV) power and concentrated solar power (CSP), showed significant worldwide growth, with electricity capacity increasing at compound annual rates of 22.7 percent (wind), 26.5 percent (CSP), and 46.1 percent (PV). In 2014, renewable energy accounted for an estimated 58.5 percent of net additions to global power-generating capacity; by the end of the year, renewable-energy facilities represented 27.7 percent of that capacity. China was the world leader in cumulative wind capacity and hydropower capacity, as well as total overall renewable electricity capacity; the United States led in both geothermal and biomass capacity; and Germany had the greatest PV capacity.

In 1990, wind power was used in only a few countries; by 2015, it played a role in the energy sources of more than one hundred countries. That year, China added more than 30 gigawatts (GW) of wind-power capacity, representing almost half of all new installations and bringing the nation's total capacity to more than 145 GW, according to the Global Wind Energy Council. However, the United States was the world leader in actual wind-powered electricity production in 2015, generating about 190 million megawatt-hours (MWh) compared to China's 185.1 MWh. Meanwhile, Argentina, Brazil, Colombia, Costa Rica, and Paraguay have become Latin America's largest biofuel producers, and other renewable technologies are being expanded as well. In 2014, the total investment in renewable energy by developing economies —including China, Brazil, and India—amounted to US$131.3 billion combined, an increase of 36 percent over the previous year. All these changes in the world energy scenario have helped to increase confidence in renewable energy. By the end of 2014, the United States had 15.5 percent of total installed capacity and 13.5 percent of total generation focused on renewable sources.

BIOMASS

The first application of biomass as a source of energy occurred in the early days of humankind, when people learned to use fire to produce heat and cook food. The first evidence that humans cooked food over controlled fires, based on the evolution of human molars, dates to 1.9 million years ago.

Biomass is organic raw material derived, directly or indirectly, from plants as a result of photosynthesis. It includes crop residues for cogeneration, forest residues, animal wastes, municipal solid waste, and energy crops such as sugarcane, switchgrass, jatropha, and corn. Photosynthesis is the chemical process that plants use to convert energy from sunlight, carbon dioxide, and water into oxygen and organic compounds, sugars (stored energy). The use of biomass instead of fossil fuels to generate energy mitigates the greenhouse effect, because even though the carbon

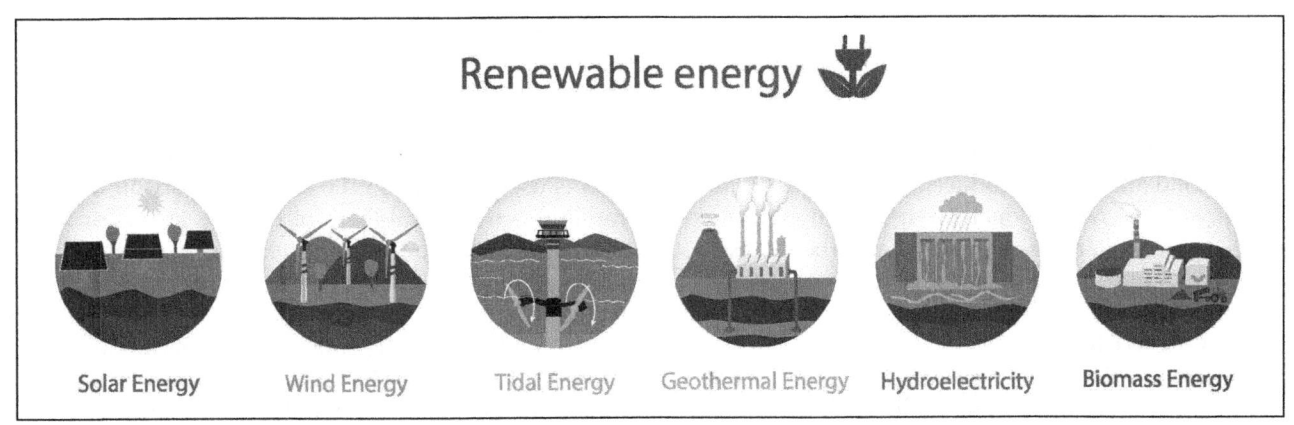

Renewable energy types. (Monica Odo via iStock)

present in the biomass will be emitted when used as fuel, it comes from plants that previously removed the carbon (as carbon dioxide) from the atmosphere; hence, the "carbon sink" role of the plant has balanced the later combustion of the fuel and emissions from the plant.

Biomass can provide raw material for multiple fuel uses, for the production of heat, electricity, and both liquid and gaseous fuels for transport. Energy is extracted from these various sources of biomass through biochemical processes, chemical reactions, and mechanical technologies to convert biomass into liquid or gaseous fuel. However, a considerable disadvantage of biomass is related to its low energy density when compared with that of fossil fuels. The processing of biomass can require significant energy inputs, which should be minimized to maximize the conversion of biomass and energy recovery.

There are three basic uses of biomass as fuel. These include thermochemical energy, biofuels, and biogas.

Biodiesel and bioethanol are the best-known biofuels available for use in automobiles and other motor vehicles. Biodiesel is produced by transesterification, which is a reversible chemical reaction of vegetable oils or animal fats with alcohols, esters (biodiesel) and glycerin. As of 2015, the most commonly used raw materials to produce biodiesel are wheat and rapeseed (predominantly in Europe) and soybean (in the United States, Brazil, and Argentina) oils. Biodiesel can be mixed with traditional diesel fuel and used in compression-ignition (diesel) engines without engine adaptations. Sugarcane, maize, wheat, sugar beets, and sweet sorghum can be used as raw materials for the production of bioethanol. These products are rich in sugars or starches, which are converted to alcohol by means of fermentation. Bioethanol production using sugarcane fermentation techniques has been commercially undertaken in Brazil since the 1980s.

Biogas can be produced from any kind of biomass by means of anaerobic microbes (bacteria that live in the absence of oxygen). Pigs, cattle, and chickens reared in confined areas produce a considerable concentration of organic waste matter with high moisture content, which can be used for biogas production. Biogas contains mainly methane and carbon dioxide, along with small amounts of other gases, such as hydrogen sulfide, ammonia, hydrogen, and carbon monoxide, giving it a very bad smell. The wet biomass is fed into an enclosed digestion tank together with a source containing anaerobic microbes; in the tank, anaerobic reactions occur. The remaining solid and liquid residues can be used as fertilizers. The period of time that biomass should remain in the digestion tank can range from a single day to several months. In 1630, it was discovered that decomposing organic matter is capable of producing a flammable gas, and in 1808 it was discovered that the gas contained methane. Biogas can be used as a low-cost fuel for heating and cooking; it can also be converted to electricity and heat or purified and compressed, much like natural gas, to create fuel to power motor vehicles.

Finally, gasification and pyrolysis are thermochemical processes in which organic matter is degraded by thermal reactions in the presence of limited amounts of air or oxygen. The major products are biochar (charcoal), bio-oil, or a gaseous product, which can also be burned as fuel. The amount of each of the three products formed is dependent on the type and nature of the biomass input, the type of facility used, and the particular process adopted. Pyrolysis aims to obtain solid and liquid products, whereas gasification produces a gaseous product, composed mostly of hydrogen, carbon monoxide, methane, carbon dioxide, and water vapor.

HYDROPOWER

Hydropower refers to the energy generated through the use of flowing water. For millennia, hydropower has been used for irrigation; notable engineering of water channels has been found in the ancient remains of Egyptian and Mayan civilizations, and the hydraulic engineering feats of the Roman civilization are well documented. There are many ways to harness the potential and kinetic energy of water to perform work; some examples include the use of waterwheels, water mills, run-of-the-river hydropower plants, hydroelectric dams, tidal power, and wave power.

Hydroelectric power is the electricity generated when water flows through a turbine to a lower level. These turbines, which are flow-controlling blades mounted on rotating shafts, are usually located within the dams. The potential energy is stored by dams as a volume of water located behind the dam. As long as water is being released through the dam, it rotates the turbines, which are coupled to generators that then supply electricity to transmission lines. Hydropower plants play a major role in the world's capacity to generate electricity. For example, the Three Gorges Dam in China—the greatest hydropower plant in the world—has more than twice the capacity of the Kashiwazaki-Kariwa Nuclear Power Plant in Japan, which is the nuclear plant with greatest capacity in the world.

Tidal power is the energy that can be extracted from the rise and fall of ocean tides. Extraction of tidal power is simple in theory. A tidal dam, termed a "barrage," is built across an estuary, creating an enclosed basin for storing water at high tide. Turbines in the barrage are used to convert the potential energy, resulting from the difference in water levels, into electrical energy.

Less common types of hydroelectricity include wave power, run-of-the-river hydropower, and marine (or ocean) current power.

SOLAR ENERGY

Most of the energy used on Earth has its origin in the electromagnetic radiation from the sun, including biomass, hydropower, and wind power. However, the term "solar power" in the context of energy generation is used to refer to the direct conversion of solar radiation to a useful form of energy. Forms of solar power include photovoltaic electricity, solar power tower plants, and solar thermal heating, among other forms.

Although only a very small fraction of the radiation from the Sun reaches Earth, sunlight represents a tremendous source of renewable, greenhouse-gas-free energy. Passage through the atmosphere splits the radiation reaching the surface into direct and diffuse components, reducing the total energy through selective absorption by dry air, water molecules, dust, and cloud layers, while heavy cloud coverage eliminates most direct radiation. Sunlight is intermittent, varying diurnally from day to night over a twenty-four-hour period as Earth rotates. Thus, storage of the energy is a very important factor if it is to be used efficiently and economically.

A typical procedure is to use a solar collector to absorb the solar energy and convert it to thermal energy, which is transferred by heat pipes carrying pumped fluids for low-temperature (less than 100°C) heating or storage. Therefore, the collector should be made of materials with high thermal conductivity and low thermal capacity, such as metals (copper, steel, and aluminum) and some thermal-conducting plastics. The most common collectors are flat, blackened plates, since they convert both direct and diffuse (cloud-mitigated) radiation into heat.

Direct solar radiation can also be focused by a range of concentrating solar power technologies and collected to provide medium- to high-temperature heating. These technologies for concentrating solar power are of three types: parabolic troughs, power towers, and heat engines. The heat generated by the radiation is then used to operate a conventional power cycle, generally by steam-generating techniques similar to those used in conventional power plants. Solar thermal power plants designed to use direct sunlight must be sited in regions with high direct solar radiation.

Another way to use solar power is by means of photovoltaic (PV) conversion. The PV effect is the production of electric potential and current when a system is exposed to light. The sun serves as the light source, and photovoltaic cells, also called solar cells, convert that light to energy. A PV cell consists of a semiconductor electrical junction device, which absorbs and converts the radiant energy of sunlight directly into electrical energy. Solar cells may be connected in series and/or parallel to obtain the required values of current and voltage for electric power generation needs. Most PV cells are made from single-crystal silicon and have been expensive for generating electricity, but they have found applications. Research has emphasized lowering the cost of PV cells by improving performance and by reducing the costs of materials and manufacturing. Besides their low efficiency (relative to the percentage of the incident sunlight that is converted into electrical output power) and high costs, PV cells' power generation is limited by the presence or absence of solar radiation. For some applications, the electricity can be stored (e.g., in batteries) to supply electricity on cloudy days and during the night.

WIND POWER

Windmills have been used to pump water and perform other kinds of mechanical work for centuries, but they were not used to produce electric power until the late 1800s. A wind power station consists of rotating blades attached to a generator, which is connected to transmission lines.

Wind power does not emit polluting emissions and does not produce unwanted substances that require careful disposal. There appear to be minor environmental impacts associated with the installation of wind turbines, aside from the possible disturbance of wildlife habitat and farming, which for some include the visual impact of a large multi-turbine wind farm on the natural aspect of an area. Wind turbines are not considered to be noisy machines; however, some noise is generated in their operation, and this has led to negative reactions of the public in some areas. Another issue is the coincident location of wind turbines in areas along the migration routes of birds; there have been reports of birds dying after colliding with the rotating blades of wind turbines. However, perhaps the greatest obstacles to wind farms have been that the areas where there is wind are often heavily populated and that the kind of equipment wind farms require is still too expensive.

GEOTHERMAL POWER

Geothermal energy is heat energy from within the depths of Earth. It originates from Earth's molten interior and from the radioactive decay of isotopes in underground rocks. The heat is brought near the surface by crustal plate movements, deep circulation of groundwater, and intrusions of molten magma from a great depth into Earth's crust. In some places, the heat rises to the surface in natural streams of steam or hot water, which have been used since prehistoric times for bathing and cooking. Wells can be drilled to trap this heat in supply pools, greenhouses, and power plants. The reservoirs developed to harness geothermal power to generate electricity are called "hydrothermal convection systems" and are characterized by circulation of water to depth. The driving force is convection, via the density difference between cold, downward-moving recharge water and heated, upward-moving thermal water. Hot water from a reservoir is flashed partly to steam at the surface, and this steam is used to drive a conventional turbine-generator set.

Geothermal energy tends to be relatively diffuse, which makes it difficult to trap. If it were not for the fact that Earth itself concentrates geothermal heat in certain regions—typically regions associated with the boundaries of tectonic plates—geothermal energy would be essentially useless.

Geothermal resources are renewable within the limits of equilibrium between offtake of reservoir water and natural or artificial recharge. Within this equilibrium, the energy source is renewable for a long period of time. Although geothermal energy may not be technically "renewable," the global geothermal potential represents a practically inexhaustible energy resource. The issue is not the finite size of the resource but the availability of technologies able to trap this kind of energy.

—*Leonardo Fonseca Valadares*

Further Reading

Boyle, Godfrey, ed. *Renewable Energy: Power for a Sustainable Future*. Oxford UP, 2012.

Magill, Bobby. "China, US Lead Global Boom in Wind Power." *Climate Central*. Climate Central, 2 Mar. 2016. Web. 17 May 2016.

REN21. *Renewables 2015: Global Status Report*. REN21 Secretariat, 2015. *REN21*. Web. 16 May 2016.

Ruin, Sven, and Gören Sidén. *Small-scale Renewable Energy Systems: Independent Electricity Systems for Community, Business and Home*. CRC Press, 2020.

Sørensen, Bent. *Renewable Energy: Physics, Engineering, Environmental Impacts, Economics & Planning*. 4th ed. Academic Press, 2011.

Spellman, Frank R. *The Science of Renewable Energy*. 2nd ed. CRC Press, 2016.

U.S. Department of Energy. National Renewable Energy Laboratory. *2014 Renewable Energy Data Book* N.p.: Author, 2015. *National Renewable Energy Laboratory*. Web. 16 May 2016.

Wróbel, Marek, Marcin Jewiaarz, and Andrzej Szlek, eds. *Renewable Energy Sources: Engineering, Technology, Innovation ICORES 2018*. Springer, 2020.

ROTATIONAL POWER

FIELDS OF STUDY
Electrical Engineering; Mechanical Engineering

ABSTRACT
Rotation has long been used to generate energy and transfer forces, from the treadmills of the ancient world to flywheels and some types of internal combustion engines in the present day.

KEY CONCEPTS
rotary engine: an internal combustion engine design that uses a central rotor driven by successive fuel ignitions rather than individual pistons connected to an eccentric crankshaft and driven in sequence by fuel ignitions

waterwheel: a large wheel-and-axle structure that uses the force of water flowing against paddles at the outer perimeter of the wheel to drive the wheel around and so drive the axle

windmill: a tower-like structure supporting three or more large sail-like vanes or aerodynamic blades attached to a common central shaft; the force exerted by wind against the vanes or blades drives them around, and so drives the shaft

vane: a triangular sail-like appendage on the central shaft of a windmill or other rotary device; in more high-tech constructions, aerodynamic "blades" are used in stead of vanes

WINDS AND WHEELS
As far back as 3,000 years ago, early machines used wheels powered by animals or humans in order to move heavy weights, grind grain, or raise water. They were a frequent feature in Greek and Roman engineering and were spread particularly by the Romans through the ancient world. The Greeks also developed toothed gearing and waterwheels, the main components of water mills, which use a waterwheel to transfer the force of moving water to the mechanical components of the mill in order to power it. Water

Model of a Roman water-powered grain-mill. The millstone (upper floor) is powered by an undershot waterwheel by the way of a gear mechanism (lower floor). (Wikimedia Commons)

mills are built along a body of running water or an artificial water body such as a millpond.

The flow of the water could be controlled or maintained by various means; often there was at least a channel or a pipe directing the water to the waterwheel. The force of the water striking the wheel turned the wheel, rotating the axle to which the wheel was attached and in turn driving whatever gears were also attached to the axle. In this manner, early rotational motion was harnessed to produce energy.

WATER AND ROTATIONAL POWER

The earliest known waterwheel is the Perachora wheel, on the periphery of Peloponesse, in the third century BCE. Written references to such wheels date to the same century, especially in the technical manuals of Philo of Byzantium (280-220 BCE), a Greek engineer. Both horizontal and vertical wheels were in use. Waterwheels became more sophisticated over time, connected to gearing mechanisms, aqueducts, sluices, and dams to adjust the speed of the water or the rotation of the wheel.

Water-powered sawmills operated saws that sliced not only through lumber but also through stone. Later, the Muslim agricultural revolution of the ninth to thirteenth centuries was driven in large part by the golden age of Muslim mathematics and science, and the ingenuity of Arab engineers in finding numerous applications for hydropower to mechanize many farm tasks.

Both Arabs and Europeans also experimented with tide mills, which were powered by tidal flow instead of running water. In such a mill, gates trap rising water during high tide and then drain it through sluice gates at low tide, so that the force of the draining water drives the waterwheel—a technique useful in many coastal locations but especially efficient in places where there is a particularly great difference between high and low tides.

In some parts of the world, the advantages of waterwheels were so great that the spread of their use required laws, such as those of eighth century China, which restricted their usage in order to prevent them from blocking the passage of ships on the waterways.

Gearing could also translate vertical rotation into horizontal rotation and vice versa. The water mills used for grinding kernels of grain in Britain and the United States leading up to and during the Industrial Revolution used a vertical waterwheel (and therefore a horizontal axle, creating horizontal rotation), which drove a horizontal shaft mounted to another wheel, which linked with a "wallower" on a vertical shaft in order to drive a spur wheel at the other end of the shaft, turning a small wheel called a stone nut, which was attached to the stones that ground the kernels. The same waterwheel could in this manner be connected to several stone nuts, grinding a great deal of grain in a short time. This gearing arrangement also allowed wheels to be rotated at a speed greater than the speed of the water driving the waterwheel, rather than being limited by it, as early Greek and Roman water mills had been.

Until the introduction of the steam engine, the water mill was the keystone of industry throughout Europe and the United States. Mills were used to cut timber and stone, to grind not only grains for food but also bark for use in leather tanning, to manufacture gunpowder, to spin yarn, to smelt lead, to make paper, to crush mined ore, to slit iron in order to make nails, and to shape metal by passing it through heavy rollers. Water mills were even used to power blast furnaces, and even after the steam engine was introduced, cotton mills used this rotational energy first to increase the water flow to the waterwheel, only later adopting steam as the motive power.

WIND AND ROTATIONAL POWER

Only slightly younger than the waterwheel is the windmill, which similarly uses the force of the wind to drive vanes that rotate an axle (usually vertical vanes and a horizontal axle, although other designs were common in antiquity). Early examples of wind-driven wheels are found in first century Greece and fourth century Tibet, and the technology became more common in medieval Europe. The earliest European mills were used to grind grain, with later industrial applications developing. The gears inside the windmill transfer the energy of the rotary motion of the vanes to mechanisms similar to those of the water mills. Never quite as popular as the water mill, there were about 200,000 windmills in Europe at the peak of their popularity, compared to half a million water mills.

Still in common use in dry agricultural regions—such as the American Southwest and Great Plains, South Africa and Namibia, and Australia—are wind pumps, which convert the rotary motion of the vanes to power a pump cylinder that either draws water up from wells or drains low-lying land. Windmills have been used for this purpose since the ninth century in the Middle East and China, but the modern wind pump dates to an 1854 design by Vermont engineer Daniel Halladay. In the United States alone, there are some 60,000 wind pumps, drawing water usually for agricultural purposes but sometimes for drinking water.

From the nineteenth century on, windmills have been built that generate electricity via a wind turbine. These have enjoyed a resurgence in the twenty-first century as concerns over the costs and effects of fossil fuels have motivated serious investment in other energy sources. Wind turbines that generate electricity instead of simply turning the wind's kinetic energy into mechanical energy are sometimes called wind generators for clarity's sake. High-efficiency windmills, typically with three-vaned rotors, are arranged in large numbers on wind farms. The gearbox of the turbine converts the slow rotation of the vanes into a faster rotation of wheels generating electricity.

The conventional electricity-generating windmill has a horizontal axis design, but in places with highly variable wind direction, vertical-axis wind turbines are sometimes used. This also eliminates the need for a tower and allows the generator and gearbox to be near the ground, making maintenance easier. Vertical-axis wind turbines do best when mounted on rooftops, to avoid the turbulent airflow at ground level. One common type of vertical-axis turbine is the "eggbeater" turbine, designed by French engineer Georges Darrieus, which is efficient but

unreliable. Shaped like an eggbeater whisk, this wind turbine has vanes that curve from the top to the bottom of a vertical shaft. There have been many attempts to develop airborne wind turbines with no mount, but none has yet succeeded in producing useful amounts of electricity.

STEAM AND ROTATIONAL POWER

The steam turbine was a sophisticated steam engine introduced in 1884, developed by English engineer Charles Parsons. It quickly replaced the extant reciprocating piston steam engine, which was heavier and less efficient. The steam turbine continues to be used today as an electricity generator; most of the world's electricity is in fact generated this way, with coal or petroleum serving as the fuel to generate the steam. Parsons's steam turbine was instrumental in making electricity affordable throughout the world; it brought about the twentieth century, the age of power lines and electric lighting.

Condensing turbines are the most common in power plants, and they exhaust partially condensed steam. Reheat turbines, instead of exhausting the steam, cause the steam to flow from a high-pressure section of the turbine back to the boiler, where it is heated again and cycles through the system repeatedly. Steam is expanded as it travels through the turbine, in order to maximize efficiency by generating work. Two types of turbine can accomplish this, referred to as impulse turbines and reaction turbines—even though in practice, most power plant turbines will operate as both an impulse turbine and a reaction turbine, at different stages of their operation. In the impulse-turbine stage, fixed nozzles direct the steam into high-speed jets. The turbine's rotor blades convert the kinetic energy of those jets into rotational speed. The pressure drop as the steam flows through the nozzle causes the steam to expand, contributing to its velocity. In the reaction-turbine stage, the turbine's rotors themselves form nozzles.

Gas turbines are a type of internal combustion engine, in which—as in the steam turbine—a stream of high-velocity gas is directed by a nozzle over the turbine blades in order to spin them and generate energy or mechanical output. "Air-breathing" jet engines use gas turbines that produce thrust from exhaust gases; when most of their thrust is so produced, they are referred to as turbojets, whereas those that produce thrust by connecting ducted fans to the gas turbines are turbofans. Gas turbines can also be incorporated into liquid propellant rockets. While gas turbines are sometimes used on ships and locomotives, the use of them in automobiles and buses has been in the experimental stage since the 1940s, with intended applications including luxury cars (Jaguar is a major funder of gas turbine automobile research) and hybrid electric cars. A small number of hybrid buses, such as the HEV-1, developed by AVS of Chattanooga, Tennessee, in 1999, use gas turbines.

THE WANKEL ENGINE

Another internal combustion engine using rotating motion instead of reciprocating pistons is the Wankel engine, named for German engineer Felix Wankel (1902-88). The terms *rotary engine* and *Wankel engine* are sometimes used interchangeably, although the Wankel engine is only the best-known rotary internal combustion engine. Its innovation was its compact size, and it was refined shortly after Wankel's invention of it by the Japanese car company Mazda, which continues to be the most extensive user of it in automobiles. However, the design is also used in chainsaws, single-passenger watercraft (such as Jet Skis), go-karts, and other applications where a compact engine is needed.

—*Richard D. Besel*

Further Reading

Boyce, Meherwan P. *Gas Turbine Engineering Handbook*. Gulf Professional, 2006.

Burton, Tony, David Sharpe, Nick Jenkins, and Ervin Bossanyi. *Wind Energy Handbook*. Wiley, 2001.

Hau, Erich. *Wind Turbines: Fundamentals, Technologies, Application, Economics*. Springer, 2006.

Muszynska, Agnieszka. *Rotor Dynamics*. Taylor & Francis, 2005.

Saravanamuttoo, H.I.H., G.F.C. Rogers, H. Cohen, and P.V. Straznicky. *Gas Turbine Theory*. Pearson Prentice Hall, 2008.

Rekioua, Djamila. *Wind Power Electric Systems: Modeling, Simulation and Control*. Springer, 2014.

Shlyakhin, P. *Steam Turbines: Theory and Design*. U Press of the Pacific, 2005.

Sadi Carnot

FIELDS OF STUDY
History of Science; Physics; Thermodynamics

ABSTRACT
Sadi Carnot was a French mathematician who is also referred to as the Father of Thermodynamics. He studied the transfer of heat when researching how to make a steam engine most efficient.

KEY CONCEPTS
thermodynamics: the science of the movement of heat energy

BIOGRAPHY OF SADI CARNOT
Nicolas Leonard Sadi Carnot was born in the Palais du Petit-Luxembourg, Paris, on June 1, 1796, the eldest son of Lazare Nicolas Marguerite Carnot (1753-1823), the Organizer of the Victory of the French Revolutionary Wars. His father was a member of the Directory, and a surviving member of the Committee of Public Safety. His younger brother, Lazare Hippolyte Carnot (1801-88), would become an important politician and preserved some of Sadi's notes among his papers. Lazare Carnot resigned from politics in 1807 in order to attend to the education of his sons.

Sadi was taught at home by his father, who had written on scientific and engineering problems. He had also studied the efficiency of water machines. The idea of work derived from falling water was Sadi's model for work on a heat engine caused by falling heat.

Sadi attended the Lycée Charlemagne in Paris in order to prepare for the entrance examination to the École Polytechnique, which he entered in 1812. In 1813, Sadi sought permission from Napoleon to join with other young students in defending France. In March 1814, he fought at Vincennes in the Battle of Paris. In October 1814, he graduated as a military engineer. His studies at the École Polytechnique placed him in the company of some of France's best scientists, including Siméon-Denis Poisson, Joseph Gay-Lussac, and André-Marie Ampère. Posted to the École du Genie at Metz as a student second lieutenant, he wrote several scientific papers that are no longer extant. During the Hundred Days War, Lazare Carnot served as Napoleon's minister of the interior, but was exiled in October 1815, following the Restoration. This ended the special attention Sadi's superiors had previously given to him.

When his studies ended in 1816, Sadi began serving in the Metz engineering regiment as a second

Sadi Carnot. (Wikimedia Commons)

299

lieutenant. For several years, he moved from garrison to garrison inspecting fortifications, drafting plans, and writing reports that were thereafter neglected in bureaucratic files. In 1819, he was assigned to the general staff in Paris. He immediately gained a permanent leave of absence, which allowed him time to engage in studying physics. Between 1818 and 1824, Carnot spent many hours studying the steam engine, an improved version of which had been invented by James Watt (1736-1819). Hailed as the technology of the future, it was a very inefficient machine that lost most of its heat energy. Carnot sought to improve its efficiency by seeking to understand the nature of heat.

SCIENTIFIC STUDIES

In 1824, Carnot published *Réflexions sur la puissance motrice du feu et sur les machines propres a developer cette puissance* (*Reflections on the Motive Power of Fire*). In the title word *feu* (fire), Carnot meant heat, and by *puissance motrice* (motive power) he meant work. The book was his only scientific work. Published by Bachelier, the most important French scientific publisher of the time, it was well received, formally presented to the *Academie des Sciences*, and favorably reviewed by P. S. Girard in the *Revue encyclopedique*. *Réflexions* showed that the efficiency of a heat engine—that is, an engine that changes heat into mechanical energy in order to do work-was dependent upon the difference between its hottest and coolest temperature. Using the formula $(T_1-T_2)/T_2$, Carnot expressed the efficiency between the hottest temperature, T_1, and coolest temperature, T_2. The equation is considered to be the first statement of the theory of heat movement. Consequently, Carnot is considered the father of the science of thermodynamics. The equation was universal because it applied to any heat engine and any temperature. At the end of his *Réflexions*, Carnot included a definition of work. It was defined as weight that was lifted to some height. In modern physics, work is the force that is applied to a body through a distance against resistance.

French army reorganization in 1827 put Carnot back on active duty. This prevented him from engaging in more scientific studies until 1828, when he resigned. He then spent his time studying the relationship between temperature and pressure. His researches were cut short when he contracted scarlet fever. He had not yet recovered when a cholera epidemic swept across Paris, taking his life. He died in Paris on August, 24, 1832. His scientific papers and all of his possessions were burned, as was the practice in an effort to purge the disease with fire. Carnot's work in thermodynamics was generally ignored until 1848 when William Thomson (Lord Kelvin) was influenced by it as he studied heat for an absolute temperature scale, the Kelvin scale. The German physicist Rudolf Clausius later modified Carnot's work and spread his ideas as the second law of thermodynamics: Entropy occurs because heat cannot move from a colder substance to a hotter substance.

—*Andrew J. Waskey*

Further Reading

Carnot, Sadi. *Reflection on the Motive Power of Fire*. Dover, 1960.

Carnot, Sadi. *Reflexions on the Motive Power of Fire: A Critical Edition with the Surviving Scientific Manuscripts*. Robert Fox, trans., ed. Manchester UP, 1986.

Cropper, William H. *Great Physicists: The Life and Times of Leading Physicists from Galileo to Hawking*. Oxford UP, 2001.

Feidt, Michel, ed. *Carnot Cycle and Heat Engine Fundamentals and Applications*. MDPI, 2020

Gillispie, Charles Coulston, and Raffaele Pisano. *Lazare and Sadi Carnot: A Scientific and Filial Relationship*. 2nd ed. Springer, 2014.

Hatfield, Philipp L. *Great Men of Science: A History of Scientific Progress*. Macmillan, 1933.

Mendoza, E., ed. *Reflections on the Motive Power of Fire by Sadi Carnot, and Other Papers on the Second Law of Thermodynamics by E. Clapeyron and R. Claussius*. Dover Publications, 1960.

SOLAR CONCENTRATOR

FIELDS OF STUDY

Electrical Engineering; Materials Science; Solar Engineering

ABSTRACT

Solar concentrators represent major advancements in solar electricity efficiency through concentrating the Sun's energy into a focal point to generate intense heat. Solar concentrators have been used since ancient times, especially in the form of convex lenses to increase the heat delivered from the sun.

KEY CONCEPTS

concentrator: a device that acts to focus sunlight to a central point, either by reflection or concentration through a lens

molten salt system: a system that uses a low-melting inorganic salt as its working fluid

parabolic: a shape that conforms to the general mathematical equation $y = ax^2 + b$

parabolic trough: a long, semicylindrical structure that is parabolic in cross-section

THE POWER OF SUNLIGHT

Solar concentrators are systems that use mirrors to focus the sun's energy onto a small focal point, creating intense heat. Solar concentrators have been used since ancient times, especially in the form of convex lenses to increase the heat delivered from the sun. In more recent times, solar concentrators have been used in conjunction with photovoltaic cells or to generate steam cheaply and efficiently. Modern designs typically fall under four main categories: parabolic dishes, Fresnel reflectors, troughs, and solar towers. Novel uses of concentrating solar energy include solar ovens and parabolic mirrors for starting fires.

The first mention of concentrated solar power comes from Greek legends that tell how Archimedes repelled an invading fleet by using a "burning glass" of several large mirrors to concentrate the Sun's energy to ignite the ships. However, the legitimacy of this claim has been repeatedly brought into question, and experiments conducted in 2005 proved that while it was theoretically possible under ideal conditions, it would have been far easier to ignite ships with a catapult or flaming arrows.

The first modern use of concentrating solar power into energy was in the nineteenth century, when a parabolic trough design was used to generate steam to run the world's first steam engine. Following this advancement, concentrated solar power was commonly put to use for irrigation, locomotion, and refrigeration.

MODERN DESIGNS

Four main types of solar concentrators are in use today for electricity generation and are either used in conjunction with modern silicon photovoltaic cells or steam energy to generate electricity. Solar concentrators, in order to obtain maximum efficiency, must be directly facing the sun and must therefore shift their orientation throughout the day, a hurdle in small-scale deployment of concentrated solar energy.

Greek mathematician Archimedes was famous for his supposed use of a "burning mirror" that could direct sunlight to start fires aboard enemy ships. (Wikimedia Commons)

The four types of solar concentrators are parabolic dishes, Fresnel reflectors, troughs, and solar concentrator towers. Parabolic dishes reflect sunlight back toward a photovoltaic cell in a similar fashion to satellite TV dishes. Larger parabolic dishes can be used with a small Stirling engine to generate electricity. A Fresnel reflector is a single lens with concentric prismatic rings that allows for higher concentration of light with lighter material than a simple convex lens. Solar troughs are similar to dishes but instead focus energy on a tube filled with a working fluid such as a molten salt that is heated, and the heat is used to generate electricity. One advantage of troughs is that they only have to track the sun on a single axis. Concentrated solar towers are the largest of the concentrated solar technologies, with a single tower placed in a field of solar reflectors, all of which reflect the sun's energy back to the tower, creating intense heat for electricity generation. Concentrated solar towers are the most efficient of the concentrated solar technologies and are able to generate electricity for several hours after daytime.

Parabolic dishes and Fresnel reflectors are commonly used in conjunction with photovoltaic cells to improve their efficiency by increasing the intensity of the sunlight hitting the cell. This drastically reduces the cost of photovoltaic solar arrays, as the cells are far more expensive than mirrors and lenses. Since about one-tenth as many silicon chips are needed, the cost per kilowatt-hour is dramatically reduced, putting solar energy closer to the cost of coal and natural gas electrical generation. However, with these two systems, heat is a major hurdle, as the concentrated solar energy also produces significant heat.

Solar concentrators are also used as solar ovens or cookers, often as development schemes in poorer countries to slow deforestation and desertification. Basic models are simply a foil arranged in a parabolic shape to concentrate the sun's energy, while more complex models are box-shaped with a combination of reflectors and black surfaces to generate heat. These solar ovens use no fuel, produce no smoke, and can be used in a similar manner as a slow cooker; however, they are most efficient during the hottest time of the day and take a long time to cook food, requiring advance planning.

CURRENT AND FUTURE APPLICATIONS

Since the cost of fuel for solar energy is free, the most important factor in reaching grid parity is the cost of building solar concentrators. Current technology has brought the cost per kilowatt-hour to levels roughly

Working model of luminescent solar concentrator. (Levita.lev via Wikimedia Commons)

equal to both natural gas and coal, but still roughly 10 times higher than nuclear energy. Some analysts argued that solar concentrators could lower the cost to $0.06 per kilowatt-hour by 2015, given improvements in efficiency and mass production.

The United States initially led in concentrated solar energy with the 354-megawatt solar electric generating systems (SEGS) concentrated solar energy station in the Mojave desert (comprising 99 percent of total U.S.-installed solar capacity in 1992). It was at one time the largest solar plant in the world. Another major scheme that is under development is Desertec, backed by German engineers; it is part of a larger scheme to link Europe, north Africa, and the Middle East in a renewable energy grid. Desertec will begin in Morocco and is projected to supply 15 percent of Europe's energy by 2050. The largest concentrated solar power plant in the world currently is Morocco's Ouarzazate plant, begun on February 4, 2016, and with the capacity to power more than a million households in Morocco.

Potential future developments include dyes that can diffuse and redirect the sun's energy to eliminate the need for solar tracking. Solar tracking adds a significant cost and maintenance to solar concentrators. However, the dyes are somewhat unstable and do not last the life span of typical solar panels.

—*Matthew Branch*

Further Reading

Bose, Debajyoti, Krishnam Goyal, and Vidushi Bhardwaj. *Design and Development of a Parabolic Solar Concentrator and Integration with a Solar Desalination System.* GRIN, 2016.

Green, Martin A. *Third Generation Photovoltaics: Advanced Solar Energy Conversion.* Springer, 2005.

Greene, Lori E. *High and Low Concentrator Systems for Solar Electric Applications IV.* SPIE: 2009.

Jagoo, Zafrullah. *Tracking Solar Concentrators: A Low Budget Solution.* Springer, 2013.

López, Antonio L. Luque, and Viacheslav M. Andreev. *Concentrator Photovoltaics.* 1st ed. Springer, 2010.

Olofsson, William L., and Viktor I Bengtsson. *Solar Energy: Research, Technology and Applications.* Nova Science Publishers, 2008

Rabl, A. "Comparison of Solar Concentrators." *Solar Energy*, vol. 18, no. 2, 1976.

Twidell, John, and Toney Weir. *Renewable Energy Resources.* 3rd ed. Routledge, 2015.

Solar Energy

FIELDS OF STUDY
Materials Science; Solar Power Engineering; Thermodynamics

ABSTRACT
The Sun is Earth's most important primary energy source, providing solar radiation that can be used immediately for many human needs from heating water and buildings to generating electricity, and that has provided essentially all of the energy on the planet since its formation.

KEY CONCEPTS
albedo: the amount of light and other energy reflected from a planet relative to the amount that it receives; an albedo of 0.3 indicates that the planet reflects and emits 30 percent of its incoming insolation back into space

insolation: specifically, the light and other solar emissions striking Earth from the Sun (Sol)

nuclear fusion: the joining of two atomic nuclei to produce the atomic nucleus of a third element; in the Sun, hydrogen nuclei are fused together under the force of the Sun's gravity to produce helium nuclei

ozone: an oxygen molecule consisting of three oxygen atoms as O_3 instead of the usual two as O_2

solar collector: a structure, typically of mirror-like surfaces, designed to focus sunlight to a central point in order to capture heat energy

solar thermal power plant: a power generating plant that uses a large array of mirror surfaces to capture and focus sunlight as the heat source for producing steam to drive electrical generators

EARTH, SUN AND ENERGY
The Sun is about 149.6 million kilometers (93 million miles) from Earth, and it is Earth's primary source of energy. It provides an annual energy of about 3.9×10^{24} joules, the equivalent of 1.5×10^{18} kilowatt-hours. This energy, which reaches Earth in the form of electromagnetic radiation, is called insolation and makes up about 99.98 percent of the entire energy budget of Earth. The remaining part originates from geothermal sources (heat from the interior of Earth). Solar energy is estimated to be available at a constantly renewing rate (daily) for the next 4 or 5 billion

The sun. (Jessie Eastland via Wikimedia Commons)

years—its biggest advantage compared to secondary sources of energy such as fossil fuels (coal, gas, and oil, which are the remains of once-living organisms that capture the energy of sunlight through photosynthesis, require eons to develop and hence will eventually be depleted).

SOLAR FACTS

The Sun is a star, a massive ball of hot gas, with a surface temperature of 5,800 to 6,000 K (5,500°C to 5,700°C). It is approximately 1.4 million kilometers (0.8 million miles) in diameter and consists mainly of hydrogen (73.5 percent) and helium (25 percent). The remaining elements include oxygen, carbon, and other heavier elements, up to iron. Elements with mass numbers greater than iron do not occur in stars like the Sun.

The solar magnetic field can be described best as a panel of radiating dipoles. Polarity reversal takes place in an irregular cycle of 11 years, called solar magnetic activity cycle (solar cycle for short), and is related to the activity of sunspots. Sunspots appear because of interferences in the magnetic field of the sun. The solar cycle can be understood as an internal hydromagnetic dynamo process, which has effects on the Sun's atmosphere, including its corona (as seen in coronal mass ejections), and solar wind (the outflow of particles from the sun); it also modulates the flux of short-wavelength radiation from the ultraviolet (UV) range to X-radiation. Simultaneously with the solar cycle, total solar irradiance (TSI) at Earth's surface varies about 0.1 percent, affecting Earth's regional and global climate and weather patterns.

This modulation also has an impact on Earth's ozone layer. Usually, in the stratosphere, oxygen molecules are split by solar UV radiation, thus allowing the creation of ozone molecules, which protect the Earth from overexposure to the sun's radiation. When UV radiation is modified to X-rays, the irradiance is not energetic enough to split oxygen molecules, thus resulting in a lower ozone concentration. Variations of radiation in the ozone layer cause fluctuations in the amount of UV radiation reaching the Earth's surface by about 400 percent.

Another impact of the solar cycle can be detected in Earth's ionosphere. The Sun influences interplanetary space through its magnetic field and solar wind. Solar wind is caused by the emission of particles that leave the sun at a speed of some 1 million miles per hour. Solar eruptions can fortify the speed and density of solar wind, leading to impressive polar lights on Earth but also causing interference within electronic systems and radio frequencies. Solar photons and cosmic rays (very short-wavelength radiation) ionize the ionosphere, thus affecting the propagation of radio waves and interfering with local and long-distance transmissions relevant for marine and aircraft communications as well as for radio broadcasting.

PRIMARY ENERGY SOURCE

The Sun produces solar energy through nuclear fusion, amounting to approximately 5,000 times as much energy as the current global demand, assuming that all factors that shield incoming solar energy from reaching Earth's surface are taken into account. Since measurements began, solar irradiance has been nearly constant; there are no historical references to any significant fluctuations.

The Sun's heat flux density, J, which is radiant energy per area unit and time unit, depends on the angle of incidence (b): $J = J_0 \times \sin(b)$. At a flat angle, fewer photons reach and heat the surface of Earth than at a vertical angle of incidence, J_0, which explains the diversity of climate zones from tropical to polar on Earth. Solar radiation consists of the entire electromagnetic spectrum, including long-wave radio and infrared waves, visible light, short-wave X-rays, and ultraviolet radiation. Photon frequencies are related to the radiation wavelength, l, by the equation: $l = c/v$. In this formula, c is the speed of light in a vacuum, defined as 3×10^8 meters per second.

The average insolation intensity at the edge of Earth's atmosphere is about 1,367 watts per square meter, a quantity scientists refer to as the *solar constant*. The effective solar radiation on Earth's surface, however, after subtracting for the angle of incidence at different latitudes, as well as atmospheric reflection and absorption, is about 340 watts per square meter. This in turn amounts to billions of watts worldwide. Less than one-billionth of the Sun's total energy reaches Earth's outer atmosphere, where 30 percent is immediately reflected back into space (particularly because of clouds, dust particles, and snow cover) as Earth's planetary albedo.

At Earth's surface, about 47 percent of the Sun's energy is absorbed by the atmosphere and converted to heat. About 23 percent of the incoming energy is responsible for the hydrologic cycle, driving regional weather and global climate patterns. Less than 1 percent of the energy drives the winds and ocean currents, and just 0.02 percent is captured for photosynthesis. Plants use UV radiation for photosynthesis, converting solar energy to biomass, sugars, and hence chemical energy, which in turn is consumed by animals, sustaining the food chain that supports life. (Biomass is also the basis for the fossil fuels—from peat to coal to oil and natural gas—that dominate the world's energy economy.) Solar radiation is also responsible for differences in atmospheric pressure, while about one-third of the tidal effects in oceans derive from solar gravitational force. Thus, the Sun's energy is primary, even for other so-called primary sources: All life depends on sunlight, and other sources of energy—from hydropower to tides to winds—are driven by the Sun, including the plant life (biomass) that humans can consume or burn.

USING SOLAR ENERGY

Total human primary energy consumption in 2006 was 1.0×10^{14} kilowatt-hours. The sun produces about 5,000 times as much as energy as this. A potentially very large amount of this energy could supply electrical power by means of solar technologies: solar collectors to convert electromagnetic waves into thermal energy; solar thermal power plants to produce heat and water vapor, driving turbines to produce electrical power; solar cookers and solar ovens to cook meals and sterilize medical instruments; solar photovoltaic (PV) cells to produce electrical direct current (DC); and sunlight itself to grow plants that produce biomass for processing into liquid and gaseous fuels, such as vegetable oils, ethanol and methane.

To use solar radiation most effectively, modules and collectors have to be arranged at optimal angles—much as many plants are structured in nature. In addition to the incident angle of the sun, the azimuth angle and the angle of inclination are important factors for the installation of modules and collectors. The azimuth angle measures the deviation of collec-

305

tors from an exact southward position. Earth's geometry and rotation, including latitude, declination, and solar-hour angle, are important factors for planning the installation of solar collectors. Furthermore, the local elevation, surface inclination, orientation, and shadows must be taken into account. At higher altitudes, the density of the atmosphere changes and thus has an influence on the intensity of radiation. All these factors refer to the sun's position above the horizon and can be accurately calculated using established formulas.

Atmospheric attenuation by absorption of radiation through gas molecules, such as carbon dioxide, aerosols, and water vapor, has an impact on the efficiency of solar collectors. Solar radiation reaching the surface does not have the same intensity at every point, but is modified by topographical features and atmospheric conditions. Gases in the atmosphere are mostly colorless and transparent, so gaseous attenuation usually does have a great effect on the clarity of the atmosphere, a measure known as relative optical mass. Although the air appears clear, however, the absorbing properties of some gases can be very strong. In the case of a humid atmosphere due to water vapor or pollutants, the attenuating properties are described as Linke turbidity, defining the optical density of a vaporous atmosphere in relation to a clear and dry atmosphere. It is not a precise value, considering all dynamic factors, but a simplified comparison. With regard to water vapor and its reducing impact on incoming solar radiation, clouds hold a special position due to their thickness and appearance at several atmospheric layers. Because all necessary data needed to analyze the interactions between clouds and solar radiation are seldom available, here again estimates based on empirical data are common practice.

With regard to attenuating atmospheric factors, the incoming solar radiation is classified as direct, unhindered radiation, termed beam; scattered radiation, termed diffuse radiation; and reflected radiation, for the small part of the radiation that is reflected from the ground. Together, these three types of radiation are called global radiation, mostly expressed as the solar constant.

HISTORY

Applications of solar energy date back to ancient eras, between 800 BCE and 600 CE, when concave mirrors were used to focus light to induce enough heat to start a fire. For example, the Olympic Games fire was—and still is—traditionally ignited by means of concave mirrors. In the eighteenth century, Horace Bénédict de Saussure invented the precursor of today's solar collectors. The photogalvanic effect, whereby electric currents are produced from light-induced chemical reactions, was discovered in 1839 by the French physicist Alexandre-Edmond Becquerel. It took until 1876 for William G. Adams and Richard E. Day to demonstrate this effect for selenium crystals. A few years later, in 1891, the first patent for a solar plant, a simple thermal solar collector, was obtained by the metal manufacturer Clarence M. Kemp from Baltimore. Albert Einstein described the photoelectric effect precisely in 1905 and received the Nobel Prize in Physics in 1921 for this research.

In 1954, Daryl Chapin, Calvin, and Gerald Pearson developed the first silicon solar cells, which worked with an efficiency of only 4 percent. The first technical application for this technology, to supply power for telephone amplifiers, was invented one year later. Since the late 1950s, photovoltaic cells have been used in satellite technology, selected for their ruggedness, reliability, and relative light weight. Vanguard 1 was the first satellite rigged with solar cells, entering Earth orbit on March 17, 1958. In the 1960s and 1970s, ongoing demand from aeronautics led to significant progress in the development of photovoltaic cells. Originally prompted by energy crises of the 1970s and growing environmental consciousness since then, attempts to open up a market for solar energy converters have grown, using silicon, gallium, and cadmium, which function as semiconductors, in solar cells.

The technologies in place for using solar energy are advancing and increasingly being implemented at all levels, from grassroots efforts in developing nations to large, government-sponsored programs. For example, a program sponsored by the U.S. Department of Energy, the Million Solar Roofs Initiative, ran from 1997 to 2005. In the program's retrospective report, the accomplishments noted included the installation of what amounted to more than 377,000 solar water-heating, PV, and pool-heating systems; the installation of 200 megawatts of grid-connected PV capacity, and 200 megawatt-hours of solar water-heating capacity; and a growth in the acceptance of PV technology from 8 percent (1997) to 41 percent (2005).

SOLAR ENERGY APPLICATIONS

Solar energy can heat buildings, heat water, cook food, drive pumps and refrigerators, and generate electricity. Solar heat refers to the conversion of solar energy into heat energy. The average temperature of the diffuse thermal radiation of Earth is too low to be used energetically. Some technical applications focus the sunlight by means of concave mirrors on thermal solar modules, to gain a higher temperature in the heat-absorbing medium. The simplest form is called passive solar use, which includes few or no mechanical devices. This has a range of applications. Passive solar space heating typically involves the use of insulated glass structures that allow sunlight in, but trap heat from passing back out. Building design can incorporate a variety of simple applications of this principle, as well as more complex installations that multiply effective heating through convection, conduction, thermal mass storage systems, and other add-ons.

There are even options for solar air heating and cooling, solar air-conditioning. For example, some advanced devices employ evaporation for cooling purposes, known as adiabatic cooling. This technology uses the same principle as human perspiration. One cubic meter of water is sufficient to cool a 1,000-square-meter interior room. Solar air-conditioning of this type can be incorporated into green archi-

Solar water disinfection in Indonesia. (SODIS Eawag via Wikimedia Commons)

tecture. On an industrial scale, solar thermal energy is employed through various heat transfer media—such as water and salt solutions—to produce steam, which then drives turbines to generate electrical power.

Solar-electric cells, as photovoltaic (PV) devices are sometimes called, are a semiconducting technology that converts sunlight into direct current electricity, at efficiencies ranging from less than 10 percent to levels approaching 20 percent. PVs can be applied as panels, flexible strips, roof shingles, films, and other configurations. The electric energy, in the form of direct current, can be used immediately at the site of installation, or to supply an enclosed power network, stored in accumulators, or can become inducted into public power networks.

With power converters, the direct current (DC) generated by PVs is turned into alternating current (AC), the type in general use. Because of changes in seasonal and daily weather and radiation, an appropriate storage technology is necessary to ensure a constant energy supply. Without storage, there is little possibility of responding directly to fluctuating energy demand; with storage, solar power becomes a baseline source of electricity, capable of holding its own economically and practically against the fossil fuels that up until now have provided the greatest general reliability of electrical generation at scale.

The composition of the solar spectrum, duration of sunshine, and the angle at which solar radiation reaches Earth's surface depend on time, season, and latitude. All these factors have an influence on the energy of radiation. In central Europe, solar radiation is about 1,000 kilowatt-hours per square meter per year, while in the Sahara Desert it is about 2,350 kilowatt-hours per square meter per year. Different scenarios for supplying renewable energy to the European Union have included transmitting solar energy from the Sahara to central Europe and the Middle East by means of high-voltage DC conduction. In an article titled "A Solar Grand Plan," Ken Zweibel, James Mason, and Vasilis Fthenakis suggest a similar use of solar energy in the United States, with transmission from the Great Plains states eastward to the densely populated Atlantic seaboard region.

GOVERNMENT INCENTIVES

The scalability of such applications is an important factor in the ready adaptation of solar energy technologies. While governments typically provide incentives such as feed-in tariffs (which provide for certain guarantees such as minimum quantities of electricity output purchases at guaranteed minimum rates over a multiyear period) to industrial-sized entities, the adoption of solar energy by individual homeowners, local community boards, and other smaller-scale entities benefits from a different class of incentives. These tend to include tax credits, tax deductions, technical assistance grants, in-kind funding, preferred-rate financing, and related benefits.

Such incentives have proven attractive in many places. For example, in Germany, an acknowledged leader in pursuing renewable energy sources, the Q-Cells company in 2011 completed a PV power park rated at 91 megawatts at a defunct airport near Berlin. The company's investment decision was based in part on Germany's feed-in tariff system, which in turn was partly motivated by its national policy goals in its shift toward renewables. A schedule of New York State property tax deduction incentives for PVs that ran from 2008 through 2012 offered a rebate value of up to $62,500 annually for up to four years.

Incentives such as these have begun to inject real momentum into the burgeoning solar energy field. Germany reported in 2011 that it had passed the 100,000 employee milestone for workers in the PV industry. By way of comparison, that is higher than the number of employees in the U.S. steel industry. Germany, however, is not necessarily the fastest-growing nation in solar installation. Italy reported the number of renewable energy plants in that country doubled every year from 2006 through 2010, with installed capacity surpassing 30,000 megawatts in 2010, a 65% increase since 2000. The expansion, the country declared, has been especially strong in wind farms, biofuels plants—but above all in PV installations. By 2010, Italy had more than 155,000 separate solar generating facilities from tiny to giant, with a combined electrical output capacity of 3.5 gigawatts.

Among developing countries, the trend is catching on fast. A project in Peru to build a PV installation with capacity of 44 megawatts firmed up its $145 million financing as 2011 drew to a close. Thinking big, Rajasthan state in India opened bidding in early 2012 for companies to build 100 megawatts of PV plants, plus 100 megawatts of solar thermal plants. On February 4, 2016, Morocco began construction of a 1,000 megawatt Ouarzazate solar array, and in September, 2016, China opened the 1,000 megawatt Yanchi Ningxia solar complex.

—Manja Leyk

Further Reading

Badescu, Viorel. *Modeling Solar Radiation at the Earth's Surface: Recent Advances*. Springer, 2008.

Letcher, Trevor M., and Vasilis M. Fthenakis, eds. *A Comprehensive Guide to Solar Energy Systems, with Special Focus on Photovoltaic Systems*. Academic Press, 2018.

Nemet, Gregory F. *How Solar Energy Became Cheap: A Model for Low-Carbon Innovation*. Routledge, 2019.

Palz, Wolfgang, J. Greif, and Commission of the European Communities, eds. *European Solar Radiation Atlas: Solar Radiation on Horizontal and Inclined Surfaces (Algorithms and Combinatorics)*. Springer, 1996.

Petela, Richard. *Engineering Thermodynamics of Thermal Radiation: For Solar Power Utilization*. McGraw-Hill, 2010.

Plante, Russell H. *Solar Energy. Photovoltaics and Domestic Hot Water*. Academic Press, 2014.

Sukhatme, S.P., and J.K. Nayak. *Solar Energy: Principles of Thermal Collection and Storage*. 3rd ed. Tata McGraw-Hill, 2008.

U.S. Department of Energy. *Laying the Foundation for a Solar America: The Million Solar Roofs Initiative; Final Report*. October 2006. http://www.nrel.gov/docs/fy07osti/40483.pdf.

Zweibel, Ken, James Mason, and Vasilis Fthenakis. "A Solar Grand Plan: By 2050 Solar Power Could End U.S. Dependence on Foreign Oil and Slash Greenhouse Gas Emissions." *Scientific American*, December 16, 2007. http://www.colorado.edu/physics/phys3070/phys3070_fa08/Reading/phys3070_sciamerican_solargrandplan.pdf.

Solar Thermal Systems

FIELDS OF STUDY
HVAC Trades; Physical Science; Plumbing Trades

ABSTRACT
A system that converts energy from the Sun into heat in usable form is called a "solar thermal system." With improvements in technology and the increasing cost of conventional fuels, solar thermal systems are attracting attention.

KEY CONCEPTS
central receiver: a structure placed to receive the reflected sunlight from a large symmetrical array of mirror surfaces

desalination: conversion of saltwater to salt-free water by evaporating the saline water and condensing the water vapor in a receiver separated from the salt residue; essentially a form of distillation

parabolic dish: a structure having a parabolic shape, the interior being reflective; sunlight entering the parabolic cavity is reflected to the focal point of the parabolic surface

zeolite: a mineral form in which the arrangement of atoms within the material forms interatomic cavities that are able to trap and hold water molecules

SOLAR THERMAL SYSTEMS
Energy from the sun can be harnessed directly for creation of either high temperature steam (greater than 100°C), or low temperature heat (less than 100°C) for various heat and power applications. A system that converts energy from the Sun into heat in usable form is called a solar thermal system. With improvements in technology and the increasing cost of conventional fuels, solar thermal systems are attracting attention.

HIGH TEMPERATURE SOLAR THERMAL SYSTEMS
These systems use mirrors or other reflecting surfaces to concentrate solar radiation. There are three main technologies under high temperature systems: central receiver, parabolic dish system, and solar line-focusing collector. Solar heat at high temperatures, sufficient to generate electricity, is collected and electricity is generated using a heat engine. Compared to solar photovoltaic (PV) systems, solar thermal power systems have the following benefits:

- Electricity production is not limited to only sunlight hours, unlike solar PV.
- Effective utilization of the solar spectrum.
- Conversion efficiency is high and the dusty climatic conditions do not affect the power generation.
- There is no need for storage of large amounts of electricity for usage outside nonsunlight hours via batteries.

Solar Thermal Systems

The 150 MW Andasol solar power station is a commercial parabolic trough solar thermal power plant, located in Spain. The Andasol plant uses tanks of molten salt to store solar energy so that it can continue generating electricity even when the sun isn't shining. (BSMPS via Wikimedia Commons)

Electricity generation from high temperature solar thermal systems is a reality but currently there is only a small market for the same. Its future depends a lot on the availability and success of further research and development.

LOW TEMPERATURE SOLAR THERMAL SYSTEMS

These systems collect solar radiation to heat air or water for both domestic and industrial purposes. Low temperature solar thermal systems are a proven technology and have a mature industry with a wide range of applications. These technologies can be used as passive or as active systems.

Energy is collected in passive systems through orientation, materials, and construction of the collector, and does not require the use of pumps or motors. The properties of the collector allow it to absorb, store, and use solar radiation.

Active systems make use of pumps or motors to circulate water or some other heat-absorbing fluid through solar collectors. These collectors are typically made up of copper tubes bonded to a metal plate, painted black, and encapsulated in an insulated box covered by a glass.

Among solar thermal technologies, solar water heating, solar process heat, and solar buildings have progressed far enough to contribute to the energy pool and toward combating climate change. Thus far, solar water heating is the most widely accepted technology worldwide.

SOLAR COOKING

Utilizing the Sun's heat to cook food is not a new technology: it has been used by for centuries. A device that cooks food with heat energy from the sun is called a "solar cooker" and can be easily manufactured from materials that are readily available. There are certain limitations to the technology; cooking is limited to times when the sun is shining, must take place in an open space, and cooking takes significantly longer time than using other means. One of the major developments in solar thermal technology in the twenty-first century is the introduction of concentrating solar cookers that concentrate the heat on a single focal point. These cookers have been found to be very effective.

SOLAR DRYING

Controlled solar drying is another potential solar thermal application. Everyone who has hung out wet clothes to dry in the sunshine understands the basic principle of solar drying. Drying is also a vital operation in the postharvest sector, and losses (e.g., grains) during uncontrolled drying are significant. Moreover, open-sun drying can lead to product contamination. Controlled hot-air drying minimizes loss and contamination. With the increasing difficulty of obtaining firewood and the ever-increasing prices of other sources of energy, solar drying systems, in addition to being environmentally sound, can also be more cost effective.

SOLAR AIR HEATING AND COOLING

Solar-heated air can be used for space heating during the winter or in cold regions where heating is required for most of the year. Several air-heating collectors have been developed for space heating, which not only conserve fuel but also helps reduce the indoor air pollution caused by burning fuels in a living space. Solar cooling can be achieved by passing air over solid desiccants like silica gel or zeolite, drawing moisture from the air, and creating an efficient evaporative cycle. These desiccants can then be regenerated by drying, using solar thermal systems.

SOLAR HOMES

In the past, people designed homes to take advantage of the heating power of sunlight, but the availability of relatively cheap electricity in recent centuries has caused most homes to be designed with artificial heating and cooling. Using the sun to heat or cool houses is called "passive solar design." In cold climates, homes are designed so that sunlight entering the house is stored by a solid mass through the night. These designs also include well-insulated walls and roofs; heavy curtains can also be used to trap heat.

In passive solar cooling, buildings are designed to slow the rate of heat transfer by using shading insulation, thoughtful landscaping, and removing excess heat with fans, vents, and cross-ventilation.

SOLAR WATER AND POOL HEATING

Solar water heaters, with year-round utility, are economically attractive. These designs, using heat exchangers, are widely used by fabricators and can also soften hard water. Recently, systems with evacuated tube collectors have found success in the marketplace, as they are less expensive and free from problems like scaling, corrosion, and leaking. However, improvements are still needed to make these tubes durable, to make the selective glass coating more effective, and to ensure that storage tanks are more efficient in extracting heat from the tubes. In solar pool heating, the systems are simple and relatively inexpensive. The pool systems commonly use simple and low-cost unglazed plastic collectors, and the pool itself is the thermal storage container for the system.

SOLAR DESALINATION

Solar desalination is a technique used to desalinate water using solar energy. The process is called the "solar humidification-dehumidification (HDH) process." In this technique, the natural water cycle is mimicked for a shorter time frame by evaporating and condensing water to separate the present substances. The driving force in this process is solar thermal energy, which produces water vapor that is later condensed in a separate chamber to obtain desalinated water.

—*Saurabh H. Mehta*

Further Reading

Biswas, Jagadish, and Agnimitra Biswas. *Modeling and Optimization of Solar Thermal Systems: Emerging Research and Opportunities*. IGI Global, 2021.

Build It Solar. "Solar Pool (and Hot Tub) Heating." http://www.builditsolar.com/Projects/PoolHeating/pool_heating.htm.

Chandra, Laltu, and Ambesh Dixit, eds. *Concentrated Solar Thermal Energy Technologies: Recent Trends and Applications.* Springer, 2018.

Duffie, J.A., and W.A. Beckman. *Solar Engineering of Thermal Processes.* 3rd ed. Wiley, 2006.

Howell, John R. *Solar-Thermal Energy Systems: Analysis and Design.* McGraw Hill, 1982.

Ramlow, Bob, and Benjamin Nusz. *Solar Water Heating: A Comprehensive Guide to Solar Water and Space Heating Systems.* New Society Publishers, 2010.

Sarbu, Ioan, and Calin Sebarchievici. *Solar Heating and Cooling Systems: Fundamentals, Experiments and Applications.* Elsevier, 2017.

Solar Cookers World Network. http://solarcooking.org.

STEAM AND STEAM TURBINES

FIELDS OF STUDY
Electrical Engineering; Mechanical Engineering; Steam Power Engineering

ABSTRACT
A steam boiler converts the chemical energy in fuel into the thermal energy of steam. A steam turbine converts this thermal energy into the mechanical energy of a rotating shaft. This shaft can drive an electric generator or other device.

KEY CONCEPTS
rotor: the inner part of a turbine or generator; the rotor is the part that turns

throttle valve: a valve designed to control the amount of steam flowing into the generator by opening to admit a greater flow of input steam, or closing to decrease the flow of input steam

BACKGROUND
Fossil fuels such as oil and coal contain chemical energy. Uranium contains nuclear energy. Either of these forms of energy can be converted into thermal energy (heat), and this thermal energy can be used to make steam in a boiler. A steam turbine can be used to convert the thermal energy of steam into the mechanical energy of a rotating shaft. When the turbine shaft is used to drive an electric generator, electricity is produced. Although electric generators can be driven by diesel engines, gas turbines, and other devices, most electricity is generated using steam turbines.

PRINCIPLES OF TURBINE OPERATION
High-pressure, high-temperature steam enters a steam turbine through a throttle valve. Inside the turbine the steam flows through a series of nozzles and rotating blades. As it flows through a nozzle, the pressure and temperature of the steam decrease, and its speed increases. The fast-moving steam is directed against rotating blades, which work something like the blades on a pinwheel. The steam is deflected as it passes over the rotating blades, and in response the steam pushes against the blades and makes them rotate. As the steam flows over the rotating blades its speed decreases.

Large turbines are composed of many stages. Each stage has a ring of nozzles followed by a ring of rotating blades. The slow-moving steam leaving the rotating blades of one stage enters the nozzles of the next stage, where it speeds up again. This arrangement is called pressure compounding. The energy of the steam is converted to mechanical work in small steps. Less of the steam's thermal energy is wasted or lost if it is converted in small steps.

The amount of power produced by a turbine depends on the amount of steam flowing through it and on the inlet and outlet steam pressures. Steam flow is constantly regulated by the throttle valve, but the steam pressures are fixed by the design of the system. Inlet steam pressure is determined by the operating pressure of the boiler that supplies it. Outlet pressure is determined by where the steam goes when it leaves the turbine. If the steam simply escapes into the atmosphere, the outlet pressure is atmospheric. If the outlet steam pressure is made lower than atmospheric pressure, the turbine produces more power. This is accomplished by having the steam leaving the turbine flow into a condenser.

Cooling water passing through tubes inside the condenser removes heat from the steam flowing around the tubes and causes it to condense and become liquid water. Since water occupies a much smaller volume as a liquid than as steam, condensing creates a vacuum. When a turbine is connected to a condenser, the outlet steam pressure can be far below one atmosphere.

DETAILS OF TURBINE CONSTRUCTION

Inside the steel turbine casing, stationary partitions called diaphragms separate one turbine stage from the next. Each diaphragm has a hole at the center for the rotor shaft to pass through. Nozzle passages are cut through the diaphragms near their outer rims, and the steam is forced to pass through these nozzles to get to the next stage.

The rotor of a turbine is made up of solid steel disks that are firmly attached to a shaft. Rotating blades are mounted around the rims of the disks. Where the shaft extends from the casing at each end, it is supported by journal bearings and a thrust bearing. The journal bearings are stationary hollow cylinders of soft metal that support the weight of the rotor. A thrust bearing consists of a small disk on the shaft of the turbine that is trapped between two stationary disks supported by the casing. If the rotor tries to move forward or back along its own axis, the rotating disk presses against one of the stationary disks. Thrust and journal bearings must be lubricated by a constant flow of oil that forms a thin film between the rotating and stationary parts of the bearing and prevents them from making direct contact. Without this film of oil, the bearing would wear out in a few seconds.

A seal must be provided where the shaft of the turbine passes through the casing. At one end of the casing the steam pressure inside is high. Outside the casing the air pressure is only one atmosphere. If there were no seal, steam would rush out through the space between the casing and the shaft. At the other end of the turbine, the pressure inside may be below atmospheric. Here air would rush in if there were no seal around the shaft.

ELECTRIC POWER GENERATION

Most electric power is produced by steam turbines driving electric generators. This is true whether the source of the steam is a nuclear reactor or a boiler burning fossil fuel. The turbines in power stations are extremely large. In nuclear plants the turbines may produce as much as 1,300 megawatts of power. Power stations are often located near rivers so that water from the river can be used as cooling water in the condensers that receive steam from the large turbines.

—*Edwin G. Wiggins*

Further Reading

Avallone, Eugene A., Theodore Baumeister III, and Ali M. Sadegh. "Steam Turbines." *Marks' Standard Handbook for Mechanical Engineers.* 11th ed. McGraw-Hill, 2007.

Bloch, Heinz P. *Steam Turbines: Design, Applications, and Rerating.* 2nd ed. McGraw-Hill, 2009.

How Stuff Works. "How Steam Technology Works." http://science.howstuffworks.com/steam-technology.htm.

Peng, William W. "Steam Turbines." *Fundamentals of Turbomachinery.* J. Wiley, 2008.

Tanuma, Tadashi, ed. *Advances in Steam Turbines for Modern Power Plants.* Woodhead Publishing, 2017.

STEAM ENGINES

FIELDS OF STUDY

Mechanical Engineering; Stationary Engineering; Thermodynamics

ABSTRACT

A steam boiler converts the chemical energy in fuel into the thermal energy of steam. A steam engine converts this thermal energy into the mechanical energy of a rotating shaft. This shaft can drive an electric generator or pump.

KEY CONCEPTS

combustible: describes carbon-based material that can be readily oxidized in air to release its contained chemical energy as heat and light

constant-pressure steam: steam that is produced, maintained and used at a constant temperature rather than being allowed to cool during the functional operation of the machine

dewatering pump: pumps used to remove water from working coal mines in order to prevent them from flooding

pivoted beam: essentially an application of the lever and fulcrum principle, in which one end of the beam is driven downward by a powered rod to raise a load attached to the other end of the beam, which then returns to its previous position when the load is removed

BACKGROUND

The chemical energy that is contained within fossil fuels such as oil and coal and other combustible fuels can be converted into thermal energy (heat) by burning the fuel. This thermal energy can be used to create steam in a boiler. A steam engine converts the thermal energy of steam into the mechanical energy of a rotating shaft, and this shaft can drive a pump, a ventilating fan, a ship's propeller, and many other devices.

HISTORY

Although there were attempts to use steam to drive mechanical devices as early as 60 CE by Hero of Alexandria, the first real steam engine was designed and built by Thomas Newcomen in 1712. That year Newcomen successfully used a steam engine to pump water from a coal mine near Dudley Castle, England. In 1765, as he walked across Glasgow Green in the city of Glasgow, Scotland, James Watt conceived the idea of connecting the steam engine to a separate condenser. The first full-size engines based on this concept were built in 1776: one at John Wilkinson's blast furnace near Broseley, England, and the other at Bloomfield coal mine near Tipton, England. Newcomen's design and Watt's early designs used steam at constant pressure. Over the course of his life, Watt invented many improvements to the steam engine, including rotary engines, a device for measuring engine performance, and engines in which the steam expanded during the piston stroke. Expanding steam engines soon drove the earlier type off the market, because the fuel consumption associated with the boiler of an expanding steam engine is far less than that of a constant pressure engine. While recent steam engines operate at much higher pressure than Watt's, they are similar in design.

Having been almost completely supplanted by internal combustion diesel engines and the decline of the coal industry, steam engines have essentially become little more than nostalgia items in modern society, although there are occasional applications such as cinema where authenticity demands an actual steam engine. As such, there is no new development of steam engine technology and little new information to be found, although there is an abundance of eighteenth and early nineteenth century contemporary treatises on steam engines to be found in the technical literature.

PRINCIPLES OF OPERATION

Early steam engines would be considered upside-down by modern standards. The piston was connected to a rod that emerged from the top of the engine, and steam was fed into the cylinder below the piston. A chain connected the piston rod to one end of a pivoted beam suspended above the engine, and the other end of the beam was connected to a pump that drew water up from the bottom of a mine. The weight of the pump rod was sufficient to pull the pump end of the pivoted beam downward, which caused the other end of the beam to rise and lift the piston upward. As the piston rose, steam at just above atmospheric pressure flowed from the boiler into the growing space below the piston. When the piston reached the top of its stroke, the valve between boiler and cylinder closed, and in Newcomen's engine water was sprayed into the cylinder. As the water absorbed heat from the steam, the steam condensed, which created a partial vacuum. This vacuum, combined with atmospheric pressure acting on the upper side of the piston, caused the piston to move downward. When the piston reached the bottom of its stroke, the steam valve opened again. The steam pressure balanced the atmospheric pressure on the other side of the piston, and the weight of the pump rod again raised the piston to the top of its stroke.

Watt recognized that spraying cold water directly into the cylinder not only condensed the steam but also cooled off the cylinder itself. On the next stroke some incoming steam was wasted in reheating the cylinder. The separate condenser in Watt's engine condensed the steam without chilling the cylinder. This resulted in a dramatic improvement in fuel consumption. Watt also closed off the upper side of the piston and provided a hole just big enough for the piston rod to pass out. Constant-pressure steam was admitted to the space above the piston, and this steam provided the pressure previously supplied by the atmosphere. Watt's original purpose here was to eliminate the cooling effect of the atmosphere, but he soon realized that there was another benefit. Instead of continuing to admit steam at constant pressure during the entire downward stroke, the steam valve could be closed and the steam could be allowed to expand. This further reduced fuel consumption and paved the way for modern expansion steam engines.

APPLICATIONS OF THE STEAM ENGINE

The first applications were to drive dewatering pumps in mines and to supply pressurized air for blast furnaces that produced cast iron. It was soon realized that rotary steam engines could be used to drive all kinds of machinery. Without the invention of the steam engine, the Industrial Revolution would not have occurred in the time and place that it did. Steam engines drove spinning and weaving machines in the textile industry. Ships and railroad locomotives powered by steam engines revolutionized transportation. There were steam-powered farm tractors, automobiles, and construction machines. Early electric generators were also driven by steam engines. Many of these applications are now powered by electric motors, gasoline and diesel engines, and steam turbines, but it was the steam engine that showed the way.

—*Edwin G. Wiggins*

Further Reading

Barton, D.B. *The Cornish Beam Engine: A Survey of Its History and Development in the Mines of Cornwall and Devon from Before 1800 to the Present Day, with Something of Its Use Elsewhere in Britain and Abroad.* New ed. Author, 1966.

Bray, Stan. *Making Simple Model Steam Engines.* Crowood, 2005.

Colburn, Zerah. *The Construction and Operation of Steam Engines for Locomotives.* Salzwasser Verlag, 2010.

Collier, James Lincoln. *The Steam Engine.* Marshall Cavendish, 2006.

Crump, Thomas. *A Brief History of the Age of Steam: The Power That Drove the Industrial Revolution.* Carroll & Graf, 2007.

How Stuff Works. "How Steam Engines Work." http://www.howstuffworks.com/steam.htm.

Marsden, Ben. *Watt's Perfect Engine: Steam and the Age of Invention.* Columbia UP, 2002.

Rose, Joshua. *Modern Steam Engines.* H.C. Baird, 1886. Reprint. Astragal Press, 2003.

Savery, Thomas. *The Miner's Friend.* Outlook Verlag, 2020.

Steingress, Frederick M., Harold J. Frost, and Daryl R. Walker. *Stationary Engineering.* 3rd ed. American Technical, 2003.

Tredgold, Thomas. *The Steam Engine.* J. Taylor, 1827. Reprint. Cambridge UP, 2014.

STIRLING, ROBERT

FIELDS OF STUDY
Mechanical Engineering; Thermodynamics

ABSTRACT
Scottish engineer Robert Stirling invented the thermodynamically efficient four-step external combustion Stirling-cycle engine, featuring a heat regenerator that stores and reuses excess heat.

KEY CONCEPTS

external combustion engine: an engine in which fuel combustion takes places outside of the engine works and is not an integral part of the engine's mechanical operation

heat exchanger: a device designed to transfer the heat given off by the spent product of the operation to the unheated input of the operation; in the Stirling engine, heat is recovered from the output stage of the operating cycle and used to heat the material for the input stage of the operating cycle

thermal efficiency: the percentage of available heat energy that is converted to usable work relative to the heat energy that is lost as waste

BACKGROUND

Robert Stirling was a Scottish engineer as well as an ordained minister of the Church of Scotland. Stirling was known for the practical nature of his experiments in radiant heat and engine design, carried out with the aid of his younger brother, James. Stirling emerged as one of the leading engineers of his time after developing the four-cycle external combustion engine that bears his name. The Stirling engine introduced the use of a heat regenerator that recovered the excess heat produced during the engine's cycle, allowing it to be reused during the next cycle. Despite the engine's thermodynamic efficiency, it never enjoyed widespread use because of its cost and the complexity of its manufacture.

Stirling was born into a farm family in Perthshire, Scotland, on October 25, 1790. Stirling followed his basic education at the University of Edinburgh with divinity studies at University of Glasgow. After his ordination, he received a parish at Kilmarnock. He also showed an early aptitude for science and engineering, conducting scientific experimentation both during and after his education. His interests included scientific instrument design, radiant heat, and engine design. Manufacturer Thomas Norton provided Stirling with a workshop in which to carry out his practical experiments. He married Jean Rankin Stirling in 1819; the couple had seven children.

English inventor James Watt had launched a revolution with the development of a practical steam engine in the late eighteenth century, but the engines had a number of difficulties. Stirling became interested in designing an engine that used hot air rather than steam to reduce the danger of explosions and increase thermal efficiency and fuel economy. Stirling received an 1816 patent for a series of methods and parts designed to improve fuel economy in glass melting furnaces, breweries, and factories. He also included the design for an external combustion hot-air engine utilizing his innovations. Air under pressure passes through a cycle of heating, expansion, cooling, and compression.

The work for which Stirling received his 1816 patent represented the forerunner of what became known as the Stirling or Stirling-cycle engine. Robert worked with his brother James, also an engineer, to improve the engine through additional research and experimentation on model engines. James had provided the suggestion that the engine should utilize compressed air with higher pressures rather than air at atmospheric pressure. The Stirling engine is an external combustion engine utilizing pistons and cylinders. The use of two separate cylinders for fluid expansion and compression marked a change from steam engines, whose design features only one cylinder. The engine works through a four-step process generating a closed cycle similar to that of a steam engine.

The Stirling engine can be manufactured using a variety of metals and runs on a gas, often air, hydrogen, or helium, rather than on steam. The air or gas is heated outside the main cylinder. Inside, the working fluid absorbs this heat and expands to move the pistons. One of the Stirling engine's unique features is its use of a regenerative heat exchanger rather than a condenser. Stirling called his exchanger the economizer. This exchanger allows the engine to store the excess heat generated during each thermodynamic cycle rather than simply losing that heat in a condenser. This allows the excess heat to be used in the next cycle.

Robert Stirling. (Wikimedia Commons)

The exchanger thus greatly enhances the Stirling engine's thermal efficiency.

The brothers had less success in the manufacture of the engines for actual use, although several of their improvements were patented in subsequent years. Stirling, however, was never an active businessman and was relatively unconcerned with the profit motive or the need to protect his patents from competitors in the marketplace. A few Stirling engines were in use, including the one that powered the foundry where James worked from 1843 until 1847. William Thomson (Lord Kelvin) employed one of Stirling's model engines to demonstrate the reversible cycle while he was a professor at the University of Glasgow.

Stirling engines were not widely adopted in the nineteenth century for a number of reasons. In addition to the complexity of their design and manufacture mentioned above, the cast iron in use at the time was unable to withstand the high temperatures produced by the working engine; moreover, technological difficulties allowed for the production of engines with only limited horsepower. Although steam engines posed more danger given their high rates of boiler explosions, they would remain in widespread use throughout the nineteenth century.

Stirling engines did not disappear, however; they found limited uses in the twentieth and twenty-first centuries, despite being overshadowed by the development of the more cost-effective four-step Otto-cycle internal combustion engine. Modern uses for the Stirling engine include the production of liquefied air in research laboratories and the powering of weather and spy satellites and submarines. The fact that Stirling engines have the highest possible thermodynamic efficiency may outweigh their expense and complexity of design and manufacture as environmental conservation becomes an increasing concern.

Robert Stirling died on June 6, 1878, at Galston, Scotland. He was not widely known for his work with engines in the years following his death, despite the fact that his engine carried his name. All his sons followed in his footsteps: Four became railroad engineers and the fifth joined the clergy. Stirling's only two surviving early engine models are displayed at the Royal Scottish Museum and the University of Glasgow.

—*Marcella Bush Trevino*

Further Reading
Ibrahim, Mounir B., and Roy C. Tew, Jr. *Stirling Convertor Generators*. CRC Press, 2012.
Organ, A.J. *Thermodynamics and Gas Dynamics of the Stirling Machine*. Cambridge UP, 1980.
Organ, Allan J. *Stirling Cycle Engines: Inner Workings and Design*. John Wiley & Sons, 2013.
Sier, Robert. *Rev. Robert Stirling, D.D.* L.A. Mair, 1995.
Sier, Robert. "Stirling and Hot Air Engine Homepage." http://www.stirlingengines.org.uk.
Vineeth C.S. *Stirling Engines: A Beginner's Guide*. Vineeth CS, 2012.
Walker, Graham. *Stirling Engines*. Oxford UP, 1980.

Sugar Beet Ethanol

FIELDS OF STUDY
Agriculture; Biotechnology; Fermentation Engineering

ABSTRACT
Ethanol (ethyl alcohol) is increasingly used not only as an energy carrier in the transportation sector but also in a wide variety of other applications. It can be produced from a variety of renewable feedstocks, including sugar beets (and molasses, a byproduct of sugar production). Sugar beets can be grown in a wide range of soil and climate conditions.

KEY CONCEPTS
diffuser: a processing stage in which thinly sliced sugar beets are left to soak in water, allowing their sugar content to diffuse out from the beet slices and dissolve in the water
grain alcohol: ethanol produced by fermentation of grains rather than sugar beets and sugarcane
molasses: the thick, sweet, strong flavored syrup residue remaining after sugar crystallization and separation from the crude sugar solution extracted from various source plants for sugar production; contains a fairly high percentage of remnant glucose sugar mixed with flavonoid compounds and other sugars
vinasse: the solid residue remaining from distillation of product ethanol from fermented sugar beet extract

SUGAR BEET ETHANOL IN THE WORLD
Ethanol (ethyl alcohol) is a volatile, flammable, and colorless liquid. It is increasingly used not only as an energy carrier in the transportation sector but also in a wide variety of other applications: in beverages, industrial feedstocks, and solvents. Also called bioethanol, it can be produced from a variety of renewable feedstocks, including sugar beets (and molasses, a byproduct of sugar production). Sugar beets can be grown in a wide range of soil and climate conditions, with world production concentrated in Europe, Russia, Turkey, the eastern United States, and Chile. In 2009, France, Germany, Poland, the United Kingdom, and the Netherlands accounted for more than 75 percent of the total European Union (EU) production.

HISTORY OF BIOETHANOL
Bioethanol has been known as a motor fuel for more than a century. Its use for transport started when Henry Ford designed the Model T in the expectation that bioethanol produced by American farmers would

be used as its main fuel. The car was produced from 1903 to 1926 and was able to run on bioethanol starting with the 1908 model year. At the same time, several farmers also used self-produced bioethanol to fuel their own agricultural machinery (and themselves).

By the mid-1930s, every European country was using gasoline blended with ethanol, and some countries even mandated its use. In France, a compulsory incorporation rate of 10 percent alcohol in all gasoline was imposed, whereas in Germany a blend of gasoline with potato-derived ethanol was sold under the name *Reichskraftsprit*. In Britain, gasoline was blended with 25 percent grain alcohol under the Discol brand. Discol was seen as a high-performance fuel and was often used in motor racing, aviation, and power-boat racing. During World War II, the use of bioethanol continued, helping to overcome oil shortages.

Nevertheless, oil became the major source of transportation fuel because of its better compatibility with the materials used to construct engines as well as the growing supply of cheaper fuel oil. The use of ethanol as a transportation fuel was mostly abandoned after World War II and was not resumed until the early 1970s, with the first world oil crisis. In 1975, Brazil started the Pró-Álcool (National Alcohol) program, a government-supported plan aimed at promoting the production and use of ethanol fuel from sugarcane. In 1978, large-scale production of corn-based bioethanol started in the United States. Nowadays, the United States and Brazil represent approximately 88 percent of the total worldwide production of bioethanol fuel: The United States produced more than 49 billion liters (13 billion gallons) in 2010, followed by Brazil, with more than 26 billion liters (nearly 7 billion gallons). In Europe, ethanol fuel production started in 1992, but the sector soared after 2004, and it nearly septupled between then and 2009, reaching 4.45 billion liters (1.18 billion gallons) by 2010.

PRODUCTION OF BIOETHANOL FROM SUGAR BEETS

Sugar beet cultivation includes several steps: soil preparation (plowing), fertilizer application, sowing, weed control, and harvesting. After transportation to a processing plant, the beets are washed and sliced into chips. Slicing maximizes the efficiency of the

Sugar Beet. (Barcin via iStock)

next step, diffusion, in which the chips are passed into a hot water solution to extract the sweet raw juice. Beet pulp is the most important product of beet processing and, after pressing and drying, can be sold as animal-feed concentrate or burned for process heat. Approximately 75 kilograms of dried beet pulp are produced per ton of sugar beet processed.

The raw juice exiting the diffuser can be used for bioethanol or sugar production. Market conditions drive the determination of sugar and bioethanol production shares. Each ton of sugar beets processed, at 16 percent sugar content, yields about 100 liters of ethanol. Vinasse (stillage) is a byproduct from ethanol distillation and, after concentration, can be sold as an additive for animal feed or fertilizer. Per liter of bioethanol produced, nearly 0.6 kilogram of concentrated vinasse is obtained.

ENERGY AND ENVIRONMENTAL ISSUES

In addition to security of energy supply and rural development support, climate change is another rationale driving the promotion of bioethanol as a transport fuel. According to the EU renewable energy directive, 2009/28/EC, typical savings in greenhouse gas (GHG) emissions attributable to sugar beet ethanol would amount to 61 percent.

However, this figure does not take into account soil carbon emissions from (direct and indirect) land-use change (LUC). LUC is an emergent topic in the literature, as it may have a significant impact on the GHG balance of biofuels. Bioethanol fuel production from sugar beets is constrained by the agricultural land available, which may trigger competition with the food sector. Even though the technology is not yet mature at a commercial scale, production of ethanol from cellulosic biomass (so-called second-generation ethanol)—namely straw, grass, and wood—seems very attractive for ensuring a plentiful and low-cost supply of such feedstock.

—*Fausto Freire*

Further Reading

Clark, James, and Fabien Deswarte, eds. *Introduction to Chemicals from Biomass*. 2nd ed. Wiley, 2015.

Drapcho, Caye M., Nghiem Phu Nhuan, and Terry H. Walker. *Biofuels Engineering Process Technology*. 2nd ed. McGraw-Hill, 2020.

Ingledew, W.M. *The Alcohol Textbook: A Reference for the Beverage, Fuel, and Industrial Alcohol Industries*. 5th ed. Nottingham UP, 2009.

Malça J., and F. Freire. "Renewability and Life-Cycle Energy Efficiency of Bioethanol and Bio-Ethyl Tertiary Butyl Ether (bioETBE): Assessing the Implications of Allocation." *Energy*, vol. 31, no. 15, 2006.

SenterNovem. *Bioethanol in Europe: Overview and Comparison of Production Processes*. Rapport 2GAVE0601. SenterNovem, 2006.

Walker, G.M. *Bioethanol: Science and Technology of Fuel Alcohol*. Ventus, 2010.

Sugarcane Ethanol

FIELDS OF STUDY
biotechnology; refinery operations; agriculture;

ABSTRACT
Ethanol can be made by the fermentation of sugars produced by sugarcane. It is used primarily as a transport fuel, either in pure form or as a gasoline additive. Brazil, the world's largest ethanol producer, commonly makes ethanol from sugarcane.

KEY CONCEPTS

bagasse: the cellulosic residue (plant matter) left after the juice has been squeezed out from the prepared sugarcane stalks

flex-fuel vehicles: cars and trucks with engines designed to operate efficiently with fuels containing any amount of ethanol

gasohol: common name for gasoline blends containing less than 25 percent ethanol in Brazil

greenhouse gases (GHGs): commonly refers to carbon dioxide, but includes a number of other gases capable of absorbing heat energy radiating from Earth's surface and then emitting much of that heat energy back into the atmosphere

ETHANOL, AND SUGAR, FROM SUGARCANE

The interest in sugarcane-based ethanol was reinforced at the beginning of the twenty-first century by the increasing volatility and rapid surge in crude oil prices. The high dependence of the majority of the world's developed and developing countries on a handful of oil and gas producers in the Middle East and West Asia, plagued by spare capacity concerns and geopolitical tensions, has made it almost inevitable that these dependent nations must diversify their energy resources. Ethanol from sugarcane provides an attractive alternative, with its potential technical, social, economic, and environmental benefits as compared to gasoline and ethanol produced from other sources. However, its commercial viability and competitiveness is largely dependent on prolonged government nurturing and support of the sugarcane industry. Furthermore, large-scale production of ethanol from sugarcane may trigger a food-fuel trade-off, leading to increased food prices and loss of biodiversity as scarce land resources become increasingly dedicated to biofuel-producing sugarcane crops.

ETHANOL VS. OIL

The interest in sugarcane-based ethanol as an alternative fuel for transport, especially in oil-importing developing nations, is motivated by the need to diver-

Sugarcane. (bianliang via iStock)

sify energy sources and lower exposure to the price volatility of the international oil market. Sugarcane-based ethanol is especially attractive for countries that import oil heavily (for instance, landlocked countries) that pay high prices for delivered petroleum but already have a comparative advantage in production of sugarcane. Furthermore, production of ethanol from sugarcane holds the promise of contributing to rural development by creating jobs in production, manufacture, transport, and distribution of feedstock and products.

Brazil uses pure sugarcane-based ethanol in about 20 percent of its vehicles; in the rest of the nation's vehicles, it uses at least 25 percent ethanol blend. Some developing countries, including China, use 10 percent ethanol blend; India has mandated 5 percent ethanol blend. Brazil's National Alcohol Program (PRO-ALCOOL) started as a response to the Middle East oil embargo of 1973. At that time, Brazil's dependence on foreign oil made it even more vulnerable than the United States. Brazil's program has been extremely successful, although its development has not come without hitches. The feedstock costs account for 58 to 65 percent of the cost of ethanol production in Brazil. Thus, commercial viability of ethanol is critically dependent on the cost of cane production. However, the center-south region of Brazil has no parallel in terms of productivity, and the ethanol produced in the region is the cheapest in the world. This could be attributed to the country's rain-fed cane cultivation, availability of large amounts of unused land, constant support through decades of research, and commercial cultivation.

ETHANOL ECONOMICS IN BRAZIL

Most distilleries in Brazil belong to sugar mill/distillery complexes that are capable of changing the production ratio of sugar to ethanol. This capability enables plant owners to take full advantage of fluctuations in the relative prices of sugar and ethanol, as well as benefit from the much higher price that can be fetched by converting molasses into ethanol. Besides, flex-fuel vehicles, which were introduced in Brazil in 2003, have further increased the attractiveness of building hybrid sugar-ethanol complexes. More than 60 percent of all motor vehicles produced in Brazil are now flex-fuel: able to run on any mixture of alcohol and gasoline, as well as on 100 percent alcohol. These engines can also operate with regular gasoline alone if there is a shortage of biofuels.

A critical issue is whether Brazil's success in achieving self-sufficiency and a commercially competitive sugarcane-based ethanol industry can be replicated elsewhere. In this context, it deserves to be mentioned that Brazil's success was preceded by more than 20 years of government support. The country continues to maintain a significant tax differential between gasohol (80 percent gasoline and 20 percent ethanol) and hydrous ethanol. A large number of countries around the world are growing sugarcane, but none has been able to match Brazil's sugarcane cost structure. Thus, continued support in the form of subsidies—indirect, direct, or both—may be needed to launch and maintain a sugarcane-based biofuels industry in most developing countries. Although a rapid surge in crude prices is expected to make ethanol production and consumption viable in the near future, governmental policy making will remain hinged to uncertainty and hence may lack adequate clarity with regard to the extent of support necessary to render a burgeoning ethanol industry self-sustaining.

ENVIRONMENTAL CONSIDERATIONS

Ethanol in Brazil provides significant reductions in greenhouse gas (GHG) emissions compared to those from gasoline, a result of the relatively energy-efficient nature of sugarcane production, the use of bagasse (left over after the juice has been squeezed out of sugarcane stalks) as process energy, and the advanced state of sugar farming and processing. Thus, the savings in net energy value or balance (that is, the difference between the energy in the fuel product, or output energy, and the energy needed to produce the product, or input energy) and associated GHG emissions from blending of sugarcane-based ethanol with gasoline, on a life-cycle basis, may turn out to be much greater than ethanol produced from other sources, such as corn.

In order to ensure that ethanol is produced from sugarcane in a sustainable manner and enhances savings in net energy value and GHG emissions, ethanol plants should be produced using biomass and not fossil fuels; cultivation of annual feedstock crops should be avoided on land rich in carbon (above- and below-ground), such as peat soils used as permanent grassland; byproducts should be utilized efficiently in order to maximize their energy and GHG benefits; nitrous oxide emissions should be kept to a minimum by means of efficient fertilization strategies; and the commercial nitrogen fertilizer used for crops should be produced in plants that have facilities to mitigate the nitrous oxide gases generated during production.

—*Kaushik Ranjan Bandyopadhyay*

Further Reading

Basso, Thalita Peixoto, and Luiz Carlos Basso, eds. *Fuel Ethanol Production from Sugarcane*. IntechOpen, 2019.

Borjesson, Pal. "Good or Bad Bioethanol from a Greenhouse Gas Perspective: What Determines This?" *Applied Energy*, vol. 86, 2009.

Cortez, Luís A.B., Manoel Regis L.V. Leal, and Luiz A. Horta Nogueira, eds. *Sugarcane Bioenergy for Sustainable Development. Expanding Production in Latin America and Africa*. Routledge, 2019.

Dias de Moraes, M.A.F., and David Zilberman. *Production of Ethanol from Sugarcane in Brazil from State Intervention to a Free Market*. Springer, 2014.

Energy Sector Management Assistance Program. "Potential for Biofuels for Transport in Developing Countries." Report 312/05. Washington, DC: World Bank, 2005.

Gerdal, B. Hahn-Ha, M. Galbe, M.F. Gorwa-Grauslund, G. Liden, and G. Zacchi. "Bio-ethanol: The Fuel of Tomorrow from the Residues of Today." *Trends in Biotechnology*, vol. 24, 2006.

Santos, Fernando, Sarita Candida Rabelo, Mario de Matos, and Paulo Eichler, eds. *Sugarcane Biorefinery, Technology and Perspectives*. Academic Press, 2020.

Sorda, Giovanni, Martin Banse, and Claudia Kemfert. "An Overview of Biofuel Policies Across the World." *Energy Policy*, vol. 38, 2010.

Sun Day

FIELDS OF STUDY
Economics; Energy and Energy Resources; Public Policy

ABSTRACT
Sun Day, May 3, 1978, was successful in that it helped policymakers gather input for the development of national solar energy policies and raised awareness of the benefits of solar power, but progress in solar technologies continued at a relatively slow pace after the event, and the United States continued to be heavily dependent on fossil fuels.

KEY CONCEPTS
embargo: banning of sale or trade of goods to one country by another country or international coalition
renewable energy: energy that does not come from finite resources
solar energy: energy emanating from the Sun in the form of electromagnetic radiation and subatomic particles

THE NEED
During the 1960s and 1970s expanding demands for energy, increasing concerns regarding environmental quality, and limited domestic capacity to meet energy demands with traditional fossil fuels brought many Americans to the realization that the United States needed to place a higher priority on renewable sources of energy, particularly solar energy. The urgency of the problem was dramatically impressed upon the leadership of the nation with the Middle East oil embargo in 1973. It became clear to the American public that while the oil embargo would eventually pass, the nation, and even the world, could never again operate under the assumption that the traditional dependence on fossil fuels and other existing sources of energy could continue.

The need for a comprehensive program aimed at developing solar energy as a viable contributor to the future energy supply in the United States led to the creation of the Solar Energy Research Institute (SERI) in Denver, Colorado, in 1977 and the designation of May 3, 1978, as Sun Day. Solar energy awareness and development were emphasized throughout the week of May 1 through May 7, 1978. SERI provided technical support to the federal Sun Day Committee in its efforts to generate large volumes of information on solar energy for the public. A SERI-produced slide show on the technology and potential of solar energy was distributed throughout the nation to be shown at regular intervals in larger cities during the week. At the U.S. Customs House in Bowling Green, New York, solar energy displays were open to public view from May 3 to May 7.

On Sun Day, President Jimmy Carter visited SERI and gave an address on the future of solar energy in the United States in which he requested that every federal government agency consider more ways to help solar energy become a part of everyday American life. Carter pointed out the importance of developing renewable and essentially inexhaustible sources of energy in the future, particularly placing new emphasis on the importance of solar energy in the country's coming energy transition. He concluded that the costs associated with solar power technologies must be reduced so that solar power could be used more widely and would help establish a cap on rising fossil-fuel prices. In addition, Carter stated that he had just provided the U.S. Department of Energy with an additional $100 million for expanded efforts in solar research, development, and demonstration projects.

Following Carter's Sun Day address, a series of well-attended forums were conducted across the country. Participants included congressional representatives; state and local government officials; representatives of industries, labor organizations, public utilities, and special interest groups; and members of the general public. These public forums identified citizen groups interested in solar energy and provided input for the development of national solar energy policies.

—*Alvin K. Benson*

Further Reading
Laird, Frank N. *Solar Energy, Technology Policy, and Institutional Values.* Cambridge UP, 2001.
Carter, Jimmy. *Public Papers of the Presidents of the United States Jimmy Carter 1978 January 1 to June 30, 1978.* https://books.google.ca/books?id=mVPVAwAAQBAJ&pg=PA574&dq=sun+day+Jimmy+Carter&hl=en&sa=X&ved=2ahUKEwjPybbUwavuAhUOFVkFHbXZDrIQ6AEwAXoECAcQAg#v=onepage&q=sun%20day%20Jimmy%20Carter&f=false.
Scheer, Hermann. *A Solar Manifesto.* 2nd ed. James & James, 2001.

T

Tesla, Nikola

FIELDS OF STUDY
Electrical Engineering; Electronics Engineering Technology

ABSTRACT
Nikola Tesla was an inventor as well as a mechanical and electrical engineer who contributed significantly to the development and distribution of commercial electricity. He harnessed the alternating current mode for electricity transmission while developing patents for fluorescent lighting, the radio, and hundreds of other products still used today.

KEY CONCEPTS
alternating current: electrical current in which the direction of flow of the current reverses sinusoidally between positive and negative maximum values; normal house current oscillates at a frequency of 60 Hertz (Hz)

direct current: electrical current in which the current flows at a constant value and in one direction only

induction motor: an electrical motor that uses a system of wire coils (the stator) to generate a rotating magnetic field to drive the rotation of a central permanent magnet on a shaft (the rotor)

stepping up/down: transferring a voltage from a low value to a high value, or from a high value to a lower value, by means of transformers

BACKGROUND
Nikola Tesla was an inventor, as well as a mechanical and electrical engineer, who contributed significantly to the development and distribution of commercial electricity. The son of a Serbian Orthodox priest, Tesla immigrated to America, where he harnessed the current mode for electricity transmission—alternating current—while developing patents for fluorescent lighting, the radio, and hundreds of other products still used today.

Born an ethnic Serb in 1856 in the Croatian Military Frontier in the Austrian Empire (modern Croatia), Tesla was constantly constructing ideas in his mind and building devices from an early age. He absorbed languages and mathematics easily. After completing his elementary education in Croatia, he studied at the Polytechnic School in Graz and the University of Prague, followed by employment as an electrical engineer in Germany, Hungary, and France.

While walking with a friend in a city park in Budapest, Hungary, in January 1881, Tesla was suddenly

Nikola Tesla. (Wikimedia Commons)

excited by an inspiration. As his friend attempted to get him to rest for fear of his health, Tesla began drawing a diagram of an idea in the dirt. In this moment he envisioned the principle of the rotating magnetic field, which would later be used to introduce the world to his revolutionary induction motor. Tesla said, "No more will men be slaves to hard tasks; my motor will set them free. It will do the work of the world."

In 1884, Tesla arrived in New York City with four cents in his pocket and a letter of recommendation to Thomas Alva Edison from Charles Batchelor, a European business associate of Edison. Upon meeting Edison, Tesla presented his idea for the induction motor and alternating current, which Edison declared too dangerous. However, Edison agreed to give Tesla a job.

Tesla's position was short-lived, however. Differences arose between him and Edison over Tesla's vision for using alternating current (AC) and Edison's commitment to direct current (DC). Tesla left, igniting a longstanding hostility between the two men. Tesla formed his own company, Tesla Electric Light and Manufacturing, but he was allegedly cheated out of patents by partners, leading him to work as a laborer in New York for a period. In 1885, George Westinghouse (Westinghouse Electric Company) bought patent rights to Tesla's AC system, but their partnership was not confirmed for a few years.

Tesla's laboratory was established in New York City in 1887 where he produced experiments of shadowgraphs (which would later lead to the discovery of

Cutaway view through stator of induction motor. (S.J. de Waard via Wikimedia Commons)

324

X-rays), a carbon button lamp, various types of lighting, and exhibitions allowing electricity to flow through his body to light lamps. He was invited in 1888 to lecture before the American Institute of Electrical Engineers. Westinghouse was impressed and a partnership was formed.

A "war of the currents" was launched between Edison and the Tesla-Westinghouse team. Edison had a head start, with numerous stations already operating in the United States, but the AC system seemed to be favored by technology experts because of its ability to be "stepped up" to high voltage between cities and "stepped down" for distribution to users. DC required power stations every two miles, whereas AC allowed for long-distance transmission. AC was more dangerous than DC, as a shock could easily kill. Edison launched a campaign against AC, which was dismissed when Westinghouse successfully used Tesla's system to light the World Columbian Exposition, Chicago's world fair of 1893. This success allowed Tesla to pursue his childhood dream of harnessing the power of Niagara Falls. With funding support from J. P. Morgan, the Vanderbilts, and others, an electric capacity of 100,000 horsepower (75 megawatts) was generated and the power transmitted to Buffalo, New York, 20 miles away.

By 1891, when he became a U.S. citizen, Tesla was experiencing great success and rapidly turning out patents and products, such as the induction motor, new types of generators and transformers, AC power transmitters, and a new type of steam turbine. He also experimented with a wireless broadcasting tower and turbines, but much remained only in notebooks, given Tesla's lack of funds. He developed a basic design for a radio in 1892 and patented a radio-controlled robot boat in 1898. Guglielmo Marconi claimed the first patents for radio, but a U.S. Supreme Court decision in 1943 recognized Tesla as the true creator, and Marconi's patents were deemed invalid.

Tesla was tall and thin; he lived among members of high society when he could afford it. He believed that inventors should never marry, yet he was linked romantically to several heiresses. His phobias also prevented intimacy: He could not touch a person's hair and had a strong aversion to earrings, a germ phobia, and numerical fixations. He had few close friends, one of whom was author Mark Twain (Samuel Langhorn Clemens). Tesla's last days were spent at the Hotel New Yorker (paid for by Westinghouse), where he conducted various electrical experiments and befriended pigeons. He died of heart failure on January 7, 1943, with more than 700 patents to his name.

—*Stacy Vynne*

Radio sets became the first mass-produced consumer electronics good in the 1920s. (Wikimedia Commons)

Further Reading
Bailey, Ronald. "The Wizard Who Electrified the World." *American History*, June 2010.
Carlson, W. Bernard. "Inventor of Dreams." *Scientific American*, vol. 292, no. 3, 2005.
Carlson, W. Bernard. *Tesla: Inventor of the Electrical Age*. Princeton UP, 2013.
Jatras, Stella L. "The Genius of Nikola Tesla." *The New American*, vol. 19, no. 15, 2003.

Martin, Thomas Commerford. *The Inventions, Researches and Writings of Nikola Tesla.* Outlook, 2020.

Public Broadcasting Service. "Master of Lightning." http://www.pbs.org/tesla/.

Tesla Memorial Society of New York. "Tesla's Biography." http://www.teslasociety.com.

Tesla, Nikola, and Ben Johnston. *My Inventions: The Autobiography of Nikola Tesla.* Arcturus Publishing, 2019.

THERMOCHEMISTRY

FIELDS OF STUDY
Chemical Engineering; Physical Chemistry; Thermodynamics

ABSTRACT
The study of the heat and energy associated with physical transformations and chemical reactions, thermochemistry focuses on the principles and processes involving calories, enthalpy, entropy, the heat of combustion, the heat of formation, and heat capacity.

KEY CONCEPTS
calorimeter: a laboratory device used to measure changes of heat energy

heat of combustion: the specific amount of heat energy released when a material is completely combusted with oxygen

heat of formation: the specific heat (enthalpy) change in the formation of a specific material from its component atoms, with both reactants and products being in their standard states

phase change: the change of a material in conversions from liquid to gas, gas to liquid, liquid to solid, solid to liquid, solid to gas, or gas to solid

thermochemistry: the study of chemical processes relative to heat

BACKGROUND
Thermochemistry is the study of the heat and energy associated with physical transformations and chemical reactions. Thermochemistry studies calories, enthalpy, entropy, the heat of combustion, the heat of formation, and heat capacity. In particular, thermochemistry is founded on two laws: Antoine-Laurent de Lavoisier and Pierre-Simon Laplace's law, formulated in 1780, states that the energy change accompanying any transformation is equal and opposite to the energy change accompanying the reverse of that process. Germain Henri Hess's law, formulated in 1840, says that the energy change accompanying any transformation is the same, regardless of the number of steps involved in the process of the transformation. These laws later influenced the formulation of the first law of thermodynamics (by Rudolf Clausius), in

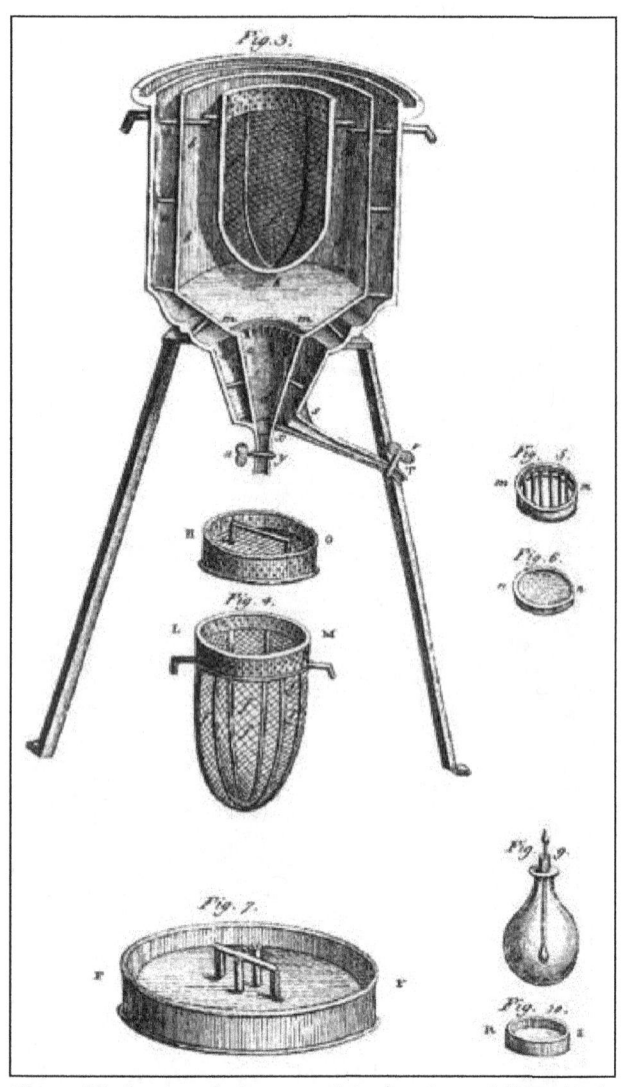

The world's first ice-calorimeter, used in the winter of 1782–83, by Antoine Lavoisier and Pierre-Simon Laplace, to determine the heat evolved in various chemical changes; calculations which were based on Joseph Black's prior discovery of latent heat. These experiments mark the foundation of thermochemistry. (Wikimedia Commons)

1850, which stated that the internal energy of a system is equal to the heat supplied to the system minus the work done by the system.

Thermochemistry studies systems, such as specific chemical processes or objects. A system refers to the part of the world that is being studied, with everything else considered as its environment. A system may be closed, in which case it can exchange energy with its environment, not matter; it may be open, in which case both matter and energy can be exchanged; or it may be isolated, meaning that neither energy nor matter may be exchanged. Processes may be described as adiabatic (without heat exchange), isobaric (without pressure change), or isothermal (without temperature change).

Thermochemistry pays particular attention to the energy exchanges of phase changes, chemical reactions, and the formation of solutions. It studies four major state functions: internal energy (the total energy contained by a system and necessary to create the system), enthalpy (internal energy plus the energy required to make room for a system by displacing its environment), entropy (a measure of energy dispersal at a specific temperature), and Gibbs free energy (the work obtainable from the system).

ENDOTHERMIC AND EXOTHERMIC PROCESSES

The two types of reactions studied by thermochemistry are endothermic and exothermic.

Endothermic reactions absorb heat from the surroundings, as occurs when a chemical cold pack is activated (by breaking a seal separating two chemicals, such as ammonium chloride with water) to initiate a chemical reaction that absorbs heat in order to turn it into chemical bond energy. Photosynthesis is another endothermic process, whereby plants create chemical energy (in the form of carbohydrates) by exposure to carbon dioxide, light, and water.

Perhaps the best-known endothermic process is the evaporation of water. Because a molecule must have sufficient kinetic energy to overcome liquid-phase intermolecular forces and kinetic energy is proportional to temperature, the warmest molecules evaporate first, which lowers the average temperature of the molecules remaining in a process called evaporative cooling. Humans sweat because it cools us off: Heat from the body's surface is transferred to the sweat, which then "carries it away" by evaporating.

Exothermic processes release heat from a system. The burning of a candle and the combustion of fuels such as wood, coal, and petroleum are obvious exothermic processes. The condensation of water vapor is also an exothermic process, as is the setting of cement, concrete, and epoxy. In chemical reactions, heat released by an exothermic reaction is usually absorbed in the form of electromagnetic energy, and as kinetic energy is lost via reacting electrons, light is released.

CALORIMETRY

Calorimetry is the process of measuring heat changes. Modern thermochemists use the differential scanning calorimeter, which measures the amount of heat required to increase the temperature of a sample being tested and a reference with a well-defined heat capacity, as a function of temperature. As the sample goes through phase transitions and other physical transformations, either more or less heat will be required to keep it at the same temperature as the reference material, depending on whether the process is exothermic or endothermic. A sample undergoing exothermic processes will require less heat, whereas a sample undergoing endothermic processes will require more. The difference in heat flow is observed while maintaining the sample and the reference at the same temperature. Differential scanning calorimetry is widely used in industry as a way to test the purity of sample materials.

—*Bill Kte'pi*

Further Reading

Brown, Robert C., ed. *Thermochemical Processing of Biomass: Conversion into Fuels, Chemicals and Power.* 2nd ed. Wiley, 2011.

Crocker, Mark, ed. *Thermochemical Conversion of Biomass to Liquid Fuels and Chemicals.* RSC Publishing, 2010.

Lele, Armand Fopah. *A Thermochemical Heat Storage System for Households: Combined Investigations of Thermal Transfers Coupled to Chemical Reactions.* Springer, 2016.

McCray, Tama. *An Introduction to Thermochemistry.* White Word Publications, 2012.

Mirskiy, Anton G. *Thermochemistry and Advances in Chemistry Research.* Nova Science Publishers, 2009.

Perrot, Pierre. *A to Z of Thermodynamics.* Oxford UP, 1998.

Rosendahl, Lasse, ed. *Direct Thermochemical Liquefaction for Energy Applications*. Woodhead Publishing, 2018.

Stacy, Angelica M., Janice A. Coonrod, and Jennifer Claesgens. *Fire: Energy and Thermochemistry*. Key Curriculum Press, 2005.

THERMODYNAMICS AND ENERGY

FIELDS OF STUDY
Physical Chemistry; Physics; Thermodynamics

ABSTRACT
Thermodynamics means movement of heat, the principles of which are at the foundation of some very important processes in the universe, from the movement and formation of stars and planets to the conversion of the food we eat into muscular exertions that allow us to live and work.

KEY CONCEPTS
absolute zero: the temperature at which no energy exists and from which no energy can flow

closed system: any system that has not input of energy or other resource from any external source and so is entirely self-contained

enthalpy: the energy of a system expressed as heat

entropy: the energy associated with the dissipation of energy from a state of higher energy intensity to one of lower energy intensity

free energy: the amount of energy within a system that is available to be converted to work

negentropy: the creation of order or increased energy density within a system

open system: any system that can receive an input of energy or other resource from any external source and so is not entirely self-contained

thermodynamics: the movement of energy within a system; the term originally referred to the movement of heat within a system

THE FOUR LAWS OF THERMODYNAMICS
There are four laws of thermodynamics: the first and second are often called, respectively, "the conservation and entropy laws"; the third law of thermodynamics pertains to the minimum or absence of entropy; and the fourth law—which is formally called the "zeroth law," since it is the most fundamental yet was discovered after the first three—explains the thermal equivalence between systems. Having withstood scientific scrutiny, the laws of thermodynamics and their myriad applications and related studies inform us about the evolution of systems based on energy flow.

THE ZEROTH LAW
The zeroth law of thermodynamics was formulated by Ralph Fowler in 1931. The law states a transitive property between systems in thermal equilibrium: If system A and system B are in thermal equilibrium with each other, and system A is in thermal equilibrium with system C, then systems B and C are also in thermal equilibrium. This thermal equivalence between systems means that if they come in contact, there will be no net exchange of energy between them. The heat contents of the systems in thermal equilibrium are therefore the same; the value on a thermometer reflects this equilibrium point.

THE FIRST LAW
The first law of thermodynamics—again, along with the second and third laws, found earlier than the zeroth law—covers the conservation of energy. It was formulated independently by Julius Robert von Mayer, James Joule, and Hermann von Helmholtz in the 1840s. The law, through various statements, concludes that the sum of all energy in the known universe is fixed and can be neither created nor destroyed. Energy can take different forms, such as heat, light, and also matter, and it can be transformed, such as by combusting fossil fuels to generate electricity, but there is no net loss or gain of total energy in the universe or, on a smaller scale, a system and its surroundings.

THE SECOND LAW
The second law of thermodynamics, formulated by Rudolf Clausius in the 1850s, governs the transformation of energy among its various forms and the corresponding heat dispersion. The law states that as energy is used in work, the amount of available, or free, energy decreases as the energy intensity is degraded and heat is produced. (The first law of thermodynamics states that the energy does not disappear, so the sum total of energy remains the same.) Clausius used the term *entropy* for the value of

the degradation, or disorder, that occurs. As free/available energy is transformed, there is a change from order toward disorder in the system; for energy, it is the decrease in the amount of energy available to perform work as heat flows from the system into its surroundings. For example, as the energy in coal is transformed into electricity, some of the energy in coal is transformed into a less intense state, heat exhaust that warms the environment.

The second law also reveals the irreversibility of energy flow and, even more abstract, of time. Energy dissipates from high-potential to low-potential states, but the reverse does not occur—the amount of potential work that can be completed by an energy source does not increase as it is used. This occurs whenever there is a thermal gradient in the system or between the system and its surroundings.

Steam escapes from a kettle of boiling water into the cooler air surrounding the stove, yet the steam does not contain the same energy intensity as the water, nor does the steam move from the cooler air back into the warmer water. Time can be measured by assessing the amount of entropy; the presence of more entropy in the system and its surroundings indicates a progression in time. Yet entropy does not increase steadily, as specific events, such as a volcanic eruption, can alter the pace of entropy's change.

THE THIRD LAW

The third law of thermodynamics, formulated by Walther Nernst in 1906, states that entropy is present at every heat level above absolute zero. (Absolute zero is minus 273.15°C, or minus 459.67°F, or 0 Kelvin.) However, it is impossible for a system, such as a machine, to function at absolute zero, as there would be no energy flow; the heat produced will always be above absolute zero. Conversely, maximum entropy, occurring when there is no more free (available) energy, is a situation in which there is no thermal gradient, as the amount of heat has dispersed evenly throughout the entire system and its surroundings. This is also a point when the system would cease to function; that is, additional work could not be done.

SYSTEM CONTEXTS

Thermodynamics operate differently based on the type of the system in which energy is flowing. In a closed system, such as in the universe as a whole, or in a smaller, isolated system, the amount of energy is indeed fixed. However, in an open system, such as Earth within the solar system, the amount of energy is not fixed, since radiation from the sun is (for Earth) an external source of free energy. As long as the sun continues to furnish additional energy to the system on Earth, order (negentropy) can be created, and maximum entropy—and thus thermal equilibrium—will not be reached. The total entropy of the entire environment encompassing Earth's system—that is, the system and its surroundings—will increase as energy dissipates, but the additional energy entering the system from its surroundings prevents equilibrium from being reached in the system. The study of nonequilibrium thermodynamics, the processes in open systems and closed systems not at equilibrium, was first formulated by Edwin Schrödinger in 1944.

The study of thermodynamics has been extended to living and nonliving entities, societies and other systems, and to the universe. The study of thermodynamics also occurs as three main branches: classical, which analyzes the macroenvironment; chemical, which analyzes energy flow in chemical reactions; and, together with our increasing understanding of atoms and molecules, molecular or statistical thermodynamics, which analyzes microlevel behavior.

Bioenergetics, including the study of human metabolism, has been an active area of research. Alfred Lotka's law of maximum energy—that the evolutionary success of organisms depends on the amount of useful energy harnessed—rests on thermodynamic principles, as does Charles Elton's pyramid of numbers, which explains the relationship between the quantities of different organisms in an ecosystem. Moreover, Albert Einstein's equation $E = mc^2$, equating mass with energy, may be the best-known application of thermodynamics.

—*Kirk S. Lawrence*

Further Reading

Atkins, Peter. *Four Laws That Drive the Universe.* Oxford UP, 2007.

Ben-Naim, Arieh. *The Four Laws That Do Not Drive the Universe: Elements of Thermodynamics for the Curious and Intelligent.* World Scientific, 2018.

Chaisson, Eric J. *Cosmic Evolution: The Rise of Complexity in Nature.* Harvard UP, 2001.

Müller, Ingo. *A History of Thermodynamics: The Doctrine of Energy and Entropy.* Springer, 2007.
Smil, Vaclav. *Energy in Nature and Society: General Energetics of Complex Systems.* MIT Press, 2008.
Tosun, Ismail. *Thermodynamics Principles and Applications.* World Scientific, 2015.

THOMPSON, BENJAMIN (COUNT RUMFORD)

FIELDS OF STUDY
Materials Science; Physics; Thermodynamics

ABSTRACT
Benjamin Thompson, also known as Count Rumford, was a British officer, Bavarian politician, physicist, and inventor. He had a significant impact on the development of thermodynamics and was one of the most significant physicists of the late eighteenth century.

KEY CONCEPTS
caloric theory: the notion that "heat" was contained in materials as a kind of invisible fluid that could only be experienced when it was released from within the material
friction: the interaction of two surfaces in motion relative to each other; the interaction generates heat in accord with the degree of interaction
kinetic theory: the notion that "heat" is produced by motion
thermodynamics: the study of the active nature and movement of heat energy

BACKGROUND
Benjamin Thompson was born March 26, 1753, in Woburn, Massachusetts. As a young man, he developed a great interest in mathematics and science and experimented with fireworks and the nature of heat. During the American Revolution, he joined the British army as a spy and left for London in 1776. There he joined the Royal Society in 1779 and published his observations concerning the speed of artillery shells and the explosive power of gunpowder. From 1781 to 1783, Thompson lived in America for the last time, as a lieutenant colonel. Back in Britain, he was promoted to the rank of colonel, but he saw no future for himself in a military career in London after the American Revolutionary War. At first, Thompson wanted to build a career in Vienna, but during his journey there he became acquainted with a relative of Bavaria's elector, Karl Theodor. He decided to go back to England, where he was knighted by King George III, who allowed him to enter the elector's service. In Munich, Thompson became a member of the Bavarian Academy of Sciences and was made a grand chamberlain in 1785 and a privy councilor in 1787. His first major task was the reorganization of the Bavarian army.

In this position, he also became a social reformer: Because soldiers were malnourished and poorly dressed, he had potato fields cultivated for better food supply, creating the nutritional and inexpensive "Rumford soup," and experimented with the thermal conductivity of textiles such as wool and cotton to create appropriate uniforms. Moreover, Thompson built almshouses, workhouses, and factories using his scientific knowledge to improve the living situation of Munich's citizens. In 1792, he was made a count

Benjamin Thompson (Count Rumford). (Wikimedia Commons)

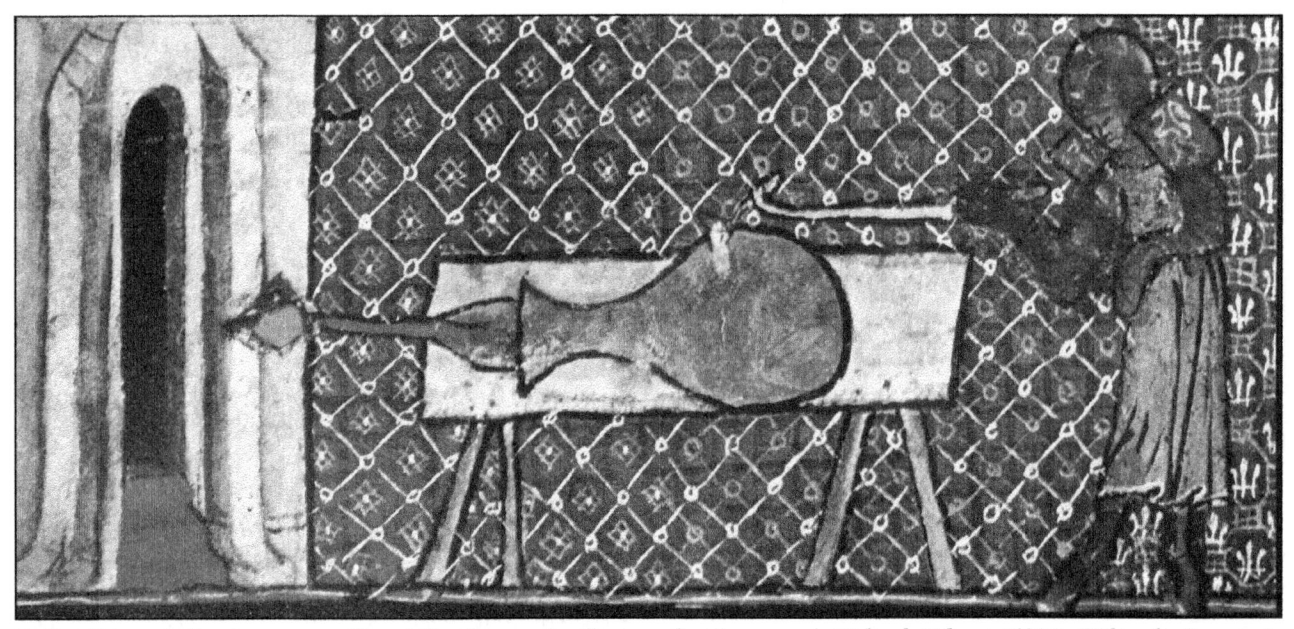

Earliest picture of a European cannon, De nobilitatibus, sapientiis et prudentiis regum, *by Walter de Milemete (1326). (Wikimedia Commons)*

(*Reichsgraf*) of the Holy Roman Empire and named himself after the town of Rumford, where he had been married.

A MAN OF SCIENCE

Thompson was one of the most significant pioneers of thermodynamics and the modern theory of heat. Following his early experiments with explosives, he became interested in the general nature of heat. He constructed lamps, chimneys, and the famous Rumford stove, an energy-saving and thermally efficient kitchen range that consumed only half as much fuel as an ordinary open fireplace. He published his findings on the improvement of the open fireplace in his popular essay "On Chimney Fireplaces" in 1796.

THERMODYNAMICS AND THE MODERN THEORY OF HEAT

In the late eighteenth century, the kinetic and the caloric theories of heat were being discussed, and Thompson's research would help elucidate heat theory, or the study of thermodynamics. During his most famous experiments in Munich's armory in 1797, he observed the heat produced while drilling out cannons. He discovered that they remained hot as long as the friction of drilling continued and that the heat could even melt the cannon.

This contradicted the theory of the French chemist Antoine-Laurent de Lavoisier, whose caloric theory was widely accepted at the time. Lavoisier thought of a thermal element called *caloricum*, an invisible liquid form of matter. Thompson's experiments showed that the amount of heat produced during the drilling stayed the same. He discovered that such a big amount of heat could not be contained in the metal of the cannon tube. Thompson stated that mechanical friction is an inexhaustible source of heat.

This disproved the theory that the heat stored in the cannon tube that had been released through the drilling should have been depleted eventually. He concluded from his results that heat was not some sort of matter but was rather a form of kinetic energy, the basis for the modern theory of heat as a form of energy and the law of the conservation of energy, although Thompson's calculations for the mechanical equivalent of heat were not as accurate as those by James Joule decades later. In 1798, Thompson presented his findings to the Royal Society and published them in *An Experimental Inquiry Concerning the Source of the Heat Which Is Excited by Friction*.

With regard to the nature of heat, Thompson also concluded that gases could not conduct heat, an argument he later extended to liquids. These bold ideas

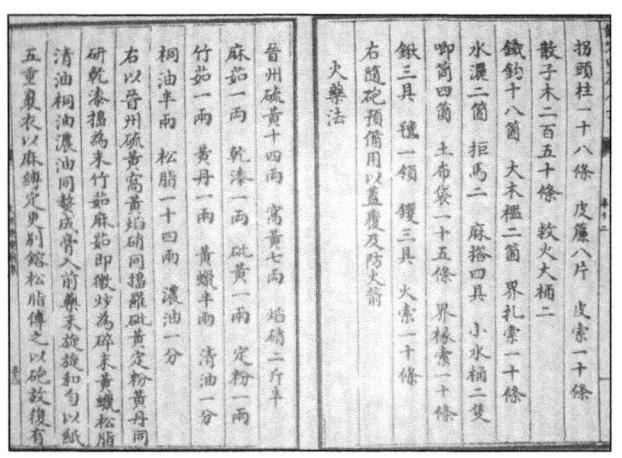

Earliest known written formula for gunpowder, from the Wujing Zongyao of 1044 CE. (Wikimedia Commons)

were attacked by various scientists of his time and were soon proved wrong.

HONORS AND AWARDS
Thompson received the Royal Society's Copley Medal in 1792. He endowed the Rumford Medal of the Royal Society (the first honoree of which he became in 1800 "for his various discoveries respecting heat and light") and the Rumford Prize of the American Academy of Arts and Sciences. He also encouraged the foundation of Great Britain's Royal Institution in 1799, an association for research and education, the first director of which was the chemist Sir Humphry Davy.

In 1804, Thompson moved to Paris, where he joined the Institut de France. In 1805, after his first wife had died in 1792, he married Lavoisier's widow, Marie Anne, but they were divorced four years later. He moved to Auteuil, today a district of Paris, where he dedicated himself to further research and inventions, among them a drip coffee pot. He died on August 21, 1814, in Auteuil, having endowed a part of his estate to Harvard College, where the Rumford Professorship was established in 1816.

—*Sophie Gerber*

Further Reading
Brown, George I. *Count Rumford: The Extraordinary Life of a Scientific Genius; Scientist, Soldier, Statesman, Spy.* Sutton, 1999.
Graf von Rumford, Benjamin. *The Complete Works of Count Rumford.* HardPress, 2020.
Sparrow, W.J. *Knight of the White Eagle: A Biography of Sir Benjamin Thompson, Count Rumford, 1753-1814.* Hutchinson, 1964.
Thompson, Benjamin, of Rumford. *Collected Works of Count Rumford.* 5 vols. Edited by Sanborn C. Brown. Harvard UP, 1968-1970.

THOMSON, JOSEPH JOHN

FIELDS OF STUDY
Analytical Chemistry; Electromagnetism; Electronics; Physics

ABSTRACT
British physicist Joseph John Thomson was the Cavendish Professor of Experimental Physics at Cambridge and a Nobel laureate in physics noted for his interest in the discharge of electricity and atomic structure. In 1897, he identified electrons and protons.

KEY CONCEPTS
canal rays: streams of positively charged particles produced as a byproduct of the formation of cathode rays, therefore representing the massive, positively charged remnants of the atoms that had emitted the cathode ray particles
cathode rays: streams of negatively charged particles-electrons-that have been forcibly ejected from neutral atoms through electrical stimulation of those atoms
magnetic field: a three-dimensional force field that nevertheless has a definite directional component surrounding a magnet
"plum pudding" model: a nineteenth-century model of atomic structure in which the structure of the atom is analogous to that of a "plum pudding" (or perhaps a raisin-bran muffin), with the newly discovered electrons scattered randomly throughout the whole of the atom

BACKGROUND
British physicist Joseph John Thomson, commonly known as J.J. Thomson, was born near Manchester, England, on December 18, 1856. He became the Cavendish Professor of Experimental Physics at Cambridge in 1884. Thomson's research centered on the

conduct of electric charges in gases as well as atomic structure. His work on atomic structure led to his 1897 discovery of the electron, which demonstrated for the first time that the atom was not the fundamental building block of matter, but was instead composed of smaller subatomic particles. His work was widely published, and he received numerous awards, including the Nobel Prize in Physics in 1906.

Thomson's family wished him to pursue engineering, but his father's death when he was 16 forced him to abandon an engineering apprenticeship and seek another career path. Thomson was educated at Owens College, Manchester, and Trinity College, Cambridge, where he received a Master's degree in mathematics in 1880. He served on the faculty of Trinity and was named the Cavendish Professor of Experimental Physics at Cambridge in 1884; he would achieve the rank of Master in 1918. Thomson married Rose Elisabeth Paget in 1890. The couple had one son, George Paget Thomson, and one daughter, Joan Paget Thomson.

SCIENTIFIC RESEARCHES

In his role as Cavendish Professor, Thomson supervised the Cavendish Laboratory's experiments in such fields as electromagnetism and atomic structure. Although he resigned the position in 1918, he remained affiliated with Cambridge and his noted student Ernest Rutherford throughout his career. Rutherford succeeded him as Cavendish Professor. Thomson himself researched the discharge of electricity in gases and developed his own theory of atomic structure. In the late nineteenth century, Thomson began studying electric discharge in empty glass cathode-ray tubes in an effort to resolve scientific questions regarding the nature of cathode rays. He was able to measure the bending of cathode rays in the presence of magnetic fields and their energy.

In 1897, Thomson recognized that cathode rays were formed of streams of small, negatively charged subatomic particles, which he initially named corpuscles but which quickly came to be known as electrons. He also correctly theorized that electrons are one of the atom's basic material components, although he incorrectly theorized that electrons are the atom's only material component. In the production of cathode rays, a second type of "ray" called "canal rays" is produced. Thomson reasoned, quite logically, that if the cathode rays were composed of negatively charged particles emitted from neutral atoms, then the canal rays must consist of the remaining positively charged portion of those atoms. Canal ray particles were indeed determined to be positively charged, but the behavior of canal ray particles indicated they were much more massive than cathode ray particles and therefore must represent the major mass of the atoms. Hence, they were called "protons," for their primary role in the mass of the atom. Thomson first publicly presented his findings before the Royal Institution on April 30, 1897.

His findings proved revolutionary: He was the first to propose that the atom itself is not the basic unit of matter, as previously believed. Years of additional research and experimentation by a variety of scientists

Joseph John Thomson. (Wikimedia Commons)

tested his initial theories. Thomson next turned his attention to determining how the particles he had discovered were incorporated into the structure of an atom. The Thomson atomic model, also known as the "plum pudding model," theorized that an atom consisted of a positively charged cloud or sphere filled with negatively charged electrons. Cathode rays thus consisted of electron "raisins" that had been forcibly ejected from within the "plum pudding" atoms.

Thomson was author or coauthor of numerous important scientific publications, including *A Treatise on the Motion of Vortex Rings* (1883), *Applications of Dynamics to Physics and Chemistry* (1888), *Lectures on Properties of Matter* (1890), *Elements of the Mathematical Theory of Electricity and Magnetism* (1895), *The Discharge of Electricity Through Gases*(1898), *Conduction of Electricity Through Gases* (1903), *Structure of Light* (1925), *The Corpuscular Theory of Matter* (1907), *Rays of Positive Electricity and Their Application to Chemical Analyses* (1913), *The Electron in Chemistry* (1923), *Notes on Recent Researches in Electricity and Magnetism* (1893), and his autobiography, *Recollections and Reflections* (1936). He travelled to the United States on two occasions: to deliver lecture series at Princeton University in 1896 and Yale University in 1904.

AWARDS AND HONORS
Thomson was elected a fellow and president of the Royal Society and president of the British Association for the Advancement of Science. He also held honorary doctorates from numerous U.S. and European universities. His numerous awards included the Order of Merit (1912), the Hughes (1902) and Copley (1914) Medals, the Smithsonian Institution's Hodgkins Medal (1902), and the Institute of Civil Engineers' Faraday Medal (1938). He was knighted in 1908, and his most notable award, the 1906 Nobel Prize in Physics, was awarded for his work regarding the discharge of electricity in gases. Thomson's son, Sir George Paget Thomson, also became a distinguished scientist and professor, coauthoring works with his father and winning his own Nobel Prize in Physics in 1937.

John Joseph Thomson died on August 30, 1940. Later scientific research would both build upon and disprove his theories. His student Rutherford developed a different atomic model in 1911, proposing that the atom consisted of a positively charged nucleus circled by negatively charged electrons. Subsequent research would lead to the introduction of newer atomic models, as well as to the discovery that subatomic elements such as protons and neutrons were formed by a number of additional fundamental particles, including photons, muons, and quarks. Albert Einstein's quantum theory showed that electrons could act as either particles or waves, under different conditions.

Understanding of atomic structure was key to the future generation of nuclear energy and its applications not only in war but also for peaceful generation of power. Other advances based on Thomson's work include the development of the mass spectrograph, the ability to use positive rays to separate atoms and molecules, and the discovery of isotopes. His work also opened important avenues of research and development that led to modern advances in communications and computation, including the invention of the television and the computer.

—*Marcella Bush*

Further Reading
Navarro, Jaume. *A History of the Electron. J.J. and G.P. Thomson*. Cambridge UP, 2012.
Staley, Kent W. "The Discovery of the Electron." American Institute of Physics. http://www.aip.org/history/electron/jjhome.htm.
Thomson, George Paget, Sir. *J.J. Thomson and the Cavendish Laboratory*. Nelson, 1964.
Thomson, J.J. "Award Ceremony Speech." *Nobel Lectures: Physics, 1901-1921*. Elsevier, 1967. http://www.nobelprize.org/nobel_prizes/physics/laureates/1906/press.html.
Thomson, J.J. *Recollections and Reflections*. G. Bell, 1936.
Thomson, J.J. *Rays of Positive Electricity and Their Application to Chemical Analyses* Creative Media Partners LLC, 2018.

THOMSON, WILLIAM (LORD KELVIN)

FIELDS OF STUDY
Electrical Engineering; Physics; Thermodynamics

ABSTRACT
William Thomson was an Irish-born mathematician, physicist, and inventor who formulated the first two laws of thermodynamics, discovered absolute zero, and made important

contributions to the study of thermodynamics, mathematics, and electromagnetism.

KEY CONCEPTS
absolute zero: the temperature at which there is absolutely no energy present; the temperature at which the volume of an ideal gas, decreasing as temperature decreases, becomes non-existent

electrical conductivity: a measure of a material's ability to conduct electrical current, the inverse of electrical resistance, stated in Mhos (Ω^{-1})

electrical resistance: a measure of a material's requirement for work to be done by an electrical current passing through the material, stated in Ohms (Ω)

kinetic energy: the energy associated with a physical object in motion

CHILD PRODIGY
William Thomson, Lord Kelvin, was a physicist who made numerous, fundamental contributions to our understanding of heat, work, and energy. Moreover, he was instrumental in explaining the nature of electricity and magnetism. Thomson was born in Belfast, Ireland, on June 26, 1824, and died on December 17, 1907, in Ayrshire, Scotland. A child prodigy in mathematics, he had a knack for finding solutions to both practical and scientific problems. In 1846, he became the professor of natural philosophy at the University of Glasgow, a post that he acquired at the age of 22 and retained until he retired in 1899. The study of natural philosophy was the precursor of modern physics. Over the course of his lifetime, he published more than 650 papers, applied for 70 patents, and, through his textbook and work with students, helped to turn the study of physics into the modern scientific discipline that we know today.

DISCOVERIES AND INVENTIONS
Thomson believed that all forms of energy were interrelated, and he made great strides in formulating a unified theory of thermodynamics (the study of conversion of different forms of energy). Building on the work of Sadi Carnot, Thomson proved that absolute zero was equal to minus 273.15°C, which came to be denoted as 0 kelvin (with no degree sign), based on the absolute scale of temperature or thermodynamic tem-

William Thomson (Lord Kelvin). (Wikimedia Commons)

perature scale now named for him. He also proved that heat is equivalent to work. His name was also given to a number of his other discoveries and inventions that are important to our understanding and application of energy. These include the Thomson (thermoelectric) effect, the Kelvin-Holmholtz timescale (related to the physics of stars), the Kelvin water dropper (an electrostatic generator), the Kelvin bridge (a device used for measuring low levels of electrical resistance), the Kelvin circulation theorem (in fluid mechanics), and the Kelvin (Ampere) balance.

Thomson discovered an accurate method of measuring electrical conductivity or electrical resistance, including the quadrant electrometer for measuring static electricity. He helped to pioneer the study of electromagnetism, helped lay the first transatlantic submarine cable, and developed myriad maritime navigation instruments, including an adjustable compass and a pressure-recording device for taking depth soundings while a ship was moving—a device still used today. He coined the term 'kinetic energy' and was the first person to teach experimental science, whereby students do experiments as a means of learning. While not named after him, Stokes's theorem, which is a proof in vector calculus, was also Thomson's work—one of his many contributions to the study of mathematics and the presentation of mathematical analyses of physics-related processes, including electromagnetic phenomena. In addition, along with Peter Guthrie Tait, Thomson wrote the multivolume *Treatise on Natural Philosophy* (1867), the first modern textbook on physics.

GREAT ADVANCES IN SCIENCE

Over the course of his career, Thomson exchanged ideas and was friends with many of the great scientists of his day, including James Clerk Maxwell, Hermann von Helmholtz, Michael Faraday, James Joule, and the mathematician George Gabriel Stokes. The free exchange of ideas between these men contributed to many of the great advances made in science throughout the second half of the nineteenth century. For example, Thomson was inspired by Faraday's work on electromagnetism, and in turn Maxwell was inspired by Thomson's work in the same field. Thomson and Joule often collaborated, with Thomson offering critiques of Joule's work. Their collaboration led to the acceptance of the theory of kinetic energy. While Thomson is credited with establishing the groundwork for the study of thermodynamics and for composing the first two laws of thermodynamics, he readily admitted that he built upon the work of other researchers. He credited the first law of thermodynamics (the conservation of energy) to Joule and the second law (thermodynamic entropy) to both Sadi Carnot and Rudolf Clausius. A well-known figure in both England and America, Thomson travelled and lectured in the United States. He headed the commission that designed Niagara Falls hydroelectric power station and was instrumental in having alternating current (AC) selected as the standard method for power transmission.

While Thomson made many great strides in advancing our understanding of physics, he also made some mistakes. For example, using calculations based on energy conservation and heat loss, he calculated Earth to be between 20 million and 400 million years old. (It is now believed that Earth is approximately 4.54 billion years old.) Although his calculations were accurate, the basic assumptions on which he based his work were faulty, one being that Earth had consistent thermal properties throughout. Consequently, he obtained an incorrect result.

AWARDS AND HONORS

Thomson was the recipient of many awards and honors for his scientific work, including the Royal Medal (1856), the Keith Medal (1864), the Copley Medal (1883), the Gunning Victoria Jubilee Prize (1887), the Knight Grand Cross of the Victorian Order (1896), and the John Fritz Medal (1905). He was knighted by Queen Victoria in 1866 for his work in helping to lay the first undersea cable and for inventing two devices, the mirror galvanometer and siphon recorder, which helped to ensure a steady transmission rate for data via undersea cables. Thomson was also the first scientist to be elevated to the British peerage. In 1892, he became the first and only Baron Kelvin of Largs, thanks not only to his scientific endeavors but also to the contributions that he made to Britain's commercial enterprises. Although he married twice, he had no heirs; the title of Baron Kelvin died with Thomson.

—*Rochelle Caviness*

Further Reading

Collins, M.W., R.C. Dougal, C.S. König, and I.S. Ruddock, eds. *Kelvin, Thermodynamics and the Natural World*. WIT Press, 2016.

Flood, Raymond, Mark McCartney, and Andrew Whitaker, eds. *Kelvin: Life, Labours, and Legacy*. Oxford UP, 2008.

Gray, Andrew. *Lord Kelvin*. Outlook, 2020.

Lindley, David. *Degrees Kelvin: A Tale of Genius, Invention, and Tragedy*. Joseph Henry Press, 2004.

Thompson, Silvanus P. *The Life of William Thomson, Baron Kelvin of Largs*. 2 vols. 1910. Cambridge UP, 2011.

TIDAL POWER GENERATION

FIELDS OF STUDY

Electrical Engineering; Ecology; Marine Construction; Oceanography

ABSTRACT

Tidal power is the harnessing of power from the cyclical ebb and flow of the ocean's tides and is considered a renewable

energy resource. The use of tidal power as a means of generating electricity began in the middle of the twentieth century with the first large-scale construction of a tidal power plant. As a result of growing concerns over oil and coal availability alongside carbon dioxide and other greenhouse gas (GHG) emissions, there is currently growing interest in developing improved technologies and implanting schemes to harness tidal energy to produce electricity.

KEY CONCEPTS

barrage: a structure that bars and controls the tidal inflow and outflow of water in a river channel or other coastal intrusion

diurnal: describes a change that produces two states in one day, such as day and night

neap tide: the tidal change that occurs with the least difference in height between high and low tides

semidiurnal: describes a change that produces twice as many states in one day as a diurnal change, such as two periods of high and low tides in one day

spring tide: the tidal change that occurs with the greatest difference in height between high and low tides

TIDES AND TIDAL POWER GENERATION

The use of tidal power as a means of generating electricity began in the middle of the twentieth century with the first large-scale construction of a tidal power plant. As a result of growing concerns over oil and coal availability alongside carbon dioxide and other greenhouse gas (GHG) emissions, there is currently growing interest in developing improved technologies and implanting schemes to harness tidal energy to produce electricity.

Tidal power is a form of hydropower, water-based power, which harnesses the cyclical rising and falling of the Earth's tides. Tides result from a combination of Earth's rotation about its axis in conjunction with the gravitational attraction between the mass of seawater on the surface of Earth and the mass of the Sun and Moon. The tidal currents ebbing and flowing between high and low tides can be used by turbines to generate electricity. Tidal patterns globally fall within a cycle of 24 hours and 50 minutes. The predominant tidal pattern globally is known as semidiurnal, resulting in two periods of alternating high tide and low tide, with each period lasting 12 hours and 25 minutes within the general tidal pattern, about two per day. Tidal patterns may vary in frequency and intensity as a result of coastline features such as headlands and peninsulas, which can increase tidal current velocities, making such tides yield increased energy.

PREDICTABLE BUT VARIABLE

Of the various forms of renewable energies, including wind and solar energy, tidal power is considered one of the most predictable, although still variable. While tides occur regularly and thus are in that sense predictable, they will fluctuate during cycles of approximately 14 days between spring tide and neap tide ranges. Spring tide range is the maximum of the range of differences between high and low tide, while the neap tide is the minimum of the range of differences between high and low tide. The spring tide and neap tide ranges correspond to the relative location of the moon with respect to both Earth and Sun. During a given tidal period, there will be time when the energy generated from the tide will be zero and time when the energy will be at a peak. This variation, while predictable, will often not correspond with daily fluctuations in consumption of electrical power.

As a result, tidal power generation must also include strategies and technologies for storing electricity, which can then be tapped when energy derived from tides is at its minimum, and for buffering or smoothing out peaks, which pose potential problems when energy derived from the tides is at its maximum. Additionally, local climate and weather patterns, such as strong winds, will affect tidal variations and thus the potential amount of electricity that can be derived. There are two major categories of tidal power electricity generation: tidal barrage schemes and tidal current turbines.

HISTORY

The history of harnessing tidal power by humans dates back to the Middle Ages and even ancient Rome. Tide mills were specialized water mills driven by tidal changes. By using a dam with a sluice gate and a reservoir, rising tidal waters could flow into a reservoir through the gate. From there, water would be stored in the reservoir as the tide level receded. The sluice gate would close automatically, storing water in the reservoir. The stored water would then drive a waterwheel when released. Tidal barrage schemes operate in a similar fashion. The simplest tidal barrage

Sihwa Lake Tidal Power Station (South Korea) is the world's largest tidal power installation, with a total power output capacity of 254 MW. (Wikimedia Common)s

schemes employ large dams, similar to those used for hydroelectric power, and sluice gates that direct tidal water into a reservoir where it is stored.

When the tide falls, the water in the reservoir is released across turbines to produce electrical power through what is called "ebb generation." Such schemes are typically constructed across the span of bays, rivers, estuaries, and other semienclosed coastal water features that are subject to tides. Tidal barrage schemes are most effective where tidal amplitude is high. Ebb generation produces power only when the tide moves from high to low tide, in a single direction. Although two-way tidal power generation can produce energy from the tidal flow both into and out of the reservoir, it is costlier than single-direction systems. This is a result of necessarily more complex and robust turbines and support structures that are capable of generating power in two directions.

ENVIRONMENTAL CONCERNS

Several environmental concerns are associated with tidal barrage schemes. As with most large-scale construction projects that affect rivers and other waterways, such as dams, local ecosystems are affected by both the construction and operation of the barrage scheme. The Barrage-de-la-Rance scheme, built on the Rance River in France in 1967, was the first tidal power station built and took several years to construct. During this time, the river was cut off from the ocean, which led to the deaths of many species of animals and plants. After the completion of the Barrage-de-la-Rance scheme and reconnection of the Rance River to the ocean, the local ecology began to recover.

The current state of the La Rance basin suggests a recovered ecosystem. Operation of the tidal barrage scheme, however, produces additional concerns. Silting and accumulation of particles, including human-made chemicals such as fertilizers, can be exacerbated by the enclosed reservoir. The barrage itself makes movement of organisms between the ocean and the body of water enclosed by the reservoir and around the barrage difficult, particularly with respect to turbine motion. Because tidal barrage facilities span rivers and other semienclosed coastal waters, such geographies become unavailable for shipping, trade, and transportation use.

Barrage-de-la-Rance is currently the world's largest tidal power plant. The first North American tidal power plant, the Annapolis Royal Generating Station, was established in 1984 in the Bay of Fundy of Nova Scotia, Canada, and is the world's second-largest tidal power plant. Two other notable tidal barrage power plants include the Jiangxia Tidal Power Station, located in Leqing Bay, China, and the Kisalaya Guba power plant, in Russia at the Barents Sea.

TURBINES

The second major form of tidal power generation is derived from tidal current turbines, also called "tidal stream generators," in a way that is similar to the way that wind turbines harness wind to produce electricity. Because of the relatively smaller size of the turbines that operate without the need for water-impounding dams, tidal current turbines as a means of generating electricity from tidal power are more flexible, less expensive, and less ecologically damaging than tidal barrage systems. Turbines used to generate electricity from tidal currents are conventionally categorized by whether the axis of rotation of the turbine is horizontally or vertically mounted.

Vertical axis tidal current turbines always rotate in the same direction, regardless of the direction of the tidal stream or current itself. Those oriented along a horizontal axis of rotation, on the other hand, must be able to accommodate the reversal of the turbine blades to mirror the alternating direction of the tide. A distinction that can be made between tidal current turbines centers on whether the electrical energy is produced directly from the tidal water's contact with turbine blades or indirectly, through devices that employ the Venturi effect, which converts tidal flow into airflow that then turns a generator. Tidal current turbines are most effective in locations where fast currents occur, usually as a result of tidal waters that are concentrated between natural obstructions such as at the entrances to rivers and bays, around headlands, or between neighboring landmasses such as islands. Currently, several tidal current turbine farms are distributed throughout the world along the coastlines of Norway, Canada by Nova Scotia, South Korea, and China, among other locales. India, the United Kingdom, and the Philippines have proposed establishing tidal current turbine farms in the future.

Unlike tidal barrage schemes, which generate power only from large heads of water, tidal current turbines produce electricity from freely flowing currents or streams of water. Because of the reduced size and cost of turbines, multiple tidal current turbines can be more flexibly and strategically placed in high-velocity stream and current paths that would be impractical to harness for power through a tidal barrage scheme. Additionally, tidal current turbine farms do not impound water and are consequently unaffected by issues associated with the construction and operation of tidal barrage schemes. Moreover, tidal current turbines are much easier to establish and less capital intensive in initial installation, since there is no need to erect a water-impounding dam.

—*Josef Nguyen*

Further Reading

Charlier, Roger H., and Charles W. Finkl. *Ocean Energy: Tide and Tidal Power.* Springer, 2009. http://www.springerlink.com/content/x7u431/#section=30137&page=1.

Greaves, Deborah, and Gregorio Iglesias, eds. *Wave and Tidal Energy.* Wiley, 2018.

Hodge, B.K. "Ocean Energy." *Alternative Energy Systems and Applications.* John Wiley and Sons, 2010.

Khare, Vicas, Cheshta Khare, Savita Nema, and Prashant Baredar. *Tidal Energy Systems, Design, Optimization and Control.* Elsevier, 2019.

Lyatkher, Victor. *Tidal Power: Harnessing Energy from Water Currents.* Scrivener, 2014.

Millar, Dean L. "Wave and Tidal Power." *Energy ... Beyond Oil*, edited by Fraser Armstrong and Katherine Blundell. Oxford UP, 2007.

Units of Measurement

FIELDS OF STUDY
General Sciences; Metrology; Physics

ABSTRACT
Units of measure are standardized forms of measurement of quantities such as mass or weight, time, length or distance, and other quantities. There are many systems of measurement, including the British Gravitational System (often called the "Imperial System") and the International System of Units (or SI system, after its French initials). Other systems of measurement, including derived units, also exist.

KEY CONCEPTS
ampere: the flow of about 6.28 X 1018 electrons, or one coulomb of charge, per second in a conductor

base unit: a unit of measurement that is not constructed or derived from other units of measurement

candela: a measurement of the light intensity of a light source directly, rather than the amount of light at a distance from that source

derived unit: a unit of measurement that is a composite of other units of measurement

kelvin: the unit of measurement for temperature; one kelvin has the same magnitude as one Celsius degree, but the kelvin temperature scale starts at absolute zero rather than the freezing point of water such that 0°C = 273.15 k

metric: refers to the measurement of properties relative to appropriate standard and defined units of measurement

mole: an absolute measure of the amount of matter present based on the number and type of atoms in the matter, based on the rule that one mole of any pure substance contains exactly the same number of atoms or molecules as one mole of any other pure substance, that number being approximately 6.022597×10^{23}, amounting to the weight in grams equivalent to the molecular weight of the material; for example, 238 g of uranium-238 contains exactly the same number of atoms as 1 g of hydrogen atoms or 12 g of carbon-12 atoms

supplementary unit: dimensionless units for the measurement of solid and plane angles; the term is now obsolete as both supplementary units have been formalized as derived SI units

HOW BIG, HOW MANY, ETC.
Units of measure are standardized forms of measurement of quantities such as mass or weight, time, length or distance, and other quantities. Some examples include the kilogram or pound (both units of mass), the second or minute (units of time), and the meter, kilometer, inch, and mile (all measures of distance).

SI UNITS
All systems of weights and measures, metric and nonmetric, are linked through a network of international agreements supporting the International System of Units. The International System is also called the "SI," using the first two initials of the system's French name, *Système International d'Unités*. The key agreement is the Treaty of the Meter (*Convention du Mètre*), signed in Paris on May 20, 1875. The United States is a charter member of this metric club, having signed the original document, although the United States persists in using nonmetric units to the present day.

At the heart of the SI is a short list of base units defined in an absolute way without referring to any other units. The base units are consistent with the part of the metric system called the "MKS system." The MKS system is so named because it uses the meter, kilogram, and second as base units. The SI unit is fully consistent, and there is only one recognized unit for each physical quantity (variable).

Three types of units are used: base units, supplementary units, and derived units. Base units are dependent on accepted standards. In all, there are seven SI base units: the meter (which measures distance, or length), the kilogram (which measures mass), the second (which measures time), the ampere (which measures electric current), the kelvin (which measures temperature), the mole (which measures the amount of a substance), and the candela (which measures light intensity).

Supplementary units included the radian and the steradian when the International System was introduced, but because the plane angle was expressed as a ratio between two lengths and a solid angle as the ratio between an area and the square or a length, the International System came to recognize the radian (plane angle) and the steradian (solid angle) as dimensionless derived quantities, and in 1995 supplementary units as a class of SI units was removed, leaving only two classes: the base and the derived units.

Derived units are defined algebraically in terms of these fundamental units. Currently, there are 22 derived SI units. For example, the SI unit of force, the newton, is defined to be the force that accelerates a mass of 1 kilogram at the rate of 1 meter per second per second. This means the newton is equal to 1 kilogram meter per second squared, so the algebraic relationship is N = kg × m × s².

Each SI unit is represented by a symbol, not an abbreviation. These symbols are the same in every language of the world. However, the names of the units themselves vary in spelling according to national conventions.

Base Units of the International System of Units (SI System)

Quantity	Name	Unit Symbol
Amount	mole	mol
Electric current	ampere	A
Length/distance	meter	m
Luminous intensity	candela	cd
Mass	kilogram	kg
Thermodynamic temperature	kelvin	k
Time	second	s

ENGLISH UNITS

In the system of British or English units known as the Imperial System, short-distance units are based on the dimensions of the human body. The inch represents the width of a thumb. The foot (12 inches) was originally the length of a human foot, although 12 inches is now shorter than most people's feet. The yard (3 feet) got its start in England as the name of a 3-foot measuring stick, but it is also understood to be the distance from the tip of the nose to the end of the middle finger of the outstretched hand. Finally, if one's arms are stretched out to the sides as far as possible, the total "arm span," from one fingertip to the other, is a fathom (6 feet).

Land was traditionally measured by the gyrd or rod, an old Saxon unit probably equal to 20 "natural feet." In the Saxon land-measuring system, 40 rods make a furlong (*fuhrlang*), the length of the traditional furrow (*fuhr*) as plowed by ox teams on Saxon farms. The rod and the furlong are still used today with essentially no change. Longer distances in England are traditionally measured in miles. The mile is a Roman unit, originally defined to be the length of 1,000 paces of a Roman legion. A pace was two steps, right and left, or about 5 feet, so the mile is a unit of roughly 5,000 feet. In terms of area, in English-speaking countries, land is traditionally measured by the acre, a very old Saxon unit. There are references to the acre at least as early as the year 732. The word *acre*

The former Weights and Measures office in Seven Sisters, London. (Nico Hogg via Wikimedia Commons)

also meant field, and as a unit an acre was originally a field the size that a farmer could plow in a single day.

The basic traditional unit of weight or mass, the pound, originated as a Roman unit and was used throughout the Roman Empire. The Roman pound was divided into 12 ounces, but many European merchants preferred to use a larger pound of 16 ounces, perhaps because a 16-ounce pound is conveniently divided into halves, quarters, and eighths.

The names of the traditional volume units are the names of standard containers. Until the eighteenth century, it was difficult to measure the capacity of a container accurately in cubic units, so the standard containers were defined by specifying the weight of a particular substance—for example, wheat or beer. The gallon, the English unit of volume, was originally the volume of eight pounds of wheat. There were a multiplicity of units, as different commodities were carried in containers of slightly different sizes. Gallons are always divided into 4 quarts, which are further divided into 2 pints each. For larger volumes of dry commodities, there are 2 gallons in a peck and 4 pecks in a bushel. Larger volumes of liquids were carried in barrels or other containers, whose size in gal-

Some SI Derived Units

Quantity Name	Unit	Symbol	SI Units	Other Units
Catalytic activity	katal	kat	$mol\ s^{-1}$	—
Celsius temperature	degree Celsius	°C	K	—
Electric capacitance	farad	F	$A^2s^4kg^{-1}m^{-2}$	C/V
Electric charge	coulomb	C	As	—
Electric conductance	siemen	S	$A^2s^3kg^{-1}m^{-2}$	A/V
Electric potential force	volt	V	$kgm^2s^{-3}A^{-1}$	W/A
Electric resistance	ohm	Ω	$kgm^2s^{-3}A^{-2}$	V/A
Energy, work, quantity of heat	joule	J	kgm^2s^{-2}	N m
Force	newton	N	$kgms^{-2}$	J/m
Frequency	hertz	Hz	s^{-1}	—
Illuminance	lux	lx	$cdsrm^{-2}$	lm/m^2
Inductance	henry	H	$kgm^2s^{-2}A^{-2}$	Wb/A
Luminous flux	lumen	lm	cdsr	—
Magnetic flux	weber	Wb	$kgm^2s^{-2}A^{-1}$	V • s
Magnetic induction, flux density	tesla	T	$kgs^{-2}A^{-1}$	Wb/m^2
Plane angle	radian	rad	mm^{-1}	dimensionless
Power	watt	W	kgm^2s^{-3}	J/s
Pressure, stress	pascal	Pa	$kgm^{-1}s^{-2}$	N/m^2
Radiation dose: equivalent, biological risk	sievert	Sv	m^2s^{-2}	—
Radiation dose: absorbed	gray	Gy	m^2s^{-2}	J/kg
Rates of radioactivity and other random events	becquerel	Bq	C	—
Solid angle	steradian	sr	m^2m^{-2}	dimensionless

lons tended to vary with the commodity, with wine units being different from beer and ale units, for example.

THE METRIC SYSTEM

The metric system replaces all the traditional units, except the units of time and of angle measure. One fundamental unit is defined for each quantity. These units are now defined precisely in the International System of Units.

In the metric system of measurement, designations of multiples and subdivisions of any unit may be arrived at by combining the name of the unit with the prefixes *deka*, *hecto*, and *kilo*, meaning, respectively, 10 (that is, 10 times the base unit), 100, and 1,000, and *deci*, *centi*, *and milli*, meaning, respectively, one-tenth, one-hundredth, and one one-thousandth. Other multiples and fractions exist as well. In certain cases, for example (particularly in scientific usage), it becomes convenient to provide for multiples larger than 1,000 and for subdivisions smaller than one one-thousandth, such as *mega*, *giga*, *and tera*, meaning, respectively, 10^6, 10^9, and 10^{12}.

Further Reading

Crowell, Benjamin. *Light and Matter*. Light and Matter, 2003. http://www.lightandmatter.com/html_books/2cl/ch01/ch01html#Section1.2.

Gupta, S.V. *Units of Measurement. History, Fundamentals and Redefining the SI Base Units*. 2nd ed. Springer, 2020.

Hebra, Alex. *Measure for Measure: The Story of Imperial, Metric and Other Units*. Johns Hopkins UP, 2003.

National Institute of Standards and Technology. *NIST Handbook 44: Specifications, Tolerances, and Other Technical Requirements for Weighing and Measuring Devices*. 2010.

Rowlett, Russ. *A Dictionary of Units of Measurement*. Center for Mathematics and Science Education, U of North Carolina. http://www.unc.edu/~rowlett/units.

U.S. Department of Energy (DOE)

FIELDS OF STUDY

Environmental Science; Intergovernmental Affairs; Policy Studies

ABSTRACT

The U.S. Department of Energy (DOE) regulates and supports energy industries and research in the United States, combining economic, environmental, and energy security missions. The DOE promotes scientific and technological innovation for reliable, clean, and affordable energy production and the environmental cleanup of the national nuclear weapons complex. Energy sources supported by the U.S. Department of Energy include bioenergy from wood, wood waste, straw, manure, sugarcane, and products from agricultural processes; fossil fuels from coal, natural gas, and oil; and renewable energies from rain, tides, geothermal, hydropower, solar power, wind, and nuclear power.

KEY CONCEPTS

cybersecurity: protection of networked computer systems from intrusion and unauthorized access by foreign and domestic activity

fusion reactor: a nuclear reactor that operates on the principle of nuclear fusion rather than nuclear fission

nuclear arsenal: the aggregate collection of all nuclear weapons possessed or controlled by the United States

nuclear fusion: in fusion, small nuclei are combined to form larger nuclei accompanied by the release of a large amount of energy; in fission large nuclei are split into smaller nuclei accompanied by the release of an amount of energy that is much less than that released by fusion

nuclear physics: the study of the structural characteristics, behavior and possible interactions of atomic nuclei

RESPONSIBILITIES OF THE DOE

The U.S. Department of Energy (DOE) regulates and supports energy industries and research in the United States, combining economic, environmental, and energy security missions. The DOE promotes scientific and technological innovation for reliable, clean, and affordable energy production and the environmental cleanup of the national nuclear weapons complex. The DOE is the largest supporter of basic research in the physical sciences in the United States, providing more than 40 percent of total federal funding in high-energy physics, nuclear physics, and the fusion energy sciences. Its national security priorities include ensuring the "integrity and safety" of the country's nuclear weapons; promoting international nuclear safety; advancing nuclear nonproliferation; and providing safe, efficient, and effective nuclear power plants for the United States Navy.

The proposed fiscal 2013 budget for the DOE provided about $27.4 billion in discretionary funds, representing a 3.2 percent increase over the enacted 2012 budget. (The DOE budget also includes mandatory funds, consisting mainly of direct loan financing for innovative technology development.) Of the $27.4 billion plan, the breakout by the DOE unit was roughly: nuclear security and defense, 44 percent; environmental management, 20 percent; science, 18 percent; energy resource development, 16 percent; and other programs, 2 percent.

The fiscal 2020 budget request for the DOE was $31.7 billion, an increase of $1.1 billion from the 2019 request. Of this amount, 7.25 percent was to go towards energy independence, energy security and new energy sources, 17.35 percent for support of research, and 75 percent toward modernizing the nuclear arsenal, nuclear clean-up and grid cybersecurity.

SCOPE

The DOE manages more than 130,000 employees and contractors and operates, coordinates, or funds a network of laboratories, industrial and military test sites, and other facilities independently and in partnership with corporations, universities, state and local governments, and various other federal agencies and bodies. The DOE maintains an active interest in every form of energy generation and conservation, and all types of fuel sources, whether conventional, transitional, or experimental. Programs include such long-running, publicly promoted concepts as the Energy Star program, launched in 1992 to identify and grade energy-efficient household appliances with an eye on conservation. An even longer-run DOE program was started by the Nuclear Waste Policy Act of 1982, promulgated by Congress to assign the DOE Office of Civilian Radioactive Waste Management to oversee the safe handling and storage of the waste products of nuclear weapons production, nuclear-powered naval vessels, and the commercial nuclear electric power industry complex.

The DOE Office of Energy Efficiency and Renewable Energy (EERE) invests in technologies "that strengthen the economy, protect the environment, and

Seal of the U.S. Department of Energy. (Wikimedia Commons)

reduce dependence on foreign oil," according to the DOE mission statement for the division. EERE programs cover the development of energy conservation efforts for homes and other buildings, transportation, and manufacturing. On the production side, EERE makes key efforts in solar, wind, water, biomass, geothermal, and hydrogen and fuel cell technology.

HIGH ENERGY PHYSICS

The Office of High Energy Physics (HEP) supports scientific energy research on the constituents and architecture of the universe, large concentrations of particles in nature, and natural space sources of charged particles. Research areas include proton and electron accelerator-based physics, nonaccelerator physics, theoretical physics, and applied physics such as investment in and development of advanced technology. The Tevatron, the largest particle accelerator ever built in the United States, operated from 1983 through 2011. Located at DOE's Fermilab in Batavia, Illinois, experiments and discoveries at the Tevatron helped advance global scientific understanding of a number of advanced high energy physics problems. The 2011 shutdown came after the Tevatron's capabilities were overtaken by such newer facilities as the Large Hadron Collider in Europe.

Programs of the U.S. Department of Energy

Joint Initiatives	Focus
Biomass Research and Development Initiative: Department of Energy (DOE) and the Department of Agriculture	Biobased products, bioenergy research
Carbon Sequestration Leadership Forum: 21-nation climate change initiative	Transport, storage, and capture of carbon dioxide
Clean Coal Power Initiative (CCPI): U.S. government and industry	Coal-based power, carbon capture, and storage
Clean Energy Technology Exports Initiative: developing world	Access to efficient energy services
Climate Voluntary Innovative Sector Initiatives: Opportunities Now (VISION): public-private industry partnership	Energy efficiency and greenhouse gas emissions
Energy Star®: DOE and the Environmental Protection Agency (EPA)	Seal of energy efficiency on 1.5 billion products
FreedomCAR and Fuel Partnership: industry-government research initiative	Petroleum-free cars and light trucks
Fueleconomy.gov: DOE and the EPA	Practical information on fuel and vehicles
Generation IV: international	Availability, climate change, air quality, security
Hawaii Clean Energy Initiative: DOE and the state of Hawaii	Renewable resources by the year 2030
Nuclear Power 2010: U.S. government agencies and industries	Identify sites for new nuclear power plants, test regulatory processes
Science.gov: U.S. government agencies	Research and development information
Solar America Cities: DOE, Solar America Initiative, U.S. cities	Solar energy technologies
State Energy Efficiency Action Network: U.S. states	Energy efficiency improvements

The DOE's HEP program features physicists who study the behavior of subatomic particles at high energies. Although sometimes regarded as highly theoretical, the discipline has resulted in technologies with applications for everyday life, including medicine, advanced engineering, materials science, electronics, and energy production.

NUCLEAR PHYSICS

Nuclear physics entails the study of atomic nuclei interactions, and their applications for science, power generation, and weaponry. Nuclear physics has diverse applications, for example it is used in medicine for magnetic resonance imaging (MRI); in materials engineering for ion implantation; and in archaeology for radiocarbon dating. The DOE's Office of Science nuclear physics program supports fundamental research on the nature of matter and energy and develops scientific knowledge, technologies, and trained manpower. The DOE uses its peer-reviewed research and development for nuclear-related national security, energy, and environmental quality.

Nuclear research focuses on atomic reactions, radioactivity, and isotope and heavy element synthesis. Quarks and gluons, cosmic rays and neutrinos, and supernovae are of interest in this DOE program. The Nuclear Energy Advisory Committee (NEAC), formerly the Nuclear Energy Research Advisory Committee (NERAC), began in 1998 to provide independent advice to the Office of Nuclear Energy (NE) on complex science and technical issues that arise in the planning, managing, and implementation of DOE's nuclear energy program. The NEAC advises the NE regarding priorities and strategies in science and engineering development efforts. National policy concerning nuclear energy research can be requested from the Secretary of Energy or the assistant secretary for nuclear energy. The committee includes representatives from higher education, industries, foreign nationals, and government laboratories.

FUSION ENERGY

The application of nuclear fusion for electricity generation remains experimental, but if the physical and cost obstacles to its controlled use can be overcome, it would have tremendous potential. DOE Fusion Energy Sciences (FES) methods have long-range promise for fusion as an abundant, clean source of energy, recommended within the National Energy Policy. Fusion power is released by sustained burning of plasma, a state of matter essentially comprised of a gas with an abundance of charged particles; nuclear fusion is the process at work in the core of the sun, where immense pressures break down the natural repulsion of atomic nuclei and compress hydrogen atoms into helium—a reaction that emits tremendous amounts of energy in the form of heat, light, and the full spectrum of radioactive particles. The FES has implemented U.S. support of the International Thermonuclear Experimental Reactor (ITER), now under construction in Cadarache, France. The most ambitious fusion project in history, ITER was projected to come on line in 2019, after an estimated construction cost of $20 billion over a 12-year period. Costs are borne by the seven ITER members: the European Union, China, India, Japan, South Korea, Russia, and the United States.

The ITER project goal is to operate a fusion energy power plant that will be capable of producing a heat energy output up to 10 times its energy input. The fu-

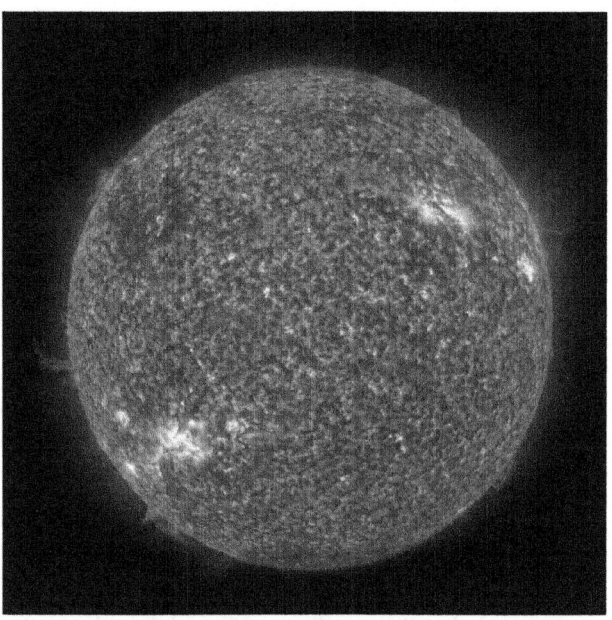

The Sun is a main-sequence star, and thus generates its energy by nuclear fusion of hydrogen nuclei into helium. In its core, the Sun fuses 500 million metric tons of hydrogen each second. (Wikimedia Commons)

sion process is obviously favored for this idea, as well as its zero emissions profile in terms of greenhouse gas. However, the obstacles to achieving practical application on a commercial scale are extremely steep; realization on that scale remains decades away. Experimental fusion reactors to date have produced bursts of energy for only fractions of a second. Since applied research began in the 1950s, no controlled fusion reactor has produced a self-sustained reaction, known as "break-even."

The immense cost of ITER, and other fusion reactor experiments, stems both from the intensively engineered structures needed to contain a reaction that reaches a temperature similar to that of the sun's core, and from the massive amounts of electricity needed to contain the plasma, either through extremely powerful magnetic fields or lasers.

HISTORY OF DOE

The DOE has its roots in the Manhattan Project, the secret crash program that developed the atomic bombs used against Japan at the end of World War II. The U.S. Army Corps of Engineers confined top-secret nuclear research within the Manhattan Engineer District, giving the project its code name. Following the war, Congress passed the Atomic Energy Act of 1946, creating the Atomic Energy Commission (AEC); its function was to maintain civilian government control over the field of atomic research and development, a hotly debated issue of the time. As the Cold War proceeded between the United States and Western European nations, on one hand, and the Soviet Union and nations in its sphere of influence on the other, the AEC supported the design and production of nuclear weapons as well as reactors for nuclear submarine propulsion. The 1954 Atomic Energy Act ended exclusive government use of the atom, inaugurating commercial nuclear power as an industry to enhance U.S. energy infrastructure. The AEC continued to regulate development and testing of nuclear power for U.S. consumers.

With the Energy Reorganization Act of 1974, the AEC became two new agencies: the Nuclear Regulatory Commission, which would regulate U.S. nuclear power, and the Energy Research and Development Administration, which would manage all nuclear weapons, the naval reactors, and development programs. During this period, a lengthy oil-driven energy crisis prompted interest in alternative, lower-cost fuel resources, unifying energy research, development, and planning for a sustainable energy future within the United States. The Department of Energy Organization Act of 1977 consolidated federal government agencies and programs around those goals, launching the DOE. The newly formed DOE began administration of the established Federal Energy Administration, the Energy Research and Development Administration, the Federal Power Commission, and many other energy-related government agencies. Within a decade, research into nuclear weapons, development, and production rose in importance at DOE.

Beginning in the 1990s, the DOE was tasked with environmental cleanup of nuclear weapons domestically and abroad, nonproliferation and wise stewardship of nuclear materials, energy efficiency and conservation by citizens, technology transfer and industrial competitiveness.

The DOE supports the Atomic Testing Museum in Las Vegas, Nevada; and historic facility tours of the Experimental Breeder Reactor 1 at DOE's Idaho National Engineering and Environmental Laboratory; the Oak Ridge National Laboratory; the Y-12 National Security Complex; the East Tennessee Technology Park; the U.S. Army White Sands Missile Range; and the Trinity Site, where the world's first atomic test device was exploded in 1945. Day tours of the Nevada test site preserve its scientific and technological value.

—*Qingjiang Yao*

Further Reading

Committee to Review the U.S. ITER Science Participation Planning Process. *A Review of the DOE Plan for U.S. Fusion Community Participation in the ITER Program.* National Academies Press, 2009.

Fehner, Terrence R., and Francis G. Goslin. *Coming in From the Cold: Regulating U.S. Department of Energy Nuclear Facilities, 1942-1996.* American Society for Environmental History and the Forest History Society in association with Duke UP, 1996.

Holl, Jack M., and Terrence R. Fehner. *Department of Energy, 1977-1994: A Summary History.* Office of Scientific and Technical Information, 1994.

U.S. Department of Energy. "About Us." http://energy.gov/about-us.

U.S. Department of Energy. "Department of Energy FY 2020 Budget Request Fact Sheet." 11 Mar. 2019, https://www.energy.gov/articles/department-energy-fy-2020-budget-request-fact-sheet#:~:text=The%20President's%20Budget%20Request%20for,from%20the%20FY%202019%20request.

V

VOLTA, ALESSANDRO

FIELDS OF STUDY
Automotive Engineering; Bioelectronics; Electrical Engineering

ABSTRACT
Alessandro Volta was an Italian physicist and chemist renowned for his pioneering work in chemistry of gases, electrochemistry, and electromagnetism. Today he is best known for his invention of the battery, or voltaic pile, in 1800.

KEY CONCEPTS
animal electricity: the notion that the electrical energy causing muscle response when muscle tissue was stimulated was stored within the muscle and was released by the stimulus

calorimetry: the measurement of the amount of heat energy either absorbed or given off in a specific reaction or in a specific physical change

capacitance: the amount of electrical charge that can be temporarily stored in a capacitor; the standard unit of capacitance is the farad (a contraction of faraday)

condenser: early name for the charge storage device now called a "capacitor"

electrostatic machines: devices that used friction to build up a static electrical charge that could be drawn off as electrical current while the machine was in operation

induction: the generation of an electrical current in a conductor by the action of a moving magnetic field

quantity of electricity (Q): the calculated amount of charge in a capacitor as the product of its capacitance (C) and the applied or resultant voltage (T), as $Q = C \times T$

serial connection: connection of electrical components end to end "in series" rather than side-by-side to common connection points "in parallel"

tension: an early term for voltage, still in use in regard to capacitors and transformers

BACKGROUND
Alessandro Volta was born in Como, Italy, on February 18, 1745, to an aristocratic family. From 1758 to 1760, he attended the city's Jesuit school and pursued his studies in philosophy, displaying great intelligence and ability in many disciplines. He subsequently began to take an interest in electric phenomena and studied the works of Petrus van Musschenbroek, Jean-Antoine Nollet, and Giambattista Beccaria, the three greatest scientists of the time in the field of electricity.

ELECTRICITY AND OTHER THINGS
In 1769, when he was only 24, Volta published his first work on electricity, titled *De vi ac attractiva ignis electrici ac phaenomenis independentibus* (On the Forces of Attraction of Electric Fire). In 1774, he was appointed super-

Alessandro Volta. (Wikimedia Commons)

intendent and regent at the Royal School of Como. The following year, he informed English chemist and theologian Joseph Priestley of his first notable invention, which he called the "perpetual electrophorus." This device could provide electricity without the need for continuous friction, as was the case for electrostatic machines of that time. For the first time, induction was applied to the systematic, abundant, and sustainable production of electricity, and the device was immediately adopted by many European laboratories. In October 1775, Volta became chair of experimental physics at the Royal School of Como.

In 1776, during his summer holiday on Lake Maggiore, Volta observed gas bubbles rising to the surface. He collected them, analyzed the gas, and came to the conclusion that it was an inflammable gas, which he labeled "inflammable air native to the swamps"; he had found methane. Volta noticed that it was possible to cause an explosion of a mixture of gases through a spark (as now occurs when air and fuel are mixed and then sparked in car engines), and he built a device called "Volta's pistol," which he used to measure the force of explosion of inflammable gases. Volta's pistol became the eudiometer, an instrument used to measure the amount of oxygen in the air.

That same year, Volta planned to broadcast from Como to Milan an electrical signal through a long wire. The wire had to be kept isolated from the ground with wooden stakes, and the electrical signal was to be generated in Como and received in Milan through the explosion of the gas pistol. This idea predated the invention of the telegraph. In November 1778, Volta was appointed professor of physics at the University of Pavia; at that time, he was working on the condenser and was able to implement a device capable of measuring weak electric charges, which he called a "condenser electroscope."

In 1780, Volta went to Florence to visit the Royal Museum of Physics and Natural History; in 1781, he visited Switzerland, Germany, the Netherlands, and Belgium and, toward the end of December, arrived in Paris, where he stayed for four months to work with Antoine-Laurent de Lavoisier and Pierre-Simon Laplace. In 1782, Volta arrived in London, where on March 14 he read to the Royal Society his essay on the condenser. This paper also correctly stated the concepts of quantity of electricity (Q), capacitance (C), and tension (T), and enunciated the fundamental relationship of the condenser, $Q = C \times T$, a law that is still discussed in physics textbooks. In 1783, he was appointed rector of the University of Pavia.

In the years from 1786 to 1792, Volta's research focused on electricity and meteorology, specifically on physicochemical properties of gaseous elements, which led him to determine the law of uniform expansion of the air. The research carried out during those years in electrolysis, electrical meteorology, calorimetry, geology, and the chemistry of gases solidified his reputation as one of the most famous scientists in Europe.

VOLTA AND GALVANI...AND A FROG

In 1792, Volta became aware of Luigi Galvani's experiments on "animal electricity" after an experiment in which two different metals were connected in series with a frog's leg and to one another. Volta started reproducing Galvani's experiments, and after extensive studies, he concluded that the frog's contractions were not due to an electricity of animal origin; instead, electricity apparently present in the dead frog was external to the animal's body and was caused by the contact of the two metals. The role of the frog in the experiment was that of an extremely sensitive electroscope. Galvani and the supporters of animal electricity refuted the idea, and thus began a scientific controversy that affected the whole European scientific community.

The dispute grew until Volta, exploiting the difference in potential due to the juxtaposition of two different metals, was able to achieve, by introducing a third conductor, a serial connection able to harness the contributions of the individual elements. This led to the battery that would take his name. On March 20, 1800, Volta presented to the Royal Society the invention of the voltaic pile, or battery, which he originally called "artificial electric organ." The importance of this invention and the applications it spurred in only a few months seemed to resolve the controversy in Volta's favor. However, the idea of animal electricity did not prove to be wrong. Volta's invention gave birth to electrochemistry and electromagnetism, as well as modern applications of electricity, while Galvani's research prompted the birth of electrophysiology and modern biology.

Volta presented the battery to the Institut de France in 1801, in the presence of First Consul Napoleon Bonaparte, who awarded him a gold medal and, four years later, an annual pension and an appointment as a knight of the Legion of Honor. Despite his enormous success, Volta remained a modest and down-to-earth person, attached to the calm of domestic life. In 1819, he retired to private life in a country house in Camnago, where he died March 5, 1827, at the age of 82.

Further Reading

Blondel, Christine, and Bertrand Wolff. "*Electricité animale ou électricité métallique? La Controversé*; Galvani-Volta et l'invention de la pile." Mar. 2007. http://www.ampere.cnrs.fr/parcourspedagogique/zoom/galvanivolta/controverse/index.php.

Dibner, Bern. *Alessandro Volta and the Electric Battery*. F. Watts, 1964.

Pancaldi, Giuliano. *Volta: Science and Culture in the Age of Enlightenment*. Princeton UP, 2003.

Pera, Marcello. *The Ambiguous Frog: The Galvani-Volta Controversy on Animal Electricity*. Princeton UP, 2014.

Schlesinger, Henry. *The Battery: How Portable Power Sparked a Technological Revolution*. HarperCollins, 2010.

Trasalti, S. "1799-1999: Alessandro Volta's 'Electric Pile'; Two Hundred Years, but It Doesn't Seem Like It." *Journal of Electroanalytical Chemistry*, vol. 460, no. 1, 1999.

Waste Heat Recovery

FIELDS OF STUDY
Boiler and Pipe Fitting Trades; Electrical Engineering; Physical Chemistry; Thermodynamics

ABSTRACT
Waste heat can be recovered from many sources and distributed by district heating and cooling systems to improve the efficiency of energy systems and reduce greenhouse gases. Recently developed technologies also allow electricity to be generated from waste heat.

KEY CONCEPTS
combined heat and power (CHP) system: a power generation system that captures and utilizes spent heat as a second output

district heating system: a system set up in and around a local power generating station or industrial production facility to distribute waste heat within that local district for heating and cooling purposes

organic Rankine cycle: a Rankine cycle power generating process used when high-temperature steam is not available for use, but low temperature steam is available; the cycle uses a high molecular weight organic liquid that boils at a significantly lower temperature than water as the working fluid

WASTED HEAT = WASTED ENERGY
Traditional methods of generating electricity and air conditioning produce large amounts of waste heat. In many cases, this heat can be captured and reused to heat and cool buildings, usually in district heating and cooling systems. This practice, known as combined heat and power (CHP), was widely practiced in the early twentieth century, when electric generating plants were built in cities and connected to steam and hot-water district heating systems, many of which were already in place distributing waste heat from factories. As these small electric generating plants have been closed over time and replaced by large electric generating plants in remote locations in North America, along with the closure of urban factories, these legacy heat distribution systems have often been abandoned, resulting in the loss of efficient capture of waste heat from electric generation.

European cities responded to the energy crisis of the 1970s by building modern district heating systems and updating existing ones as a way to use waste energy and reduce oil imports. There are new opportunities to use waste heat in district heating and cooling systems from renewable energy sources, such as wood, both to address the need for energy and to reduce greenhouse gas (GHG) emissions that lead to climate change.

EARLY USES OF WASTE HEAT
The wide-scale use of steam engines beginning in the late eighteenth century soon resulted in the use of waste heat from these engines. A patent was awarded to Oxford brewer Sutton Thomas Wood in 1784 for the beneficial use of waste heat from steam engines. The adoption of high-pressure steam engines in the early nineteenth century greatly increased the amount and temperature of exhaust heat, and as early as 1813 Oliver Evans visited a factory near Baltimore where the exhaust from one of his high-pressure engines was efficiently heating the factory. Many factories in England also used waste heat for heating. Another application of waste heat was to the compound steam engine, which used the exhaust steam from a high-pressure engine to run a low-pressure steam engine. This method was first used in 1771, and over time engines (and later turbines) incorporated multiple pressure reductions within a single unit.

MORE RECENT DISTRICT HEATING SYSTEMS
The notion of large-scale industrial and urban waste heat recovery and distribution emerged at the end of

the nineteenth century in North America with the production of electricity from reciprocating engines and with engineers recognizing potential profit from the sale of so-called waste steam. The United States witnessed a proliferation of large-scale waste heat recovery and its use in district energy systems. These included steam distribution systems in cities such as New York, Detroit, Boston, St. Louis, Philadelphia, and Milwaukee. An extensive hot-water system, which included sidewalk snow melting, was developed in Toledo, Ohio, where in 1909 the International District Energy Association was founded. The work to capture and distribute waste heat in North America was paralleled in Europe, in cities such as Hamburg, Stockholm, and St. Petersburg.

DECLINE OF DISTRICT HEATING IN THE UNITED STATES

In the United States, the scale and the value of recapturing waste heat decreased after World War II. Urban planning practices discouraged proximity of power generation to urban centers and condemned deployment of waste heat in the United States to a legacy of mostly inefficient steam distribution. Policies by the federal Department of Housing and Urban Development that promoted individual boiler systems in buildings during the 1960s and 1970s led to further erosion of waste heat recapture, according to the National Research Council. Today in the United States, approximately 50 percent of electricity is produced in large, remote coal-burning power plants that consume far more energy to make unused heat than to generate useful electricity.

These plants are characterized by high conversion losses and inefficiencies of up to 69 percent because of the lack of waste heat recapture and lack of proximity to urban regions, according to John Randolph and Gilbert Masters. It is estimated that 45 percent of all energy use in Arlington, Virginia, is attributable to lost heat in electricity conversion and transmission. This is nearly equal to the amount of energy used in the buildings, cars, and trucks in the county. District heating systems face another challenge; even if power plants were closer to urban areas, there is little energy distribution infrastructure available to use the waste heat. A very small handful of systems, such as the dis-

A Regenerative thermal oxidizer (RTO) is an example of a waste heat recovery unit that utilizes a regenerative process. (Combustion2016 via Wikimedia Commons)

trict heating and cooling system serving St. Paul, Minnesota's, central business district, are being extended and upgraded to modern standards.

DISTRICT HEATING IN EUROPE

The story has been different In Europe. Since the 1970s, many European (and with greater frequency, Asian) cities have successfully built energy systems in which waste heat from large utilities or industrial sources has been captured and channeled efficiently via cogeneration (and increasingly via microcogeneration) plants and district-heating systems. Urban planning policies in European countries such as Germany encourage the development and expansion of district energy systems by promoting density and the inclusion of larger-scale energy sources within an urban area and appropriate energy supply zoning policies. As a result, a growing percentage of homes and buildings in Europe are served by district heating (and cooling) systems. For example, a new wood-chip-fired district heating system has been installed in the town of Ostfildern outside Stuttgart, Germany, at a former air force base that was redeveloped into housing and businesses.

District heating is intensively used in Denmark, with more than 60 percent of space and water heating supplied by district heating. District heating supplies approximately 50 percent of demand in other countries, such as Poland, Sweden, and Finland.

District heating is increasingly recognized as a priority in public policy. In 2004, the European Union enacted Cogeneration Directive 2004/8/EC. The directive calls on member states to remove barriers to cogeneration, through guarantees that electricity from CHP systems will be transmitted and distributed in a nondiscriminatory manner, and to implement other policies to promote its use. In 2007, at the Meseberg Summit, Germany became one of the first countries to elevate heat recovery to a national strategic priority.

Waste heat recovery in many European cities has traditionally used conventional nonrenewable fuels such as coal, gas, and oil. Heat recovery using CHP sources is increasingly being combined with renewable energy sources, including biomass and solar thermal. Industrial waste heat is also finding its way into the district energy source mix. District energy systems, combined with multiple waste heat sources, are seen as a critical technology to reach challenging greenhouse gas reduction and energy efficiency targets. This need is not lost on Asia: Many of the major urban developments in South Korea and China are rapidly deploying large-scale district energy systems served by CHP sources.

Recent developments in waste heat technology allow electricity to be generated from temperatures down to 91°C (200°F). Organic Rankine cycle turbines are used for this purpose and are similar to Rankine cycle steam turbines but use fluids suitable for use in low-temperature turbines.

—*Peter Garforth*

Further Reading

Al-Bahadly, Ibrahim H., ed. *Energy Conversion. Current Technologies and Future Trends.* IntechOpen, 2019.

Danish Board of District Heating. "Danish District Heating History." http://www.dbdh.dk/artikel.asp?id=464&mid=24.

Declaye, Sébastien. *Design, Optimization, and Modeling of an Organic Rankine Cycle for Waste Heat Recovery.* June 2009. http://www.labothap.ulg.ac.be/cmsms/uploads/File/TFE_SD090623.pdf.

Euroheat and Power. "District Heating and Cooling." http://www.euroheat.org/Default.aspx?ID=4.

Infosource Europe. "EU Energy Policy: Cogeneration Directive." December 2008. http://www.inforse.dk/europe/eu_cogen-di.htm.

International District Energy Association. "What Is District Energy?" http://www.districtenergy.org/what-is-district-energy/.

Lasala, Silvia, ed. *Organic Rankine Cycles for Waste Heat Recovery-Analysis and Applications* IntechOpen, 2020.

Stehlik, Petr, and Zdenek Jegla. *Heat Exchangers for Waste Heat Recovery.* MDPI AG, 2020.

University of Rochester. "District Heat Library." http://www.energy.rochester.edu/.

Wang, Enhua, ed. *Organic Rankine Cycle Technology for Heat Recovery.* IntechOpen, 2018.

WATT, JAMES

FIELDS OF STUDY

Fluid Dynamics; Mechanical Engineering; Pipe Fitting Trades

ABSTRACT
Often mistakenly credited for inventing the steam engine, Watt is best known for making improvements to the steam engine of Thomas Newcomen with the addition of a separate condenser. By increasing the steam engine's efficiency, Watt paved the way for the advances of the Industrial Revolution.

KEY CONCEPTS
condenser: in steam systems, a chamber in which spent steam is allowed to cool and condense back into liquid water that can then be returned to the boiler to produce live steam

external combustion engine: an engine in which fuel combustion takes place outside or apart from the inner workings of the engine itself

steam engine: the quintessential external combustion engine, in which fuel is burned in a firebox and the hot combustion gases are circulated through pipes in a separate boiler to produce steam that is subsequently used to drive the operation of piston-cylinder arrangements

watt: the standard International System of Units (SI) unit of power, defined electrically as the flow of one ampere of electrical current across a potential difference (a voltage) of one volt

BACKGROUND
Inventor, mechanic, engineer, and chemist James Watt was born on January 19, 1736, in Greenock, a small town outside Glasgow, Scotland. Often mistakenly credited for inventing the steam engine, Watt is best known for making improvements to the steam engine of Thomas Newcomen with the addition of a separate condenser. By increasing the steam engine's efficiency, Watt paved the way for the advances of the Industrial Revolution.

Watt was born into a fairly prosperous family. His father, also named James Watt, was a shipwright and merchant. Homeschooled during his early youth by his mother, Agnes Muirhead, Watt excelled in his studies but was often plagued by health problems and a weak constitution. Upon completing grammar school, Watt was allowed to have a work space in his father's shop. Although the makings of a gifted craftsman were evident, Watt had not yet taken on an official apprenticeship. Shortly after his mother's death in 1753, Watt decided to become a mathematical instrument maker, perhaps influenced by the occupation of his grandfather, Thomas Watt, who was a mathematics instructor. Temporarily living with his relatives in Glasgow, Watt was later able to secure an unofficial apprenticeship with the London-based instrument maker John Morgan. Shortly after the completion of an exhaustive year of study, Watt sought to open a shop of his own upon his return to Glasgow in 1754. However, because Watt had not completed a formal apprenticeship, the local craft guild hindered Watt from practicing his trade in the town. Taking refuge at the University of Glasgow, Watt was able to practice his trade beyond the reach of the local guild, as the university was in need of a skilled craftsman who could repair various scientific instruments. There, Watt took full advantage of the opportunities offered by the academic setting and first encountered a model of a Newcomen steam engine.

ENGINEERING THE INDUSTRIAL REVOLUTION AND THE FUTURE
By 1758, Watt had his own shop at the university and had befriended the chemist Joseph Black. It was

James Watt. (Wikimedia Commons)

through his conversations with Black that Watt learned a great deal about chemistry and heat. In 1763, Watt discovered that the university had access to a nonfunctional model of a Newcomen steam engine that was in London being repaired. Upon learning about Watts's interest in the engine, John Anderson, a professor of natural history, had the engine returned and given to Watt. Watt was able to repair the device but discovered that it was inefficient. In 1764, Watt married his cousin, Margaret Miller.

Upon realizing that a significant portion of heat from the steam was being lost, Watt ingeniously thought of condensing the steam in a separate compartment rather than in the main chamber, as the Newcomen engine was designed to do. Watt had a working model by 1765 and agreed to partner with John Roebuck to produce the commercial versions of the engine, but the patent was not granted until 1769. Prior to Watt's addition, the Newcomen engine was used only in a limited range of mining operations. With its increased efficiency, however, the new engine had applicability in a variety of industries. The prototype for a mining operation, at Kinneil, was constructed in 1769.

By 1773, Watt's business partner, Roebuck, was encountering financial troubles. With two children to support and the death of his wife in the same year, Watt set aside the development of his new engine, instead taking up civil engineering and surveying work. A year later, Watt entered into a new partnership with Matthew Boulton. In 1775, Boulton used his political and business connections to extend the 1769 patent to the year 1800 by an act of Parliament. The new Boulton and Watt engines were soon used throughout the mining operations of Cornwall, replacing the older and less efficient Newcomen engines. Boulton and Watt both achieved a measure of financial success well into their later lives.

During the 1780s, Watt engaged in a number of additional creative activities beyond improving the steam engine, although none was as commercially successful. In 1780, Watt patented a means of copying documents via the use of a press and thin paper. In 1784, Watt submitted a patent for the development of a steam locomotive. In 1786, he dabbled with chemical bleaching. Watt finally retired from the steam engine business in 1800, turning over control of the company to his and Boulton's sons. He died on August 25, 1819, at his home in Handsworth, Birmingham.

The SI contains a measure of power, the watt, that bears his name. Streets have been named after him, statues in his likeness have been erected, and monuments in his honor are scattered throughout the world. The British Institution of Mechanical Engineering also awards the James Watt International Gold Medal to outstanding mechanical engineers. Arguably the single most influential inventor of the Industrial Revolution, James Watt left a legacy that is still felt today.

—*Richard D. Besel*

Further Reading

Dick, Malcolm, and Caroline Archer-Parré, eds. *James Watt (1736-1819): Culture, Innovation, and Enlightenment.* Liverpool UP, 2020.

Dickinson, H.W. *James Watt: Craftsman and Engineer.* Augustus M. Kelley, 1936.

Ericson, Robert, and Albert Edward Musson. *James Watt and the Steam Revolution: A Documentary History.* Augustus M. Kelley, 1969.

Hart, Ivor Blashka. *James Watt and the History of Steam Power.* H. Schuman, 1949.

Miller, David Philip. *James Watt, Chemist: Understanding the Origins of the Steam Age.* Routledge, 2016.

Russell, Ben. *James Watt: Making the World Anew.* Reaktion Books, 2014.

Watt

FIELDS OF STUDY
Classic Mechanics; Electrical Engineering

ABSTRACT
The watt is the standard unit for measuring power, which is defined as work or energy transfer over time. One watt is equivalent to one joule of energy transferred or consumed per second. Everyday appliances such as toasters use wattage to indicate the power needed to operate the device, the power it outputs, or the maximum power it can utilize without damage.

KEY CONCEPTS
derived unit: a unit created by combining two or more other units

displacement field: in electrodynamics, the electric field produced solely by free charges (e.g., free electrons)

International System of Units (SI): a standardized set of units and measures used by scientists worldwide, based on and largely synonymous with the metric system

joule: the SI derived unit of energy, work, or heat

Ohm's law: an empirical law stating that the current, or flow of electrical charge, between two points is directly proportional to the voltage, or difference in electric potential, between those points

power: the work done or energy transferred over time

power rating: the maximum electrical power a device can use without being damaged

revolution: describes circular motion wherein an object circles an internal axis (e.g., Earth spinning about its axis); contrast to rotation, wherein the axis is external (e.g., Earth orbiting the Sun)

QUANTIFYING ENERGY TRANSFER OVER TIME

Watts (W) are the International System of Units (SI) unit for measuring power—that is, work done or energy transferred over time. The power rating of a device indicates the maximum wattage it can use without damage. One watt is equivalent to one joule (J) of energy transferred per second. Because it is based on a combination of two or more SI units, the watt is considered a derived unit. It is named after Scottish engineer James Watt (1736-1819), who famously developed radical improvements to the steam engine. Wattage is most commonly used to indicate the energy consumption of electronics.

HOW ELECTRICAL POWER IS GENERATED

Electrical power is generated by the transfer of electrical energy, usually in the form of electrons, between atoms. Electrons flow down the atoms of a copper wire because there is a large difference between the electric potential of one terminal of the source (e.g., a battery) and that of the destination, the opposite terminal of the source. The energy carried by the electrons is either transferred or transformed in order to do work. An incandescent bulb works by transforming the energy carried by the electrons into heat that produces light, which is a form of electromagnetic energy. The electric field generated by free electrons and other free charges moving through conductive materials is called a displacement field.

CALCULATING WATTAGE

Electric current, or the rate at which electric charge is transferred across a material, obeys Ohm's law, named after German physicist Georg Ohm (1789-1854). The SI unit of electrical resistance, the ohm (O), is also named after him. Ohm's law states that current (I) is directly proportional to the voltage (V), or difference in electrical potential across the material, divided by the resistance (R) of that material.

$$I = (V/R)$$

Alternatively, the Ohm's law relationship states that the voltage is the product of the current and the resistance, as

$$V = I \times R$$

The term "wattage" simply means power (P) as measured in watts. Power can be expressed in a variety of ways, depending on which values are known:

$$P = I^2 R$$
or
$$P = (V^2/R)$$

Given a nine-volt battery connected to itself by a loop of copper wire with a resistance of 18 ohms, one can determine the amount of electrical power produced by the battery sending a current through the wire:

$$P = (9^2/18) = 4.5 \text{ W}$$

WATTAGE TODAY

Wattage is a vital parameter in the design and use of high-performance electronics, which often have very sensitive components. Understanding wattage and its relationship to voltage and resistance will only become more useful in an increasingly electric world.

—*Kenrick Vezina, MS*

Further Reading

"DC Circuit Theory." *Basic Electronics Tutorials*. Wayne Storr, n.d. Web. 8 June 2015.

"Electric Current." *Encyclopaedia Britannica*. Encyclopaedia Britannica, 10 Sept. 2014. Web. 8 June 2015.

Gibilisco, Stan. *Electricity Demystified*. 2nd ed. McGraw, 2012.

Henderson, Tom. "Ohm's Law." *The Physics Classroom*. Physics Classroom, n.d. Web. 1 June 2015.

Hughes, John M. *Practical Electronics: Components and Techniques*. O'Reilly, 2015.

Kelly, P.F. *Electricity and Magnetism*. CRC, 2015.

WAVE PROPERTIES

FIELDS OF STUDY
Acoustics; Classical Mechanics; Electromagnetism

ABSTRACT
Waves, whether mechanical or electromagnetic, share certain common properties. All waves have a wavelength, a frequency, and an amplitude, which together determine the ability of a wave to transmit energy and to displace the medium through which it travels.

KEY CONCEPTS
diffraction: a change in the direction of a wave as it passes around an obstruction or through an opening

interference: how waves interact when they meet in the same medium. Waves whose crests align will reinforce one another (constructive interference); waves where one's crests align with the other's troughs will dampen one another (destructive interference)

longitudinal wave: a type of wave wherein the medium is displaced in a direction parallel to the movement of energy, as in the case of sound waves

reflection: the bouncing back of a wave after it hits a barrier, as when light reflects off a mirror

refraction: the alteration of a wave's path, speed, and wavelength when it passes from one medium to another

transverse wave: a type of wave wherein the medium is displaced in a direction perpendicular to the movement of energy, as in the case of waves on the surface of water

TYPES OF WAVES
There are two major ways to categorize waves. One way is based on whether the wave can transmit energy through a vacuum. Waves that can are classified as electromagnetic waves, such as light, while waves that depend on a physical medium such as air or water are mechanical waves, such as sound waves or ocean waves. The other major categorization is based on the movement of the wave relative to the direction in which energy is transferred. Transverse waves move perpendicular to the direction of energy transfer. An example of this is when water moves up and down in ripples on a pond, while the waves travel horizontally across the surface. Longitudinal waves, also called "compression waves," move parallel to the direction of energy transfer. An example of this is when a Slinky is stretched out horizontally and one end is quickly pushed and then pulled, sending waves along the toy's length. In a transverse wave, the highest and lowest points of the waves are called crests and troughs, respectively. In a longitudinal wave, the points of maximum compression are called compressions, and points of minimal compression (i.e., maximum spread) are called rarefactions.

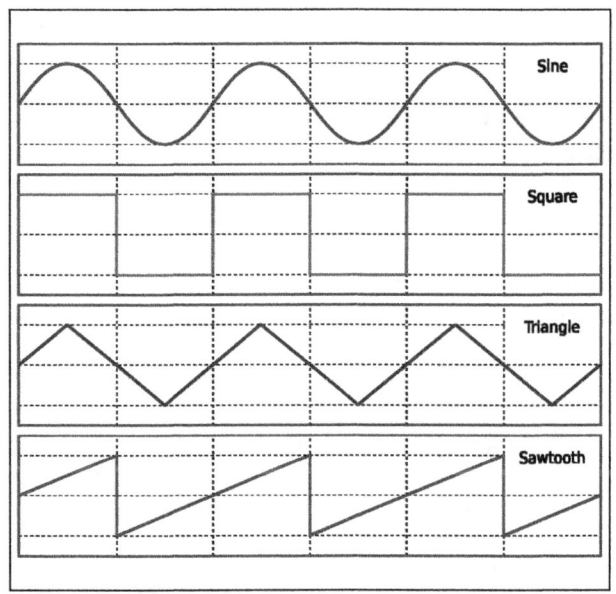

Artist's depiction of different waveforms. (Omegatron via Wikimedia Commons)

WAVELENGTH, FREQUENCY, AND AMPLITUDE

All waves are described by three properties: wavelength, frequency, and amplitude. In a transverse wave, wavelength is the distance between two peaks or two troughs; in a longitudinal wave, it is the distance between two compressions or two rarefactions. Frequency measures how often a complete wave cycle occurs in a given period of time, typically one second. Frequency and wavelength are inversely related to each other. Waves with high frequencies have short wavelengths, and waves with low frequencies have long wavelengths. The energy-transmission ability of a wave is also related to both of these properties: high-energy waves have short wavelengths and high frequencies; low-energy waves have long wavelengths and low frequencies.

Amplitude is a measure of the intensity of a wave in terms of how much it deforms the medium through which it is traveling. In a transverse wave, it is the distance between the resting equilibrium of the medium (the baseline) and the top of a crest. (An alternate way to measure amplitude is the vertical distance from the crest to the trough, called peak-to-peak amplitude.) In a longitudinal wave, amplitude is measured as the displacement of the particles of the medium at rest relative to a compression.

REFLECTION, DIFFRACTION, AND REFRACTION

When waves encounter barriers or pass from one medium to another, several phenomena can occur. Reflection occurs when a wave hits a barrier and is bounced back. Examples of this include a sound wave echoing off the side of a cliff and light bouncing off a mirror. Diffraction occurs when a wave passes by an obstacle or through an opening and bends its path as a result, as when ripples in a pond bend around rocks obstructing their path. Refraction occurs when a wave passes from one medium to another, causing its direction, speed, and wavelength to change. An example of this is light passing from air into water. This is why a straw sticking out of a glass of water looks as though it is bent at the surface of the water. Dispersion is a special type of refraction that splits a wave into several wavelengths, as when a prism refracts white light and produces a rainbow.

WAVE INTERACTIONS

Waves may also interact with one another when traveling in the same medium. This is interference, which may be constructive or destructive. If the waves align so that their peaks and troughs are paired, they merge to become a single wave with greater amplitude. This is called constructive interference. If the waves are aligned so that the peak of one wave aligns to the trough of another, they merge and dampen each other, reducing or even canceling their amplitudes. This is called destructive interference.

—*Kenrick Vezina, MS*

Further Reading

"Anatomy of an Electromagnetic Wave." *Mission: Science.* NASA, 13 Aug. 2014. Web. 29 June 2015.

"Frequency and Period." *SparkNotes: SAT Physics.* SparkNotes, 2011. Web. 29 June 2015.

Fritzsche, Hellmut. "Electromagnetic Radiation." *Encyclopædia Britannica.* Encyclopædia Britannica, 26 Nov. 2014. Web. 29 June 2015.

Ishimaru, Akira. *Electromagnetic Wave Propagation, Radiation and Scattering from Fundamentals to Applications.* 2nd ed., Wiley/IEEE Press, 2017.

Russell, Daniel A. "Longitudinal and Transverse Wave Motion." *Acoustics and Vibration Animations.* Pennsylvania State U, 18 Feb. 2015. Web. 29 June 2015.

"Waves." *The Physics Classroom.* Physics Classroom, n.d. Web. 29 June 2015.

Zielinski, Ellen, Courtney Faber, and Marissa H. Forbes. "Lesson: Waves and Wave Properties." *TeachEngineering.* Regents of the U of Colorado, 18 July 2014. Web. 29 June 2015.

Wavelength

FIELDS OF STUDY
Communications; Electromagnetism; Electronics; Physics

ABSTRACT
Physical and electromagnetic waves are defined by their wavelength and frequency. Waves, whether physical or electromagnetic, are described as either longitudinal or transverse. The time required for one cycle of any wave is the period of the wavelength. Wavelengths are related to both the

frequency and the speed at which waves travel through a medium.

KEY CONCEPTS

aperture: an adjustable opening in a barrier through which light or other electromagnetic emission can pass

crest: the highest point of a wave from its neutral value

electromagnetic spectrum: the continuous range, or continuum, of the frequencies of electromagnetic waves

frequency: the number of cycles of a property that occur in a certain amount of time

phase: a stage in a wave property; typically used to describe the relationship of two or more waves

resolution: the ability of a detector to differentiate or separate different wavelengths

trough: the lowest point of a wave from its neutral value

velocity: the speed and direction of motion

WAVE CHARACTERISTICS

All waves can be thought of as cyclic disturbances in some property. The number of times that the cycle occurs in a certain amount of time is the frequency of the wave. When there is no disturbance, a property (such as a wave) has a neutral value. For waves, the value of the property alternates continuously between equal positive and negative variations. The crest (high point) and trough (low point) seen in water waves are classic examples. The neutral value of a water wave is a perfectly smooth surface. As a water wave progresses, the level rises to a maximum value at the crest, then falls to an equal distance below the neutral point in the trough. Such a "sinusoidal" wave (one that is shaped like a sine curve) is at its neutral value three times in each cycle: at the beginning, the middle, and the end. Wavelength is the distance between adjacent crests or troughs.

Sound waves behave in a similar manner. Sound waves and water waves are examples of "longitudinal" waves. Their particles displace in either the same or the opposite direction (depending on the type) of the wave. A medium's effect on a wave can be seen in the compression-rarefaction sequence. Compression displaces matter from its neutral value. Rarefaction displaces the matter in the opposite sense. Graphs of displacement in both sound waves and water waves exhibit sinusoidal shapes.

The velocity of longitudinal waves depends on the density of the medium through which they travel. The denser the medium, the faster longitudinal waves travel through. The speed of sound in air, for example, is 331 meters per second (1,087 feet per second) at 0°C (32°F) and 342 meters per second (1,123 feet per second) at 18°C (65°F). In water, the speed of sound is about 1,140 meters per second (4,724 feet per second). The density of liquid water is about 770 times that of air.

The waves of the electromagnetic spectrum are described as "transverse" waves because the direction of their displacement is at right angles to the direction of their spread. Electromagnetic (EM) waves are not as simply described as longitudinal or physical waves because they are independent of matter and do not spread in the same way. An EM wave is perhaps best thought of as the vector combination of an electric value and a magnetic value with a single velocity. All EM waves travel at the speed of light. The magnitudes of the electric and magnetic components of an EM wave exhibit are sinusoidal, as is typical of other kinds of waves and wavelike cyclic behaviors.

THE ELECTROMAGNETIC SPECTRUM

The EM spectrum is a continuum of wavelengths, or frequencies, ranging from zero to infinity. The complete absence of any EM wave is the zero point, having neither frequency nor wavelength. At the opposite extreme, frequencies and wavelengths are theorized to become so compacted as to be indistinguishable from solid matter. Visible light, detectable by the human eye, makes up just one small part of the EM spectrum, with wavelengths ranging from 770 nanometers to 400 nanometers. (A nanometer is one-billionth of a meter.) The corresponding frequencies range from about 10^{13} to 10^{16} hertz (Hz). (One hertz is one cycle per second.) By comparison, X-rays have frequencies of about 10^{18} Hz, and gamma rays have frequencies of 10^{20} Hz and beyond.

The frequency (f) and the wavelength (λ) of EM radiation are inversely related. The greater the frequency is, the shorter the wavelength. The common factor for all EM radiation is its velocity (v), the speed of light. Unlike longitudinal waves, the velocity of EM waves is constant rather than dependent on the medium. As a result, considering the speed of EM waves is generally not useful in practical applications. Instead, an EM wave is referred to by its frequency in hertz or its wavelength in meters. In spectroscopy, the designation of "wave number" is commonly used. The wave number of an EM frequency is the reciprocal of the wavelength, normally stated in 1/cm, or cm^{-1}.

APPLICATIONS

EM radiation has a number of applications determined by the wavelength range of the appropriate frequencies. Analytical applications rely on the interaction between the particular EM radiation and the electron cloud in atoms and molecules. Lower frequencies have longer wavelengths and can be used in communications. Adjusting amplitude and phase relations of EM signals enables the speed-of-light transmission of information by radio waves and microwaves. The wavelengths of this region of the EM spectrum are mostly too large to interact effectively with atoms and molecules. Such interaction cannot happen until the energy and wavelengths of the EM waves become similar in dimension to those of the atoms and molecules. Thus, microwaves effectively transmit energy into materials placed in a microwave oven by stimulating the vibrational energy of water molecules and certain kinds of chemical bonds in the material.

Wavelength continues to decrease into and across the infrared and visible region of the EM spectrum. All wavelengths of EM radiation from this point on have application in analytical methods because they can interact with matter in specific ways. Infrared, visible, and ultraviolet wavelengths are readily absorbed and emitted by atoms and molecules. Electrons' patterns of absorbing or emitting EM wavelengths are routinely analyzed and used for identification and monitoring methods. Above the ultraviolet range of wavelengths are the X-ray wavelengths. These wavelengths are too short to interact with electrons effec-

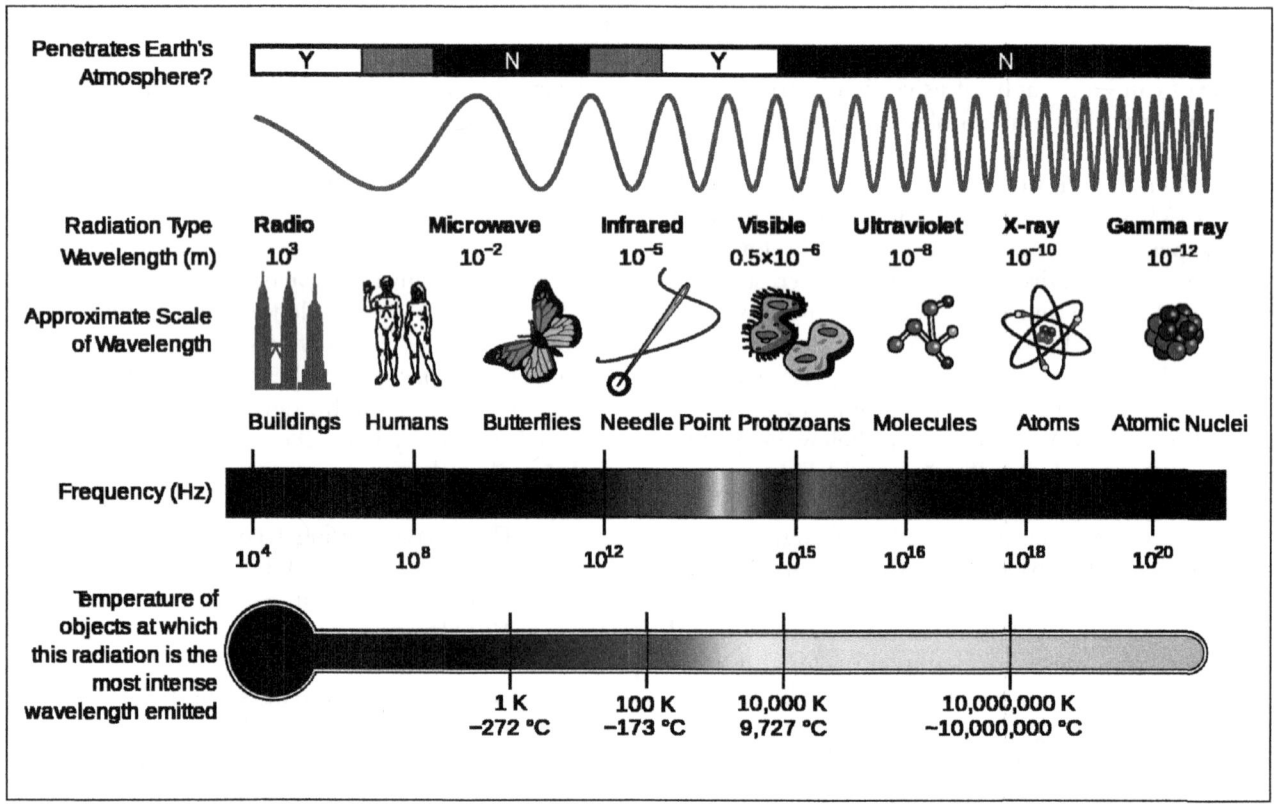

A diagram of the electromagnetic spectrum, showing various properties across the range of frequencies and wavelengths. (Inductiveload/NASA via Wikimedia Commons)

tively. However, they are closer in size to atomic nuclei and can interact with the nuclear structure of atoms. Thus, they are able to pass through solid matter fairly readily and can be diffracted by the nuclei of atoms. X-ray diffraction is used to analyze crystal structures and solid surfaces. Above the X-ray range of wavelengths are the gamma-ray wavelengths. Because of their extremely high energy and short wavelength, gamma rays destroy matter, and their use is very limited outside of astronomy.

WAVELENGTH AND RESOLUTION

The wavelengths of EM radiation are typically detected by electronic devices that convert the analog frequency of the radiation into a digital equivalent. Such detectors use an aperture to restrict the EM radiation that enters the device, where it is then resolved into component wavelengths. The resolution of the devices depends strictly on the ability of gratings and other components to differentiate the wavelengths of the light or other EM radiation that they receive. The finer the resolution, the more precise the incoming information is. High-definition cameras, microscopes, and telescopes, for example, provide clearer images than devices with lower resolution.

—Richard M. Renneboog, M.Sc.

Further Reading

"An Introduction to Waves." *GCSE Bitesize*. BBC News, 2014. Web. 25 Feb. 2015.

Anderson, Rosaleen J., David J. Bendell, and Paul W. Groundwater. *Organic Spectroscopic Analysis*. Royal Society of Chemistry, 2004.

Kirkland, Kyle. *Light and Optics*. Facts On File, 2007.

Ishimaru, Akira. *Electromagnetic Wave Propagation, Radiation and Scattering from Fundamentals to Applications*. 2nd ed., Wiley/IEEE Press, 2017.

Kumar, B.N. *Basic Physics for All*. University Press of America, 2009.

Nave, C.R. "Traveling Wave Relationship." *HyperPhysics*. Department of Physics and Astronomy, Georgia State U, 2014. Web. 25 Feb. 2015.

Pain, H.J., and P. Rankin. *Introduction to Vibrations and Waves*. Wiley, 2015.

Rogers, Alan. *Essentials of Photonics*. 2nd ed. CRC, 2008.

Shipman, James T., Jerry D. Wilson, and Charles A. Higgins Jr. *An Introduction to Physical Science*. 14th ed. Brooks/Cole, 2015.

Smith, Brian C. *Quantitative Spectroscopy: Theory and Practice*. Elsevier, 2002.

Tilley, Richard. *Colour and the Optical Properties of Materials*. 2nd ed. Wiley, 2011.

WESTINGHOUSE, GEORGE

FIELDS OF STUDY

Electrical Engineering; Power Generation Engineering

ABSTRACT

George Westinghouse is one of the most significant figures in the industrialization of the United States. He had always dabbled in the use of electricity, but in 1885 he concentrated his attention on electric lighting. Westinghouse Electric Company was formed in 1886 to produce AC equipment. Westinghouse's success at the World's Columbian Exposition earned him the contract for harnessing the water power of Niagara Falls and the success of this project made western New York the center of the U.S. electrochemical industry, including all aluminum production, for many years.

KEY CONCEPTS

AC and DC: acronyms for alternating current and direct current, respectively

electrochemical industry: industrial production concerns that require a steady, and often large, supply of electricity as part of the production process, particularly for the production of aluminum and other metals

polyphase generator: an electrical generator designed to produce an AC output current that oscillates as a number of overlapping sinusoidal frequencies to provide a more consistent availability of power under load; the overlap of the frequencies ensures that the peak values of the current flow occur three or more times per second than with a single-phase current

polyphase motor: an electric motor that requires a polyphase input current in order to operate at optimum power efficiency

A MAN FOR THE PEOPLE

George Westinghouse was an important leader in transforming the United States into an industrial society between the Civil War and World War I. George Westinghouse is one of the most significant figures in the industrialization of the United States. He founded several companies, hiring talented inventors and engineers to whom he always gave full credit for their innovations. With his fertile and imaginative mind, engineering skills, and use of science, he was a premier inventor with 361 patents. More important, he managed to attract capital and organize companies—more than 60 in all—to exploit his inventions.

As a result, he was instrumental in the development and continual improvement of the products carrying more than 3,000 patents during his lifetime. His concern for employees was unique for the time. He paid salaries far higher than what the market required and provided worker's compensation, pensions, life insurance, and health and education benefits; he also instituted a five-and-a-half-day workweek at nine hours per day when his colleagues in the steel mills required seven 12-hour days. Westinghouse sponsored family parks and recreation centers and built comfortable housing that his employees could afford to buy. He was universally admired by his employees; after his death, 50,000 of them contributed to fund an elaborate monument to him, which still stands in a public park.

MAN OF INDUSTRY

Acquiring steel components for an early invention, an apparatus for replacing derailed railroad cars, brought Westinghouse to Pittsburgh, where he centered his lifelong work. In the late 1860s, when railroads were expanding across the country, stopping a long train was a bumpy and hectic process requiring a stretch of more than a mile, with a brakeman in each car attempting to synchronize their application of mechanical brakes. Several schemes to brake all cars directly from the locomotive were proposed involving steam, a vacuum, and electrical systems, but Westinghouse devised a compressed-air transmission system, involving couplings of pipes between cars with valves that automatically opened when they were joined and closed when they separated.

In 1869, he organized the Westinghouse Air Brake Company (WABCO), and over the next three years he received more than 100 patents to improve the system. He arranged highly successful demonstrations, and by 1879 there were 36,000 air brakes installed in the United States. By 1886, electrical systems were developed to compete and it was widely believed that the air brake was doomed, but Westinghouse continued to improve his system, and in 1888 his air brake was accepted as the industry standard. He established manufacturing plants in England, France, Germany, Belgium, and Russia with combined production for outfitting 300 locomotives and 1,200 cars per month with air-brake systems. To manufacture his various products, he eventually founded companies in 20 different countries.

Westinghouse's interest in railroads led to the invention of a friction draft gear for dissipating the energy stored when cars are pressed together in stopping. This innovation grew out of his scientific studies of friction and resulted in very important savings by eliminating injuries, lost time, and damage to equipment. He became interested in the very complex problem of railroad traffic control, and in 1881

George Westinghouse. (Wikimedia Commons)

he organized Union Switch and Signal Company to apply his genius to it. He produced systems for automatically controlling signals and for interlocking them so they could not direct trains into danger. These inventions went far to revolutionize railroad traffic control.

In the 1880s, natural gas was coming into use, but it was plagued by pipe leaks, valve failures leading to deaths by asphyxiation, and explosions. Moreover, distribution systems were highly inefficient. In 1884, Westinghouse formed a company and set to work on these problems, involving 38 patents. By 1888, his systems were providing gas to 25,000 homes and 700 factories, replacing 40,000 tons of coal per day; coal use in Pittsburgh dropped from 3 million to 1 million tons per year, and the price of coal dropped by 40 percent, putting 10,000 miners out of work. However, air quality in Pittsburgh improved (until the 1890s, when the manufacture of steel greatly increased coal use).

WESTINGHOUSE, TESLA, AND EDISON

Westinghouse had always dabbled in the use of electricity, but in 1885 he concentrated his attention on electric lighting. The direct current (DC) system used by Thomas Alva Edison could not practically be transmitted over a distance of more than a mile, and the higher voltages needed for transmission over greater distances were not appropriate for lighting. That disparity could be overcome by using alternating current (AC), with voltages stepped up for transmission and down for lighting by use of transformers. Westinghouse secured an option on a crude transformer then available, his assistants redesigned it mechanically and electrically, and Westinghouse Electric Company was formed in 1886 to produce AC equipment.

Within that year, he had installed demonstration lighting systems in a New England town and in sections of Pittsburgh and Buffalo, and he had orders for 27 commercial installations. Further progress required an AC meter, developed by one of his assistants in 1888, and an AC motor. Nikola Tesla had invented an AC polyphase motor, and Westinghouse purchased the rights to it and hired Tesla. After four years of development involving a polyphase generator and transmission system and reducing the frequency from 133 to 60 cycles per second, AC was ready to be implemented.

This development started the Battle of the Currents against Edison, who advocated for the use of DC. Edison claimed that the high voltage used to transmit AC over long distances was dangerous and that insulation could not contain it. An AC electric chair was introduced to execute criminals, leading to the designation of AC as a "man killer." The battle culminated in the 1892 competition to light the 1893 Columbian Exposition in Chicago. Westinghouse's bid was one-third that of Edison, and he won the contract. To comply, he had to develop a new lightbulb that required a new vacuum pump and new vacuum techniques, and he built a glass factory with new machines for grinding glass. In the one year he had available to complete the project, he produced 250,000 lightbulbs for the exposition. He also produced a dozen 75-ton polyphase generators to power them, as well as several other AC demonstrations.

The World's Columbian Exposition was a major victory in the Battle of the Currents, but the battle against Edison General Electric, the company Edison had founded and controlled by financier J.P. Morgan, continued for many years, involving bitter fights over patents and competition for markets. Westinghouse was the only holdout preventing Morgan's banks from establishing a monopoly on electric power. However, Westinghouse Electric Company always needed to borrow cash to finance research and development, and as a result, Morgan gained control of it in 1907, a personal tragedy from which George Westinghouse never recovered.

Westinghouse's success at the World's Columbian Exposition earned him the contract for harnessing the water power of Niagara Falls, including lighting the city of Buffalo, 26 miles away. He had to engineer and build three 5,000-horsepower polyphase generators, but the success of this project made western New York the center of the U.S. electrochemical industry, including all aluminum production, for many years.

Westinghouse had always been interested in steam turbines. He acquired the rights to a promising one, introduced important improvements, and adapted it to driving AC generators. In 1900, he provided a power system for Hartford, Connecticut, driven by a steam turbine four times more powerful than any other in the world; his steam turbines soon became the principal power source for generating electricity.

Streetcars were driven by DC until 1905, when Westinghouse developed an AC locomotive. This allowed streetcars and electrified railroads to operate over large distances. It required very extensive development of new equipment, but it was successful and soon became widely used, keeping coal-burning locomotives out of cities.

STEAM TURBINES AND OTHER TECHNOLOGIES

His next project was applying the steam turbine to ship propulsion, replacing the reciprocating engine then used. The problem was that turbines operate at high rotational speeds, whereas ship propellers work at low speeds. A gear system for matching these speeds had very difficult vibration problems, but Westinghouse overcame them and developed a system to drive a 30,000-ton U.S. Navy vessel. Its success led the Navy to use turbine drives on all future ships.

Other technologies to which Westinghouse made important contributions include engines fueled by natural gas, refrigeration and heat pumps, fluorescent lights, electric motors of all sizes (with self-lubricating bearings for them), home appliances (including electric fans, toasters, and coffee makers), electric cars that for a time competed with those driven by gasoline engines, and shock absorbers for automobiles. His engineering was of the highest quality and was universally admired in the technical community.

—*Bernard Cohen*

Further Reading

Jonnes, Jill. *Empires of Light: Edison, Tesla, Westinghouse, and the Race to Electrify the World.* Random House, 2003.

Leupp, Francis E. *George Westinghouse: His Life and Achievements.* Reprint. Creative Media Partners, 2019.

McNichol, Tom. *AC/DC: The Savage Tale of the First Standards War.* Jossey-Bass, 2006.

Moran, Richard. *Executioner's Current: Thomas Edison, George Westinghouse, and the Invention of the Electric Chair.* Alfred A. Knopf, 2002.

Prout, Henry G. *A Life of George Westinghouse.* 1921. Reprint. Cosimo Classics, 2005.

Skrabec, Quentin R, Jr. *George Westinghouse: Gentle Genius.* Algora Publishing, 2007.

WHEAT ETHANOL

FIELDS OF STUDY
Agriculture; Biotechnology; Chemical Engineering; Fermentation Technology

ABSTRACT
Bioethanol used as transportation fuel is produced from several feedstocks. Wheat is the most common cereal used for bioethanol production in Europe.

KEY CONCEPTS

bioethanol: ethanol that is produced through fermentation processes based on glucose, starches, and cellulose from organic matter

ethanol: the second-simplest organic alcohol, it has the chemical formula C_2H_6O, sometimes written as H_3CCH_2OH or "shorthanded" as EtOH (the simplest organic acid is methanol, H_3COH)

hammer mill: a processing machine that uses heavy weights, or "hammers," to pulverize grain kernels as it operates; the hammers are typically large steel balls that act to crush the grain as they tumble together within the machine

saccharification: an enzyme-mediated process that cleaves starch molecules to convert them into sugar molecules

volume-to-volume (v/v) percent: the volume of pure ethanol in a specific volume of raw ethanol product; for example, 10 mL of pure ethanol in 100 mL of raw product is 10 percent v/v, also stated as 10 percent by volume

ETHANOL FROM GRAINS

Bioethanol is an alcohol produced from the biodegradable fraction of products and residues from agriculture and forestry. In 2009, the United States was the world's largest bioethanol producer (at 54 percent, mainly from corn), followed by Brazil (34 percent, mainly from sugarcane) and Europe (5 percent). In the European Union (EU), starch and molasses from sugar beets are the main feedstocks (in 2009, starch accounted for 63 percent and beets/molasses 34 percent). There has been a sharp increase in starch-based ethanol, with wheat as the major feedstock. EU bioethanol fuel production in 2009 corresponded to nearly 3.7 billion liters (1,873 kilotons of oil equivalent, or ktoe). In the past, ethanol production was mainly devoted to beverage, industrial, and technical applications. This trend was reversed in 2006, due to increasing demand for fuel purposes. Recent statistics show that European ethanol production for nonfuel applications has remained stable in recent years, whereas fuel applications are becoming considerably more significant, rising from 9.2 percent in 1998 to over 70 percent in 2009. Production for 2010 is projected at 4,508 million liters (or 2,281 ktoe). Wheat-based bioethanol has experienced sustained growth in recent years, with typical growing rates well over 30 percent.

The EU consumption of bioethanol fuel has been on a sharp rise, with annual rates greater than 30 percent between 2005 and 2010. Germany and France are the main consumers, accounting for more than 44 percent of total consumption. There is a gap between consumption and production levels in the EU, which is filled by bioethanol imports, more than 70 percent of which are from Brazil. Bioethanol is used as a transportation fuel in several blend compositions. The most common include (1) up to 5 volume-volume percent (v/v percent, and more recently 10 v/v percent is allowed) of bioethanol in gasoline for conventional vehicles; (2) up to 85 v/v percent bioethanol in gasoline for modified vehicles (flex-fuel vehicles); and (3) 95 v/v percent bioethanol in diesel fuel, used in captive fleets of Swedish buses.

PRODUCTION OF BIOETHANOL FROM WHEAT

Wheat cultivation includes several steps, namely soil preparation (plowing), fertilization, sowing, weed control, and harvesting. After transportation to the processing plant, the raw material is cleaned to re-

Wheat. (Lakov Kalinin via iStock)

move any debris, such as soil and stones. The grain goes through a grinding process in a hammer mill in order to increase the grain surface and maximize the efficiency of the subsequent steps. The milled grain is then mixed with preheated water and liquefaction enzymes, forming a mash and releasing the starch from the cell material. After cooling down, a mixture of amylase enzymes is added to break down the starch into simple sugars (saccharification). Sugars are fermented to ethanol using yeasts, in a process that releases carbon dioxide and yields a solution of 8 to 10 percent milligrams per milliliter alcohol. This alcohol concentration is increased up to 95-96 v/v percent by distillation. If ethanol is to be mixed with gasoline for transportation purposes, further dehydration of 99.7 v/v percent or more is required, which is usually achieved through molecular sieve technology. On average, 1 liter of ethanol is produced for each 2.84 kilograms of wheat processed. After fermentation and distillation, the leftover residue (whole stillage) is pressed and dried to form distiller's dried grains with solubles (DDGS). Approximately 350 kilograms of DDGS are produced per ton of wheat. DDGS has both high protein and high fiber content and can be sold as feed for ruminants.

ENERGY AND ENVIRONMENTAL ISSUES

Several factors affect the energy and greenhouse gas (GHG) balance of ethanol production from wheat: production and application of fertilizers and pesticides, fossil energy inputs, soil emissions due to land use, and (direct and indirect) land-use change (LUC). According to the EU renewable energy directive 2009/28/EC, typical GHG emission savings of wheat-based ethanol over gasoline amount to 45 percent, although not including soil carbon emissions from LUC. Recent reviews of (wheat-based) ethanol studies show that the GHG balance over the life cycle (that is, from wheat cultivation to delivery of bioethanol to the end user) may vary significantly, and if lands with high stocks of carbon are converted to biofuel production, the total GHG balance may become critical. This aspect must be balanced against the recognized advantages of biofuels promotion, namely security of energy supply and rural development.

—*Fausto Freire*

Further Reading

Drapcho, Caye M., Nghiem Phu Nhuan, and Terry H. Walker. *Biofuels Engineering Process Technology*. 2nd ed. McGraw-Hill, 2020.

Goldstein, Walter E. *The Science of Ethanol*. CRC Press, 2017.

Ingledew, W.M. *The Alcohol Textbook: A Reference for the Beverage, Fuel, and Industrial Alcohol Industries*. 5th ed. Nottingham UP, 2009.

Malça J., and F. Freire. "Renewability and Life-Cycle Energy Efficiency of Bioethanol and Bio-Ethyl Tertiary Butyl Ether (bioETBE): Assessing the Implications of Allocation." *Energy*, vol. 31, no. 15, 2006.

Ray, Ramesh C., and S. Ramachandran, eds. *Bioethanol Production from Food Crops. Sustainable Sources, Interventions and Challenges*. Academic Press, 2019.

SenterNovem. *Bioethanol in Europe: Overview and Comparison of Production Processes*. Rapport 2GAVE0601. SenterNovem, 2006.

Smith, T.C., D.R. Kindred, J.M. Brosnan, R.M. Weightman, M. Sheperd, and R. Bradley. "Wheat as a Feedstock for Alcohol Production." *Research Review*, no. 61, Dec. 2006.

Walker, G.M. *Bioethanol: Science and Technology of Fuel Alcohol*. Ventus, 2010.

WILLIS CARRIER AND AIR CONDITIONING

FIELDS OF STUDY

HVAC Trades; Mechanical Engineering; Thermodynamics

ABSTRACT

Willis Carrier was an American engineer known as the father of the modern air conditioner for his patented Rational Psychometric Formulae Apparatus for Treating Air, and low-pressure centrifugal refrigeration machine. Although Carrier was not the first to design and manufacture cooling systems, he was the first to successfully create a system that could artificially control temperature and humidity as opposed to simply venting away hot air.

KEY CONCEPTS

dielene: a chemical compound used as a refrigerant liquid, designated R-1130, with the proper name *cis*-1,2-dichloroethylene

humidity: the amount of water vapor in ambient air, relative to the maximum amount of water vapor possible in air at specific temperatures and pressures (100 percent humidity); for example, 25 percent humidity means the amount of water vapor present in the air is 25 percent of the maximum amount possible

nonflammable: any compound or material that will not support combustion as a fuel, though it may be 'burned' (decomposed by heat to release combustible components) when combined with other combustible fuels

WILLIS CARRIER

Willis Haviland Carrier was born on November 26, 1876, in Angola, New York. He was raised on a family farm, but developed an interest in engineering from a young age. He attended Cornell University on a scholarship, receiving his master's degree in electrical engineering in 1901. Carrier achieved his first major scientific breakthrough in 1902 at the age of 25 while employed at the Buffalo Forge company. The company produced heating and air exhaust systems. Carrier's early work earned him a promotion to head of the company's experimental engineering department.

EARLY INVENTION

The mechanical system Carrier created to artificially control temperature and humidity within the plant is considered the world's first modern air conditioner. Carrier's machine forced air through a filter and over refrigerated coils chilled by coolant in order to lower both the humidity and the temperature of an enclosed space. Cool, dehumidified air would then flow into the room while heated air was expelled outside the building through ventilation. Carrier also noted that alterations to the temperature of the coils and the rate of airflow allowed for the creation of machines of different sizes and capacities. His machine could also lower the humidity level.

Although Carrier was not the first to design and manufacture cooling systems, he was the first to successfully create a system that could artificially control temperature and humidity as opposed to simply venting away hot air. Carrier claimed that his breakthrough in the scientific understanding of the relationship between temperature, humidity, and dew point, and the significance of that understanding

Willis Carrier. (Wikimedia Commons)

to artificially controlling indoor temperature and humidity, came to him on a foggy evening while he was waiting for a train. Carrier's spray-driven machine, known as an "Apparatus for Treating Air," received its U.S. patent in 1906. By 1911, Carrier was ready to present his Rational Psychometric Formulae to the American Society of Mechanical Engineers.

Carrier cofounded the Carrier Engineering Corporation in 1915. His partners were the engineers Edmund Heckel, Ernest Lyle, Alfred Stacey Jr., Logan Lewis, Irvine Lyle, and Edward Murphy. The corporation specialized in air-conditioning technology and systems. The partners contributed most of the approximately $35,000 in start-up capital from their own pockets. The New Jersey-based corporation would relocate its headquarters to Syracuse, New York, and open a Japanese division in 1930. Notable sites housing Carrier air-conditioning systems include the White House and the U.S. Senate and House of Representative buildings in Washington, D.C., and Madison Square Garden in New York City. One of Willis Carrier's first installed private residential air-conditioning systems was located in the Minneapolis, Minnesota, home of Charles Gates. One of his earliest residential air-conditioning systems was known as the Weathermaker.

IMPROVEMENTS
Carrier continued to experiment with his designs over the next several decades, receiving several more patents, and installing air conditioning systems in a variety of businesses. One of Carrier's key developments during this period was the low-pressure centrifugal refrigeration machine, or centrifugal chiller, which was patented in 1921. The centrifugal system increased both safety and efficiency. Carrier also improved the safety of refrigerants. Early cooling systems relied on toxic refrigerants such as ammonia. Carrier introduced the use of nontoxic, nonflammable refrigerants such as dielene. He was even able to make air conditioners smaller in size and more affordable.

The J.L. Hudson Department Store in Detroit, Michigan, and the Rivoli movie theater in New York City were among the first commercial buildings to have centrifugal chiller air-conditioning systems installed in the 1920s. Commercial sales boomed throughout the decade as air conditioning emerged as a successful advertising technique. Consumer demand dropped during the Great Depression and World War II, but military sales took over. The Carrier Corporation was heavily involved in military production during U.S. involvement in World War II. The corporation's air conditioning and refrigeration equipment was utilized in wartime factories, vessels, airplanes, and storage facilities. He also designed a simulation system to train pilots for the cold temperatures of high-altitude flight.

Carrier died at the age of 73 in New York City on October 9, 1950, but his legacy remained. Industrial and scientific research facilities and museums benefited from artificial temperature and humidity control through better products and working conditions. Human health and comfort was improved. The development of modern air conditioning and soaring consumer sales after World War II opened southern and Sun Belt states to further development and population growth by making them more attractive places to live in the hot summer months.

Carrier's Rational Psychometric Formulae has since remained the scientific basis for the fundamental calculations underlying the air-conditioning industry, while the Carrier Engineering Corporation remains a global leader in the field. Carrier has designed air-conditioning systems for various modes of transportation, such as buses, railroad cars, planes, submarines, and space vehicles.

—*Marcella Bush Trevino*

Further Reading
Cooper, Gail. *Air-Conditioning America: Engineers and the Controlled Environment, 1900-1960*. Johns Hopkins UP, 2002.
Hundy, G.F., A.R. Trott, and T.C. Welch. *Refrigeration, Air Conditioning and Heat Pumps*. 5th ed. Elsevier, 2016.
Public Broadcasting System. "Willis Carrier." http://www.pbs.org/wgbh/theymadeamerica/whomade/carrier_hi.html.
Yergen, Daniel. *The Quest: Energy, Security, and the Remaking of the Modern World*. Penguin Books, 2012.

WIND ENERGY

FIELDS OF STUDY
Electrical Engineering; Environmental Science; Mechanical Engineering

ABSTRACT
Wind energy has been a significant energy resource in human history, from its use by ancient sailboats to its employment in modern large-scale electrical generators. In the twenty-first century, it is one of the fastest-growing energy sources in the world. Wind energy is a clean, renewable, and free power source that is a viable alternative to fossil fuels, which pollute the environment and promote global warming.

KEY CONCEPTS
distributed energy system: a network of interconnected local electrical grids accepting input from a number of different suppliers for distribution to common use

wind farm: an array of wind turbines established to harvest wind energy, analogous to an agricultural operation that harvests various crops

windmill: a wind-powered structure that converts the kinetic energy of wind to mechanical energy to perform mechanical work

wind turbine: a wind-powered structure that converts the kinetic energy of wind to electrical energy

BACKGROUND

The energy of wind was captured by humans as early as the fourth century BCE, when the Egyptians used wind to propel sailboats on the Nile River. Soon, there were sailing vessels in the Mediterranean Sea. Windmills, machines that convert wind into mechanical power, were used to pump water and mill grain in ancient times. In a windmill, wind blows on sails or blades radiating from a wind shaft (cylindrical part on which the sails turn), which produces mechanical energy when it rotates. The first documented wind device was a Persian windmill shown in drawings from about 500 CE. This windmill was the horizontal axis type, with a horizontal wheel holding the sails, a vertical windshaft, and vertical sails made of bundles of reeds or wood. These windmills spread throughout the Middle East and into China.

During the twelfth century, the first European windmills appeared in England and France. These wooden machines were called "post mills," because they had a central vertical post. They had a horizontal windshaft, had a revolving platform atop the post, and were rotated by hand. In the fourteenth century, a larger and sturdier windmill called the tower mill was developed. The tower mill had a stationary body that supported a rotatable wooden cap, to which the rotor was attached. The blades faced the wind, and there was storage space for the grain at the base. This windmill was popular in Holland, where windmills were used for land drainage. Between 1300 and 1850 CE, windmills provided about 25 percent of Europe's industrial power. They were used for grinding dyes, spices, and paint pigment as well as for irrigation and grain milling.

In the nineteenth and early twentieth centuries, windmills played an essential role in the development of the western and Great Plains regions of the United States. In 1854, Daniel Halladay designed small, multibladed, and inexpensive windmills that were sturdy enough for the Great Plains. American farmers and homesteaders used windmills to pump underground water to the surface. Windmill-generated water was crucial for livestock, human use, crop irrigation, and steam locomotives. Windmills made the arid Great Plains bloom, opening up the West to towns, farms, and most important, the transcontinental railroad. Windmills were responsible for changing cattle ranching from a nomadic to a stable business and transforming the Great Plains into a breadbasket. Between 1880 and 1930, approximately 6 million windmills were installed in the western United States and the Great Plains.

As Europe and the United States became industrialized, the steam engine gradually replaced water-pumping windmills in Europe, and electricity

Photo of a windmill in Utrecht, Netherlands. (Kattenkruid via Wikimedia Commons)

replaced wind power in rural America. The Rural Electrification Act of 1936 provided low-cost federal loans for bringing electricity into rural areas. Low-cost power from town and regional electric generators became available, and power lines were extended to remote areas of the country. In 1935, only 11 percent of the farms in the United States had electric service, but by the early 1970s, about 98 percent did.

Meanwhile, scientists developed larger windmills, called wind turbines, that could generate electricity. In 1890, in Denmark, P. LaCour built the first windmill capable of generating electricity. In 1941, Palmer Putnam built the world's largest wind turbine on a windy mountaintop called Grandpa's Knob in Vermont. This nearly 230-tonne generator served an entire town by feeding electric power into the existing local utility grid.

Energy from fossil fuels (coal, natural gas, and oil) was in abundance and had essentially displaced wind power. The Atomic Energy Act of 1954 allowed private companies to develop nuclear energy for peaceful purposes. However, Europe was developing more wind-power technologies. For instance, from 1956 to 1957 in Denmark, Johannes Juul built the world's first alternating current (AC) wind turbine, the very efficient Gedser wind turbine. The 1973 oil crisis, the environmental movement, and the dangers of atomic energy led to renewed interest in wind energy, which is a renewable energy source. Electricity generated by renewable energy sources (solar, wind, hydro, geothermal, and biomass energy) is called "green" power. During the 1990s, wind power was one of the fastest-growing sources of energy.

WIND ENERGY TECHNOLOGY

When the Sun warms areas of Earth at different rates and the various surfaces absorb or reflect the radiation differently, there are differences in air pressure. As hot air rises, cooler air comes in to replace it. Wind, or air in motion, is the result. Air has mass, and moving air contains kinetic energy, the energy of that motion.

Windmills convert wind energy into mechanical power or electricity. The modern electricity windmills are called wind turbines or wind generators. In the wind turbine, wind turns two or three propeller-like rotor blades, which are the sails of the system. When the blades move, energy is transferred to the rotor. The wind shaft is connected to the rotor's center, so both the rotor and shaft spin. The rotational energy is thus transferred to the shaft, which spins an electrical generator at the other end.

The ability to generate electricity is measured in units of power called watts. A kilowatt represents 1,000 watts, a megawatt is 1 million watts, and a gigawatt represents 1 billion watts. Electricity consumption and production are described in kilowatt-hours. Multiplying the number of kilowatts by the number of hours equals the kilowatt-hours. One kilowatt-hour equals the energy of one kilowatt produced or used for a period of one hour.

The turbine's size and the speed of the wind through the rotor determine the output of the tur-

The 1973 oil crisis contributed to increasing interest in renewable resources. Here, an Oregon gas station lists the new rules under which gas was available during the crisis. (Wikimedia Commons)

bine. As of July 2016, the world's largest turbine was the MHI Vestas V164, the first wind turbine with 8-megawatt-plus rated power. Wind turbines can generate electricity for an individual building or for widespread distribution by connecting to an electricity grid or network.

WIND FARMS

The world's largest onshore wind farm is Gansu Wind Farm in China, which had a capacity of 6,000 megawatts (MW) as of 2012—four times the capacity of the next largest, India's Muppandal Wind Farm. The largest wind farm in the United States is Alta Wind Energy Center in California, with a capacity of 1,320 MW as of 2013. Along with Germany and Spain, these three countries were the global leaders in wind power capacity as of 2014. At this time, wind power accounted for 21.1 percent of Spain's energy consumption, 8.9 percent of Germany's, and 8.5 percent of India's. Despite the higher capacities of the United States and China, wind power accounted for less than 5 percent of each country's energy consumption due to the countries' higher overall energy demands. All five countries focused their wind power efforts primarily on onshore wind farms.

Wind farms can also be sited offshore in the shallow waters of the oceans in order to capture the stronger winds. The United Kingdom is the world leader in offshore wind farms, and as of 2013, the world's two largest offshore wind farms (the London Array, with a capacity of 630 MW, and Greater Gabbard, 504 MW) were located in the country. Denmark, in second place internationally for offshore wind farming, hosts the third-largest offshore farm, Anholt (400 MW). The United States began construction of its first offshore wind farm off the coast of Rhode Island in 2015. Offshore wind farms have been somewhat more controversial than their onshore counterparts due to en-

Power generating wind turbines in the countryside. (jovan_epn via iStock)

vironmental concerns about their impact on sea life and birds and to complaints about their appearance ruining the scenery, which may affect tourism in some areas. In the United States, efforts to build offshore wind farms have also been fought by the wealthy oil industry, a factor which contributed to the delay and ultimate demise of Massachusetts's ambitious Cape Wind project.

ADVANTAGES AND DISADVANTAGES

The first advantage of wind energy is that the fuel is free. The main costs of generating electricity from wind are those of installation, operation, and maintenance. Earth has an abundant supply of wind power that can help promote energy independence from expensive imported energy worldwide and thus reduce national economic and security risks. Since 1973, the United States has spent more than $7 trillion on foreign oil. The wind industry has also created jobs and helped stabilize electricity costs. Since 1980, the cost of wind energy has dropped more than 80 percent to the present day.

Wind energy has significant long-term benefits for the environment, human health, and global climate change. Wind is a clean, renewable energy resource that is inexhaustible and easily replenished by nature. Wind power plants do not pollute the air or need waste cleanups like fossil-fuel and nuclear-generation plants, and wind turbines do not emit greenhouse gases or cause acid rain.

A significant disadvantage is that wind is inconsistent and intermittent. It is variable power that does not always blow at times of electrical demands. To be cost-effective, wind sites are located where both strong winds and land are available, usually in remote locations, far from large population centers where consumer demand is the greatest. For instance, China's major wind-energy resources are located in the Northern China wind belt, including the sparsely populated Xinjiang Uygur and the windy grasslands of Nei Monggol. Even in locations where winds are strong, there are wide differences in wind velocities over relatively short distances.

To meet these challenges, storage of surplus wind energy and electrical distribution systems to transmit this energy to consumers are necessary. Wind power can be stored in batteries, and technology already exists that can convert wind energy into fuels such as ethanol and hydrogen. However, economic feasibility is a major consideration. Enhanced electrical transmission systems improve reliability for consumers, relieve congestion in existing systems, and provide access to new and remote wind-generation sources. Typically, large wind plants are connected to the local electric utility transmission network. In the European Union, there is a proposal for a super grid of interconnected wind farms in Western Europe, including Denmark, England, Ireland, and France.

On a smaller scale, distributed energy is a viable solution. Consumers can make their own wind power with private wind turbine and batteries as backup. As more communities or individual consumers use distributed energy, they lower the costs of central wind power plants and transmission systems. Established in 1987, Southwest Windpower in Arizona is the world's leading producer of small wind turbines. Applications include offshore platform lighting, remote homes and cabins, utility-connected homes and businesses, water pumping, and telecommunications. Their wind generators often work as part of a hybrid wind-solar battery-charging system. Also, small domestic turbines can complement power from a larger electrical power system, and utility companies buy back any surplus electricity.

There are also aesthetic and environmental concerns surrounding the large-scale implementation of wind energy. Local residents and public advocacy groups have often opposed wind farms because of the rotor noise, visual impact, and potential harm to property values and local wildlife and its habitats. For instance, in 2001, residents concerned about the environment opposed the proposal of the Cape Wind Project, the first offshore wind farm in the country. In many cases, technological advances and the appropriate siting of the wind generators away from populated areas have mitigated the problems.

THE FUTURE OF WIND ENERGY

Wind energy is already one of the fastest-growing energy sources, and the market is forecast to expand in the future. Wind power is affordable, readily available, and renewable. Wind energy technology has developed to the point that it can compete successfully with conventional power generation technologies, such as oil, nuclear, coal, and most natural gas-fired generation.

Utility companies have increased investments in wind farms and wind technology. In 2005, General Electric's turbine business doubled, and by 2009, it was the leading U.S. wind turbine supplier and a world leader, with more than ten thousand wind turbine installations worldwide, comprising more than 15,000 megawatts of capacity. GE operates wind power manufacturing and assembly facilities in Spain, Canada, China, Germany, and the United States. In 2008, GE passed the $4 billion mark in investments in wind farms.

Vestas, a Danish company, is the only global energy company to focus solely on wind energy, and is another market leader in turbines and wind energy. They have installed more than 54,000 turbines in seventy-three countries, generating more than 90 million megawatt-hours of energy per year. As of 2015, nearly a quarter of those turbines were in the United States, though Vestas also has a significant presence in Europe.

Significantly, there has been increasing federal government support of wind power. In 2008, the U.S. Department of Energy released its groundbreaking technical report "Twenty Percent Wind Energy by 2030: Increasing Wind Energy's Contribution to U.S. Electricity Supply". More than one hundred people from government, industry, utilities, and nongovernment organizations worked on this report, which supports a scenario in which by 2030, wind power would supply 20 percent of U.S. electricity. Other benefits would be to reduce emissions of greenhouse gases by 25 percent, avoid consumption of 15 trillion liters of water, cut electric sector water consumption by 17 percent, create $2 billion in local annual revenues through jobs and other economic benefits, and reduce nationwide natural gas use by 11 percent with savings of $86-$214 billion for gas consumers. The Department of Energy has also researched the use of wind energy for hydrogen production, water treatment and irrigation, and hydropower applications.

The American Recovery and Reinvestment Act of 2009 provided measures to benefit renewable energy, including a Treasury Department grant program for renewable energy developers, increased funding for research and development, and a manufacturing tax credit. The ARRA also includes an extension of the wind-energy production tax credit to December 31, 2012. Consumers are allowed federal tax credits for energy efficiency, including tax credits of 30 percent of the cost of residential small wind turbines placed in service before December 31, 2016.

—*Thomas W. Weber and Alice Myers*

Further Reading

Bartmann, Dan, and Dan Fink. *Homebrew Wind Power: A Hands-on Guide to Harnessing the Wind.* Buckville, 2008.

Chiras, Dan, Mick Sagrillo, and Ian Woofenden. *Power from the Wind: Achieving Energy Independence.* New Society, 2009.

Craddock, David. *Renewable Energy Made Easy: Free Energy from Solar, Wind, Hydropower, and Other Alternative Energy Sources.* Atlantic, 2008.

Ding, Yu. *Data Science for Wind Energy.* CRC Press, 2020.

Foster, Robert, ed. *Wind Energy: Renewable Energy and the Environment.* CRC Press, 2009.

Gipe, Paul. *Wind Energy Energy for the Rest of Us: A Comprehensive Guide to Wind Power and How to Use It.* Wind Works, 2018.

Letcher, Trevor M., ed. *Wind Energy Engineering: A Handbook for Onshore and Offshore Wind Turbines.* Academic Press, 2017.

Nelson, Vaughn. *Wind Energy: Renewable Energy and the Environment.* 3rd ed. CRC Press, 2018.

Newton, David E. *Wind Energy: A Reference Handbook.* ABC-CLIO, 2015.

Stiebler, Manfred. *Wind Energy Systems for Electric Power Generation.* Springer, 2008.

"10 of the Biggest Turbines." *Windpower Monthly*, 26 July 2016, www.windpowermonthly.com/10-biggest-turbines. Accessed 16 Nov. 2016.

Valentine, Katie. "First Offshore Wind Farm in the US Kicks Off Construction." *ClimateProgress.* ThinkProgress, 28 Apr. 2015. Web. 13 May. 2015.

WORK AND ENERGY

FIELDS OF STUDY
Physics; Mechanics

ABSTRACT
Work is the act of changing the energy of a particle, body, or system. The energy of a mass represents the capacity of the mass to do work.

KEY CONCEPTS

foot-pound: typically a torque measurement due to a mass of one pound acting at a radius of one foot

inch-pound: typically a torque measurement due to a mass of one pound acting at a radius of one foot

joule: a force of one Newton acting through a distance of one meter

mass: an intrinsic property of matter arising from the cumulative atomic weights of the atoms composing the mass

Newton: the force required to accelerate a mass of one kilogram by one meter per second

scalar property: a property that has magnitude but no associated direction, such as mass and speed

vector property: a property that has both a scalar magnitude and an associated direction, such as velocity (speed in a specific direction)

MASS AND ENERGY

The energy of a mass represents the capacity of the mass to do work. Such energy can be stored and released. There are many forms that it can take, including mechanical, thermal, electrical, and magnetic. Energy is a positive, scalar quantity, although a change in energy can be either positive or negative. The total energy of a body can be calculated from its mass, m, and the specific energy, U (that is, the energy per unit mass):

$$E = mU.$$

Typical units of mechanical energy are foot-pounds and joules. A joule is equivalent to the units of $N \cdot m$ and $kg \cdot m^2 s^{-2}$. In countries that use traditional English units, the British thermal unit (Btu) is used for thermal energy, whereas the kilocalorie (kcal) is still used in some applications in countries that use the International System of Units (SI units). Joule's constant, or the Joule equivalent (778.26 ft-lbf/Btu), is used to convert between English mechanical units and thermal energy units, where ft-lbf indicates foot-pound force.

$$\text{Energy in Btu} = \text{energy in ft-lbf}/J$$

LAW OF CONSERVATION OF ENERGY

The law of conservation of energy says that energy cannot be created or destroyed. However, energy can be converted into different forms. Therefore, the sum of all energy forms is constant:

$$\Sigma E = \text{constant}$$

WORK

Work is the act of changing the energy of a particle, body, or system. For a mechanical system, external work is done by an external force, whereas internal work is done by an internal force. Work is a signed, scalar quantity. Typical units are inch-pounds, foot-pounds, and joules. Mechanical work is seldom expressed in British thermal units or kilocalories.

For a mechanical system, work is positive when a force acts in a direction of motion and helps a body move from one location to another. Work is negative when a force acts to oppose motion. Friction, for example, always opposes the direction of motion and can only do negative work. The work done on a body by more than one force can be found by superposition.

From a thermodynamic standpoint, work is positive if a particle or a body does work on its surroundings. Work is negative if the surroundings do work on the object. An example would be inflating a tire, which represents positive work being done to the tire by the air as it goes into the tire, but negative work by the tire as it resists the incoming air. This is consistent with the law of conservation of energy, since the sum of the negative work and positive energy increase is zero (that is, there is no net energy change in the system).

POTENTIAL ENERGY OF A MASS

Potential energy (gravitational energy) is a form of mechanical energy possessed by a body due to its relative position in a gravitational field. Potential energy is lost when the elevation of a body decreases. The lost potential energy usually is converted to kinetic energy or heat:

$$E_{potential} = mgh$$
$$\text{or}$$
$$E_{potential} = mgh/gc$$

In the absence of friction and other nonconservative forces, the change in potential energy of a body is equal to the work required to change the elevation of the body:

$$W = E_{potential}$$

KINETIC ENERGY OF A MASS
Kinetic energy is a form of mechanical energy associated with a moving or rotating body. The kinetic energy of a body moving with instantaneous linear velocity, v, is:

$$E_{kinetic} = (½) mv^2$$
or
$$E_{kinetic} = mv^2/2\, gc$$

The work-energy principle states that the kinetic energy is equal to the work necessary to initially accelerate a stationary body or to bring a moving body to rest:

$$W = E_{kinetic}$$

A body can also have rotational kinetic energy:

$$E_{rotational} = (½) I\omega^2$$
or
$$E_{rotational} = I\omega^2/2gc$$

SPRING ENERGY
A spring is an energy storage device, since the spring has the ability to perform work. In a perfect spring, the amount of energy stored is equal to the work required to compress the spring initially. The stored spring energy does not depend on the mass of the spring. Given a spring with spring constant (stiffness), k, the spring energy is as follows:

$$E_{spring} = (½) kx^2$$

where x is the distance the spring is compressed or extended.

INTERNAL ENERGY OF A MASS
The total internal energy, usually given the symbol U, of a body increases when the body's temperature increases. In the absence of any work done on or by the body, the change in internal energy is equal to the heat flow, Q, into the body. Q is positive if the heat flow is into the body and negative otherwise.

$$U2 - U1 = Q$$

The property of internal energy is encountered primarily in thermodynamics problems. Typical units are British thermal units, joules, and kilocalories.

WORK-ENERGY PRINCIPLE
As energy can be neither created nor destroyed, external work performed on a conservative system goes into changing the system's total energy. This is known as the work-energy principle (or principle of work and energy):

$$W = E = E_2 - E_1$$

The term *work-energy principle* is limited to use with mechanical energy problems, such as the conversion of work into kinetic or potential energy. When energy is limited to kinetic energy, the work-energy principle introduces some simplifications into many mechanical problems:

- It is not necessary to calculate or know the acceleration of a body to calculate the work performed on it.
- Forces that do not contribute to work—for example, are normal to the direction of motion—are eliminated.
- Only scalar quantities are involved.
- It is not necessary to individually analyze the particles of component parts in a complex system.

CONVERSION BETWEEN ENERGY FORMS
Conversion of one form of energy into another form of energy does not violate the law of conservation of energy. Most problems involving conversion of energy are really just special cases of the work-energy principle. An example is a falling body that is acted upon by a gravitational force. The conversion of potential energy into kinetic energy can be interpreted as equating the work done by the constant gravitational force to the change in kinetic energy.

POWER
Power is the amount of work done per unit time. It is a scalar quantity:

$$P = (W/t)$$

For a body acted upon by a force or torque, the instantaneous power can be calculated from the velocity.

$$P = Fv \text{ (linear systems)}$$
$$\text{or}$$
$$P = T\omega \text{ (rotational systems)}$$

Basic units of power are ft-lbf/sec and watts (J/s), although horsepower is also widely used. Some useful power conversion formulas include:

$$1\text{hp} = 550 \text{ (ft-lbf/sec)} = 33{,}000 \text{ (ft-lbf/min)} = 0.7457 \text{ kW} = 0.7068 \text{ (Btu/sec)}$$

$$1 \text{ kW} = 737.6 \text{ (ft-lbf/sec)} = 44{,}250 \text{ (ft-lbf/min)} = 1.341 \text{ hp} = 0.9483 \text{ (Btu/sec)}$$

$$1 \text{ (Btu/sec)} = 778.26 \text{ (ft-lbf/sec)} = 46{,}680 \text{ (ft-lbf/min)} = 1.415 \text{ hp}$$

EFFICIENCY

For energy-using systems (such as cars, electrical motors, and televisions), the energy-use efficiency, ?, of a system is the ratio of an ideal property to an actual property. The property used is commonly work, power, or, for thermodynamic problems, heat. When the rate of work is constant, either work or power can be used. Except in rare instances, the numerator and denominator of the ratio must have the same units:

$$\eta = (P_{ideal}/P_{actual}) = 1.0 \; (P_{actual} = P_{ideal})$$

For energy-producing systems (such as electrical generators, prime movers, and hydroelectric plants), the energy-production efficiency is:

$$\eta = (P_{actual}/P_{ideal}) \; (P_{ideal} \neq P_{actual})$$

The efficiency of an ideal machine is 1.0 (100 percent). However, all real machines have efficiencies of less than 1.0 due to friction and other energy losses.

—*Joanta Green*

Further Reading

Crowell, Benjamin. *Conservation Laws*. Light and Matter, 2003. http://www.lightandmatter.com/html_books/2cl/ch01/ch01.html#Section1.2.

Kanoglu, Mehmet, Yunus A. Çengel, and Ibrahim Dinçer. *Efficiency Evaluation of Energy Systems*. Springer, 2012.

Ross, John S. "Work, Power, Kinetic Energy." Project PHYSNET, Michigan State U, 23 Apr. 2002. http://www.physnet.org/modules/pdf_modules/m20.pdf.

Serway, Raymond A., and John W. Jewett. *Physics for Scientists and Engineers*. 8th ed. Brooks/Cole Cengage Learning, 2010.

Smil, Vaclav. *Energy in Nature and Society: General Energetics of Complex Systems*. MIT Press, 2008.

Tipler, Paul. *Physics for Scientists and Engineers: Mechanics*. 3rd ed. W.H. Freeman, 2008.

WORK-ENERGY THEOREM

FIELDS OF STUDY
Classical Mechanics; Thermodynamics

ABSTRACT
The work-energy theorem describes the relationship between work performed on an object and the kinetic energy of that object. It states that when some net amount of work is performed on an object, that object's kinetic energy will change. This theorem can be written mathematically to relate an object's mass and the change in its velocity to the amount of work performed—a useful way of connecting the motion and mass of an object to its capacity for work.

KEY CONCEPTS
conservation of energy: the principle that energy in the universe can be neither created nor destroyed, only transformed and transferred

displacement: the absolute distance an object moves from its starting point, regardless of the path it travels

kinematics: a subfield of classical mechanics that studies the motion of objects without reference to the forces that cause this motion

kinetic energy: the energy contained in an object due to its motion

net force: the sum of all forces acting on an object

potential energy: the energy stored within an object or system due to its position or configuration relative to the forces acting on it

total mechanical energy: the sum of the kinetic energy and the potential energy an object possesses as a result of work done on it

work: the successful displacement of an object caused by the application of a force

WORK AND KINETIC ENERGY

The work-energy theorem describes the relationship between kinetic energy (the energy of an object in motion) and work (the displacement of an object by a force). It states that when work is performed on an object, the kinetic energy of that object will change. When the kinetic energy of an object changes, it moves. So, in simple terms, performing work on an object causes it to move.

Because the work-energy theorem is concerned only with masses and velocities, not with forces, it is considered part of the field of kinematics. Kinematics is a subfield within classical mechanics that studies the motion of objects without regard for the forces causing the motion. Classical mechanics, in turn, is the branch of physics concerned with the physical laws that govern both the motion of objects and the forces that move them. Isaac Newton (1642-1727) laid the foundations for modern classical mechanics with his three laws of motion, published in the late seventeenth century.

WORK, ENERGY, AND FORCE

Mathematically, the work-energy theorem is represented by the following equation, where W is the total work performed on an object, ΔK is the change in the object's kinetic energy, m is the mass of the object, and v_i and v_f are its initial and final velocities, respectively:

$$W = \Delta K = \tfrac{1}{2}mv_f^2 - \tfrac{1}{2}mv_i^2$$

Thus, the total work done is equal to the total change in kinetic energy. In this sense, work can be thought of as the transfer of energy, if that transfer of energy results in displacement. Indeed, in the International System of Units (SI), the unit for both work and energy is the joule (J).

Consider a game of billiards. When the cue ball is in motion, it has kinetic energy. When it collides with another ball, it transfers some of its kinetic energy to the second ball. The force of the collision performs work on the second ball, causing it to move. This interaction underlines the relationship between energy, work, and force. In SI units, one joule represents the amount of work done or energy transferred when one newton (N) of force acts over a distance of one meter (m). In other words, if the cue ball exerts one newton of force on the second ball, causing it to be displaced by one meter, then one joule of energy has been transferred from the cue ball to the second ball, and one joule of work has been performed.

Forces in physics are interactions. According to Newton's second law, the net force (F; sum of the forces) acting on an object is equal to the object's mass (m) times its resulting acceleration (a):

$$F = ma$$

In turn, the work (W) done by that force is equal to the net force (F) applied times the resulting displacement (s) of the object:

$$W = Fs$$

Displacement is the absolute distance and direction an object has moved from its starting position, ignoring the path taken. Therefore, a car that drove in a perfect circle and stopped exactly where it started would have a displacement of zero, no matter how large the circle it traveled. Similarly, a car that drove ten miles east, made a U-turn, and drove back five miles west would only have a displacement of five miles east, even though it traveled fifteen miles total.

The equation for work reveals that in order for a force to have performed work on an object, the displacement of that object must have a nonzero value. In other words, the object has to have moved. If an applied force does not result in displacement, no work has been done.

CONSERVATION OF ENERGY

The law of conservation of energy states that in an isolated system, energy is conserved. An isolated system is one from which neither matter nor energy can escape. The universe is, in theory, the ultimate isolated system. Thus, according to this law, energy in the universe is never created or destroyed; it can only be transformed or transferred. The work-energy theorem is an extension of the law of conservation of energy, rewritten in a usable form.

Not all energy is kinetic energy. Energy can exist in a variety of forms. One such form is potential energy, which is energy that is stored in an object or system until it can be converted to another form of energy to do work. Potential energy itself comes in different forms, such as gravitational potential energy and chemical potential energy. The human body makes use of the chemical potential energy that exists in food due to its molecular configuration. When food is digested, it undergoes chemical reactions that break down its molecules and convert some of this chemical potential energy into the thermal energy of body heat and the kinetic energy of moving limbs and beating hearts. A combustion engine similarly converts the chemical potential energy of the fuel into the kinetic energy of the moving pistons that transfer the power in the engine.

The principle of conservation of energy is useful when examining any isolated system. Consider the billiards example again. The billiards table can be treated as an isolated system, because the balls stay on the table and any energy from the environment (heat from overhead lights, the kinetic energy in a gust of air) has such a small effect that it can be ignored. Therefore, when two balls collide, the total amount of energy in the system must remain the same before and after the impact. Kinetic energy is simply transferred from one ball to the other. A miniscule amount might be converted to thermal energy due to friction with the table.

Often, when considering some kinematic interaction, it is useful to know the total mechanical energy of the objects at play. The total mechanical energy of an object or system is simply the sum of its potential and kinetic energies. In the real world, total mechanical energy is not typically conserved, because friction must be taken into account. Consider a driver in a speeding car who suddenly slams on the brakes. As the car's tires stop rotating and start sliding across the surface of the pavement, they generate friction. Friction converts kinetic energy into thermal energy, which is not a form of mechanical energy. This energy then dissipates away from the tire tracks into the surrounding environment.

THE WORK-ENERGY THEOREM IN EVERYDAY LIFE

The work-energy theorem is useful whenever the effect of work on the motion of an object is of interest. For example, understanding how the chemical potential energy in a fuel source performs work when it is released and converted into kinetic energy is an essential part of engineering efficient combustion engines. The fuel has to contain enough energy to move the pistons without breaking them.

Countless other devices in modern life also convert potential energy into kinetic energy in order to perform work. Everyday examples include vacuum cleaners, clocks, and fans. By expressing the relationship between energy transfer (work) and the motion (kinetic energy) of these objects in easy-to-measure terms (mass and velocity), the work-energy theorem makes engineering these devices possible.

—*Kenrick Vezina, MS*

Further Reading

Allain, Rhett. "What's the Difference between Work and Potential Energy?" *Wired*. Condé Nast, 1 July 2014. Web. 22 Sept. 2015.

Boleman, Michael. "Experiment # 6: Work-Energy Theorem." *Mr. Boleman's Course Information*. U of South Alabama, n.d. Web. 22 Sept. 2015.

"Energy, Kinetic Energy, Work, Dot Product, and Power." *MIT OpenCourseWare*. MIT, 13 Oct. 2004. Web. 22 Sept. 2015.

Henderson, Tom. *Kinematics*. N.p.: Physics Classroom, 2013. Digital file.

Nave, Carl R. "Work, Energy and Power." *HyperPhysics*. Georgia State U, 2012. Web. 22 Sept. 2015.

Shankar, Ramamurti. "Lecture 5: Work-Energy Theorem and Law of Conservation of Energy." *Open Yale Courses*. Yale U, 2006. Web. 22 Sept. 2015.

Simanek, Donald E. "Kinematics." *Brief Course in Classical Mechanics*. Lock Haven U, Feb. 2005. Web. 22 Sept. 2015.

Bibliography

"9 Out of 10 People Worldwide Breathe Polluted Air, but More Countries Are Taking Action." World Health Organization, 2 May 2018, www.who.int/news-room/detail/02-05-2018-9-out-of-10-people-worldwide-breathe-polluted-air-but-more-countries-are-taking-action. Accessed 9 Oct. 2018.

"10 of the Biggest Turbines." *Windpower Monthly*, 26 July 2016, www.windpowermonthly.com/10-biggest-turbines. Accessed 16 Nov. 2016.

Abdel Rahman, Rehab O., and Hosam El-Din Em Saleh, eds. *Principles and Applications in Nuclear Engineering-Radiation Effects, Thermal Hydraulics, Radionuclide Migration in the Environment.* IntechOpen, 2018.

Abersek, B., and J. Flašker. *How Gears Break.* WIT, 2004.

Ableman, Michael. *From the Good Earth: A Celebration of Growing Food Around the World.* H.N. Abrams, 1993.

Aguado, Edward, and James E. Burt. *Understanding Weather and Climate.* 5th ed. Prentice Hall, 2009.

Akli, Michaël, Patrick Bayer, S.P. Harish. and Johannes Urpelainen. *Escaping the Energy Poverty Trap: When and How Governments Power the Lives of the Poor.* MIT Press, 2018.

Alami, Abdul Hai. *Mechanical Energy Storage for Renewable and Sustainable Energy Resources.* Springer, 2020.

Al-Bahadly, Ibrahim H., ed. *Energy Conversion. Current Technologies and Future Trends.* IntechOpen, 2019.

Allain, Rhett. "What's the Difference between Work and Potential Energy?" *Wired.* Condé Nast, 1 July 2014. Web. 22 Sept. 2015.

Allen, Robert C. *The British Industrial Revolution in Global Perspective.* Cambridge UP, 2009.

Al-Megren, Hamid A., and Rashid H. Altamimi, eds. *Advances in Natural Gas Emerging Technologies.* IntechOpen, 2017.

Alonso, Gustavo, Edmundo del Valle, and Jose Ramon Ramirez. *Desalination in Nuclear Power Plants.* Woodhead, 2020.

American Industrial Hygiene Association. *Radio-Frequency and Microwave Radiation.* 3rd ed. American Industrial Hygiene Association, 2004.

American Institute of Physics. "Josiah Willard Gibbs, 1839-1903." http://www.aip.org/history/gap/Gibbs/Gibbs.html.

American Society of Civil Engineers. *Guidelines for Electrical Transmission Line Structural Loading.* ASCE, 2020.

Amidpour, Majesh, and Mohammad Hasan Khoshgoftar Manesh. *Cogeneration and Polygeneration Systems.* Academic Press, 2021.

Ampère, André-Marie, and Henriette Cheuvreux. *The Story of His Love: Being the Journal and Correspondence of André-Marie Ampère, with His Family Circle During the First Republic, 1793-1804.* R. Bentley and Son, 1873.

"Anatomy of an Electromagnetic Wave." *Mission: Science.* NASA, 13 Aug. 2014. Web. 29 June 2015.

Anderson, Rosaleen J., David J. Bendell, and Paul W. Groundwater. *Organic Spectroscopic Analysis.* Royal Society of Chemistry, 2004.

Annaratone, Donatello. *Transient Heat Transfer.* Springer, 2011.

Annual Energy Outlook 2017, with Projections to 2050. U.S. Energy Information Administration, 5 Jan. 2017, www.eia.gov/outlooks/aeo/pdf/0383(2017).pdf. Accessed 10 Oct. 2018.

Atkins, Peter. *Four Laws That Drive the Universe.* Oxford UP, 2007.

Attenberg, Roger H., ed. *Global Energy Security.* Nova Science Publishers, 2009.

Avallone, Eugene A., Theodore Baumeister III, and Ali M. Sadegh. "Steam Turbines." *Marks' Standard Handbook for Mechanical Engineers.* 11th ed. McGraw-Hill, 2007.

Aziz, Muhammad, ed. *Exergy and Its Application-Toward Green Energy Production and Sustainable Environment.* InTech Open, 2019.

Babarit, Aurélien. *Ocean Wave Energy Conversion: Resource, Technologies and Performance.* Elsevier, 2017.

Badescu, Viorel. *Modeling Solar Radiation at the Earth's Surface: Recent Advances.* Springer, 2008.

Bailey, Regina. "All About Cellular Respiration." *ThoughtCo.*, 6 May 2019, www.thoughtco.com/cellular-respiration-process-373396. Accessed 3 Mar. 2020.

Bailey, Ronald. "The Wizard Who Electrified the World." *American History*, June 2010.

Bain, Don. "We Should Have Celebrated Rudolph Diesel's Birthday Yesterday." *Torque News*, 19 Mar. 2012, http://www.torquenews.com/397/we-should-have-celebrated-rudolph-diesel%e2%80%99s-birthday-yesterday.

Bainbridge, David, and Ken Haggard. *Passive Solar Architecture: Heating, Cooling, Ventilation, Daylighting, and More Using Natural Flows.* Chelsea Green, 2011.

Baird, Davis, R.I.G. Hughes, and Alfred Nordmann. *Heinrich Hertz: Classical Physicist, Modern Philosopher*. Kluwer Academic, 1998.

Baker, Keith, and Gerry Stoker. *Nuclear Power and Energy Policy: The Limits to Governance*. Palgrave, 2015.

Baker, R.J., and R.K. Chesser. "The Chernobyl Nuclear Disaster and Subsequent Creation of a Wildlife Preserve." *Environmental Toxicology and Chemistry*, vol. 19, no. 5, 2000.

Barber, James. "Biological Solar Energy." *Energy ... Beyond Oil*. Ed. Fraser Armstrong and Katherine Blundell, 137-55. Oxford UP, 2007.

Barry, Roger Graham, and Richard J Chorley. *Atmosphere, Weather, and Climate*. Routledge, 2010.

Bartis, James T., et al. *Oil Shale Development in the United States: Prospects and Policy Issues*. Rand Institute, 2005.

Bartmann, Dan, and Dan Fink. *Homebrew Wind Power: A Hands-on Guide to Harnessing the Wind*. Buckville, 2008.

Bartnik, Ryszard, and Zbigniew Buryn. *Conversion of Coal-Fired Power Plants to Cogeneration and Combined-Cycle: Thermal and Economic Effectiveness*. Springer, 2011.

Barton, D.B. *The Cornish Beam Engine: A Survey of Its History and Development in the Mines of Cornwall and Devon from Before 1800 to the Present Day, with Something of Its Use Elsewhere in Britain and Abroad*. New ed. Author, 1966.

Basso, Thalita Peixoto, and Luiz Carlos Basso, eds. *Fuel Ethanol Production from Sugarcane*. IntechOpen, 2019.

Bastian-Pinto, Carlos, et al. "Valuing the Switching Flexibility of the Ethanol-Gas Flex Fuel Car." *Annals of Operations Research*, vol. 176, 2010.

Basu, Prabir, and Scott A. Fraser. *Circulating Fluidized Bed Boilers Design and Operations*. Butterworth-Heineman, 2013.

Batchelor, Tony, and Robin Curtis. "Geothermal Energy." *Energy: Beyond Oil*. Ed. Fraser Armstrong and Katherine Blundell. Oxford UP, 2007.

Baxter, R. *Energy Storage: A Non-Technical Guide*. PenWell Corporation, 2006.

Beith, Robert, ed. *Small and Micro Combined Heat and Power (CHP) Systems. Advanced Design, Performance, Materials and Applications*. Woodhead Publishing, 2011.

Benjamin, David S. *Embodied Energy and Design: Making Architecture Between Metrics and Narratives*. Columbia UP, 2017.

Ben-Naim, Arieh. *Entropy Demystified: The Second Law Reduced to Plain Common Sense with Seven Simulated Games*. Expanded ed. World Scientific, 2008.

———. *The Four Laws That Do Not Drive te Universe: Elements of Thermodynamics for the Curious and Intelligent*. World Scientific, 2018.

Bennion, K., and M. Thornton. "Fuel Savings from Hybrid Electric Vehicles." National Renewable Energy Council Technical Report. Mar. 2009. http://www.nrel.gov/docs/fy09osti/42681.pdf.

Berger, Lars T., and Krzysztof Iniewski. *Smart Grid Applications, Communications, and Security*. John Wiley & Sons, 2012.

Bergman, T.L. *Fundamentals of Heat and Mass Transfer*. Wiley, 2011.

Berlow, Lawrence. "Battery." *How Products Are Made*. 2012.

Berners-Lee, Mike, and Duncan Clark. *The Burning Question: We Can't Burn Half the World's Oil, Coal, and Gas. So How Do We Quit?* Greystone, 2013.

Bhandari, V.B. *Design of Machine Elements*. 3rd ed. Tata/McGraw-Hill, 2010.

Biberian, Jean-Paul, ed. *Cold Fusion: Advances in Condensed Matter Nuclear Science*. Elsevier, 2020.

Bilgili, Faik, and Ilhan Ozturk. "Biomass Energy and Economic Growth Nexus in G7 Countries: Evidence from Dynamic Panel Data." *Renewable and Sustainable Energy Reviews*, vol. 49, 2015, pp. 132-38.

"Biomass Explained." *US Energy Information Administration*. Department of Energy, 17 July 2015. Web. 13 May 2016.

Biswas, Gautam, Amaresh Dalal, and Vijay K. Dhir. *Fundamentals of Convective Heat Transfer*. CRC Press, 2019.

Biswas, Jagadish, and Agnimitra Biswas. *Modeling and Optimization of Solar Thermal Systems: Emerging Research and Opportunities*. IGI Global, 2021.

Bitterly, J.G. "Flywheel Technology: Past, Present, and 21st Century Projections." *Aerospace and Electronic Systems Magazine*, vol. 13, no. 8, Aug. 1998.

Black, Edwin. *Internal Combustion: How Corporations and Governments Addicted the World to Oil and Derailed the Alternatives*. St. Martin's Press, 2006.

Blewitt, Laura, and Mario Parker. "Think Crude's Cheap? Biodiesel's Going for Free in Some Places." *Bloomberg*. Bloomberg, 2 Feb. 2016. Web. 13 May 2016.

Bloch, Heinz P. *Steam Turbines: Design, Applications, and Rerating*. 2nd ed. McGraw-Hill, 2009.

Blondel, Christine, and Bertrand Wolff. "Electricité animale ou électricité métallique? La Controversé; Galvani-Volta et l'invention de la pile." Mar. 2007. http://www.ampere.cnrs.fr/parcourspedagogique/zoom/galvanivolta/controverse/index.php.

Bodansky, David. *Nuclear Energy: Principles, Practices, and Prospects*. 2nd ed. Springer, 2004.

Boden, David R. *Geologic Fundamentals of Geothermal Energy*. CRC Press, 2017.

Bodner Research Web. "Gibbs Free Energy." http://chemed.chem.purdue.edu/genchem/topicreview/bp/ch21/gibbs.php.

Boleman, Michael. "Experiment # 6: Work-Energy Theorem." *Mr. Boleman's Course Information*. U of South Alabama, n.d. Web. 22 Sept. 2015.

Borjesson, Pal. "Good or Bad Bioethanol from a Greenhouse Gas Perspective: What Determines This?" *Applied Energy*, vol. 86, 2009.

Bose, Debajyoti, Krishnam Goyal, and Vidushi Bhardwaj. *Design and Development of a Parabolic Solar Concentrator and Integration with a Solar Desalination System*. GRIN, 2016.

Boyce, Meherwan P. *Gas Turbine Engineering Handbook*. Gulf Professional, 2006.

Boyle, Godfrey, ed. *Renewable Energy: Power for a Sustainable Future*. Oxford UP, 2012.

Brabec, Christoph, Ulrich Scherf, and Vladimir Dyakonov, eds. *Organic Photovoltaics: Materials, Device Physics and Manufacturing Technologies*. 2nd ed. John Wiley & Sons, 2014.

Bray, Stan. *Making Simple Model Steam Engines*. Crowood, 2005.

Breeze, Paul. *Combined Heat and Power*. Academic Press, 2018.

———. *Fuel Cells*. Academic Press, 2017.

———. *Power System Energy Storage Technologies*. Academic Press, 2018.

Brown, George I. *Count Rumford: The Extraordinary Life of a Scientific Genius; Scientist, Soldier, Statesman, Spy*. Sutton, 1999.

Brown, Robert C., ed. *Thermochemical Processing of Biomass: Conversion into Fuels, Chemicals and Power*. 2nd ed. Wiley, 2011.

Buchmann, Isidor. *Batteries in a Portable World: A Handbook on Rechargeable Batteries for Non-Engineers*. 3rd ed. Cadex Electronics Inc.

Buchwald, Jed Z. *The Creation of Scientific Effects: Heinrich Hertz and Electric Waves*. U of Chicago P, 1994.

Buckeridge, Marcos Silveira, and Gustavo Henrique Goldman. *Routes to Cellulosic Ethanol*. Springer, 2011.

Build It Solar. "Solar Pool (and Hot Tub) Heating." http://www.builditsolar.com/Projects/PoolHeating/pool_heating.htm.

Bumpus, Adam G., James Tansey, Blas L. Pérez Henriquez, and Chukwumerije Okereke, eds. *Carbon Governance, Climate Change and Business Transformation*. Routledge, 2016.

Bumstead, H.A., and R.G. Van Name, eds. *The Scientific Papers of J. Willard Gibbs*. 2 vols. Longmans, Green and Co., 1906.

Bunch, Bryan. *The History of Science and Technology*. Houghton Mifflin, 2004.

Bundschuh, Jochen, and Barbara Tomaszewska, eds. *Geothermal Water Management*. CRC Press, 2018.

Bunn, James H. *The Natural Law of Cycles: Governing the Mobile Symmetries of Animals and Machines*. Transaction Publishers, 2014.

Burton, Tony, David Sharpe, Nick Jenkins, and Ervin Bossanyi. *Wind Energy Handbook*. Wiley, 2001.

Bynyard, Peter. "Pursuing the Energy Efficiency-Conservation Path." *Energy Policy*, 1988.

Cabeza, Luisa F., ed. *Advances in Thermal Energy Storage Systems: Methods and Applications*. 2nd ed. Woodhead Publishing, 2020.

Cahan, David, ed. *Hermann von Helmholtz and the Foundations of Nineteenth-Century Science*. U of California P, 1994.

Cahan, David. *Helmholtz: A Life in Science*. U of Chicago P, 2018.

Campbell, A. Malcolm, and Christopher J. Paradise. *Cellular Respiration*. Momentum Press, 2016.

Cangeloso, Sal. *LED Lighting: A Primer to Lighting the Future*. O'Reilly, 2012.

Cao, Jing. *China's Energy Intensity: Past Performance and Future Implications*. Economy and Environment Program for Southeast Asia, 2010.

Cardwell, Donald S.L. *From Watt to Clausius: The Rise of Thermodynamics in the Early Industrial Age*. Cornell UP, 1971.

———. *James Joule: A Biography*. Manchester UP, 1989.

Carlisle, Juliet E., et al. "The Public's Trust in Scientific Claims Regarding Offshore Oil Drilling." *Public Understanding of Science*, vol. 19, no. 5, 2010.

Carlson, W. Bernard. "Inventor of Dreams." *Scientific American*, vol. 292, no. 3, 2005.

———. *Tesla: Inventor of the Electrical Age*. Princeton UP, 2013.

Carnot, Sadi. *Reflection on the Motive Power of Fire*. Dover, 1960.

———. *Reflexions on the Motive Power of Fire: A Critical Edition with the Surviving Scientific Manuscripts*. Robert Fox, trans., ed. Manchester UP, 1986.

Carter, James F., and Lesley A. Chesson, eds. *Food Forensics: Stable Isotopes as a Guide to Authenticity and Origin*. CRC Press, 2017.

Carter, Jimmy. *Public Papers of the Presidents of the United States Jimmy Carter 1978 January 1 to June 30, 1978.* https://books.google.ca/books?id=mVPVAwAAQBAJ&pg=PA574&dq=sun+day+Jimmy+Carter&hl.

Casazza, Jack, and Frank Delea. *Understanding Electric Power Systems: An Overview of Technology, the Marketplace, and Government Regulations.* 2nd ed. IEEE Press/Wiley, 2010.

"Cellular Respiration." *IUPUI Department of Biology,* February 2004, www.biology.iupui.edu/biocourses/N100/2k4ch7respirationnotes.html. Accessed 3 Mar. 2020.

Chaichian, Masud, Hugo Perez Rojas, and Anca Tureanu. *Basic Concepts in Physics: From the Cosmos to Quarks.* Springer, 2014.

Chaisson, Eric J. *Cosmic Evolution: The Rise of Complexity in Nature.* Harvard UP, 2001.

Chamberlin, J. Edward. *Horse: How the Horse Has Shaped Civilizations.* BlueBridge, 2006.

Chandra, Laltu, and Ambesh Dixit, eds. *Concentrated Solar Thermal Energy Technologies: Recent Trends and Applications.* Springer, 2018.

Charles River Associates. *Primer on Demand-Side Management with an Emphasis on Price-Responsive Programs.* CRA D06090. World Bank, 2005. http://siteresources.worldbank.org/INTENERGY/Resources/PrimeronDemand-SideManagement.pdf.

Charlier, Roger H., and Charles W. Finkl. *Ocean Energy: Tide and Tidal Power.* Springer, 2009. http://www.springerlink.com/content/x7u431/#section=30137&page=1.

Chastko, Paul. *Developing Alberta's Oil Sands: From Karl Clark to Kyoto.* U of Calgary P, 2004.

Chaudhuri, Uttam Ray. *Fundamentals of Petroleum and Petroleum Engineering.* CRC Press, 2011.

Chenevix Trench, Charles. *A History of Horsemanship.* Doubleday, 1970.

Childs, Peter R.N. *Mechanical Design Engineering Handbook.* Elsevier, 2014.

Ching, Francis D.K., and Ian M. Shapiro. *Green Building Illustrated.* Wiley, 2014.

Chiras, Dan, Mick Sagrillo, and Ian Woofenden. *Power from the Wind: Achieving Energy Independence.* New Society, 2009.

Chopey, Nicholas P., et al. *Handbook of Chemical Engineering.* McGraw-Hill, 1984.

Chudnovsky, Bella. *Electrical Power Transmission and Distribution: Aging and Life Extension Techniques.* CRC Press, 2013.

Clark, James, and Fabien Deswarte, eds. *Introduction to Chemicals from Biomass.* 2nd ed. Wiley, 2015.

Clarke, Tony. *Tar Sands Showdown: Canada and the New Politics of Oil in an Age of Climate Change.* J. Lorimer, 2008.

Clayton, Blake C. *Market Madness: A Century of Oil Panics, Crises, and Crashes.* Oxford UP, 2015.

Clean Up Australia, Ltd. "Battery Recycling Fact Sheet Clean Up." May 2010. http://www.cleanup.org.au/PDF/au/cua_battery_recycling_factsheet.pdf.

Cleveland C. "Energy Quality, Net Energy, and the Coming Energy Transition." http://www.oilcrisis.com/cleveland/EnergyQualityNetEnergyComingTransition.pdf.

———. "Net Energy from Oil and Gas Extraction in the United States, 1954-1997." *Energy,* vol. 30, 2005.

"Climate Neutrality." http://www.firstclimate.com/climate-neutral-services/climate-neutrality.html.

Clutton-Brock, Juliet. *A Natural History of Domesticated Mammals.* 2nd ed. Cambridge UP, Natural History Museum, 1999.

———. *Domesticated Animals from Early Times.* U of Texas P, 1981.

———. *Horse Power: A History of the Horse and the Donkey in Human Societies.* Harvard UP, 1992.

"Coal & Electricity." *World Coal Association.* World Coal Association, 22 Oct. 2015. Web. 13 May 2016.

"Coal. International Energy Agency." OECD/IEA, 28 May 2015. Web. 13 May 2016.

Coiro, Domenico P., and Tonio Sant. *Renewable Energy from the Ocean: From Wave, Tidal and Gradient Systems to Offshore Wind and Solar,* Institution of Engineering and Technology, 2019.

Colborn, T., C. Kwiatkowski, K. Schultz, and M. Bachran. "Natural Gas Production from a Public Health Perspective." *International Journal of Human and Ecological Risk Assessment,* vol. 17 no. 5, 2011.

Colburn, Zerah. *The Construction and Operation of Steam Engines for Locomotives.* Salzwasser Verlag, 2010.

Collier, James Lincoln. *The Steam Engine.* Marshall Cavendish, 2006.

Collins, M.W., R.C. Dougal, C.S. König, and I.S. Ruddock, eds. *Kelvin, Thermodynamics and the Natural World.* WIT Press, 2016.

Combined Heat and Power Association. "History of the CHPA." http://www.chpa.co.uk.

Committee to Review the U.S. ITER Science Participation Planning Process. *A Review of the DOE Plan for U.S.*

Fusion Community Participation in the ITER Program. National Academies Press, 2009.

"Communications on Energy: Household Energy Conservation." *Energy Policy*, 1984.

Conaway, Charles F. *The Petroleum Industry: A Nontechnical Guide.* PennWell Books, 1999.

Conejo, Antonio J., Luis Baringo, S. Jala Kazampour, and Afzal S. Siddiqui. *Investment in Electricity Generation and Transmission: Decision Making Under Uncertainty.* Springer, 2016.

Conlin, R. "On the Path to Climate Neutrality." *Yes*, vol. 10, 2010. http://www.yesmagazine.org/blogs/richard-conlin/on-the-path-to-climate-neutrality.

Cook, N. *Refrigeration and Air-Conditioning Technology.* Macmillan, 1995.

Cook, Peter J. *Clean Energy, Climate and Carbon.* CSIRO Publishing, 2012.

Cooper, Gail. *Air-Conditioning America: Engineers and the Controlled Environment, 1900-1960.* Johns Hopkins UP, 2002.

Cortez, Luís A.B., Manoel Regis L.V. Leal, and Luiz A. Horta Nogueira, eds. *Sugarcane Bioenergy for Sustainable Development. Expanding Production in Latin America and Africa.* Routledge, 2019.

Costanza R. "Embodied Energy and Economic Valuation." *Science*, vol. 210, no. 4475, 12 Dec. 1980.

Costanza, Robert, and Cutler Cleveland. "Net Energy/Full Cost Accounting: A Framework for Evaluating Energy Options and Climate Change Strategies." http://summits.ncat.org/docs/EROI.pdf.

Cottrell, Michelle. *Guidebook for the LEED Certification Process.* Wiley, 2011.

Cox, Lydia. "The Surprising Decline in US Petroleum Consumption." *World Economic Forum*. World Economic Forum, 10 July 2015. Web. 16 May 2016.

Craddock, David. "Catching a Wave: The Potential of Wave Power." *Renewable Energy Made Easy: Free Energy from Solar, Wind, Hydropower, and Other Alternative Energy Sources.* Atlantic, 2008.

Craddock, David. *Renewable Energy Made Easy: Free Energy from Solar, Wind, Hydropower, and Other Alternative Energy Sources.* Atlantic, 2008.

Crawford, Robert H. *Licycle Assessment in the Built Environment.* Spon Press, 2011.

Crocker, Mark, ed. *Thermochemical Conversion of Biomass to Liquid Fuels and Chemicals.* RSC Publishing, 2010.

Cropper, William H. *Great Physicists: The Life and Times of Leading Physicists from Galileo to Hawking.* Oxford UP, 2001.

Crowell, Benjamin. *Conservation Laws*. Light and Matter, 2003. http://www.lightandmatter.com/html_books/2cl/ch01/ch01.html#Section1.2.

———. *Light and Matter*. Light and Matter, 2003. http://www.lightandmatter.com/html_books/2cl/ch01/ch01html#Section1.2.

Crowther, J.G. *Famous American Men of Science.* Books for Libraries Press, 1969.

Crump, Thomas. *A Brief History of the Age of Steam: The Power That Drove the Industrial Revolution.* Carroll & Graf, 2007.

Cruz, João, ed. *Ocean Wave Energy: Current Status and Future Perspectives.* Springer, 2008.

Cubitt, Sean, et al. "Does Cloud Computing Have a Silver Lining?" *Media Culture and Society*, vol. 32, no. 6, Nov. 2010.

Cundiff, John S., and Michael F. Kocher. *Fluid Power Circuits and Controls: Fundamentals and Applications.* 2nd ed. CRC Press, 2020.

Curley, Robert, ed. *Fossil Fuels: Energy Past, Present, and Future.* Britannica Educational, 2012.

Cusick, Marie, and Susan Phillips. "U.S. Proposes New Safety Rules for Natural Gas Pipelines." *StateImpact: Pennsylvania*. WITF, 18 Mar. 2016. Web. 23 May 2016.

Da Rosa, Aldo Vieira. *Fundamentals of Renewable Energy Processes.* 3rd ed. Elsevier, 2012.

Daines, James R. *Fluid Power: Hydraulics and Pneumatics.* Goodheart-Wilcox Company, 2012.

D'Andrade, Brian W., ed. *The Power Grid: Smart, Secure, Green and Reliable.* Academic Press, 2017.

Danish Board of District Heating. "Danish District Heating History." http://www.dbdh.dk/artikel.asp?id=464&mid=24.

Darlngton, Roy, and Keith Strong. *Stirling and Hot Air Engines: Designing and Building Experimental Model Stirling Engines.* Crowood Press, 2005.

Dawson, Todd E., and Rolf T.W. Siegwolf, eds. *Stable Isotopes as Indicators of Ecological Change.* Elsevier, 2007.

"DC Circuit Theory." *Basic Electronics Tutorials*. Wayne Storr, n.d. Web. 8 June 2015.

De Jong, Wiebren, and J. Ruud van Ommen, eds. *Biomass as a Sustainable Energy Source for the Future: Fundamentals of Conversion Processes.* Wiley, 2015.

De La Rosa, Francisco C. *Harmonics, Power Systems and Smart Grids.* 2nd ed. CRC Press, 2015.

De Pree, Christopher G. *Physics Made Simple.* Three Rivers, 2013. Digital file.

"The Death of the Internal Combustion Engine." *The Economist*, 12 Aug. 2017, www.economist.com/news/

leaders/21726071-it-had-good-run-end-sight-machine-changed-world-death. Accessed 23 Aug. 2017.

Declaye, Sébastien. *Design, Optimization, and Modeling of an Organic Rankine Cycle for Waste Heat Recovery.* June 2009. http://www.labothap.ulg.ac.be/cmsms/uploads/File/TFE_SD090623.pdf.

Deffeyes, Kenneth S. *Hubbert's Peak: The Impending World Oil Shortage.* Princeton UP, 2008.

Delbeke, Jos, and Peter Vis, eds. *Towards a Climate-Neutral Europe: Curbing the Trend.* Routledge, 2019.

DeMarco, C.L. "Trends in Electrical Transmission and Distribution Technology." *Transmission and Distribution Technology*, 1 Feb. 2006. http://www.mrec.org/pubs/06%20MREC%20DeMarco%20PSERC.pdf.

Demirel, Yasar. *Energy: Production, Conversion, Storage, Conservation, and Coupling.* Springer, 2012.

Denby, David. "More Heat Than Light." *New York Magazine*, 2 Apr. 1979.

Desmond, Kevin. *Innovators in Battery Technology. Profiles of 95 Influential Electrochemists.* McFarland & Company, 2016.

Dias de Moraes, M.A.F., and David Zilberman. *Production of Ethanol from Sugarcane in Brazil from State Intervention to a Free Market.* Springer, 2014.

Dibner, Bern. *Alessandro Volta and the Electric Battery.* F. Watts, 1964.

Dick, Malcolm, and Caroline Archer-Parré, eds. *James Watt (1736-1819): Culture, Innovation, and Enlightenment.* Liverpool UP, 2020.

Dickinson, H.W. *James Watt: Craftsman and Engineer.* Augustus M. Kelley, 1936.

Dickson, Mary H., and Mario Fanelli, eds. *Geothermal Energy: Utilization and Technology.* 1995. Reprint. Earthscan, 2005.

———. *Geothermal Energy: Utilization and Technology.* United Nations Educational, Scientific and Cultural Organization, 2003.

Digital Communities. "21st-Century Smart Grids Update: U.S. Electric Grid." http://www.digitalcommunities.com/templates/gov_print_article?id=99859074.

Dimitrov, Alexander V. *Energy Modeling and Computations in the Building Envelope.* CRC Press, 2016.

Dincer, Ibrahim, and Marc A. Rosen. *Exergy: Energy Environment and Sustainable Development.* 2nd ed. Elsevier, 2013.

Dincer, Ibrahim, and Osamah Siddiqui. *Ammonia Fuel Cells.* Elsevier, 2020.

Ding, Yu. *Data Science for Wind Energy.* CRC Press, 2020.

DiPippo, Ronald. *Geothermal Power Generation: Developments and Innovation.* Woodhead Publishing/Elsevier, 2016.

Dixit, M.K., J.L. Fernandez-Solis, S. Lavy, and C.H. Culp. "Identification of Parameters for Embodied Energy Measurement: A Literature Review." *Energy and Buildings*, vol. 42, 2010.

Donateo, Teresa, ed. *Hybrid Electric Vehicles.* IntechOpen, 2017.

Dowling, John. "The China Syndrome." *Bulletin of the Atomic Scientists*, vol. 35, no. 6, 1979.

Drapcho, Caye M., Nghiem Phu Nhuan, and Terry H. Walker. *Biofuels Engineering Process Technology.* 2nd ed. McGraw-Hill, 2020.

Drbal, L., K. Westra, and P. Boston. *Power Plant Engineering.* Springer, 1996.

Duffie, J.A., and W.A. Beckman. *Solar Engineering of Thermal Processes.* 3rd ed. Wiley, 2006.

Duffield, John S. *Fuels Paradise: Seeking Energy Security in Europe, Japan, and the United States.* Johns Hopkins UP, 2015.

Duhé, Sonya. "Crisis Communication." *Encyclopedia of Science and Technology Communication*, edited by S.H. Priest. Sage, 2010.

Durakovic, Benjamin. *PCM-Based Building Envelope Systems: Innovative Energy Solutions for Passive Design.* Springer, 2020.

Eastop, T.D., and A. McConkey. *Applied Thermodynamics for Engineering Technologists.* 5th ed. Longman, 1993.

Eckermann, Erik. *World History of the Automobile.* Society of Automotive Engineers International, 2001.

Eggink, John. *Managing Energy Costs: A Behavioral and Non-Technical Approach.* Fairmont Press, 2007.

Ehsani, Mehrdad, et al. *Modern Electric, Hybrid Electric, and Fuel Cell Vehicles, Fundamentals, Theory, and Design.* CRC Press, 2005.

Einstein, Albert. "Die Grundlage der allgemeinen Relativitatstheorie." *Annalen der Physik*, vol. 354, no. 7, 1916.

Eisentraut, Anselm. *Sustainable Production of Second-Generation Biofuels: Potential and Perspectives in Major Economies and Developing Countries.* International Energy Agency, 2010.

Elbashir, Nimir O., Mahmoud M. El-Halwagi, Ioannis G. Economou, and Kenneth R. Hall, eds. *Natural Gas Processing from Midstream to Downstream.* Wiley, 2019.

Elcock, D. *Baseline and Projected Water Demand Data for Energy and Competing Water Use Sectors.* U.S. Department of Energy, National Energy Technology Library, and Argonne National Laboratory, 2008.

"Electric Current." *Encyclopaedia Britannica*. Encyclopaedia Britannica, 10 Sept. 2014. Web. 8 June 2015.

"Electric Potential." *Encyclopaedia Britannica*. Encyclopaedia Britannica, 3 Apr. 2014. Web. 19 June 2015.

"Electric Potential Energy." *Khan Academy*. Khan Acad., n.d. Web. 19 June 2015.

"Electrical Power." *BBC Bitesize*. BBC, n.d. Web. 5 May 2015.

El-Sharkawi, Mohamed A. *Electric Energy: An Introduction*. 3rd ed. CRC Press, 2013.

Embassy of the People's Republic of China in the United States of America. "The Three-Gorges Project: A Brief Introduction." http://www.china-embassy.org/eng/zt/sxgc/t36502.htm.

Energy Quest. "The Energy Story." http://www.energyquest.ca.gov/story/chapter07.html.

Energy Sector Management Assistance Program. "Potential for Biofuels for Transport in Developing Countries." Report 312/05. Washington, DC: World Bank, 2005.

Energy Star. "Light Bulbs (CFLs)." http://www.energystar.gov/index.cfm?fuseaction=find_a_product.showProductGroup&pgw_code=LB.

"Energy, Kinetic Energy, Work, Dot Product, and Power." *MIT OpenCourseWare*. MIT, 13 Oct. 2004. Web. 22 Sept. 2015.

Energy.gov. "Energy Storage." http://www.oe.energy.gov/storage.htm.

Enescu, Diana. *Green Energy Advances*. IntechOpen, 2019.

Ericson, Robert, and Albert Edward Musson. *James Watt and the Steam Revolution: A Documentary History*. Augustus M. Kelley, 1969.

Erjavec, Jack, and Jeff Arias. "Propane/LPG Vehicles." *Hybrid, Electric, and Fuel-Cell Vehicles*. Thomson, 2007.

Erlichson, H. "André-Marie Ampère, the 'Newton of Electricity' and How the Simplicity Criterion Resulted in the Disuse of His Formula." *Physis*, vol. 37, no. 1, 2000.

"Ethanol Benefits and Considerations." *Alternative Fuels Data Center*, Energy Efficiency & Renewable Energy, U.S. Department of Energy, 28 Mar. 2018, www.afdc.energy.gov/fuels/ethanol_benefits.html. Accessed 10 Oct. 2018.

Euroheat and Power. "District Heating and Cooling." http://www.euroheat.org/Default.aspx?ID=4.

European Commission. "Communication from the Commission to the European Parliament, the Council, the European Economic and Social Committee and the Committee of the Regions: Energy 2020, A Strategy for Competitive, Sustainable, and Secure Energy."

EUR-Lex. European Union, 1998-2016. Web. 16 May. 2016. http://eur-lex.europa.eu/LexUriServ/LexUriServ.do?uri=CELEX:52010DC0639:EN:NOT.

European Commission. *Combining Heat and Power: Using Cogeneration to Improve Energy Efficiency in the European Union*. European Commission, Directorate-General for Energy and Transport, 2005.

European Environment Agency. "Climate Change and a Low-Carbon European Energy System." Report 1/2005. 29 June 2005. http://www.eea.europa.eu/publications/eea_report_2005_1.

———. "EN 20 Combined Heat and Power (CHP)." http://www.eea.europa.eu/data-and-maps/indicators/en-20-combined-heat-and.

Evans, Robert L. "Clean Coal Processes." *Fueling Our Future: An Introduction to Sustainable Energy*. Cambridge UP, 2007.

Evans, Stephen G., Reginald L. Hermanns, Alexander Strom, and Gabriele Scarascia-Mugnozza, eds. *Natural and Artificial Rockslide Dams*. Springer, 2011.

Faraday, Michael. *The Correspondence of Michael Faraday: 1811-December 1831*. Edited by Frank A.J.L. James. Institution of Electrical Engineers, 1991.

Farmer, Edward J. *Detecting Leaks in Pipelines*. ISA, 2017.

Farr, James R., ed. *The Industrial Revolution in Europe, 1750-1914*. Gale Group, 2002.

Faruqui, Ahmad, Ryan Hledik, Sam Newell, and Johannes Pfeifenberger. "The Power of Five Percent: How Dynamic Pricing Can Save $35 Billion in Electricity Costs." The Brattle Group Discussion Paper, 16 May 2007. http://www.brattle.com/_documents/Upload Library/Upload574.pdf.

Faure, Gunter, and Teresa M. Mensing. *Isotopes: Principles and Applications*. 3rd ed. Wiley, 2005.

Fehner, Terrence R., and Francis G. Goslin. *Coming in From the Cold: Regulating U.S. Department of Energy Nuclear Facilities, 1942-1996*. American Society for Environmental History and the Forest History Society in association with Duke UP, 1996.

Feidt, Michel, ed. *Carnot Cycle and Heat Engine Fundamentals and Applications*. MDPI, 2020

Fell, Robin, Patrick MacGregor, David Stapledon, Graeme Bell, and Mark Foster. *Geotechnical Engineering of Dams*. 2nd ed. CRC Press, 2015.

Fen, Wu-Chun. *Green Computing: Large-Scale Energy Efficiency*. CRC Press, 2011.

Fergus, Jeffrey W., Rob Hui, Xianguo Li, David P. Wilkinson, and Jiujun Zhang, eds. *Solid Oxide Fuel Cells: Materials, Properties and Performance*. CRC Press, 2009.

Ferreira, Alex Luiz, et al. "Flex Cars and the Alcohol Price." *Energy Economics*, vol. 31, 2009.

Fijalkowski, B.T. *Automotive Mechatronics: Operational and Practical Issues*. Vol. II. Springer, 2011.

Firk, Frank W.K. "Essential Physics I." http://www.physicsforfree.com/essential.html.

Fisanick, Christina, ed. *Eco-architecture*. Greenhaven Press, 2008.

Fischhoff, Baruch. "The Nuclear Energy Industry's Communication Problem." *Bulletin of the Atomic Scientists*, 11 Feb. 2009.

Flin, David. *Cogeneration: A User's Guide*. The Institution of Engineering and Technology, 2010.

Flood, Raymond, Mark McCartney, and Andrew Whitaker, eds. *Kelvin: Life, Labours, and Legacy*. Oxford UP, 2008.

Forbes, Nancy, and Basil Mahon. *Faraday, Maxwell and the Electromagnetic Field: How Two Men Revolutionized Physics*. Prometheus Books, 2014.

Ford, Daniel F. *The Cult of the Atom: The Secret Papers of the Atomic Energy Commission*. Simon & Schuster, 1984.

"Fossil." *Energy.gov*, U.S. Department of Energy, www.energy.gov/science-innovation/energy-sources/fossil. Accessed 18 Aug. 2020.

Foster, Robert, ed. *Wind Energy: Renewable Energy and the Environment*. CRC Press, 2009.

Foulds, Chris, and Rosie Robison, eds. *Advancing Energy Policy. Lessons on the Integration of Social Sciences and Humanities*. Palgrave/MacMillan, 2018.

Franklin, Benjamin. *Experiments and Observations on Electricity Made at Philadelphia in America*. 1769. Reprint. Cambridge UP, 2019.

Freese, Barbara. *Coal: A Human History*. Perseus, 2003.

"Frequency and Period." *SparkNotes: SAT Physics*. SparkNotes, 2011. Web. 29 June 2015.

Fritzsche, Hellmut. "Electromagnetic Radiation." *Encyclopædia Britannica*. Encyclopædia Britannica, 26 Nov. 2014. Web. 29 June 2015.

Fry, Brian. *Stable Isotope Ecology*. Springer, 2006.

Funabashi, Yoichi, and Kay Kitazawa. "Fukushima in Review: A Complex Disaster, a Disastrous Response." *Bulletin of the Atomic Scientists*, vol. 68, no. 2, 2012.

Gabbar, Hassam A., ed. *Energy Conservation in Residential, Commercial, and Industrial Facilities*. John Wiley & Sons, 2018.

Galbraith, Kate. "Why Is a Utility Paying Customers?" *New York Times*, 23 Jan. 2010.

Gallagher, Ed. "Making Biofuels Sustainable." *Our Planet: The Magazine of the United Nations Environment Programme*, Dec. 2008.

http://www.unep.org/pdf/Ourplanet/2008/dec/en/OP-2008-12-en-FULLVERSION.pdf.

Gamow, George. *Thirty Years That Shook Physics: The Story of Quantum Theory*. Anchor Books, 1966.

Ganapathy, V. *Steam Generators and Waste Heat Boilers for Process and Plant Engineers*. CRC Press, 2015.

Gardiner, K.R. "André-Marie Ampère and His English Acquaintants." *British Journal for the History of Science*, vol. 2, 1964/65.

Gears and Stuff. "Different Types of Gears." http://www.gearsandstuff.com/types_of_gears.htm.

Gears Manufacturers. "Gears History." http://www.gears-manufacturers.com/gears-history.html.

General Electric Consumer and Industrial Lighting. "Compact Fluorescent Light Bulbs." http://www.gelighting.com/na/home_lighting/ask_us/faq_compact.htm#how_work.

Geology.com. "Marcellus Shale-Appalachian Basin Natural Gas Play." http://geology.com/articles/marcellus-shale.shtml.

Georgia State University. "Kinetic Energy." http://hyperphysics.phy-astr.gsu.edu/hbase/ke.html.

Gerdal, B. Hahn-Ha, M. Galbe, M.F. Gorwa-Grauslund, G. Liden, and G. Zacchi. "Bio-ethanol: The Fuel of Tomorrow from the Residues of Today." *Trends in Biotechnology*, vol. 24, 2006.

Gerrsen-Gondelach, Sarah J. "Performance of Batteries for Electric Vehicles on Short and Longer Term." *Journal of Power Sources*, vol. 212, no. 15, 2012.

Ghenai, Chaouki, and Tareq Salameh, eds. *Sustainable Air Conditioning Systems*. IntechOpen, 2018.

Giambattista, Alan, and Betty McCarthy Richardson. *Physics*. 2nd ed. McGraw, 2010.

Giambattista, Alan, Betty McCarthy Richardson, and Robert C. Richardson. *Physics*. 3rd ed. McGraw, 2016.

Gibbs, J. Willard. *Elementary Principles in Statistical Mechanics Developed with Special Reference to the Rational Foundation of Thermodynamics*. Reprint. Creative Media LLC, 2019.

———. *The Collected Works of J. Willard Gibbs*. [1906]. Yale UP, 1948.

Gibilisco, Stan. "Propulsion with Methane, Propane, and Biofuels." *Alternative Energy Demystified*. McGraw-Hill, 2007.

———. *Electricity Demystified*. 2nd ed. McGraw, 2012.

Gilbert, P.U.P.A., and W. Haeberli. *Physics in the Arts*. Elsevier, 2008.

Gillispie, Charles Coulston, and Raffaele Pisano. *Lazare and Sadi Carnot: A Scientific and Filial Relationship*. 2nd ed. Springer, 2014.

Gipe, Paul. *Wind Energy Energy for the Rest of Us: A Comprehensive Guide to Wind Power and How to Use It*. Wind Works, 2018.

Glanz, James. "Quake Threat Leads Swiss to Close Geothermal Project." *New York Times*, 10 Dec. 2009. http://www.nytimes.com/2009/12/11/science/earth/11basel.html?_r=2.

Glasby, G.P. "Abiogenic Origin of Hydrocarbons: An Historical Overview." *Resource Geology*, vol. 56, no. 1, 2006.

Glashow, Sheldon L. "Partial-Symmetries of Weak Interactions." *Nuclear Physics*, vol. 22, no. 4, 1961. (1961).

Glassley, William E. *Geothermal Energy. Renewable Energy and the Environment*. 2nd ed. CRC Press, 2015.

Global Energy Network Institute. "The International Grid." http://www.geni.org/globalenergy/library/technical-articles/transmission/new-scientist/the-international-grid/index.shtml.

Goettemoeller, Jeffery, and Adrian Goettemoeller. *Sustainable Ethanol: Biofuels, Biorefineries, Cellulosic Biomass, Flex-Fuel Vehicles, and Sustainable Farming for Energy Independence*. Prairie Oak, 2007.

Goettemoeller, Jeffrey. *Sustainable Ethanol: Biofuels, Biorefineries, Cellulosic Biomass, Flex-Fuel Vehicles, and Sustainable Farming for Energy Independence*. Prairie Oak Publications, 2007.

Goldfarb, Veniamin, Evgenii Trubachev, and Natalya Barmina, eds. *New Approaches to Gear Design and Production*. Springer Nature, 2020.

Goldstein, Walter E. *The Science of Ethanol*. CRC Press, 2017.

Gonzalez, George A. *American Empire and the Canadian Oil Sands*. Palgrave, 2016.

Goodell, J. *Big Coal*. Houghton Mifflin, 2006.

Gorbaty, Martin L., John W. Larsen, and Irving Wender, eds. *Coal Science*. Academic, 1982.

Goswami, D. Yogi, and Frank Kreith, eds. *Energy Efficiency and Renewable Energy Handbook*. 2nd ed. CRC Press, 2016.

Gou, Bei, Woon Ki Na, and Bill Diong. *Fuel Cells: Modeling, Control and Applications*. CRC Press, 2010.

Graf von Rumford, Benjamin. *The Complete Works of Count Rumford*. HardPress, 2020.

Gray, Andrew. *Lord Kelvin*. Outlook, 2020.

Greaves, Deborah, and Gregorio Iglesias, eds. *Wave and Tidal Energy*. Wiley, 2018.

The Green Grid. "About the Green Grid." http://www.thegreengrid.org/about-the-green-grid.

Green, John A.S., ed. *Aluminum Recycling and Processing for Energy Conservation and Sustainability*. ASM International, 2007.

Green, Martin A. *Third Generation Photovoltaics: Advanced Solar Energy Conversion*. Springer, 2005.

Greene, Ann Norton. *Horses at Work: Harnessing Power in Industrial America*. Harvard UP, 2008.

Greene, Lori E. *High and Low Concentrator Systems for Solar Electric Applications IV*. SPIE: 2009.

"Greenhouse Gas." *The Encyclopedia of Earth*, edited by Cutler J. Cleveland. National Council for Science and the Environment. http://www.eoearth.org/article/Greenhouse_gas?toxic=49554.

GreenSource Magazine. *Emerald Architecture: Case Studies in Green Building*. McGraw-Hill, 2008.

Griffiths, D.J. *Introduction to Elementary Particles*. John Wiley, 1987.

Grigas, Agnia. *The New Geopolitics of Natural Gas*. Harvard UP, 2017.

Grünewald, Philipp. "The Role of Large-Scale Storage in a GB Low Carbon Energy Future: Issues and Policy Challenges." *Energy Policy*, vol. 39, no. 9, 2011.

Gunther, Leon. *The Physics of Music and Color*. Springer, 2012.

Gupta, Harsh K., and Sukanta Roy. *Geothermal Energy: An Alternative Resource for the Twenty-first Century*. Elsevier, 2007.

Gupta, S.V. *Units of Measurement. History, Fundamentals and Redefining the SI Base Units*. 2nd ed. Springer, 2020.

Hager, Willi H., Anton J. Schleiss, Robert M. Boes, and Michael Pfister. *Hydraulic Engineering of Dams*. CRC Press, 2021.

Hahn, Barbara. *Technology in the Industrial Revolution*. Cambridge UP, 2020.

Hall, Charles, et al. "What Is the Minimum EROI That a Sustainable Society Must Have?" *Energies* 2 (2009).

Halliday, David, Robert Resnick, and Jearl Walker. *Fundamentals of Physics*. John Wiley & Sons, 2009.

Hammond, G., and C. Jones. "University of Bath, Inventory of Carbon and Energy." http://www.bath.ac.uk/mech-eng/sert/embodied/.

Hansen, James. *Storms of My Grandchildren: The Truth About the Coming Climate Catastrophe and Our Last Chance to Save Humanity*. Bloomsbury Press, 2009.

Hart, Ivor Blashka. *James Watt and the History of Steam Power*. H. Schuman, 1949.

Harvey, L.D. Danny. *A Handbook on Low-Energy Buildings and District-Energy Systems: Fundamentals, Techniques and Examples*. Earthscan, 2006.

Hatfield, Philipp L. *Great Men of Science: A History of Scientific Progress*. Macmillan, 1933.

Hau, Erich. *Wind Turbines: Fundamentals, Technologies, Application, Economics*. Springer, 2006.

Heath, Garvin A. *Life Cycle Assessment of the Energy Independence and Security Act of 2007*. National Renewable Energy Laboratory, 2009.

Hebner, R. "Flywheel Batteries Come Around Again." *Spectrum*, vol. 39, no. 4, Apr. 2002.

Hebra, Alex. *Measure for Measure: The Story of Imperial, Metric and Other Units*. Johns Hopkins UP, 2003.

Heinberg, Richard, and Daniel Lerch, eds. *The Post Carbon Reader: Managing the 21st Century's Sustainability Crises*. Watershed, 2010.

Henderson, Tom. *Kinematics*. N.p.: Physics Classroom, 2013. Digital file.

———. "Ohm's Law." *The Physics Classroom*. Physics Classroom, n.d. Web. 1 June 2015.

Hewlett, Richard G., Oscar E. Anderson, and Francis Duncan. *A History of the United States Atomic Energy Commission*. U of California P, 1990.

Hiereth, Hermann, and Peter Prenninger. *Charging the Internal Combustion Engine (Powertrain)*. Springer, 2010.

Higginbottom, Adam. *Midnight in Chernobyl: The Untold Story of the World's Greatest Nuclear Disaster*. Simon & Schuster, 2020.

Higgs, Peter W. "Broken Symmetries and the Masses of Gauge Bosons." *Physical Review Letters*, vol. 13, no. 16, 19 Oct. 1964.

High Beam Business. "Primary Batteries, Dry and Wet: Industry Report." 2012. http://business.highbeam.com/industry-reports/equipment/primary-batteries-dry-wet.

Hill, Jason, et al. "Environmental, Economic, and Energetic Costs and Benefits of Biodiesel and Ethanol Biofuels." *Proceedings of the National Academy of Sciences*, vol. 103, no. 30, 2006.

Hirsch, Tim. *A Case for Climate Neutrality: Case Studies on Moving Towards a Low Carbon Economy*. United Nations Environment Programme, 2009.

Hirshfeld, A.W. *The Electric Life of Michael Faraday*. Bloomsbury Publishing, 2009.

Hobson, Keith A., and Leonard I. Wassenaar, eds. *Tracking Animal Migration with Stable Isotopes*. Academic Press, 2008.

Hodge, B.K. "Ocean Energy." *Alternative Energy Systems and Applications*. John Wiley and Sons, 2010.

Hodgson, Peter E. *Energy, the Environment and Climate Change*. Imperial College Press, 2010.

Hoffman, J.A. "Climate Change as a Cultural and Behavioral Issue: Addressing Barriers and Implementing Solutions." *Organizational Dynamics*, vol. 39, 2010, pp. 295-305.

Hofmann, James R. *André-Marie Ampère*. Blackwell, 1996.

Holl, Jack M., and Terrence R. Fehner. *Department of Energy, 1977-1994: A Summary History*. Office of Scientific and Technical Information, 1994.

Hoogers, G. *Fuel Cell Technology Handbook*. CRC Press, 2002.

How Stuff Works. "How Steam Engines Work." http://www.howstuffworks.com/steam.htm.

Howell, John R. *Solar-Thermal Energy Systems: Analysis and Design*. McGraw Hill, 1982.

Howse, Jennifer. *Levers*. Weigl, 2010.

Hu, Zhaoguang, and Zheng Hu. *Electricity Economics: Production Functions with Electricity*. Springer, 2013.

Huang, Haibo, Stephen Long, and Vijay Singh. "Techno-economic Analysis of Biodiesel and Ethanol Co-production from Lipid-producing Sugarcane." *Biofuels, Bioproducts & Biorefining*, vol. 10, 2016, pp. 299-315. Print.

Huang, Zujian. *Application of Bamboo in Building Envelope*. Springer, 2019.

Huggins, R.A. *Energy Storage*. Springer, 2010.

Hughes, John M. *Practical Electronics: Components and Techniques*. O'Reilly, 2015.

Huizenga, J.R. *Cold Fusion: The Science Fiasco of the Century*. Oxford UP, 1994.

Hundy, G.F., A.R. Trott, and T.C. Welch. *Refrigeration and Air Conditioning*. Elsevier, 2008.

———. *Refrigeration, Air Conditioning and Heat Pumps*. 5th ed. Elsevier, 2016.

Hyman, Mark, M.D. "The Impact of Mercury on Human Health and the Environment." *Alternative Therapies*, vol. 10, no. 6, 2004.

Ibrahim, Mounir B., and Roy C. Tew, Jr. *Stirling Convertor Generators*. CRC Press, 2012.

Inage, Shin-Ichi. "Prospects for Large-Scale Energy Storage in Decarbonised Power Grids." International Energy Agency Working Paper, http://www.iea.org/papers/2009/energy_storage.pdf.

Inderwildi, Oliver, and David King. "Quo Vadis Biofuels." *Energy and Environmental Science*, vol. 2, 2009.

Infosource Europe. "EU Energy Policy: Cogeneration Directive." December 2008. http://www.inforse.dk/europe/eu_cogen-di.htm.

Ingledew, W.M. *The Alcohol Textbook: A Reference for the Beverage, Fuel, and Industrial Alcohol Industries*. 5th ed. Nottingham UP, 2009.

International Atomic Energy Agency. "Thyroid Cancer Effects in Children." Aug. 2005.

http://www.iaea.org/NewsCenter/Features/Chernobyl-15/thyroid.shtml.

International Atomic Energy Agency. *IAEA Nuclear Energy Series Establishment of Uranium Mining and Processing Operations in the Context of Sustainable Development*. IAEA Nuclear Energy Series NF-T-1. International Atomic Energy Agency, 2009.

International District Energy Association. "What Is District Energy?" http://www.districtenergy.org/what-is-district-energy/.

International Energy Agency, Energy Technology Network. *Coal Information 2010*. International Energy Agency, 2010.

———. *Comparative Study on Rural Electrification Policies in Emerging Economies: Keys to Successful Policies*. International Energy Agency, 2010.

———. Energy Access Outlook 2017: From Poverty to Prosperity-World Energy Outlook Special Report. OECD/IEA, 2017, www.iea.org/publications/freepublications/publication/WEO2017SpecialReport_EnergyAccessOutlook.pdf. Accessed 9 Oct. 2018.

———. "Energy Conservation Through Energy Storage." 2012. http://www.iea-eces.org.

———. *Hybrid and Electric Vehicles: The Electric Drive Advances*. Paris, International Energy Agency, 2010. http://www.ieahev.org/pdfs/2009_annual_report.pdf.

———. "IEA Advanced Fuel Cells: Implementing Agreement," 2012. http://www.ieafuelcell.com.

———. *Key World Energy Statistics 2015*. OECD/IEA, 2015. PDF file.

———. *Natural Gas Information 2016*. Organisation for Economic Co-operation and Development, 2016.

———. "Technology Agreements: Energy Storage." 2011. http://www.iea.org/techno/iaresults.asp?id_ia=13.

———. *Technology Roadmap: Solar Photovoltaic Energy*. 2014 ed. Author, 2014. *International Energy Agency*. Web. 13 May 2016.

———. *World Energy Outlook 2009*. International Energy Agency, 2009.

International Geothermal Association. "What Is Geothermal Energy?" http://www.geothermal-energy.org/314,what_is_geothermal_energy.html.

International Panel for Sustainable Resource Management and UNEP. "Metal Stocks in Society: Scientific Synthesis." 2010. http://www.unep.fr/shared/publications/pdf/DTIx1264xPA-Metal%20stocks%20in%20society.pdf.

International Science Panel on Renewable Energies. *Research and Development on Renewable Energies: A Preliminary Global Report on Biomass*. International Science Panel on Renewable Energies, 2009.

International Standards Organization. "Technical Energy Systems: Methods for Analysis, Part 1: General." ISO 13602-1 2002. http://www.iso-standard.org/13626.html.

Intratec U.S. *Cellulosic Ethanol from Switchgrass Report Ethanol E81A Basic Cost Analysis*. Intratec, 2019.

———. *Cellulosic Ethanol from Wood Chips Report Ethanol E52A Basic Cost Analysis*. Intratec, 2019.

———. *Ethylene Production via Cracking of Ethane/Propane: Report Ethylen E21A Basic Cost Analysis*. Intratec Solutions, 2019.

———. *Propylene Production from Propane-Cost Analysis: Propylene E33A*. Intratec, 2016.

"Introduction to Circuits and Ohm's Law." *Khan Academy*. Khan Acad., n.d. Web. 19 June 2015.

"An Introduction to Waves." *GCSE Bitesize*. BBC News, 2014. Web. 25 Feb. 2015.

Isaacson, Walter. *Benjamin Franklin: An American Life*. Simon & Schuster, 2004.

Ishimaru, Akira. *Electromagnetic Wave Propagation, Radiation and Scattering from Fundamentals to Applications*. 2nd ed., Wiley/IEEE Press, 2017.

Ismail, Basel I., ed. *Advances in Geothermal Energy*. IntechOpen, 2016.

Isogawa, Yoshihito. *Gears*. No Starch Press, 2010.

Jacobson, Mark Z. "How Renewable Energy Could Make Climate Treaties Moot." *Scientific American*. Nature America, 23 Nov. 2015. Web. 6 Jan. 2016.

Jacobson, Mark. "Review of Solutions to Global Warming, Air Pollution, and Energy Security." *Energy and Environmental Science*, vol. 2, 2009.

Jagoo, Zafrullah. *Tracking Solar Concentrators: A Low Budget Solution*. Springer, 2013.

Jamasb, Tooraj, and Michael G. Pollitt. *The Future of Electricity Demand: Customers, Citizens and Loads*. Cambridge UP, 2011.

James, Frank A.J.L. *Michael Faraday. A Very Short Introduction*. Oxford UP, 2010.

Jatras, Stella L. "The Genius of Nikola Tesla." *The New American*, vol. 19, no. 15, 2003.

Jawad, Maan H. *Stress in ASME Pressure Vessels, Boilers and Nuclear Components*. Wiley, 2018.

Jha, A.R. *Next-Generation Batteries and Fuel Cells for Commercial, Military and Space Applications*. CRC Press, 2016.

Joerissen, Ludwig, et al. "Possible Use of Vanadium Redox Flow Batteries for Energy Storage in Small Grids and Stand-Alone Photovoltaic Systems." *Journal of Power Sources*, vol. 127, no. 1-2, 2004.

Johnson, Alex. "Shining a Light on Hazards of Fluorescent Bulbs." http://www.msnbc.msn.com/id/23694819/ns/us_news-environment/t/shining-light-hazards-fluorescent-bulbs.

Johnson, James. *Introduction to Fluid Power*. Delmar Thomson Learning, 2002.

Johnston, David, and Scott Gibson. *Green from the Ground Up: A Builder's Guide-Sustainable, Healthy, and Energy-Efficient Home Construction*. Taunton, 2008.

Jon, Chua Kian, Md Raisul Islam, Ng Kim Choon, and Muhammad Wakil Shahzad. *Advances in Air Conditioning Technologies: Improving Energy Efficiency*. Springer, 2021.

Jonnes, Jill. *Empires of Light: Edison, Tesla, Westinghouse, and the Race to Electrify the World*. Random House, 2003.

Jorgensen, Timothy. J. *Strange Glow: The Story of Radiation*. Princeton UP, 2016.

Josephs, Leslie. "Long before the Combustion Engine, the Hybrid Car Is Facing Obsolescence." *Quartz*, 14 July 2017, qz.com/1029464/what-percent-of-us-car-sales-are-hybrids. Accessed 10 Oct. 2018.

Josephson, Paul R. *Motorized Obsessions: Life, Liberty, and the Small-Bore Engine*. Johns Hopkins UP, 2007.

Kalaiselvam, S., and R. Parameshwaram. *Thermal Energy Storage Technologies for Sustainability: Systems Design, Assessment and Applications*. Elsevier, 2014.

Kalof, Linda, and Brigitte Resl, eds. *A Cultural History of Animals*. 6 vols. Berg, 2007.

Kamimura, Arlindo, et al. "The Effect of Flex Fuel Vehicles in the Brazilian Light Road Transportation." *Energy Policy*, vol. 36, 2008.

Kammen, Daniel. "The Rise of Renewable Energy." *Scientific American*, Sept. 2006.

Kanoglu, Mehmet, Yunus A. ?engel, and Ibrahim Din?er. *Efficiency Evaluation of Energy Systems*. Springer, 2012.

Kao, Jimmy C.M., Wen-Pei Sung, and Ran Chen, eds. *Green Building, Materials and Civil Engineering*. CRC Press, 2015.

Kapelevich, Alexander L. *Direct Gear Design*. CRC Press, 2013.

Kaplan, Stan Mark. *Smart Grid: Modernizing Electric Power Transmission and Distribution; Energy Independence, Storage and Security; Energy Independence and Security Act of 2007 (EISA); Improving Electrical Grid Efficiency, Communication, Reliability, and Resiliency; Integrating New and Renewable Energy Sources*. TheCapitol.Net, 2009.

Kartha, Sivan, and Eric D. Larson. *Bioenergy Primer: Modernised Biomass Energy for Sustainable Development*. United Nations Development Programme, 2000.

Karwatka, Dennis. "William Rankine and Engineering Education." *Tech Directions*, vol. 69, no. 3, Oct. 2009.

Kaviany, Massoud. *Essentials of Heat Transfer*. Cambridge UP, 2011.

Kelly, P.F. *Electricity and Magnetism*. CRC, 2015.

Kempos, Chris, David H. Wolpert, and Peter F. Stadler. *The Energetics of Computing in Life and Machines*. Santa Fe Institute Press, 2018.

Kenward, Michael. *Potential Energy: An Analysis of World Energy Technology*. Cambridge UP, 1976.

Kessides, Ioannis, and David Wade. "World Bank Working Research Paper: Toward a Sustainable Global Energy Supply Infrastructure." http://ideas.repec.org/p/wbk/wbrwps/5539.html.

Kessler, Günter. *Sustainable and Safe Nuclear Fission Energy: Technology and Safety of Fast and Thermal Breeder Reactors*. Springer, 2012.

Khajepour, Amir, Saber Fallah, and Avesta Goodarzi. *Electric and Hybrid Vehicles. Technology, Modeling and Control: A Mechatronic Approach*. Wiley, 2014.

Khaligh, Alireza, and Omer C. Onar. *Energy Harvesting: Solar, Wind and Ocean Energy Conversion Systems*. CRC Press, 2010.

Khan, B.H. *Non-Conventional Energy Resources*. Tata/McGraw-Hill, 2006.

Khan, Sal. "Work and Energy (Part 2)." *Khan Academy*. Khan Acad., 2015. Web. 30 Apr. 2015.

Khare, Vicas, Cheshta Khare, Savita Nema, and Prashant Baredar. *Tidal Energy Systems, Design, Optimization and Control*. Elsevier, 2019.

Khazaii, Javad. *Energy-Efficient HVAC Design: An Essential Guide for Sustainable Building*. Springer, 2014.

Kim, Albert S., and Hyeon-Ju Kim. *Ocean Thermal Energy Conversion (OTEC): Past, Present and Progress*. IntechOpen, 2020.

Kimball, John W. "Cellular Respiration." *Kimball's Biology Pages*, 1 Aug. 2019, www.biology-pages.info/C/CellularRespiration.html. Accessed 2 Mar. 2020.

Kirkland, Kyle. *Light and Optics*. Facts On File, 2007.

Klette, Patrick J. *Fluid Power Systems*. American Technical Publishers, 2014.

Knier, Gil. "How Do Photovoltaics Work?" *NASA Science*. NASA, 6 Apr. 2011. Web. 13 May 2016.

Koerth-Baker, Maggie. *Before the Lights Go Out: Conquering the Energy Crisis Before It Conquers Us*. John Wiley & Sons, 2012.

Kohan, A. *Boiler Operators Guide*. 4th ed. McGraw-Hill, 1997.

Kolanowski, Bernard F. *Small-Scale Cogeneration Handbook*. Fairmont, 2011.

Kolesnikov, Anton, et al. "Methane-Derived Hydrocarbons Produced Under Upper-Mantle Conditions." *Nature Geoscience*, vol. 2, 2009.

Kondepudi, Dilip, and Ilya Prigogine. *Modern Thermodynamics from Heat Engines to Dissipative Structures.* 2nd ed. Wiley, 2015.

Konis, Kyle, and Stephen Selkowitz. *Effective Daylighting with High-Performance Facades: Emerging Design Practices.* Springer, 2017.

Koomey, Jonathan G. "Estimating Total Power Consumption by Servers in the U.S. and the World." Final Report. Released 15 Feb. 2007, http://blogs.business2.com/greenwombat/files/serverpowerusecomplete-v3.pdf.

Kornhauser, Alan A. *Internal Combustion Engines: The Engineering Handbook.* 2nd ed., CRC Press, 2004.

Kozima, Hideo. *The Science of the Cold Fusion Phenomenon: In Search of the Physics and Chemistry Behind Complex Experimental Data Sets.* Elsevier, 2006.

Krarti, Moncef. *Weatherization and Energy Efficiency Improvement for Existing Homes.* CRC Press, 2012.

Kruger, Abe, and Carl Seville. *Green Building: Principles and Practices in Residential Construction.* Delmar, 2013.

Kruger, Paul. *Alternative Energy Resources: The Quest for Sustainable Energy.* John Wiley, 2006.

Kularatna, Nihal. *Energy Storage Devices for Electronic Systems: Rechargeable Batteries and Supercapacitors.* Academic Press, 2015.

Kumar, Anil, Om Prakash, and Prashant Singh. Chauhan. *Energy Management: Conservation and Audits.* CRC Press, 2021.

Kumar, B.N. *Basic Physics for All.* University Press of America, 2009.

Kunstler, James Howard. *The Long Emergency: Surviving the Converging Catastrophes of the Twenty-first Century.* Atlantic Monthly Press, 2005.

Kurokawa, Kosuke, Keichi Komoto, Peter van der Vleuten, and David Faman, eds. *Energy from the Desert: Practical Proposals for Very Large-Scale Photovoltaic Systems.* Earthscan, 2007.

Laird, Frank N. *Solar Energy, Technology Policy, and Institutional Values.* Cambridge UP, 2001.

Lamba, Bhawna Yadav, et al. *Preparation and Characterization of Biodiesel from Vegetable Oils.* Lambert Academic, 2011.

Lang, Susan S. "Cornell Ecologist's Study Finds That Producing Ethanol and Biodiesel from Corn and Other Crops Is Not Worth the Energy." *Cornell University News Service*, 2005. http://www.news.cornell.edu/stories/july05/ethanol.toocostly.ssl.html.

Langwith, Jacqueline, ed. *Renewable Energy.* Greenhaven, 2009.

Lankford, Ronald D. *Greenhouse Gases.* Greenhaven Press, 2009.

Larminie, James, and John Lowry. *Electric Vehicle Technology Explained.* Wiley, 2003.

Lasala, Silvia, ed. *Organic Rankine Cycles for Waste Heat Recovery-Analysis and Applications* IntechOpen, 2020.

Laxer, Gordon. *After the Sands: Energy and Ecological Security for Canadians.* Douglas & McIntyre, 2015.

Leclercq, Ludovic, Benoit Robyns, and Jean-Michel Grave. "Control Based on Fuzzy Logic of a Flywheel Energy Storage System Associated with Wind and Diesel Generators." *Mathematics and Computers in Simulation*, vol. 63, no. 3-5, Nov. 2003.

Lee, Sunggyu. "Energy from Biomass Conversion." *Handbook of Alternative Fuel Technologies*, edited by Sunggyu Lee, James G. Speight, and Sudarshan K. Loyalka. CRC, 2007.

Lee, Yuh-Ming, and Yun-Ern Tzeng. "Development and Life-Cycle Inventory Analysis of Wind Energy in Taiwan." *Journal of Energy Engineering*, vol. 134, no. 53, 2008.

Leggett, M., and M. Finlay. "Science, Story, and Image: A New Approach to Crossing the Communication Barrier Proposed by Scientific Jargon." Public Understanding of Science, vol. 10, no. 2, 2001.

Lele, Armand Fopah. *A Thermochemical Heat Storage System for Households: Combined Investigations of Thermal Transfers Coupled to Chemical Reactions.* Springer, 2016.

Lemons, Don S. *A Student's Guide to Entropy.* Cambridge UP, 2013.

Lenk, Ron, and Carol Lenk. *Practical Lighting Design with LEDs.* John Wiley & Sons, 2017.

Leslie, Jacques. *Deep Water: The Epic Struggle Over Dams, Displaced People, and the Environment.* Farrar, Straus and Giroux, 2005.

Letcher, Trevor M., and Vasilis M. Fthenakis, eds. *A Comprehensive Guide to Solar Energy Systems, with Special Focus on Photovoltaic Systems.* Academic Press, 2018.

Letcher, Trevor M., ed. *Wind Energy Engineering: A Handbook for Onshore and Offshore Wind Turbines.* Academic Press, 2017.

Leupp, Francis E. *George Westinghouse: His Life and Achievements.* Reprint. Creative Media Partners, 2019.

Lewis, Christopher J.T. *Heat and Thermodynamics.* Greenwood, 2007.

Lienhard, John H. *A Heat Transfer Textbook.* Dover, 2011.

Lindley, D. *Degrees Kelvin: A Tale of Genius and Invention.* Joseph Henry Press, 2004.

Ling, Samuel J., Jeff Sanny, and William Moebs. *University Physics Vol. 1.* Samurai Media Limited, 2017.

Liszka, John. "Are You Sure, Mr. Carnot? A Re-examination of the Thermodynamic Principles as Formulated by Nicolas Carnot and William Rankine Over One Hundred Years Ago Might Lead to Greater Efficiency in Electrical Power Generating Stations, Together with Reduced Emissions." *Engineering Digest*, vol. 38, no. 3, June 1992.

Logan, Michael. *Coal*. Greenhaven, 2008.

López, Antonio L. Luque, and Viacheslav M. Andreev. *Concentrator Photovoltaics*. 1st ed. Springer, 2010.

Lovell, Jenny. *Building Envelopes: An Integrated Approach*. Princeton Architectural Press, 2010.

Lovelock, James. *The Revenge of Gaia: Earth's Climate Crisis and the Fate of Humanity*. Basic Books, 2006.

Lovins, Amory. "Energy Strategy: The Road Not Taken?" *Foreign Affairs*, Oct. 1976.

Lun, Y.H. Venuss, and S.L. Dennis Tung. *Heat Pumps for Sustainable Heating and Cooling*. Springer, 2020.

Lund, John W. "Characteristics, Development and Utilization of Geothermal Resources." *Geo-Heat Center Quarterly Bulletin*, vol. 28, no. 2, 2007.

Lund, John W., and Tonya L. Boyd. "Direct Utilization of Geothermal Energy 2015 Worldwide Review." *Proceedings of the World Geothermal Conference 2015*. Melbourne, Australia. International Geothermal Association, 2015. PDF file.

Lützen, J. *Mechanistic Images in Geometric Form: Heinrich Hertz's Principles of Mechanics*. Oxford UP, 2005.

Lyatkher, Victor. *Tidal Power: Harnessing Energy from Water Currents*. Scrivener, 2014.

Macey, Susan M., and Marilyn A Brown. "Demonstrations as a Policy Instrument with Energy Technology Examples." *Science Communication*, vol. 11, no. 3, 1990.

Machamer, P. "Galileo Galilei." *The Stanford Encyclopedia of Philosophy*, edited by E.N. Zalta. Metaphysics Research Lab, Center for the Study of Language and Information, Stanford U, 2010. http://plato.stanford.edu/entries/galileo/.

MacKay, David J.C. *Sustainable Energy-Without the Hot Air*. UIT Cambridge, 2009.

Madehow.com. "Rudolf Christian Karl Diesel Biography (1858-1913)." http://www.madehow.com/inventorbios/11/Rudolf-Christian-Karl-Diesel.html.

Madureira, Nuno Luis. *Key Concepts in Energy*. Springer, 2014.

Magill, Bobby. "China, US Lead Global Boom in Wind Power." *Climate Central*. Climate Central, 2 Mar. 2016. Web. 17 May 2016.

Maher, Neil M. *Nature's New Deal: The Civilian Conservation Corps and the Roots of the American Environmental Movement*. Oxford UP, 2009.

Malça J., and F. Freire. "Renewability and Life-Cycle Energy Efficiency of Bioethanol and Bio-Ethyl Tertiary Butyl Ether (bioETBE): Assessing the Implications of Allocation." *Energy*, vol. 31, no. 15, 2006.

Malin, Stephanie A. *The Price of Nuclear Power: Uranium Communities and Environmental Justice*. Rutgers UP, 2015.

Mallove, E.J. *Fire and Ice: Searching for Truth Behind the Cold Fusion Furor*. John Wiley & Sons, 1991.

Manovich, Lev. *Software Takes Command*. 20 Nov. 2008, http://softwarestudies.com/softbook/manovich_softbook_11_20_2008.pdf.

Manzella, Adele, Agnes Allansdottir, and Anna Pellizzone, eds. *Geothermal Energy and Society*. Springer, 2019.

Mara, Wil. *The Chernobyl Disaster: Legacy and Impact on the Future of Nuclear Energy*. Marshall Cavendish Corporation, 2011.

Markgraf, Bert. "Cellular Respiration in Humans." *Sciencing*, 23 Apr. 2019, sciencing.com/cellular-respiration-humans-5438875.html. Accessed 3 Mar. 2020.

Marsden, Ben. *Watt's Perfect Engine: Steam and the Age of Invention*. Columbia UP, 2002.

Martin, Richard. *Coal Wars: The Future of Energy and the Fate of the Planet*. St. Martin's, 2015.

———. *Super Fuel: Thorium-The Green Energy Source for the Future*. St. Martin's Press, 2012.

Martin, Thomas Commerford. *The Inventions, Researches and Writings of Nikola Tesla*. Outlook, 2020.

Matek, Benjamin. *2015 Annual US & Global Geothermal Power Production Report*. Geothermal Energy Association, 2015. PDF file.

Maugeri, Leonardo. *Beyond the Age of Oil: The Myths, Realities, and Future of Fossil Fuels and Their Alternatives*. Praeger, 2010.

Maxwell, James Clerk. "A Dynamical Theory of the Electromagnetic Field." *Philosophical Transactions of the Royal Society of London*, vol. 155, 1865. Reprint. Wipf and Stock, 1996.

Mayersohn, Norman. "The Internal Combustion Engine Is Not Dead Yet." *New York Times*, 17 Aug. 2017, www.nytimes.com/2017/08/17/automobiles/wheels/internal-combustion-engine.html. Accessed 23 Oct. 2017.

Mays, Larry W. *Water Resources Engineering*. John Wiley & Sons, 2005.

McCaffrey, Paul, ed. *U.S. National Debate Topic, 2008-2009: Alternative Energy*. Wilson, 2008.

McCall, Martin W. *Classical Mechanics from Newton to Einstein: A Modern Introduction*. 2nd ed. Wiley, 2011.

McCarthy, James E., Cianni Marino, Nico Costa, Larry Parker, and Gina McCarthy. *EPA Regulation of Greenhouse Gases: Considerations and Options*. Nova Science Publishers, 2011.

McCray, Tama. *An Introduction to Thermochemistry*. White Word Publications, 2012.

McDonough, William, and Michael Braungart. *Cradle to Cradle: Remaking the Way We Make Things*. North Point Press, 2002.

———. *The Hannover Principles: Design for Sustainability*. W. McDonough Architects, 2003.

McIlveen, J.F.R. *Fundamentals of Weather and Climate*. Oxford UP, 2010.

McLean-Conner, Penni. *Energy Efficiency: Principles and Practices*. PennWell, 2009.

McNichol, Tom. *AC/DC: The Savage Tale of the First Standards War*. Jossey-Bass, 2006.

MDPI. *Emerging Technologies for Electric and Hybrid Vehicles*. MDPI, 2018.

Meier-Augenstein, Wolfram. *Stable Isotope Forensics: Methods and Forensic Applications of Stable Isotope Analysis*. 2nd ed., Wiley, 2018.

Mendoza, E., ed. *Reflections on the Motive Power of Fire by Sadi Carnot, and Other Papers on the Second Law of Thermodynamics by E. Clapeyron and R. Claussius*. Dover Publications, 1960.

Mertens, Konrad. *Photovoltaics-Fundamentals, Technology and Practice*. Wiley, 2019.

Meyer, John Erik. *The Renewable Energy Transition: Realities for Canada and the World*. Springer, 2020.

Middleton, Paul. *A Brief Guide to the End of Oil*. Constable and Robinson, 2007.

Miesner, Thomas, and William Leffler. *Oil and Gas Pipelines in Nontechnical Language*. PennWell, 2006.

Millar, Dean L. "Wave and Tidal Power." *Energy ... Beyond Oil*, edited by Fraser Armstrong and Katherine Blundell. Oxford UP, 2007.

Miller, Bruce G. *Coal Energy Systems*. Elsevier, 2005.

Miller, David Philip. *James Watt, Chemist: Understanding the Origins of the Steam Age*. Routledge, 2016.

Minerals Management Service, U.S. Department of the Interior. "Ocean Current Energy."

———. "Ocean Wave Energy." http://ocsenergy.anl.gov/guide/wave/index.cfm.

Mirskiy, Anton G. *Thermochemistry and Advances in Chemistry Research*. Nova Science, 2009.

Mittica, P. *Chernobyl: The Hidden Legacy*. Trolley, 2007.

Modi, V., S. Mcdade, D. Lallement, and J. Saghir. *Energy Services for the Millennium Development Goals: Achieving the Millennium Development Goals*. World Bank and United Nations Development Programme, 2005. http://www.unmillenniumproject.org/documents/MP_Energy_Low_Res.pdf.

Mokhatab, Saeid, John Y. Mak, Jaleel V. Valappil, and David A. Wood. *Handbook of Liquefied Natural Gas Transmission*. Gulf Professional, 2014.

Mokyr, Joel. "The Second Industrial Revolution, 1870-1914." August 1998. http://faculty.wcas.northwestern.edu/~jmokyr/castronovo.pdf.

Moon, John Frederick. *Rudolf Diesel and the Diesel Engine*. Priory Press, 1974.

Moran, Richard. *Executioner's Current: Thomas Edison, George Westinghouse, and the Invention of the Electric Chair*. Alfred A. Knopf, 2002.

Morris, Craig. *Energy Switch: Proven Solutions for a Renewable Future*. New Society, 2006.

Mosey, David. *Reactor Accidents: Institutional Failure in the Nuclear Industry*. Nuclear Engineering International, 2006.

Moya, Bernardo Llamas, and Juan Pous, eds. *Greenhouse Gases*. IntechOpen, 2016.

Mulder, Machiel. *Regulation of Energy Markets: Economic Mechanisms and Policy Evaluation*. Springer, 2021.

Muller, Ingo. *A History of Thermodynamics*. Springer, 2010.

———. *A History of Thermodynamics: The Doctrine of Energy and Entropy*. Springer, 2007.

Muller, R. *Physics for Future Presidents: The Science Behind the Headlines*. Norton, 2006.

Muller, Richard A. *Physics for Future Presidents: The Science Behind the Headlines*. W.W. Norton, 2008.

Müller-Kraenner, Sascha. *Energy Security*. Earthscan, 2008.

Mulvaney, Dustin, and Paul Robbins, eds. *Green Energy: An A-to-Z Guide*. SAGE, 2011.

Munir, Kashif. *Cloud Computing Technologies for Green Enterprises*. IGI Global, 2018.

Muszynska, Agnieszka. *Rotor Dynamics*. Taylor & Francis, 2005.

Mycio, Mary. *Wormwood Forest: A Natural History of Chernobyl*. Joseph Henry Press, 2005.

Myers, Richard L. "Propane." *The One Hundred Most Important Chemical Compounds: A Reference Guide*. Greenwood, 2011.

Nalule, Victoria R. *Energy Poverty and Access Challenges in Sub-Saharan Africa*. Palgrave MacMillan, 2019.

National Aeronautics and Space Administration, Glenn Research Center. "Enthalpy."

http://www.grc.nasa.gov/WWW/K-12/airplane/enthalpy.html.

———. *Metallic Rotor Sizing and Performance Model for Flywheel Systems.* NASA, 2019.

National Fluid Power Association. "Our Industry Overview." http://www.nfpa.com/OurIndustry/OurInd_Overview.asp.

National Geothermal Collaborative. "Geothermal Direct Use." http://www.geocollaborative.org/publications/Geothermal_Direct_Use.pdf.

National Institute of Building Sciences. "Building Envelope Design Guide and Whole Building Design Guide." http://www.wbdg.org/design/envelope.php.

National Institute of Standards and Technology. *NIST Handbook 44: Specifications, Tolerances, and Other Technical Requirements for Weighing and Measuring Devices.* 2010.

National Research Council. *Hidden Costs of Energy: Unpriced Consequences of Energy Production and Use.* National Academy of Sciences Press, 2009.

Navarro, Jaume. *A History of the Electron. J.J. and G.P. Thomson.* Cambridge UP, 2012.

Nave, C.R. "Cellular Respiration." *HyperPhysics*, 2016, hyperphysics.phy-astr.gsu.edu/hbase/Biology/celres.html. Accessed 2 Mar. 2020.

———. "Electric Potential Energy." *HyperPhysics*. Georgia State U, n.d. Web. 19 June 2015.

———. "Traveling Wave Relationship." *HyperPhysics*. Department of Physics and Astronomy, Georgia State U, 2014. Web. 25 Feb. 2015.

———. "Work, Energy and Power." *HyperPhysics*. Georgia State U, 2012. Web. 22 Sept. 2015.

Neill, Simon P., and M. Reza Hashemi. *Fundamentals of Ocean Renewable Energy: Generating Electricity from the Sea.* Academic Press, 2018.

Nelson, Vaughn. *Wind Energy: Renewable Energy and the Environment.* CRC Press, 2009.

———. *Wind Energy: Renewable Energy and the Environment.* 3rd ed. CRC Press, 2018.

Nemet, Gregory F. *How Solar Energy Became Cheap: A Model for Low-Carbon Innovation.* Routledge, 2019.

Newton, David E. *Wind Energy: A Reference Handbook.* ABC-CLIO, 2015.

Newton, Isaac. *Philosophiae Naturalis Principia Mathematica.* 1687. Reprint.

Newton's Principia: The Central Argument, edited by Dana Densmore and William H. Donahue. Green Lion Press, 2010.

Nikiforuk, Andrew. *Tar Sands: Dirty Oil and the Future of a Continent.* Greystone Books, 2009.

Nobel Foundation. "The Nobel Prize in Physics 1921: Albert Einstein." http://nobelprize.org/nobel_prizes/physics/laureates/1921/einstein-bio.html.

North, John, ed. *Mid-Nineteenth Century Scientists.* Pergamon Press, 1969.

Northern Arizona University, Colorado Plateau Stable Isotope Laboratory. "What Are Stable Isotopes?" http://www.mpcer.nau.edu/isotopelab/isotope.html.

Nothiger, Andreas. "Einstein Theory." Hyperhistory, http://www.hyperhistory.com/online_n2/History_n2/index_n2/einstein_theory.html.

Nunes, Leonel Jorge Ribeiro, Joao Carlos de Oliveira Matias, and Joao Paulo da Silva Catalao. *Torrefaction of Biomass for Energy Applications from Fundamentals to Industrial Scale.* Academic Press, 2018.

Nunez, Christina. "Fossil Fuels, Explained." *National Geographic,* 2 Apr. 2019, www.nationalgeographic.com/environment/energy/reference/fossil-fuels/. Accessed 18 Aug. 2020.

"Obituary, Rudolf Diesel." *Power,* vol. 38, no. 18, 1913, p. 628. http://books.google.com/books?id=b70fAQAAMAAJ&pg=PA628&dq=rudolf+diesel

Odum, Howard T. *Environmental Accounting: Energy and Environmental Decision Making.* John Wiley and Sons, 1996.

"Oil Information: Overview." *International Energy Agency,* July 2020, www.iea.org/reports/oil-information-overview Accessed 18 Aug. 2020.

"Oil: Crude and Petroleum Products Explained." *US Energy Information Administration.* Department of Energy, 5 Nov. 2015. Web. 16 May 2016.

Olofsson, William L., and Viktor I Bengtsson. *Solar Energy: Research, Technology and Applications.* Nova Science Publishers, 2008

Organ, A.J. *Thermodynamics and Gas Dynamics of the Stirling Machine.* Cambridge UP, 1980.

Organ, Allan J. *Stirling Cycle Engines: Inner Workings and Design.* John Wiley & Sons, 2013.

Ortiz, David, and Jerry Sollinger. *E-Vision 2002: Shaping Our Future by Reducing Energy Intensity in the U.S. Economy.* RAND, 2003.

Owen, M.S., ed. *ASHRAE Handbook: Fundamentals.* American Society of Heating, Refrigerating and Air-Conditioning Engineers, 2009.

Pahl, Greg. *Biodiesel: Growing a New Energy Economy.* 2nd ed. Chelsea Green, 2008.

Pain, H.J., and P. Rankin. *Introduction to Vibrations and Waves.* Wiley, 2015.

Pais, Abraham. *"Subtle Is the Lord": The Science and the Life of Albert Einstein*. Oxford UP, 1982.

Palmer, Jerrell Dean, and John G. Johnson. "Big Inch and Little Big Inch." http://www.tshaonline.org/handbook/online/articles/dob08.

Palz, Wolfgang, J. Greif, and Commission of the European Communities, eds. *European Solar Radiation Atlas: Solar Radiation on Horizontal and Inclined Surfaces (Algorithms and Combinatorics)*. Springer, 1996.

Pancaldi, Giuliano. *Volta: Science and Culture in the Age of Enlightenment*. Princeton UP, 2003.

Pappu, Vijay, Marco Carvalho, and Panos M. Pardalos, eds. *Optimization and Security Challenges in Smart Power Grids*. Springer, 2013.

Parris, George. *Thermodynamics: Heat Capacity, Enthalpy, Entropy and Free Energy*. 3rd ed. Independent, 2019.

Pecher, Arthur, and Jens Peter Kofoed, eds. *Handbook of Ocean Wave Energy*. SpringerOpen, 2017.

Peng, William W. "Steam Turbines." *Fundamentals of Turbomachinery*. J. Wiley, 2008.

Pera, Marcello. *The Ambiguous Frog: The Galvani-Volta Controversy on Animal Electricity*. Princeton UP, 2014.

Perrot, Pierre. *A to Z of Thermodynamics*. Oxford UP, 1998.

Petela, Richard. *Engineering Thermodynamics of Thermal Radiation: For Solar Power Utilization*. McGraw-Hill, 2010.

"Petroleum & Other Liquids." *US Energy Information Administration*, www.eia.gov/petroleum/. Accessed 18 Aug. 2020.

Philibert, Cédric. *Interactions of Policies for Renewable Energy and Climate*. International Energy Agency, 2011. Digital file. http://www.iea.org/papers/2011/interactions_policies.pdf.

"Photovoltaic (PV) Systems." *Canada Mortgage and Housing Corporation*. Canada Mortgage and Housing, 2010. Web. 13 May 2016.

Pierson, Jean-Marc, ed. *Large-Scale Distributed Systems and Energy Efficiency: A Holistic View*. Wiley, 2015.

Pifer, Linda K. "The Development of Young American Adults' Attitudes About the Risks Associated with Nuclear Power." *Public Understanding of Science*, vol. 5, no. 2, 1995.

Pimentel, David, and Tad Patzek. "Ethanol Production Using Corn, Switchgrass, and Wood; Biodiesel Production Using Soybean and Sunflower." *Natural Resources Research*, vol. 14, no. 1, Mar. 2005.

Pimentel, David, ed. *Biofuels, Solar, and Wind as Renewable Energy Systems: Benefits and Risks*. Springer, 2008.

Pimentel, David. "Corn Ethanol as Energy." *Harvard International Review*, vol. 31, no. 2, 2009.

———. "Ethanol Fuels: Energy Balance, Economics, and Environmental Impacts Are Negative." *Natural Resources Research*, vol. 12, no. 2, June 2003.

Pittock, Jamie, Karen Hussey, and Stephen Dovers, eds. *Climate, Energy and Water: Managing Trade-offs, Seizing Opportunities*. Cambridge UP, 2015.

Plante, Russell H. *Solar Energy. Photovoltaics and Domestic Hot Water*. Academic Press, 2014.

Plekhanov, Vladimir G. *Isotopes in Condensed Matter*. Springer, 2013.

Pomponi, Francesco, Catherine de Wolf, and Alice Moncaster, eds. *Embodied Carbon in Buildings: Measurement, Management and Mitigation*. Springer, 2018.

Poor People's Energy Outlook. Practical Action, policy.practicalaction.org/policy-themes/energy/poor-peoples-energy-outlook.

"Powering the Cell: Cellular Respiration." *CK-12 Foundation*, www.boyertownasd.org/cms/lib/PA01916192/Centricity/Domain/743/B.%20Chapter%204-Lesson%202-Powering%20the%20Cell-Cellular%20Respiration.pdf. Accessed 3 Mar. 2020.

PowerStream Technology. "Chemical Changes in the Battery." http://www.powerstream.com/1922/battery_1922_WITTE/batteryfiles/chapter04.htm.

"A Primer." http://www.netl.doe.gov/technologies/oil gas/publications/EPreports/Shale_Gas_Primer_2009.pdf.

Prout, Henry G. *A Life of George Westinghouse*. 1921. Reprint. Cosimo Classics, 2005.

Public Broadcasting Service. "Master of Lightning." http://www.pbs.org/tesla/.

———. "Willis Carrier." http://www.pbs.org/wgbh/theymadeamerica/whomade/carrier_hi.html.

Pulkrabek, Willard W. *Engineering Fundamentals of the Internal Combustion Engine*. Pearson Prentice Hall, 2004.

Purcella, Guy. *Do It Yourself Guide to Biodiesel: Your Alternative*. Ulysses, 2008.

Pynn, Larry. "Logging with Horse Power." *Canadian Geographic*, vol. 111, no. 4, Aug.-Sept. 1991, p. 30.

Rabie, M. Galal. *Fluid Power Engineering*. McGraw-Hill, 2009.

Rabl, A. "Comparison of Solar Concentrators." *Solar Energy*, vol. 18, no. 2, 1976.

Radzevich, Stephen P. *Advances in Gear Design and Manufacture*. CRC Press, 2019.

Raikar, Santosh, and Seabron Adamson. *Renewable Energy Finance: Theory and Practice*. Academic Press, 2020.

Raman, V.V. "William John Macquorn Rankine (1820-1872)." *Journal of Chemical Education*, vol. 50, no. 4, 1973.

Ramlow, Bob, and Benjamin Nusz. *Solar Water Heating: A Comprehensive Guide to Solar Water and Space Heating Systems*. New Society Publishers, 2010.

Rathakrishnan, Ethirajan. *Elements of Heat Transfer*. CRC Press, 2012.

Rathore, M.M., and Raul Raymond Kapuno. *Engineering Heat Transfer*. Jones and Bartlett, 2011.

Ray, Ramesh C., and S. Ramachandran, eds. *Bioethanol Production from Food Crops. Sustainable Sources, Interventions and Challenges*. Academic Press, 2019.

Rayaprolum, Kumar. *Boilers for Power and Process*. CRC Press, 2009.

Raymond, Martin S., and William L. Leffler. *Oil and Gas Production in Nontechnical Language*. PennWell, 2006.

Reddy, Thomas. *Linden's Handbook of Batteries*. 4th ed. McGraw-Hill Professional, 2010.

Reeder, Linda. *Guide to Green Building Rating Systems: Understanding LEED, Green Globes, Energy Star, the National Green Building Standard, and More*. Wiley, 2010.

Reilly, Helen. *Connecting the Country: New Zealand's National Grid, 1886-2007*. Steele Roberts, 2008.

Reinders, Angèle. *Designing with Photovoltaics*. CRC Press, 2020.

Rekioua, Djamila. *Wind Power Electric Systems: Modeling, Simulation and Control*. Springer, 2014.

REN21. *Renewables 2015: Global Status Report*. REN21 Secretariat, 2015. *REN21*. Web. 13 May 2016.

"Renewable & Alternative Fuels." *US Energy Information Administration*. EIA, 2016. Web. 13 May. 2016.

Renewable Energy Policy Network for the 21st Century (REN21). *Renewables 2010 Global Status Report*. Paris: REN21 Secretariat, 2010. http://www.ren21.net/REN21Activities/Publications/GlobalStatusReport

"Renewables." *International Energy Agency*. OECD/IEA, 2016. Web. 13 May. 2016.

Rigden, John S. *Einstein 1905: The Standard of Greatness* Harvard UP, 2005.

Riznic, Jovica, ed. *Steam Generators for Nuclear Power Plants*. Woodhead, 2017.

Roaf, Sue, and Fergus Nicol, eds. *Running Buildings on Natural Energy: Design Thinking for a Different Future*. Routledge, 2018.

Rogers, Alan. *Essentials of Photonics*. 2nd ed. CRC, 2008.

Rogers, Elizabeth, and Thomas Kostigen. *The Green Book: The Everyday Guide to Saving the Planet One Simple Step at a Time*. Foreword by Cameron Diaz and William McDonough. Three Rivers Press, 2007.

Rogers, G., and Y. Mayhew. *Thermodynamics: Work and Heat Transfer*. 4th ed. Longmann Scientific, 1992.

Rose, Joshua. *Modern Steam Engines*. H.C. Baird, 1886. Reprint. Astragal Press, 2003.

Rosen, Mark A., and Seama Koohi-Fayegh. *Geothermal Energy: Sustainable Heating and Cooling Using the Ground*. Wiley, 2017.

Rosendahl, Lasse, ed. *Direct Thermochemical Liquefaction for Energy Applications*. Woodhead Publishing, 2018.

Ross, John S. "Work, Power, Kinetic Energy." Project PHYSNET, Michigan State U, 23 Apr. 2002. http://www.physnet.org/modules/pdf_modules/m20.pdf.

Roth, Kurt, and James Brodrick. "Seasonal Energy Storage." *ASHRAE Journal*, Jan. 2009.

Rowlett, Russ. *A Dictionary of Units of Measurement*. Center for Mathematics and Science Education, U of North Carolina. http://www.unc.edu/~rowlett/units.

Rufer, Alfred. *Energy Storage Systems and Components*. CRC Press, 2018.

Ruin, Sven, and Gören Sidén. *Small-scale Renewable Energy Systems: Independent Electricity Systems for Community, Business and Home*. CRC Press, 2020.

Rukeyser, Muriel. *Willard Gibbs*. Ox Bow Press, 1988.

Ruppert, Michael C. *Confronting Collapse: The Crisis of Energy and Money in a Post Peak Oil World*. Chelsea Green, 2009.

Russell, Ben. *James Watt: Making the World Anew*. Reaktion Books, 2014.

Russell, C.A. *Michael Faraday: Physics and Faith*. Oxford UP, 2000.

Russell, Daniel A. "Longitudinal and Transverse Wave Motion." *Acoustics and Vibration Animations*. Pennsylvania State U, 18 Feb. 2015. Web. 29 June 2015.

Rusu, Eugen, and Vengatesan Venugopal, eds. *Offshore Renewable Energy: Ocean Waves, Tides and Offshore Wind*. MDPI, 2019.

Ryan, V. "Levers." http://www.technologystudent.com/forcmom/lever1.htm.

Rybach, Ladislaus. "Geothermal Sustainability." *Geothermics*, vol. 32, nos. 4-6, 2003, pp. 463-70.

Rydh, Carl Johan, and Magnus Karlström. "Life Cycle Inventory of Recycling Portable Nickel-Cadmium Batteries." *Resources, Conservation and Recycling*, vol. 34, 2002.

Salam, Abdus. "Weak and Electromagnetic Interactions." *Il Nuovo Cimento*, vol. 11, no. 4, Feb. 1959.

Salam, Abdus, and N. Svartholm, eds. Elementary Particle Physics: Relativistic Groups and Analyticity. Eighth Nobel Symposium. Almqvist and Wiksell, 1968.

Sammes, N. *Fuel Cell Technology: Reaching Towards Commercialization*. Springer, 2006.

Santo Pietro, David. "Power." *Khan Academy*. Khan Acad., 2015. Web. 5 May 2015.

Santos, Fernando, Sarita Candida Rabelo, Mario de Matos, and Paulo Eichler, eds. *Sugarcane Biorefinery, Technology and Perspectives*. Academic Press, 2020.

Saravanamuttoo, H.I.H., G.F.C. Rogers, H. Cohen, and P.V. Straznicky. *Gas Turbine Theory*. Pearson Prentice Hall, 2008.

Sarbu, Ioan, and Calin Sebarchievici. *Solar Heating and Cooling Systems: Fundamentals, Experiments and Applications*. Elsevier, 2017.

Savery, Thomas. *The Miner's Friend*. Outlook Verlag, 2020.

Scheer, Hermann. *A Solar Manifesto*. 2nd ed. James & James, 2001.

Schiemann, Gregor. *Hermann von Helmholtz's Mechanism: The Loss of Certainty: A Study on the Transition from Classical to Modern Philosophy of Nature*. Springer, 2009.

Schiffler, Michael Brian. *Draw the Lightning Down: Benjamin Franklin and Electrical Technology in the Age of Enlightenment*. U of California P, 2006.

Schiller, Mark. *Simplified Design of Building Lighting*. John Wiley and Sons, 1997.

Schlesinger, Henry. *The Battery: How Portable Power Sparked a Technological Revolution*. HarperCollins, 2010.

Schmidt, Michael J., and Richard Ross. "Working Elephants: They Earn Their Keep in Asia by Providing an Ecologically Benign Way to Harvest Forests." *Scientific American*, vol. 274, no. 1, Jan. 1996, p. 82.

Schubert, E. Fred. *Light-Emitting Diodes*. 2nd ed. Cambridge UP, 2005.

Schwab, Klaus. *The Fourth Industrial Revolution*. Crown Business, 2016.

Scragg, Alan. *Biofuels: Production, Application, and Development*. CABI, 2009.

Scudder, Thayer. *The Future of Large Dams: Dealing with Social, Environmental, Institutional, and Political Costs*. Earthscan, 2005.

Segrè, Emilio. *From Falling Bodies to Radio Waves: Classical Physicists and Their Discoveries*. W.H. Freeman, 1984.

SenterNovem. *Bioethanol in Europe: Overview and Comparison of Production Processes*. Rapport 2GAVE0601. SenterNovem, 2006.

Serway, Raymond A., and John W. Jewett. *Physics for Scientists and Engineers*. 8th ed. Brooks/Cole Cengage Learning, 2010.

Serway, Raymond A., Chris Vuille, and Jerry S. Faughn. *College Physics*. 8th ed. Brooks/Cole Cengage Learning, 2009.

Serway, Raymond A., John W. Jewett, and Vahé Peroomian. *Physics for Scientists and Engineers with Modern Physics*. 9th ed. Brooks, 2014.

Shales, Stuart. "Biofuels: The Way Forward?" *Science for Environmental Policy*, no. 1, Feb. 2008. http://ec.europa.eu/environment/integration/research/newsalert/pdf/1si.pdf.

Shankar, Ramamurti. "Lecture 5: Work-Energy Theorem and Law of Conservation of Energy." *Open Yale Courses*. Yale U, 2006. Web. 22 Sept. 2015.

Sheldrick, Bill, and Sally Macgill. "Local Energy Conservation Initiatives in the UK: Their Nature and Achievements." *Energy Policy*, 1988.

Shepard, M. "Coal Technologies for a New Age." *EPRI Journal*, Jan.-Feb. 1988, pp. 4-17.

Sherwood, Dennis, and Paul Dalby. *Modern Thermodynamics for Chemists and Biochemists*. Oxford UP, 2018.

Shipman, James T., Jerry D. Wilson, and Charles A. Higgins Jr. *An Introduction to Physical Science*. 14th ed. Brooks/Cole, 2015.

Shively, Bob, and John Ferrare. *Understanding Today's Natural Gas Business*. Enerdynamics, 2009.

Shlyakhin, P. *Steam Turbines: Theory and Design*. U Press of the Pacific, 2005.

Sier, Robert. *Rev. Robert Stirling, D.D.* L.A. Mair, 1995.

———. "Stirling and Hot Air Engine Homepage." http://www.stirlingengines.org.uk.

Silveira, Semida. *Bioenergy-Realizing the Potential*. Elsevier, 2005.

Simanek, Donald E. "Kinematics." *Brief Course in Classical Mechanics*. Lock Haven U, Feb. 2005. Web. 22 Sept. 2015.

Simcock, Neil, Harriet Thomson, Saska Petrova, and Stefan Bouzarovski, eds. *Energy Poverty and Vulnerability: A Global Perspective*. Routledge, 2018.

Simon, Christopher A. "Geothermal Energy." *Alternative Energy: Political, Economic, and Social Feasibility*. Rowman, 2007.

Simonis, Doris, ed. "Clausius, Rudolf." *Scientists, Mathematicians, and Inventors: Lives and Legacies: An Encyclopedia of People Who Changed the World*. Oryx Press, 1999.

Singh, Chanan, Panida Jirutitijaroen, and Joydeep Mitra. *Electric Power Grid Reliability Evaluation*. IEEE/Wiley, 2019.

Sissine, Fred, coordinator. "Energy Independence and Security Act of 2007: A Summary of Major Provisions, CRS Report for Congress." 21 Dec. 2007. http://energy.senate.gov/public/_files/R1342941.pdf.

Skinner, Steve. *Hydraulic Fluid Power: A Historical Timeline.* Steve Skinner Productions, 2014.

Skrabec, Quentin R, Jr. *George Westinghouse: Gentle Genius.* Algora Publishing, 2007.

Slack, Kara. *U.S. Geothermal Power Production and Development Update.* Geothermal Energy Association, 2008.

Sluyterman, Keetie, and Joost Dankers. *Keeping Competitive in Turbulent Markets, 1973-2007: A History of Royal Dutch Shell.* Oxford UP, 2007.

Smil, Vaclav. *Creating the Twentieth Century: Technical Innovations of 1867-1914 and Their Lasting Impact.* Oxford UP, 2005.

———. *Energy in Nature and Society: General Energetics of Complex Systems.* MIT Press, 2008.

Smith, Brian C. *Quantitative Spectroscopy: Theory and Practice.* Elsevier, 2002.

Smith, T.C., D.R. Kindred, J.M. Brosnan, R.M. Weightman, M. Sheperd, and R. Bradley. "Wheat as a Feedstock for Alcohol Production." *Research Review*, no. 61, Dec. 2006.

Smith, Zachary, and Katrina Taylor. *Renewable and Alternative Energy Resources: A Reference Handbook.* ABC-CLIO, 2008.

Solar Cookers World Network. http://solarcooking.org.

Solomon, S., D. Qin, M. Manning, Z. Chen, M. Marquis, K.B. Avery, M. Tignor, and H.L. Miller, eds. *Contribution of Working Group I to the Fourth Assessment Report of the Intergovernmental Panel on Climate Change, 2007.* Cambridge UP, 2007.

Sorda, Giovanni, Martin Banse, and Claudia Kemfert. "An Overview of Biofuel Policies Across the World." *Energy Policy*, vol. 38, 2010.

Sorebo, Gilbert N., and Michael C. Echols. *Smart Grid Security: An End-to-End View of Security in the New Electrical Grid.* CRC Press, 2011.

Sørensen, Bent. *Hydrogen and Fuel Cells: Emerging Technologies and Applications.* 2nd ed. Academic Press, 2012.

———. *Renewable Energy: Physics, Engineering, Environmental Impacts, Economics & Planning.* 4th ed. Academic Press, 2011.

———. ed., *Solar Energy Storage.* Academic Press, 2015.

Sparrow, W.J. *Knight of the White Eagle: A Biography of Sir Benjamin Thompson, Count Rumford, 1753-1814.* Hutchinson, 1964.

Speedace.info. "Rudolp (sic) Diesel and Diesel Oil." http://speedace.info/diesel.htm.

Speight, James G. *The Chemistry and Technology of Petroleum.* 4th ed. CRC Press, 2007.

———. *The Chemistry and Technology of Petroleum.* 5th ed. CRC Press, Taylor, 2014.

———. *Deep Shale Oil and Gas.* Elsevier, 2017.

———. *Natural Gas: A Basic Handbook.* 2nd ed. Gulf Professional, 2019.

Spellman, Frank R. *The Science of Renewable Energy.* 2nd ed. CRC Press, 2016.

Spreng, Daniel T. *Net-Energy Analysis and the Energy Requirements of Energy Systems.* Praeger, 1988.

Srivastava, S.P., and Jeno Hancsók. *Fuels and Fuel Additives.* Wiley, 2014.

Stacy, Angelica M., Janice A. Coonrod, and Jennifer Claesgens. *Fire: Energy and Thermochemistry.* Key Curriculum Press, 2005.

Staley, Kent W. "The Discovery of the Electron." American Institute of Physics. http://www.aip.org/history/electron/jjhome.htm.

State of Hawaii, Department of Business, Economic Development, and Tourism. "Ocean Thermal Energy." http://hawaii.gov/dbedt/info/energy/renewable/otec.

Stearns, Peter N. *The Industrial Revolution in World History.* 4th ed. Routledge, 2018.

"The Steering Column." Nikolaus Otto and His Remarkable Compression Stroke. *Car and Driver*, vol. 39, no. 8, Feb. 1994, p. 7.

Steffens, Henry John. *James Prescott Joule and the Concept of Energy.* Science History Publications, 1979.

Stehlik, Petr, and Zdenek Jegla. *Heat Exchangers for Waste Heat Recovery.* MDPI AG, 2020.

Steingress, Frederick M., Harold J. Frost, and Daryl R. Walker. *Stationary Engineering.* 3rd ed. American Technical, 2003.

Steinle, Friedrich. *Exploratory Experiments: Ampère, Faraday, and the Origins of Electrodynamics.* Alex Levine, tr. U of Pittsburgh P, 2016.

Stephenson, Michael. *Energy and Climate Change: An Introduction to Geological Controls, Interventions and Mitigations.* Elsevier, 2018.

Stiebler, Manfred. *Wind Energy Systems for Electric Power Generation.* Springer, 2008.

Stockton, Nick. "Nuclear Power Is Too Safe to Save the World from Climate Change." *Wired.* Conde Nast, 3 Apr. 2016. Web. 16 May 2016.

Strezov, Vladimir, and Hossain M. Answar, eds. *Renewable Energy Systems from Biomass: Efficiency, Innovation and Sustainability.* CRC Press, 2019.

Sui, Daniel, and David Rejeski. "Environmental Impacts of the Emerging Digital Economy: The E-for-Environment E-Commerce?" *Environmental Management*, vol. 29, no. 2, 2002, pp. 155-63.

Sukhatme, S.P., and J.K. Nayak. *Solar Energy: Principles of Thermal Collection and Storage*. 3rd ed. Tata McGraw-Hill, 2008.

Susskind, Charles. *Heinrich Hertz: A Short Life*. San Francisco Press, 1995.

Sutherland, H.B. "Professor William John Macquorn Rankine." *Proceedings of the Institution of Civil Engineers Civil Engineering*, vol. 132, no. 4, 1999.

Sutton, Anthony C. *Cold Fusion: The Secret Energy Revolution*. Dauphin Publications, 2016.

Szargut, J. *Exergy Method: Technical and Ecological Applications*. WIT Press, 2005.

Tanchev, Ljubomir. *Dams and Appurtenant Hydraulic Structures*. 2nd ed. CRC Press, 2014.

Taner, Tolga, ed. *Energy Policy*. IntechOpen, 2020.

Tanuma, Tadashi, ed. *Advances in Steam Turbines for Modern Power Plants*. Woodhead Publishing, 2017.

Taubes, G. *Bad Science: The Short Life and Weird Times of Cold Fusion*. Random House, 1993.

Tesla Memorial Society of New York. "Tesla's Biography." http://www.teslasociety.com.

Tesla, Nikola, and Ben Johnston. *My Inventions: The Autobiography of Nikola Tesla*. Arcturus Publishing, 2019.

Thaler, Alexander, and Daniel Watzenig, eds. *Automotive Battery Technology*. Springer, 2014.

Thess, André. *The Entropy Principle: Thermodynamics for the Unsatisfied*. Springer, 2011.

Thompson, Benjamin, of Rumford. *Collected Works of Count Rumford*. 5 vols. Edited by Sanborn C. Brown. Harvard UP, 1968-1970.

Thompson, Silvanus P. *The Life of William Thomson, Baron Kelvin of Largs*. 2 vols. 1910. Cambridge UP, 2011.

Thomson, George Paget, Sir. *J.J. Thomson and the Cavendish Laboratory*. Nelson, 1964.

Thomson, J.J. "Award Ceremony Speech." *Nobel Lectures: Physics, 1901-1921*. Elsevier, 1967. http://www.nobelprize.org/nobel_prizes/physics/laureates/1906/press.html.

———. *Rays of Positive Electricity and Their Application to Chemical Analyses* Creative Media Partners LLC, 2018.

———. *Recollections and Reflections*. G. Bell, 1936.

Thumann, Albert. *Plant Engineers and Managers Guide to Energy Conservation*. 10th ed., River Publishers, 2020.

Tilley, Richard. *Colour and the Optical Properties of Materials*. 2nd ed. Wiley, 2011.

The Times. "Revealed: The Environmental Impact of Google Searches." http://technology.timesonline.co.uk/tol/news/tech_and_web/article5489134.ece.

Tipler, Paul. *Physics for Scientists and Engineers: Mechanics*. 3rd ed. W.H. Freeman, 2008.

Tiwari, G.M., and Arvind Tiwari Shyam. *Handbook of Solar Energy: Theory, Analysis and Applications*. Springer, 2016.

Tiwari, G.N., and M.K. Ghosal. "Draught Animal Power." *Renewable Energy Resources: Basic Principles and Applications*. Alpha Science International, 2005.

Tomain, Joseph P. "Nuclear Futures." *Duke Environmental Law and Policy Forum*, vol. 15, no. 221, 2005.

Torriti, Jacopo. *Peak Energy Demand and Demand Side Response*. Routledge, 2016.

Tosun, Ismail. *Thermodynamics Principles and Applications*. World Scientific, 2015.

Trasalti, S. "1799-1999: Alessandro Volta's 'Electric Pile'; Two Hundred Years, but It Doesn't Seem Like It." *Journal of Electroanalytical Chemistry*, vol. 460, no. 1, 1999.

Tredgold, Thomas. *The Steam Engine*. J. Taylor, 1827. Reprint. Cambridge UP, 2014.

Tregenza, Peter, and Michael Wilson. *Daylighting Architecture and Lighting Design*. Routledge, 2011.

Trudell, Craig. "Tesla Surges as Morgan Stanley Says Electric Cars Will Gain Market Share." Bloomberg.com, 31 Mar. 2011, http://www.bloomberg.com/news/2011-03-31/tesla-surges-as-morgan-stanley-says-electric-cars-will-gain-1-.html.

Tucker, William. "Understanding E = mc2." 21 Oct. 2009, http://www.energytribune.com/articles.cfm/2469/Understanding-E-=-mc2.

Turner, John. "A Realizable Renewable Energy Future." *Science*, vol. 285, 1999.

Twidell, John, and Toney Weir. *Renewable Energy Resources*. 3rd ed. Routledge, 2015.

"U.S. Biodiesel and Renewable Diesel Imports Increase 61% in 2015." *US Energy Information Administration*. Department of Energy, 11 Apr. 2016. Web. 13 May 2016.

U.S. Congress. Senate. Committee on Homeland Security and Governmental Affairs. *Energy Star Program: Covert Testing Shows the Energy Star Program Certification Process Is Vulnerable to Fraud and Abuse*. U.S. Government Accountability Office, 2010.

U.S. Department of Commerce. "About the Department of Energy: Origins and Evolution of the Department of Energy." http://www.energy.gov/about/origins.htm.

———. "About Us." http://energy.gov/about-us.

———. "Alternative and Advanced Fuels." http://www.afdc.energy.gov/afdc/fuels/indes.html.

———. "Alternative and Advanced Fuels: Propane."

———. *Biodiesel Fuel Handling and Use Guidelines for Users, Blenders, Distributors*. GPO, 2010.

———. "Department of Energy FY 2020 Budget Request Fact Sheet." 11 Mar. 2019, https://www.energy.gov/articles/department-energy-fy-2020-budget-request-fact-sheet.

———. "Emissions of Potent Greenhouse Gas Increase Despite Reduction Efforts." U.S. Department of Commerce, 2010. http://www.noaanews.noaa.gov/stories2010/20100127_greenhousegas.html.

———. *Laying the Foundation for a Solar America: The Million Solar Roofs Initiative; Final Report. October 2006.* http://www.nrel.gov/docs/fy07osti/40483.pdf.

———. National Renewable Energy Laboratory. *2014 Renewable Energy Data Book* N.p.: Author, 2015. *National Renewable Energy Laboratory.* Web. 16 May 2016.

———. "Ocean Thermal Energy Conversion." http://www.energysavers.gov/renewable_energy/ocean/index.cfm/mytopic=50010.

———. "Office of Science: The Atomic Energy Commissions (AEC), 1947." http://www.ch.doe.gov/html/site_info/atomic_energy.htm.

———. "What Is the Energy Payback for PV?" http://www.nrel.gov/docs/fy04osti/35489.pdf.

U.S. Department of the Interior, Bureau of Land Management. "About Oil Shale." http://ostseis.anl.gov/guide/oilshale/index.cfm.

———. Bureau of Land Management. "About Tar Sands." http://ostseis.anl.gov/guide/tarsands/index.cfm.

———. *Ocean Energy.* Minerals Management, 2007

U.S. Department of State. "The Kyoto Protocol on Climate Change." http://www.state.gov/www/global/oes/fs_kyoto_climate_980115.html.

U.S. Energy Information Administration. *Annual Energy Review 2010.* Technical Report DOE/EIA-0387(97). U.S. Energy Information Administration, 2011.

———. Electricity Explained: Your Guide to Understanding Electricity. http://www.eia.doe.gov/energyexplained.

———. "Energy Explained: Your Guide to Understanding Energy." http://www.eia.doe.gov/energyexplained/index.cfm.

———. "International Energy Statistics: Energy Intensity." http://www.eia.doe.gov/cfapps/ipdbproject/IEDIndex3.cfm?tid=92&pid=46&aid=2.

———. Oil Imports and Exports. *EIA,* U.S. Energy Information Administration, 1 May 2018, /www.eia.gov/energyexplained/index.php?page=oil_imports. Accessed 10 Oct. 2018.

———. "Power Transactions & Interconnected Networks." http://www.eia.doe.gov/cneaf/electricity/page/prim2/chapter7.html.

———. "Proved Reserves of Natural Gas (Trillion Cubic Feet)." *EIA.gov.* U.S. Department of Energy, 2015. Web. 6 Jan. 2016.

U.S. Environmental Protection Agency. "Clean Power Plan: Clean Power Plan for Existing Power Plants." *EPA.* EPA, 11 Feb. 2016. Web. 16 May. 2016.

———. *E85 and Flex Fuel Vehicles.* U.S. Environmental Protection Agency, Office of Transportation and Air Quality, 2010.

———. "Sources of Greenhouse Gas Emissions: Commercial and Residential Sector Emissions." *EPA.gov.* Environmental Protection Agency, 17 Apr. 2014. Web. 30 Jan. 2015.

U.S. Geological Survey. "Heavy Oil and Natural Bitumen: Strategic Petroleum Reserves." http://pubs.usgs.gov/fs/fs070-03/fs070-03.pdf.

———. "Natural Bitumen Resources of the United States." http://pubs.usgs.gov/fs/2006/3133/pdf/FS2006-3133_508.pdf.

U.S. Nuclear Regulatory Commission. *Backgrounder on Chernobyl Nuclear Power Plant Accident.* April 2009. http://www.nrc.gov/reading-rm/doc-collections/fact-sheets/chernobyl-bg.html.

United Kingdom, Department for Business, Energy & Industrial Strategy. *Annual Fuel Poverty Statistics Report, 2018 (2016 Data).* Crown, 2018, assets.publishing.service.gov.uk/government/uploads/system/uploads/attachment_data/file/719106/Fuel_Poverty_Statistics_Report_2018.pdf.

United Kingdom, Department for the Environment, Food and Rural Affairs. "The Government's Strategy for Combined Heat and Power to 2010." http://www.lowcarbonoptions.net/resources/Policy-&-Measures/Policies-and-Measures-United-Kingdom/chp-strategy.pdf.

United Nations Chernobyl Forum Expert Group Environment. *Environmental Consequences of the Chernobyl Accident and Their Remediation.* Aug. 2005. http://www.iaea.org/NewsCenter/Focus/Chernobyl/pdfs/ege_report.pdf.

United Nations Conference on Climate Change. "More Details about the Agreement." United Nations Conference on Climate Change. COP21, n.d. Web. 16 May. 2016.

United Nations Development Programme. "The World Energy Assessment: Overview, 2004 Update." http://www.undp.org/energy/weaover2004.htm.

United Nations Environment Programme. *The Bioenergy and Water Nexus.* Oeko-Institution and IEA Bioenergy Task 43, 28 Nov. 2011.

United Nations, Department of Economic and Social Affairs, Division for Sustainable Development. "Intensity of Energy Use." http://www.un.org/esa/sustdev/natlinfo/indicators/isdms2001/isd-ms2001economicB.htm.

United Nations. "List of Parties That Signed the Paris Agreement on 22 April." *United Nations: Sustainable Development Goals: 17 Goals to Transform Our World*. UN, Apr. 2016. Web. 16 May. 2016.

University of Rochester. "District Heat Library." http://www.energy.rochester.edu/.

University of Strathclyde, Glasgow. "Boiler Technology Adapted to Small Scale Heating." http://www.esru.strath.ac.uk/EandE/Web_sites/06-07/Biomass/HTML/boiler_technology.htm.

Valentine, Katie. "First Offshore Wind Farm in the US Kicks Off Construction." *ClimateProgress*. ThinkProgress, 28 Apr. 2015. Web. 13 May. 2015.

van Aken, Hendrik M. *The Oceanic Thermohaline Circulation: An Introduction*. Springer, 2007.

Van Basshuysen, Richard, and Fred Schafer. *Internal Combustion Engine Handbook: Basics, Components, Systems, and Perspectives*. Society of Automotive Engineers International, 2004.

Van Mierlo, Joeri, ed. *Plug-in Hybrid Electric Vehicle (PHEV)*. MDPI, 2019.

Viegas, Jennifer. *Kinetic and Potential Energy: Understanding Changes Within Physical Systems*. Rosen, 2005.

Vineeth C.S. *Stirling Engines: A Beginner's Guide*. Vineeth CS, 2012.

Vujovic, Ljubo. "Albert Einstein (1879-1955)." Tesla Memorial Society of New York, http://www.teslasociety.com/einstein.htm.

Walker, G.M. *Bioethanol: Science and Technology of Fuel Alcohol*. Ventus, 2010.

Walker, Graham. *Stirling Engines*. Oxford UP, 1980.

Walker, J. Samuel. *Three Mile Island: A Nuclear Crisis in Historic Perspective*. U of California P, 2004.

Walker, J.S. *Three Mile Island: A Nuclear Crisis in Historical Perspective*. U of California P, 2004.

Wang, Enhua, ed. *Organic Rankine Cycle Technology for Heat Recovery*. IntechOpen, 2018.

Wang, Jun-Jie. *Hydraulic Fracturing in Earth-Rock Fill Dams*. Wiley, 2014.

Wang, Xiangzeng. *Lacustrine Shale Gas: Case Study from the Ordos Basin*. Elsevier, 2017.

Wang, Xiuli, and Michael Economides. *Advanced Natural Gas Engineering*. Gulf Professional, 2009.

Watts, Martin. *Working Oxen*. Shire, 1999.

Watts, Naomi. *Heat Transfer: Fundamentals and Applications*. NY Science Press, 2020.

"Waves." *The Physics Classroom*. Physics Classroom, n.d. Web. 29 June 2015.

Weart, Spencer R. *Nuclear Fear: A History of Images*. Harvard UP.

Weinberg, Steven. "A Model of Leptons." *Physical Review Letters*, vol. 19, no. 21, 1967.

Wenar, Leif. *Blood Oil: Tyrants, Violence, and the Rules That Run the World*. Oxford UP, 2016.

"What Is a Cell?" *US National Library of Medicine*, 3 Mar. 2020, ghr.nlm.nih.gov/primer/basics/cell. Accessed 3 Mar. 2020.

Wheeler, Lynde Phelps. *Josiah Willard Gibbs: The History of a Great Mind*. Ox Bow Press, 1998.

"Why Build Green?" *EPA.gov*. Environmental Protection Agency, 9 Oct. 2014. Web. 30 Jan. 2015.

Wimberly, Jamie. *Energy Star Shining Bright? National Consumer Survey of the Energy Star Brand*. Ecoalign, 2010.

Winterton, R.H.S. *Thermal Design of Nuclear Reactors*. Elsevier, 2014.

Wise, M. Norton. *Aesthetics, Industry and Science: Hermann von Helmholtz and the Berlin Physical Society*. U of Chicago P, 2018.

Witten, Mark L., Errol Zeiger, and Glenn D. Ritchie, eds. *Jet Fuel Toxicology*. CRC Press, 2011.

World Bank. Renewable Energy Development: The Role of the World Bank Group. World Bank, 2004.

"World Energy Outlook 2019." *International Energy Agency*, Nov. 2019, www.iea.org/reports/world-energy-outlook-2019. Accessed 18 Aug. 2020.

World Nuclear Association. "Radioisotopes in Industry."
———. "What Is Uranium? How Does It Work?" http://www.world-nuclear.org/education/uran.htm.

Worldwatch Institute. *Biofuels for Transport: Global Potential and Implications for Energy and Agriculture*. Earthscan, 2007.

Wright, Glen, Sandy Kerr, and Kate Johnson, eds. *Ocean Energy: Governance Challenges for Wave and Tidal Stream Technologies*. Earthscan, 2018.

Wróbel, Marek, Marcin Jewiaarz, and Andrzej Szlek, eds. *Renewable Energy Sources: Engineering, Technology, Innovation ICORES 2018*. Springer, 2020.

Xie, Wei-Chau, Shun-Hao Ni, Wei Liu, and Wei Jiang. *Seismic Risk Analysis of Nuclear Power Plants*. Cambridge UP, 2019.

Yablokov, Alexey V., Vassily B. Nesterenko, and Alexey V. Nesterenko. *Chernobyl: Consequences of the Catastrophe for People and the Environment*. Blackwell, 2009.

Yacobucci, Brent D. "Fuel Ethanol: Background and Public Policy Issues." *Congressional Research Service*, 3 Mar. 2006.

http://www.policyarchive.org/handle/10207/bitstreams/2757.pdf.

Yam, Vivian W.W., ed. *WOLEDs and Organic Photovoltaics: Recent Advances and Applications*. Springer, 2010.

Yang, Zhaoqing, and Andrea Copping, eds. *Marine Renewable Energy: Resource Characterization and Physical Effects*. Springer, 2017.

Yergen, Daniel. *The Quest: Energy, Security, and the Remaking of the Modern World*. Penguin Books, 2012.

Yergin, Daniel. *The Prize: The Epic Quest for Oil, Money, and Power*. New ed. The Free Press, 2008.

Young, Hugh D. *College Physics*. Addison-Wesley, 2012.

Young, Hugh D., Philip W. Adams, and Raymond J. Chastain. *Sears & Zemansky's College Physics*. 10th ed. Addison, 2016.

Yudelson, Jerry. *The Green Building Revolution*. Island, 2008.

Zhao, Guangjin. *Reuse and Recycling of Lithium-Ion Power Batteries*. Wiley, 2017.

Zhou, Shelley W.W. *Carbon Management for a Sustainable Environment*. Springer, 2020.

Zielinski, Ellen, Courtney Faber, and Marissa H. Forbes. "Lesson: Waves and Wave Properties." *TeachEngineering*. Regents of the U of Colorado, 18 July 2014. Web. 29 June 2015.

Zurschmeide, Jeff. "The Truth about Ethanol in Gasoline." *Digital Trends*, 6 Aug. 2016, www.digitaltrends.com/cars/the-truth-about-ethanol-in-your-gas. Accessed 10 Oct. 2018.

Zweibel, Ken, James Mason, and Vasilis Fthenakis. "A Solar Grand Plan: By 2050 Solar Power Could End U.S. Dependence on Foreign Oil and Slash Greenhouse Gas Emissions." *Scientific American*, December 16, 2007. http://www.colorado.edu/physics/phys3070/phys3070_fa08/Reading/phys3070_sciamerican_solargrandplan.pdf.

Zycher, Benjamin. *Renewable Electricity Generation: Economic Analysis and Outlook*. AEI Press, 2011.

Glossary

absolute zero: the temperature at which there is absolutely no energy present; the temperature at which the volume of an ideal gas, decreasing as temperature decreases, becomes nonexistent

AC and DC: acronyms for alternating current and direct current, respectively

acetogens: bacteria that acts to break down cellulosic material, producing acetate (or acetic acid) as a product of the digestion process

acetone: 2-propanone, also known as dimethyl ketone

acid mine drainage: water runoff from mines in which sulfur from certain minerals has reacted with the water to produce sulfuric acids that leach toxic heavy metals from other minerals present

actinide: elements belonging to the actinide series in the periodic table, having atomic weights greater than that of actinium

activation energy: the amount of energy required to perturb a system sufficiently to render it unstable

actuator: a device that when activated performs a motion that causes a function such as extending a robotic arm or extending an aircraft's landing gear to be carried out

aerobic: processes that require the presence of oxygen

agricultural revolution: the period in which nomadic hunter-gatherer societies established agriculture in place, leading to the establishment of permanent settlements where the products of agriculture would be readily available

albedo: the amount of light and other energy reflected from a planet relative to the amount that it receives; an albedo of 0.3 indicates that the planet reflects and emits 30 percent of its incoming insolation back into space

alkaline: mineral material capable of acting as a base, as opposed to an acid

alkanes: organic molecules composed solely of carbon and hydrogen; in normal alkanes the carbon atoms are joined together linearly, in branched alkanes the carbon atoms are joined together as groups attached to a simple linear structure

alternating current: electrical current produced by magnetic oscillation, which reverses direction periodically rather than flowing continuously in one direction; for example, normal household current in North America oscillates at a frequency of 60 Hertz (60 cycles per second) and therefore reverses direction 120 times per second

alternative: indicates some feature or process that is intended to displace and be used instead of the historical standard feature or process

alternative energy sources: refers to energy sources that do not rely on unsustainable fossil fuels (petroleum, coal, natural gas) but are entirely renewable (wind power, solar photovoltaics, small-scale hydropower, etc.)

American system: a production manufacturing system that relied on specialized or single-purpose machines rather than general or multipurpose machines

ampere: the standard unit of electrical current, defined as the movement of 1 coulomb of electric charge through a potential difference of 1 volt in 1 second

anaerobic: processes that either do not require or that do not occur in the presence of oxygen

analytical engine: a programmable mechanical calculating device proposed by Joseph Babbage but never actually constructed, as a refined and advanced version of his difference engine

animal electricity: the notion that the electrical energy causing muscle response when muscle tissue was stimulated was stored within the muscle and was released by the stimulus

anode: the relatively positive terminal of an electrochemical circuit; the anode receives electrons emitted from the cathode

anthropogenic: refers to conditions and consequences that are caused by human activities

anthropogenic emissions: byproduct emissions of materials like carbon dioxide from the burning of fossil fuels caused by human activities

aromatics: organic molecules composed solely of carbon and hydrogen atoms, in which some or all of the carbon atoms are joined together to form ring structures with more energetic bonds; they are not called aromatic for their odor, but for the special nature of the interatomic bonds in the ring structures

aperture: an adjustable opening in a barrier through which light or other electromagnetic emission can pass

atomic energy: energy derived from the use of radioactive materials undergoing nuclear fission

atomic weight: generally, the weighted average of the atomic weights of the isotopes of an element according to their natural abundances; specifically, the weight of an atom as determined by the number of protons and neutrons in its nucleus

bagasse: the cellulosic residue (plant matter) left after the juice has been squeezed out from the prepared sugarcane stalks

barrage: a structure that bars and controls the tidal inflow and outflow of water in a river channel or other coastal intrusion

barrel: a benchmark volume used as a de facto standard measurement for petroleum production, with 1 barrel being standardized as a volume of 42 gallons or 159 liters

base unit: a unit of measurement that is not constructed or derived from other units of measurement

bifuel vehicle: a vehicle that can operate using either one of two entirely different types of fuel rather than a mixture of two fuels

bilateral policy: meaning two sides, a policy agreement involving two nations or governing bodies

biodiesel: fuel for use in diesel engines, produced by processing vegetable oils and animal fats through a transesterification process and refining the product for quality

bioethanol: ethanol that is produced through fermentation processes based on glucose, starches, and cellulose from organic matter

biogas: gases, typically hydrogen and methane, that are produced as fuels through the degradation and decomposition of organic matter

biofuels: fuels produced from organic matter using either biological processes such as fermentation or chemical processes such as transesterification; principal biofuels are ethanol from fermentation and biodiesel from transesterification

biogas: a mixture of gases produced by the decomposition of animal and vegetable wastes

biogenic: produced by the decomposition of organic matter under anaerobic conditions

biomass: material produced from the growth of plants, either as a crop byproduct or for direct use

biomass oils: oils that can be obtained from biomass, such as vegetable seed oils and animal fats, for processing into usable fuels

bitumen: high molecular weight hydrocarbon materials, commonly seen as roofing or paving tar

bottom-up approach: a management approach that begins from the position of the end user with regard to needs and adjusts the higher levels of management to minimize costs and maximize efficiency

breeder reactors: nuclear reactors designed to create more fuel than they use in the production of energy

British thermal unit (Btu): defined as the amount of heat required to raise the temperature of one pound of water by one Fahrenheit degree

Brownian movement: a phenomenon observed and described by Robert Brown (1773-1858) in which very small particles suspended in a still fluid jiggle without an apparent cause

building-integrated photovoltaics (BIPVs): photovoltaic systems that are designed and built into buildings, either at the time of construction or as aftermarket additions

CAFE standards: Corporate Average Fuel Economy standards, the minimum average fuel economies all automobile manufacturers are to achieve, designated by class and type of vehicle

caloric theory: an early theory that heat existed as a type of fluid in all materials and flowed from objects with a high caloric content to objects with lower caloric content, and held that the caloric content of the universe was constant

calorimeter: a laboratory device used to measure changes of heat energy

calorimetry: the measurement of the amount of heat energy either absorbed or given off in a specific reaction or in a specific physical change

canal rays: streams of positively charged particles produced as a byproduct of the formation of cathode rays, therefore representing the massive, positively charged remnants of the atoms that had emitted the cathode ray particles

candela: a measurement of the light intensity of a light source directly, rather than the amount of light at a distance from that source

capacitance: the amount of electrical charge that can be temporarily stored in a capacitor; the standard unit of capacitance is the farad (a contraction of faraday)

cap-and-trade system: one in which a maximum amount of emissions is assessed for an industry and each contributor of emissions is assessed the maximum amount of emissions allowed (the cap). Contributors who do not emit that amount are allowed to sell or trade (the trade) their remaining emissions allowance to contributors who may have to exceed their allotment in order to maintain the overall total within the specified limits of the policy

caprock: an overlying layer of dense rock that prevents petroleum and natural gas from escaping from where it has accumulated, effectively trapping it in the more porous rock below

carbon capture: trapping of carbon dioxide, hydrocarbons and other waste gases to prevent them from entering the atmosphere

carbon footprint: a measure of the amount of atmospheric carbon dioxide a product or service represents

carbon neutral: processes and activities that result in no net increase in atmospheric carbon

Carboniferous Period: the period of geological history before the evolution of land-dwelling creatures, in which only plants existed on Earth's land surfaces

Carnot limit: the upper limit for the efficiency of a heat engine; depends on the temperature difference between the working fluid and the surroundings; the higher this difference, the higher the theoretical maximum energetic efficiency.

catalyst: a material that takes part in a reaction process, serving to lower the energy barrier to the reaction and so promote its occurrence, but is not itself consumed in the reaction

cathode: the relatively negative terminal of an electrochemical circuit; the cathode emits electrons that are to be received by the anode

cathode rays: streams of negatively charged particles-electrons-that have been forcibly ejected from neutral atoms through electrical stimulation of those atoms

cell: a construct of two different half-cells that produces an electrical potential, or voltage, as the sum of the relative potentials of the two half-cells

cellulose: a principal secondary product of photosynthesis, formed by the head-to-tail concatenation of glucose molecules that are the primary product of photosynthesis

cellulosic ethanol: ethanol produced from degradation and fermentation of cellulose-based plant matter

central receiver: a structure placed to receive the reflected sunlight from a large symmetrical array of mirror surfaces

centrifugal force: an apparent force produced by the rotation of a mass about a central axis; it appears to act perpendicular to the axis of rotation in opposition to the centripetal force

centripetal force: a real force produced by the rotation of a mass about a central axis; it is formally the tension between the individual particles of the mass and the central axis proportional to the rate of rotation of the mass

cetane number: a classification of fuel oil combustibility relative to pure cetane (hexadecane, a C-16 hydrocarbon); analogous to the octane number for gasoline fuels relating gasoline combustibility to pure iso-octane (a C-8 hydrocarbon)

cetane rating: a comparative rating assigned to diesel fuel blends on the basis of their combustion properties relative to pure n-hexadecane, a normal alkane consisting of 16 carbon atoms and 34 hydrogen atoms

CFL: compact fluorescent light

chain reaction: in a nuclear fission chain reaction, a subatomic particle ejected from one nucleus strikes another nucleus, causing it to undergo fission and release another particle, the number of ejected particles increasing geometrically as the chain reaction runs away

charcoal: an essentially pure carbon source produced by burning wood at low temperatures with insufficient air to promote combustion

charge: indicates that something has a component of electricity in its fundamental composition, or to load components of electricity into an object. In modern physics, charge is a fundamental property of matter, with the electron being the fundamental unit of negative charge and the proton being the fundamental unit of positive charge.

chromatographic separation: a preparatory technique in which compounds in a mixture are separated from each other by differential adhesion to a stationary solid phase within a moving fluid phase

clathrate: a solid material in which a small molecule of a gas is enclosed within the crystal structure of a solid

climate change: a shift in the global climate over a prolonged period of time, whether by natural or anthropogenic causes; often used interchangeably with global warming

climate change adaptation: measures designed to adapt to the changing nature of the world as climate change continues

climate change mitigation: measures intended to reduce or stop further climate change, whether anthropogenic or natural

climate neutral: processes and activities that have no net effect on the climate, whether globally or locally

closed cycle: an ocean thermal difference technology that uses transfer heat energy from a warm water stream to vaporize a low-boiling working fluid that is kept in a closed system to drive a generator, and is then condensed by heat exchange with the cold water stream

closed system: one in which the working fluid (steam) is condensed back to liquid water and returned for reuse rather than being vented away

coercion: the use of punitive consequences for failing to comply with corporate and governmental directives

coffer dam: a temporary structure designed to divert water flow around the location of work being carried in the construction of a dam

cogeneration: processes that generate both electricity and recoverable heat

coke: an essentially pure carbon source produced by heating coal, without promoting its combustion, to drive out contained tars and oils

combined heat and power (CHP) system: a power generation system that captures and utilizes spent heat as a second output

combustible: describes carbon-based material that can be readily oxidized in air to release its contained chemical energy as heat and light

combustion: generally identified as burning with the formation of flames, but technically all oxidation reactions that produce carbon dioxide as a product of the reaction, with or without flames

combustion zone: the region within a boiler at which fuel combustion is taking place

commodify: to use the certifications provided as part of a program as a tradable commodity

compression ignition: compressing a gas causes its temperature to increase; in a Diesel engine the temperature increase of air due to compression is high enough to ignite the fuel

compression ratio: the ratio of the volume of an engine cylinder when fully open to its volume when fully closed; for example, an engine cylinder with a maximum volume of 34 ounces (1 liter) and a minimum volume of 3.4 ounces (100 milliliters) has a compression ratio of 10:1

compressive stress: stress on a dam structure and its surrounding terrain by the weight of water being held behind the dam

concentrator: a device that acts to focus sunlight to a central point, either by reflection or concentration through a lens

condense: to collect charge. Centuries ago, electricity was thought to be a kind of fluid that could condense or concentrate, as when an object builds up a charge of static electricity. Today's capacitor acquires a build-up of electrical charge as it functions, and was originally called a condenser.

condenser: in steam systems, a chamber in which spent steam is allowed to cool and condense back into liquid water that can then be returned to the boiler to produce live steam

conduction: the transfer of energy by physical contact between substances of different temperatures

conductor: any material that will allow the passage of an electrical current, which all materials do to a greater or lesser degree

conservation of energy: a physical law that states energy is neither created nor destroyed, but merely changes from one form to another; also stated as the total energy of any system before a change to the system must be equal to the total energy of the system after that change

constant-pressure steam: steam that is produced, maintained and used at a constant temperature rather than being allowed to cool during the functional operation of the machine

consumption: for electrical appliances, a measure of the power produced by the electrical current passing through the appliance when activated for a specified period of time; for example, a 100-watt lightbulb operating for 10 hours consumes 1 kilowatt-hour of energy

coulomb: the basic unit of charge in the International System of Units (SI)

convection: the transfer of energy in fluids as a result of changes in density according to changes in temperature

core material: the array of rods of refined uranium, or other fissionable elements, that are the source of energy in a nuclear reactor

coupling: a connection in which the axis and the direction of rotation of the two separate parts are identical

cracking: a thermal process in which hydrocarbons are made to fracture and decompose into a number of different, smaller compounds, according to specific chemical reactions

cradle-to-grave: a term representing the span of different stages of a particular material, product or service from the moment of its inception to its final disposition; similar appropriate terms may be used to indicate different stages

craft industry: nonmechanized industries that produce traditional hand-made goods such as textiles and ceramic ware

crest: the highest part of a wave from its natural value

current: the movement of electric charges from one place to another

cybersecurity: protection of networked computer systems from intrusion and unauthorized access by foreign and domestic activity

data center: a large array of memory storage devices and server units in which information is stored and served out to remote users as a centralized service rather than on individual private machines (e.g., Microsoft Cloud, Google Dropbox, etc.)

daylighting: interior lighting achieved by capturing and directing sunlight to the interior spaces of buildings

decommission: the removal of an energy-production facility, or any other production facility, from active service, and may include its demolition and disposal

decomposition: the breakdown into its molecular components of once-living organic matter, whether of animal or plant origin

degrees of freedom: the maximum number of states of a system according to the number of independent variables (e.g., temperature, pressure, volume, etc.) operating on the system

dehumidification: the artificial removal of water vapor from the air inside of a building in order to maintain human comfort level and prevent damage due to condensation and mold growth

demonstration: typically, a small-scale project designed to show the features, characteristics and usefulness of a full-scale implementation

derived unit: a unit of measurement that is a composite of other units of measurement

desalination: conversion of saltwater to salt-free water by evaporating the saline water and condensing the water vapor in a receiver separated from the salt residue; essentially a form of distillation

dewatering pump: pumps used to remove water from working coal mines in order to prevent them from flooding

diamagnetism: a small repulsive force experienced in a magnetic field by all materials

dielene: a chemical compound used as a refrigerant liquid, designated R-1130, with the proper name cis-1,2-dichloroethylene

difference engine: a mechanical calculating device proposed and partly constructed by Joseph Babbage in 1822, for carrying out reliable calculations of life expectancies

differential heating: because of the nature of water, land surfaces heat up more quickly from sunlight than water areas do; land surfaces also give up that heat much faster than bodies of water do

diffraction: a change in the direction of a wave as it passes around an obstruction or through an opening

diffuser: a device designed to scatter incoming sunlight evenly to reduce its intensity and harshness on the eyes

digital computer: an electronic programmable calculating device that uses Boolean logic to manipulate patterns of electronic bits (high and low voltage signals) for the performance of various functions

diode: an electronic component made of semiconductive material that allows current flow only in one direction when activated; assemblies of interconnected diodes are etched onto a silicon substrate to produce transistors and complex transistor combinations

direct current: electrical current that flows in one direction only, as from a relatively negative potential to a relatively positive potential

directional drilling: a deep drilling method in which the direction the drill bit cuts and travels through rock can be controlled from the surface

discharge: the removal of a build-up of electrical charge from an object

displacement: the difference between the initial position of an object and its final position, regardless of its path

displacement field: in electrodynamics, the electric field produced solely by free charges (e.g., free electrons)

distillation: separation of the components of a solution by heating to the boiling point boiling and condensation of the resulting vapors of the individual components

distributed energy system: a network of interconnected local electrical grids accepting input from a number of different suppliers for distribution to common use

district heating system: a system set up in and around a local power generating station or industrial production facility to distribute waste heat within that local district for heating and cooling purposes

diurnal: describes a change that produces two states in one day, such as day and night

diurnal changes: literal difference between day and night

domestic policy: a policy established and applicable only within that countries national jurisdiction

domestication: conversion of a wild species to one that is able to coexist with humans in a mutually beneficial relationship

draft animals: domesticated animals used primarily as traction engines to pull loads

drag: essentially, friction between moving surfaces and the surrounding air resulting in the generation of heat

dry steam: steam that consists only of water in the gas phase, unaccompanied by any liquid water

E rating: a designation of the percent ethanol content by volume in an ethanol-blended gasoline

efficiency: the ratio of the actual work output of a machine relative to the ideal, or theoretical, work output of the ideal machine

electrical conductivity: a measure of a material's ability to conduct electrical current, the inverse of electrical resistance, stated in Mhos (Ω^{-1})

electrical current: normally, the flow of electrons through a conductor under the influence of a potential difference between two points in the conductor

electrical power, P: the product (in watts) of the amount of current I (in amperes) passing through an appliance and the electrical resistance R (in ohms) of the appliance, calculated as $P = I^2R$

electrical resistance: a measure of a material's requirement for work to be done by an electrical current passing through the material, stated in Ohms (Ω)

electrification: the provision of a secure and adequate electrical supply system in a particular area or region

electrification rate: the portion of an area or region that has been provided with secure and adequate access to electricity relative to the total area requiring such access

electrify: to provide electricity to something to enable it to function

Glossary

electrochemical: a chemical reaction system that functions through the transfer of electrons from one material (the cathode) to another (the anode)

electrochemical industry: industrial production concerns that require a steady, and often large, supply of electricity as part of the production process, particularly for the production of aluminum and other metals

electrolysis: chemical reactions driven by an electric current

electrolyte: a solution or intermediary substance that contains mobile ions, such as saltwater

electromagnetic spectrum: the continuous range of frequencies of electromagnetic waves

electromagnetism: magnetism, or a magnetic field, produced by the functioning of an electrical current or field

electronic bonds: electrons in atoms occupy an orderly arrangement about the nucleus that is determined by specific energy relationships of the nuclear components; in order to achieve the most stable arrangement within those parameters, atoms can accept, or share, an electron with another atom, which binds them together

electrostatic machines: devices that used friction to build up a static electrical charge that could be drawn off as electrical current while the machine was in operation

element: a substance whose atoms contain a specific number of protons

embargo: banning of sale or trade of goods to one country by another country or international coalition

embodied energy: an accounting of the energy consumed in the existence of a material, product or service through its cradle-to-grave lifespan

emergy: a contraction of embodied energy used to indicate the embodied energy in equivalent solar energy

emissions intensity: the amount of pollution emitted relative to the amount of energy used for the production of a certain amount of economic value

end effector: a device that functions as the end user of a fluid power system to carry out a function

energy: a term describing the innate potential of a system to do work on that system and its surroundings

energy burden: term used in the case of fuel and energy products and services to indicate the embodied energy cost of the product or service relative to the energy to be derived from the product or service

energy efficiency: the value of a product or resource relative to the amount of energy that is consumed in its production

energy efficiency ratio: a rating according to the amount of heat energy removed from the air per watt-hour of electricity consumed by the particular system

energy intensity: the amount of energy consumed by the production of a product or resource relative to its economic value

energy payback: a time-based determination of the amount of energy produced by a facility that is equal to the embodied energy of the facility

energy poverty: a state of the absence of or inadequate access to a secure and adequate supply of electricity and fuel for household needs such as cooking, hygiene, heating, etc., typically in undeveloped and underdeveloped areas of the world

energy return on investment: a calculation of the overall amount of energy produced by a facility relative to the energy invested in its operations

energy waste: unnecessary energy consumption and consumption beyond the immediate needs of the operations

enthalpy: the energy of a system expressed as heat

entropy: the energy associated with the dissipation of energy from a state of higher energy intensity to one of lower energy intensity

equilibrium: the state of a dynamic system in which no net change can occur without the input of energy from an external source

ethanol: the second-simplest organic alcohol, it has the chemical formula C_2H_6O, sometimes written as H_3CCH_2OH or shorthanded as EtOH (the simplest organic acid is methanol, H_3COH)

ethylene: the simplest member of an infinite series of unsaturated hydrocarbon molecules called alkanes, in which two adjacent carbon atoms are doubly-bonded to each other; more than one such site of unsaturation may exist in alkene molecules

eukaryote: a cell that has a central nucleus structure containing the cell's DNA and maintains its life through aerobic respiration

exergy: a concept describing the amount of usable work that can be reasonably obtained from a thermodynamic system

experimentalist: a scientist who uses the observation of well-defined practical experiments to determine the apparent properties of processes and materials

external combustion engine: an engine in which fuel combustion takes places outside of the engine works and is not an integral part of the engine's mechanical operation

facade: the side of a wall, whether interior or exterior, that faces the outside; interior facades are used to capture heat energy and diffuse light entering from the exterior facade

feed-in tariff: a rated amount paid back to energy consumers who also produce energy through the installation of geothermal, solar or other type of renewable energy and deliver their excess output into the general supply

feedstock: materials employed as the principal component of a production process

fenestration: the number, types, and sizes of the windows in a building

fermentation: digestion of available sugars by yeast or other bacteria with ethyl alcohol and carbon dioxide being the usual products

field: the manifestation through space of the potential of an electric or magnetic source

filament: in incandescent lightbulbs, a thin tungsten wire that resists the flow of electricity to the point that it becomes hot enough to emit white light

first-generation ethanol: ethanol produced from food crop materials that would normally be used as food

fiscal policy: short-term policy agreements established based on annual monetary consideration of energy

fissionable: capable of undergoing nuclear fission, either spontaneously or by neutron bombardment

fissionable materials: radioactive isotopes that undergo spontaneous nuclear fission, releasing various kinds of radiation and changing their elemental identity in the process

flex-fuel vehicles: cars and trucks with engines designed to operate efficiently with fuels containing any amount of ethanol

fluidized bed: a structure in which an inert carrier such as fine sand or alumina is suspended in a column of heated gas being blown through it; used for the combustion of powdered fuels or to conduct gas-phase chemical reactions

fluorescence: the property of a material that it will absorb ultraviolet light, as from electrically excited mercury vapor in a CFL, and re-emit energy as visible light, or fluoresce

flywheel: a relatively large mass that stores energy by virtue of its rotation; a flywheel typically may acquire rotation by converting energy from linear motion or by being directly spun up to speed by an external driver; it can then be used to moderate the energy load being placed on the system

foot-pound: typically a torque measurement due to a mass of one pound acting at a radius of one foot

force: the result of an interaction between two objects that changes the pattern of motion of the objects. A force can be a pull or a push.

fossil fuels: fuels derived from the remnants of once-living organisms, particularly coal, petroleum and natural gas (methane)

fossilization: the process by which once-living biological matter, whether of plant or animal origin, is converted to a mineral form

four-stroke engine: an internal combustion engine requiring four timed piston strokes to complete one cycle of operation per cylinder: an intake stroke to take in fuel and air, a compression stroke in which the air-fuel mixture is compressed, a power stroke in which the air-fuel mixture is ignited to drive the piston down with force, and an exhaust stroke to drive the combustion products out of the cylinder

fractional distillate: petroleum (unrefined oil) is a complex mixture of different compounds; fractionation of the mixture by distillation separates the components according to their respective boiling points. Each portion according to boiling point range is a fractional distillate

free energy: also Gibbs free energy, the useful work obtainable from a thermodynamic system at constant temperature and volume

frequency: the number of complete waves or cycles that occur in one unit of time

Fresnel lens: a type of lens that uses a series of concentric edges to refract light towards a central focal point

friction: the interaction of two surfaces in motion relative to each other; the interaction generates heat in accord with the degree of interaction

fulcrum: the supporting point upon which a lever operates

fuel poverty: a state of the inability to afford to pay the cost for household utilities and fuel needs, typically in well-developed regions of the world

fusion: the joining together of two atomic nuclei to produce the nucleus of an atom of a heavier element

fusion reactor: a nuclear reactor that operates on the principle of nuclear fusion rather than nuclear fission

gasification: the process of heating organic matter with a limited amount of oxygen so that it is driven to decompose to gaseous materials

gasohol: specifically refers to gasoline-ethanol blends containing less than 25 percent ethanol by volume

geothermal: literally earth heat, or heat from below the ground surface; can refer to heat exchanged from the constant temperature within the soil at depths below all frost, or to heat from volcanic sources

geothermal gradient: the gradual and consistent increase in ambient temperature with depth below the ground surface

geothermal heat pump: devices that use the consistent temperatures that exist underground as the heat source or heat sink for heating and cooling systems above ground

geyser: an eruption of geothermally heated water from some depth below the surface in geothermally active areas

gigawatt: 1,000,000,000 watts

glazing: the type of glass used in window construction; single-glazed panes are a single sheet of glass; double-glazed panes are two sheets of glass separated by a layer of gas such as argon; triple-glazed panes are three sheets of glass with two gas-filled spaces separating them

global conveyor belt: the planet-spanning system of interconnected surface and deep ocean currents that transports an immense amount of the energy received from the Sun and from Earth's interior

global warming: the observed gradual increase over time of Earth's average atmospheric temperature, not counting seasonal variations; often used to mean climate change

grain alcohol: ethanol produced by fermentation of grains rather than sugar beets and sugarcane

graphite: one of the many forms of pure carbon, the same material that is used as pencil lead

graywater: waste water from household operations such as bathing, laundry, etc., but not including sewage water

green energy: energy resources and practices that function to reduce energy consumption and the environmental impact of energy-consuming products

green roof: a roof covered by a layer of soil and grass or other small vegetation, rather than shingles and other traditional roofing materials

greenhouse effect: the demonstrable phenomenon in which certain atmospheric gases absorb thermal energy radiating from Earth's surface level and return a significant portion of it back into the atmosphere instead of passing it directly out into space, resulting in a sustained higher-then-normal atmospheric temperature

greenhouse gas: any atmospheric gas capable of absorbing heat energy emanating from Earth's surface and subsequently acting to return that heat energy into the atmosphere

greenhouse gas emissions: the emission of gases known to contribute to the greenhouse effect from power generation facilities and operations, particularly of carbon dioxide (CO_2), hydrocarbons, and other gases

greenhouse gases (GHGs): commonly refers to carbon dioxide, but includes a number of other gases capable of absorbing heat energy radiating from Earth's surface and then emitting much of that heat energy back into the atmosphere

grid system: the regional electricity distribution network

gyre: a broadly circular flow of surface waters in the oceans of the world

gyroscopic force: a reactive force generated by a rotating mass to resist displacement from its original axis of rotation, consistent with conservation of angular momentum

Hadley cells: a large, convective air circulation system driven by the influx of solar energy at the equator; the cells rise from the equatorial region, move toward the north and south poles, then sink to the surface to return toward the equatorial regions

Hadley cell convection: atmospheric convection currents caused by the influx of solar energy along the equator; the warmed air rises up from the cooler sea-level air, moves to the north and south away from the equator, then cools at high altitudes and sinks back toward the surface as its density increases, ultimately to flow back toward the equatorial region

half-cell: a metal in contact with a solution of a metal salt, having a specific voltage potential relative to a defined standard half-cell

half-life: for any first-order reaction such a spontaneous nuclear fission, the length of time for one-half of any quantity of the material to undergo the reaction process

hammer mill: a processing machine that uses heavy weights, or hammers, to pulverize grain kernels as it operates; the hammers are typically large steel balls that act to crush the grain as they tumble together within the machine

harmonics: the study of the interaction of wave phenomena

heat engine: a system that converts heat energy into another form of energy such as mechanical or electrical by taking advantage of the difference in heat energy between two bodies, usually fluids

heat exchanger: a structure that separates a heat source from a working fluid but allows heat to transfer from one to the other

heavy water: water molecules in which the normal hydrogen atoms of H_2O have been replaced by the heavier isotope deuterium, producing D_2O

heat of combustion: the specific amount of heat energy released when a material is completely combusted with oxygen

heat of formation: the specific heat (enthalpy) change in the formation of a specific material from its component atoms, with both reactants and products being in their standard states

heat pump: the central component of a heating/cooling system that uses the constant heat content of a large mass such as Earth's soil below the depth affected by surface temperature variations. Heat pumps use a cooled liquid to accept heat from the depths, and recapture that heat above ground by passing the heated liquid through a heat exchanger that, like an air conditioner, recovers the heat and cools the liquid. Reversing the process allows excess heat to be taken from above ground to be dissipated below ground.

helical gear: a gear in which the teeth follow an angled, or helical, direction across the thickness of the gear; prone to thrust forces due to the angle of the teeth and the force applied against them

hemicellulose: a secondary product of photosynthesis, formed by the incorporation of cellulose molecules and lignin molecules into the rigid molecular structure of hemicellulose

herringbone gear: a combination gear in which helical gears with teeth angled in opposite directions are interfaced and used as a single gear; not prone to thrust forces as helical gears are

Hertz (Hz): the standard unit of frequency for cyclically repeating phenomena; one complete cycle of the phenomenon over a period of one second defines a frequency of one Hertz

heterogeneous: a system consisting of two or more phases, such as solid and liquid, etc.

higher-heating value (HHV): describes fluid gases used for heat transfer at temperatures above their combustion temperature

high-level radioactive waste: radioactive source materials such as spent nuclear fuel and other radioactive sources emitting hard radiation such as X-rays

humidification: the artificial addition of water vapor to the air inside of a building in order to maintain human comfort levels and avoid physical damage due to dehydration

humidity: the amount of water vapor in ambient air, relative to the maximum amount of water vapor possible in air at specific temperatures and pressures (100 percent humidity); for example, 25 percent humidity means the amount of water vapor present in air is 25 percent of the maximum amount possible

hybrid: a light-duty vehicle that uses a small internal combustion engine in conjunction with an electric motor as its power unit

hydraulic system: a system that uses fluid pressure to perform mechanical functions; hydraulic fluids may be compressible gases or incompressible liquids as the working fluid

hydrocarbon: compounds in which the component molecules are composed only of hydrogen atoms and carbon atoms, hence hydrocarbon

hydrofluorocarbons: molecules consisting of carbon, hydrogen and fluorine atoms

hydropower: power generated by the use of moving water

immiscible: the condition in which two different liquids do not dissolve each other and so do not mix, as for example oil and water

inch-pound: typically a torque measurement due to a mass of one pound acting at a radius of one foot

incompressible fluid: a fluid that does not change its volume when an external pressure is applied

induction: the generation of a magnetic field in a second material by the presence of a magnet; similarly, the generation of an electrical current in a conductor by a moving magnetic field

induction coil: a type of transformer in which a secondary coil of a large number of turns of thin wire is mated to a primary coil of a few turns of thick wire about an iron core. A low-voltage direct electrical current flowing through the primary coil is turned on and off at a regular frequency, inducing a high-voltage current of the same frequency in the secondary coil.

induction motor: an electrical motor that uses a system of wire coils (the stator) to generate a rotating magnetic field to drive the rotation of a central permanent magnet on a shaft (the rotor)

Industrial Revolution: the period beginning in the mid-1700s in which traditional animal and human power began to be replaced by steam power and mechanization

infrared radiation: electromagnetic waves having wavelengths longer than those of the visible red color, associated with and detectable as heat

in phase: the condition when the peaks and troughs of a wave tend to occur in the same place at the same time, thus adding to the amplitude of the resultant wave

insolation: specifically, the light and other solar emissions striking Earth from the Sun (Sol)

insulator: a material that has a high resistance to electric charges, preventing them from moving through it easily

interference: how waves interact when they meet in the same medium; waves whose crests align will reinforce one another (constructive interference); waves where one's crests align with the other's troughs will dampen one another (destructive interference)

internal combustion engine: an engine in which fuel is combusted within the engine itself in order to extract energy from the fuel as an integral part of the engine operation

International System of Units (SI): a standardized set of units and measures used by scientists worldwide, based on and largely synonymous with the metric system

involuntary park: an uninhabited wilderness area that was originally created and inhabited by humans, such as a ghost town that has been abandoned and reverted to wilderness

isotopes: atoms of an element in which all atoms contain the same number of protons but different numbers of neutrons

jet fuel: a particular fraction of petroleum distillates containing a limited variety of alkanes and aromatics of a particular molecular weight and boiling point range

joule: defined electrically as the work done by a current of one ampere through a potential difference of one volt

kelvin: the unit of measurement for temperature; one kelvin has the same magnitude as one Celsius degree, but the kelvin temperature scale starts at absolute zero rather than the freezing point of water such that 0°C = 273.15 k

kerogen: an insoluble organic material found in shale and the pores of other types of rock; kerogen must be heated in the absence of oxygen to be made suitable for processing

kilowatt: 1,000 watts

kilowatt-hour: a unit for measuring electricity consumption, equal to one thousand watts of power consumed over one hour, or 3.6×10^6 joules

kinematics: a subfield of classical mechanics that studies the motion of objects without reference to the forces that cause this motion

kinetic energy: the energy associated with a physical object in motion

kinetic theory: the notion that heat is produced by motion

lacustrine: formed in the sediment of ancient freshwater lakes

landfilling: disposing of discarded batteries and other items in landfill operations

LED lightbulbs: lightbulbs that use high-intensity light-emitting diodes (LEDs) rather than incandescent tungsten filaments or mercury vapor as the light source; LEDS typically use 3 to 5 watts of power to produce the same light output as a 100-watt incandescent bulb, and last much longer

lever: a rigid bar or bar-shaped structure that operates upon a fulcrum

life cycle: the period from the design of a building, through its construction and use, to its eventual demolition and disposal of the remnants

lifetime: the period of existence of a product, resource or facility from its original idea and design to its final stage of disposal

light shelves: essentially a series of baffles that reflect incoming light within a light tube to produce diffuse light as its output

light tube, solar tube: a device built into the roof or walls of a building to capture and direct sunlight to the interior of the building

light-water: normal H_2O; the designation is used to indicate which type of moderator is used in a particular nuclear reactor; heavy-water indicates that D_2O is used

longitudinal wave: a type of wave wherein the medium is displaced in a direction parallel to the movement of energy, as in the case of sound waves

lower-heating value (LHV): describes a fluid used for heat transfer at temperatures below the combustion temperature or boiling point of the fluid

low-level radioactive waste: materials that have been exposed to radioactive materials and so have acquired some radioactivity, as well as other radioactive sources that do not emit "hard" radiation such as X-rays

LPG: liquid propane gas which is effectively pure propane; not to be confused with LNG or liquid natural gas, which contains primarily methane and ethane

magma: molten rock within Earth's mantle layer; magma that has extruded onto the surface through fissures or volcanic eruption is called lava

magmatic intrusion: a location at which molten rock (magma) from Earth's mantle layer has moved into a location above it in Earth's crust

magnetic bearing: a type of bearing that uses magnetic repulsion rather than physical contact to stabilize the rotation of an axle

magnetic field: a three-dimensional force field that nevertheless has a definite directional component surrounding a magnet

magnetostriction: the phenomenon in which metal objects deform somewhat when they become magnetized

mass: an intrinsic property of matter arising from the cumulative atomic weights of the atoms composing the mass

mass spectrograph: or mass spectrometer; an analytical device that uses the deflection of a charged particle in a magnetic field according to its mass-to-charge ratio to identify and isolate particles according to their mass

matrix: an array of numerical factors that can be manipulated mathematically to solve a number of equations simultaneously

mechanical advantage: the force needed to carry out a task using a leverage function relative to the force required to carry out the task without the leverage function

mechanical theory of heat: the theory that heat is produced by mechanical movement of atoms and molecules within materials

megawatt: 1,000,000 watts

meltdown: the runaway release of energy from radioactive materials resulting in the production of heat well beyond that required to melt those materials

metabolism: the biochemical processes by which energy is extracted from food and oxygen in living biological systems

methane hydrate: a solid material formed when molecules of methane gas become encased within a structure of water ice; sometimes called ice that burns. Methane hydrate occurs naturally on cold ocean floors and often clogs pipes when water vapor is present in natural gas lines.

methanogens: bacteria that consume acetate or acetic acid, producing methane and carbon dioxide as products of the digestion process

metric: refers to the measurement of properties relative to appropriate standard and defined units of measurement

microgrid: an electric grid system that serves a small area such as a single town

mitochondria: organelles within eukaryotic cells, where energy is extracted from the glucose obtained from food

moderator: in a nuclear reactor core, the moderator is a material such as carbon or heavy water (D_2O) that slows the flight of ejected neutrons by absorbing an appropriate portion of their kinetic energy

molasses: the thick, sweet, strong flavored syrup residue remaining after sugar crystallization and separation from the crude sugar solution extracted from various source plants for sugar production; contains a fairly high percentage of remnant glucose sugar mixed with flavonoid compounds and other sugars

mole: an absolute measure of the amount of matter present based on the number and type of atoms in the matter, based on the rule that one mole of any pure substance contains exactly the same number of atoms or molecules as one mole of any other pure substance, that number being approximately 6.022597×10^{23}, amounting to the weight in grams equivalent to the molecular weight of the material; for example, 238 g of uranium-238 contains exactly the same number of atoms as 1 g of hydrogen atoms or 12 g of carbon-12 atoms

molten salt system: a system that uses a low-melting inorganic salt as its working fluid

momentum: the propensity of an object in motion to remain in motion or if at rest to remain at rest, in accord with Newton's First Law

Moore's law: an empirically derived law based on the observation that the ability to etch ever smaller transistor structures on silicon-based semiconductor surfaces advances at such a rate that the computing power of the resulting devices doubles approximately every two years until the physical size limit of available space will be attained, at which point no further advancement is physically possible

multifuel vehicle: a vehicle that can operate using either one of three or more entirely different types of fuel rather than a mixture of fuels

multilateral policy: meaning many sides, a policy agreement involving three or more nations or governing bodies

natural abundance: the relative proportions in which the isotopes of an element are found in nature

neap tide: the tidal change that occurs with the least difference in height between high and low tides

negentropy: meaning negative entropy, as occurs when a random collection of different components is assembled into an ordered whole structure, as is the case when a machine is assembled from its individual component parts

net energy ratio: the difference between the total energy that can be obtained from a source, and the energy expended to obtain the energy from that source

net force: the sum of all forces acting on an object

network society: the social culture of computing systems connected to centralized and distributed network systems

neutron: one of the three main subatomic particles that compose atoms; protons and neutrons are located in the positively-charged nucleus, and electrons surround the nucleus as an ordered cloud of negative charge

Newcomen engine: an external combustion engine that functioned by injecting high-pressure steam into a closed cylinder to raise a piston within the cylinder, then relied on the reduced pressure inside the cylinder, produced by the condensation of the steam, to draw the piston back down inside the cylinder before the next introduction of high-pressure steam

newton: the base unit of force, equal to one kilogram-meter per second squared ($kg \cdot m/s^2$)

noncondensable gases: in an ocean thermal energy conversion system, dissolved gases such as nitrogen and carbon dioxide that do not condense to liquid form in the cooling phase of the operating system since they have boiling points far below that of the working fluid

nonflammable: any compound or material that will not support combustion as a fuel, though it may be burned (decomposed by heat to release combustible components) when combined with other combustible fuels

nonfood residues: materials such as woody stems, inedible leaves, corn cobs, etc. that are separated from food sources and discarded

nuclear arsenal: the aggregate collection of all nuclear weapons possessed or controlled by a specific country or entity

nuclear fission: the spontaneous splitting apart of the atomic nucleus of atoms of a radioactive isotope, resulting in the release of various kinds of radiation and changing the identity of the isotope to that of a lower element; e.g. uranium is transmuted to lead through a nuclear fission process

nuclear fusion: in fusion, small nuclei are combined to form larger nuclei accompanied by the release of a large amount of energy; in fission large nuclei are split into smaller nuclei accompanied by the release of an amount of energy that is much less than that released by fusion

nuclear physics: the study of the structural characteristics, behavior and possible interactions of atomic nuclei

nucleus: the central core of an atom, containing a specific number of protons for each element, but variable numbers of neutrons

nuclide: a particular radioactive atomic nucleus

offshore drilling: drilling of oil and gas wells in the waters overlying the continental shelves rather than on land

octane rating: a comparative rating assigned to gasoline blends on the basis of their combustion properties relative to pure iso-octane (2,2,4-trimethylpentane)

odorant: a compound having a strong, easily detectable and unpleasant odor, such as the sulfur-containing mercaptans, an odor characteristic of skunks

Ohm's law: an empirical law stating that the current, or flow of electrical charge, between two points is directly proportional to the voltage, or difference in electric potential, between those points

oil sands: dense deposits of bitumen combined with sand

open cycle: an ocean thermal difference technology that uses flash vaporization of cold seawater under reduced pressure to produce steam directly to be used to drive a generator; subsequent condensation of the used steam produces desalinated or freshwater

open system: one in which the working fluid is not recaptured for reuse and must be replenished in order to maintain operation of the system; the opposite of a closed system

organic Rankine cycle: a generating process used when high-temperature steam is not available for use, but low temperature steam is available; the cycle uses a high molecular weight organic liquid that boils at a significantly lower temperature than water as the working fluid

out of phase: the condition when the peaks of one wave and the troughs of another tend to occur in the same place at the same time, producing a resultant wave with less amplitude than either of the incident waves

oxidant: also termed oxidizing agent, the substance accepts electrons in an electrochemical reaction, and so oxidizes the other material in the reaction

oxidation: any reaction in which a material combines with oxygen to produce oxides

oxidation numbers: a formal method of defining the electronic state of atoms in compounds according to the number of electrons they appear to have gained (reduction) or lost (oxidation)

oxidizer: a compound that supports the combustion (oxidation) of the fuel; most commonly atmospheric oxygen, although other materials such as nitrous oxide are used in specialty applications

oxygenates: organic molecules like alkanes but containing one or more oxygen atoms as part of their basic molecular structure

ozone: an oxygen molecule consisting of three oxygen atoms as O_3 instead of the usual two as O_2

pack animals: domesticated animals used primarily as trucks to carry loads

parabolic: a shape that conforms to the general mathematical equation $y = ax2 + b$

parabolic dish: a structure having a parabolic shape, the interior being reflective; sunlight entering the parabolic cavity is reflected to the focal point of the parabolic surface

parabolic trough: a long, semicylindrical structure that is parabolic in cross-section

particulates: solid microparticles that are produced during combustion of fuels, such as carbon soot, metal oxides and fly ash

Pascal's Law: the pressure exerted on a fluid is exerted in all directions within the fluid

peak period: the period during the day in which demand for a resource such as electricity is highest

penstock: a tubular structure that directs water flow to the turbine to drive a generator in a power station

perfluorocarbons: molecules consisting only of carbon and fluorine atoms

period: the length of time for one complete cycle of a wave or other cyclic property to occur

perpetual motion machine: the fallacious concept of a machine that would produce at least as much work or energy as that required for its operation

petroleum: also known as crude oil, bitumen and other terms, typically obtained from deep drilled wells or extraction from oil sands

petroliferous rock: rock that contains or can be used as a source of petroleum

pH: a rating of the degree of acidity of a water solution, according to the equation $pH = -\log[H^+]$

pharmoacokinetics: the time-related behavior of pharmacologically active compounds within living systems

phase: a stage in a wave property; typically used to describe the relationship of two or more waves

phase change: the change of a material in conversions from liquid to gas, gas to liquid, liquid to solid, solid to liquid, solid to gas, or gas to solid

photoelectric effect: the phenomenon in which light striking the surface of a material causes its surface atoms to emit electrons that can be harvested as an electric current

photosynthesis: the process in which chlorophyll-containing cells produce glucose from atmospheric carbon dioxide and water in the presence of sunlight

photovoltaics: electronic devices that produce electrical current by the photoelectron effect, in which incident sunlight causes surface atoms to eject electrons

piping: an intrusion of water into and through the structure of a dam, forming a slow leak that may develop into serious erosion and the failure of the dam

pivoted beam: essentially an application of the lever and fulcrum principle, in which one end of the beam is driven downward by a powered rod to raise a load attached to the other end of the beam, which then returns to its previous position when the load is removed

plum pudding model: a nineteenth-century model of atomic structure in which the structure of the atom is analogous to that of a plum pudding (or perhaps a raisin-bran muffin), with the newly discovered electrons scattered randomly throughout the whole of the atom

pneumatic pressure: the force of air being compressed in a confined space

polarization: the phenomenon in which electromagnetic waves, which oscillate radially about the axis of their direction of motion from a source, are made to oscillate in a single plane along the axis of their direction of motion

polyphase generator: an electrical generator designed to produce an AC output current that oscillates as a number of overlapping sinusoidal frequencies to provide a more consistent availability of power under load; the overlap of the frequencies ensures that the peak values of the current flow occur three or more times per second than with a single phase current

polyphase motor: an electric motor that requires a polyphase input current in order to operate at optimum power efficiency

potential energy: the energy that is stored in objects and can be converted to other forms of energy, such as kinetic energy

potential energy surface: a surface defined by its height above a gravitational reference surface

power: the work done or energy transferred over time

power rating: the maximum electrical power a device can use without being damaged

primary battery: a battery that cannot be recharged once its energy has been drained

probable maximum precipitation (PMP): the maximum amount of rainfall or snowmelt that is likely to be experienced by a particular dam serving a defined watershed

processing speed: the rate at which a particular central processing unit is capable of performing calculations and data manipulations

prokaryote: a cell that does not have a central nucleus structure containing the cell's DNA and maintains its life through anaerobic respiration

propylene: an unsaturated form of propane in which two adjacent carbon atoms are doubly bonded to each other which makes such compounds capable of undergoing a wide variety of reactions, formula is C_2H_6 or $H_2C=CH-CH_3$

pulley: essentially a gear without teeth used to facilitate the application of mechanical advantage

pumped storage hydroelectric plant: a hydroelectric plant that uses some of its generated electricity to pump water back into the reservoir for reuse

pyrolysis: the process of heating organic matter in such a way that it breaks down and decomposes to a mixture of liquids, oils, and gases

quantity of electricity (Q): the calculated amount of charge in a capacitor as the product of its capacitance (C) and the applied or resultant voltage (T), as Q = C X T

quantum (pl. quanta): the discrete particle of energy that characterizes a particular wavelength of light

radiation: the emission of subatomic particles and electromagnetic energy due to nuclear fission or fusion processes

radioactive decay: the spontaneous fission, or splitting apart, of the nuclei of certain unstable atoms, accompanied by the release of the corresponding quantity of the nuclear binding energy of those atoms

radiological hazards: the range of possible dangers to humans and the environment that may arise through exposure to nuclear radiation

radiopharmaceuticals: drugs and other biological compounds whose molecules have been constructed with a radioactive component

Rankine cycle: thermodynamic cycle employed in a steam engine involves four main stages: pumping of water, heat addition, work production, and heat removal

reactor core: the central component of a nuclear reactor, containing the fuel rods of refined uranium; the reactor core is constantly immersed in heavy water to absorb carry away heat and moderate the rate of nuclear fission within the fuel rods

real-time pricing: the cost to the consumer of a resource such as electricity adjusted continually according to the time of day the use occurs

reciprocal: the inverse of a value, calculated as 1 divided by the value.

reconditioning: repair and replacement of the components of a poorly-functioning battery in order to return it to a fully functional state

rectifier: a device that restricts the flow of water to one direction, or electrons if the rectifier is an electronic device in an electrical circuit

reflection: the bouncing back of a wave after it hits a barrier, as when light reflects off a mirror

refraction: the alteration of a wave's path, speed, and wavelength when it passes from one medium to another

regenerative braking: a system in which part of the kinetic energy lost in braking is captured during braking and used to regenerate electrical charge in the battery

regulatory policy: long-term policy agreements established in regard to regulation of resource management and energy use

renewable energy: energy resources that replenish themselves naturally and so are not considered finite resources that will be depleted through consumption

renewable identification number (RIN): an identification and tracking number assigned to a specific identifiable batch or production run of biofuel, typically ethanol or biodiesel; the RIN is provided by and registered with the organization or branch of government that oversees the biofuel industry, and must be reported to that body when the product is used in preparing a commercial fuel blend

reserve-to-production ratio (R/P): the amount of gas in a reserve relative to the annual rate at which gas is being removed from the reserve, stated in years

resolution: the ability of a detector to differentiate or separate different wavelengths

respiration: the process of extracting energy from glucose by converting it back to the carbon dioxide and water from which it was photosynthesized

retorting: heating of a material in the absence of oxygen in order to alter its characteristics

revolution: describes circular motion wherein an object circles an internal axis (e.g., Earth spinning about its axis); contrast to rotation, wherein the axis is external (e.g., Earth orbiting the Sun)

RFS: renewable fuel standard, the minimum amount of renewable fuels to be produced and used in transportation fuel blends

roof garden: a roofing design that uses a significant layer of soil as its outermost layer, or to at least include raised garden beds, providing an insulating and water absorbing layer while also allowing the roof to be used as a garden space

rotary engine: an internal combustion engine design that uses a central rotor driven by successive fuel ignitions rather than individual pistons connected to an eccentric crankshaft and driven in sequence by fuel ignitions

rotation: angular motion of a mass about a fixed axis

rotational speed: simply, the rate at which a rotating mass travels about its central axis; for all particles in the mass the angular rotation speed is equal, but the physical distance traveled by individual particles about the axis increases directly with radius

rotor: the inner part of a turbine or generator; the rotor is the part that turns

saccharification: an enzyme-mediated process that cleaves starch molecules to convert them into sugar molecules

saline: containing salt

salinity: the concentration of dissolved mineral salts in ocean waters

scalar quantity: a physical property that has magnitude that is not associated with a specific direction

scalar property: a property that has magnitude but no associated direction, such as mass and speed

scientific terminology: words and phrases specific to various scientific fields and operations, not used in general parlance

seasonal changes: weather conditions according to the season of the year in a location

Second Law of Thermodynamics: heat moves only from a body of higher temperature to one of lower temperature

secondary battery: a battery that is designed to be recharged by replenishing the energy that has been removed from it

secondary energy: energy that is derived from a primary source rather than being a source itself

second-generation ethanol: ethanol produced from food crop refuse and crops that are not used for food

semiconductive: describes a material that does not normally conduct and electrical current but that can be enabled to do so either electrically or by stimulation with electromagnetic frequencies

semiconductor: a material that does not normally conduct electricity, but that can be made conductive by stimulation either by applying an electrical biasing voltage or by light

semidiurnal: describes a change that produces twice as many states in one day as a diurnal change, such as two periods of high and low tides in one day

serial connection: connection of electrical components end to end in series rather than side-by-side to common connection points in parallel

sick building syndrome: a characteristic of buildings that are constructed with poor ventilation, insulation, lighting, etc., that causes them to be environments that are not conducive to the proper health of the people who live and work in them

sievert (Sv): the standard unit of radiation exposure that has replaced the rem, with 1 Sv being the equivalent of 100 rems; the sievert relates radiation dose and biological effectiveness

siltation: the deposition over time of waterborne silt, rocks and debris carried into a dam reservoir

sinusoidal: having a shape or pattern of behavior that can be described by a sine wave function.

smart grid: an electric grid system with built-in digital monitoring and control functions

smart meters: electrical current measurement devices designed to monitor electrical consumption and coordinate it with different time-of-use cost rates for billing purposes

solar collector: a structure, typically of mirror-like surfaces, designed to focus sunlight to a central point in order to capture heat energy

solar energy: energy emanating from the Sun in the form of electromagnetic radiation and subatomic particles

solar photovoltaics: solid state devices that rely on sunlight and the photoelectric effect to produce an electrical current output

solar thermal power plant: a power generating plant that uses a large array of mirror surfaces to capture and focus sunlight as the heat source for producing steam to drive electrical generators

sour crude: crude oil having a high sulfur content

speed: the distance traveled per unit of time

spring tide: the tidal change that occurs with the greatest difference in height between high and low tides

sprocket: a type of gear with pointed teeth designed to fit into the spaces of a flat drive chain; in an inverted design for use with link chains the chain fits into matching spaces cut into the edge of a pulley

steady state: often used to mean equilibrium when referring to dynamic systems, but a system in a steady state is not necessarily a dynamic system

steam engine: the quintessential external combustion engine, in which fuel is burned in a firebox and the hot combustion gases are circulated through pipes in a separate boiler to produce steam that is subsequently used to drive the operation of piston-cylinder arrangements

step-down: reduction, or stepping down, of a high voltage to a lower voltage by the use of a transformer

step-up: the increase, or stepping up, of a relatively low voltage to a higher voltage by the use of a transformer

stratified-charge engine: an internal combustion engine in which the fuel required for one cycle is delivered at successive points in the power stroke, rather than all at once at the beginning of the power stroke

supplementary unit: dimensionless units for the measurement of solid and plane angles; the term is now obsolete as both supplementary units have been formalized as derived SI units

sustainability: refers to processes that can be carried on for generations of time, typically based on entirely renewable resources

sweet crude: crude oil with a very low sulfur content

tar sand: a heavy, rock-like deposit of sand intermixed with bitumen

tensile strength: the extent to which a material is able to resist forces acting in opposite directions without deformation and failure; a flywheel rotating at such a rate that the centrifugal force exceeds the centripetal force will fail and fly apart

tension: an early term for voltage, still in use in regard to capacitors and transformers

terawatt: 1,000,000,000,000 watts

terawatt-hour: energy consumption at a rate equivalent to 1 terawatt in 1 hour, 1 terawatt being 10^{12} watts (cf: a 100-watt incandescent lightbulb consumes just 100 watts in 1 hour and 10 such bulbs would have to operate for 1 hour to consume 10^3 watts, or 1 kilowatt-hour of electricity...1 terawatt-hour is 10^9 times as much)

thermal efficiency: the percentage of available heat energy that is converted to usable work relative to the heat energy that is lost as waste

thermochemical heat storage: the use of excess or available energy such as solar radiation to drive the chemical synthesis of a product that can be used at a later time in a reaction that releases heat

thermochemistry: the study of chemical processes relative to heat

thermodynamics: the study of the active nature and movement of heat energy

thermogenic: produced by the thermal decomposition of carbonaceous matter

thermohaline circulation: global oceanic water circulation driven by changing temperatures and salinities as the water moves

throttle valve: a valve designed to control the amount of steam flowing into the generator by opening to admit a greater flow of input steam, or closing to decrease the flow of input steam

tidal energy: the energy of oceanic tides, derived from the gravitational influence of the Sun and Moon

time-of-use pricing: the cost to the consumer of a resource such as electricity according to the set rates for the time of day the use occurs

tonne: a mass of 1,000 kilograms, also called a metric ton, equivalent to 2,200 pounds or 1.1 standard tons

total mechanical energy: the sum of the kinetic energy and the potential energy an object possesses as a result of work done on it

train oil: oil recovered from the harvesting and rendering of whales, and especially from the head cavity of sperm whales

transesterification: a chemical reaction process in which the portion of an ester molecule derived from an alcohol is replaced by that of a different alcohol to produce a different ester molecule. In biodiesel production, the tri-alcohol 1,2,3-propanetriol (glycerol) is replaced in the tri-ester (or triglyceride) molecule by the monoalcohol methanol, resulting in the formation of three monoester molecules.

transformer: a constructed electrical device that uses the principle of induction to induce a voltage in a secondary coil by the current in a primary coil with a different number of "turns" of wire; a higher number of turns on the secondary coil results in induction of a higher voltage, while a lower voltage

translation: movement from one locus to another through space in a straight line

transmission loss: the dissipation of electricity as heat due to the friction of electrons moving through the transmission lines

transmission network: an electricity delivery system originating at a generating station, passing through any number of user nodes and interconnections, and terminating at the generation site to complete the electrical circuit

transverse wave: a type of wave wherein the medium is displaced in a direction perpendicular to the movement of energy, as in the case of waves on the surface of water

trough: the lowest point of a wave from its natural value

two-stroke engine: an internal combustion engine in which piston movement within the cylinder opens and closes intake and exhaust ports within the cylinder during operation, rather than using a camshaft timed to the rotation of the crankshaft to coordinate the opening and closing of separate intake and exhaust valves as in the four-stroke engine

ultraviolet light: electromagnetic radiation having frequencies higher than the highest visible light frequency in the electromagnetic spectrum

vane: a triangular sail-like appendage on the central shaft of a windmill or other rotary device; in more high-tech constructions, aerodynamic blades are used instead of vanes

vapors: typically, molecules of a liquid that have entered the gas phase at temperatures below the boiling point of the liquid

vector property: a property that has both a scalar magnitude and an associated direction, such as velocity (speed in a specific direction)

vector quantity: a physical property that has both magnitude and an associated direction

velocity: the speed and direction of motion

vibration: back-and-forth movement, or oscillation, of a physical mass about a central fixed point; vibrational motion may be longitudinal along a central axis, lateral across that axis, or a combination of both

vinasse: the solid residue remaining from distillation of product ethanol from fermented sugar beet extract

viscous, viscosity: the thickness of a liquid, expressed in its resistance to flowing and the extent to which it adheres to a solid surface as it flows

vitalism: an old philosophical concept that claimed a special vital force within living bodies was responsible for muscle movement, it was discredited when it was demonstrated that muscle movement was produced by electrical stimulation of muscle tissue

volume-to-volume (v/v) percent: the volume of pure ethanol in a specific volume of raw ethanol product; for example, 0.33 ounces (10 milliliters) of pure ethanol in 3.4 oz (100 mL) of raw product is 10 percent v/v, also stated as 10 percent by volume

voltage: the work done per unit charge when moving a charge against an electric field

water vapor: water that exists in the air of the atmosphere as a gas

watershed: the land area in which the natural drainage enters a specific river or the rivers within that area

waterwheel: a large wheel-and-axle structure that uses the force of water flowing against paddles at the outer perimeter of the wheel to drive the wheel around and so drive the axle

watt: named after Scottish engineer James Watt (1736-1819); one joule of energy per second, or the energy required to pass one ampere of electric current (or one coulomb of electrons) across a potential difference of one volt

wavelength: the distance from any point in a wave to the identical point in the next wave, usually measured from crest to crest.

wave-particle duality: the principle that light propagates as if it is a wave but interacts with matter as if it is composed of particles

wet steam: steam at a lower temperature than dry steam that is accompanied by liquid water

wind farm: an array of wind turbines established to harvest wind energy, analogous to an agricultural operation that harvests various crops

wind turbine: a wind-powered structure that converts the kinetic energy of wind to electrical energy

windmill: a tower-like structure supporting three or more large sail-like vanes or aerodynamic blades attached to a common central shaft; the force exerted by wind against the vanes or blades drives them around, and so drives the shaft

work: the energy used by a force to move an object over a given distance

worm gear: a type of gear in which a single tooth spirals around its central axis; typically used where the axis of rotation of the worm gear and the axis of rotation of the regular gear it meshes with are at right angles to each other

zeolite: a mineral form in which the arrangement of atoms within the material forms interatomic cavities that are able to trap and hold water molecules

Organizations

American Council for an Energy-Efficient Economy (ACEEE)
529 14th Street, NW
Suite 600
Washington, DC 20045
(202) 507-4000

American Council on Renewable Energy (ACORE)
1150 Connecticut Avenue, NW
Suite 401
Washington, DC 20036
(202) 393-0001

American Solar Energy Society (ASES)
2525 Arapahoe Avenue
Suite E4-253
Boulder, CO 80302
(303) 443-3130
info@ases.org

American Wind Energy Association (AWEA)
1501 M Street, NW
Suite 900
Washington, DC 20005
(202) 383-2500

International Energy Agency (IEA)
9 Rue de la Fèdèration
75739 Paris Cedex 15
France
33 (0)1 40 57 65 00

International Renewable Energy Agency (IRENA)
Masdar City
PO Box 236
Abu Dhabi
United Arab Emirates
97124179000
info@irena.org

Interstate Renewable Energy Council (IREC)
PO Box 1156
Latham, NY 12110-1156
(518) 624-7379
info@irecusa.org

National Hydropower Association (NHA)
601 New Jersey Avenue, NW
Suite 660
Washington, DC 20001
(202) 682-1700
info@hydro.org

National Renewable Energy Laboratory
15013 Denver W Parkway
Golden, CO 80401
(303) 275-3000

Office of Energy Efficiency and Renewable Energy
1000 Independence Avenue, SW
Forrestal Building
Washington, DC 20585
energy.gov/eere

Rocky Mountain Institute
22830 Two Rivers Road
Basalt, CO 81621
(970) 927-3851

Solar Energy Industries Association (SEIA)
1425 K Street, NW
Suite 1000
Washington, DC 20005
(202) 682-0556

United States Department of Energy (DoE)
1000 Independence Avenue, SW
Washington, DC 20585
(202) 586-5000
the.secretary@hq.doe.gov

United States Environmental Protection Agency (EPA)
1200 Pennsylvania Avenue, NW
Washington, DC 20460
(202) 564-4700

US Energy Information Administration (EIA)
1000 Independence Avenue, SW
Washington, DC 20585
(202) 586-8800
infoCtr@eia.gov

Subject Index

2014 Renewable Energy Data Book, 290

absolute zero, 328
AC generators, 365
acetone, 72
acoustics, 159, 359
adenosine triphosphate (ATP), 2
aerobic cellular respiration, 1-4
agricultural revolution, 14, 15, 295
agriculture, 14, 26, 29, 317, 319, 366
air pollution, 115, 126, 129, 131, 156, 199, 214, 285, 286, 311
alcohol fuel, 81
Almanack, 158
alternating current (AC), 100, 273, 287, 308, 324, 336, 365, 372
alternative energy, 7-12
Alternative Motor Fuels Act of 1988, 147
American Academy of Arts and Sciences, 332
American Institute of Electrical Engineers, 325
American Recovery and Reinvestment Act of 2009 (ARRA, also Stimulus), 214, 288, 375
American Revolution, 157, 330
American Society of Testing and Materials (ASTM), 27
ampere, 12, 341
Ampère, André-Marie, 12-14
anaerobic respiration, 1
analytical chemistry, 332
animal husbandry, 14
animal power, 14-18
anode, 162
anthropology, 14
aquaculture (fish farming), 26, 179, 184
arch dams, 86
archaeology, 224, 227, 346
Archimedes, 239, 301
architecture, 38, 89, 120, 187
Arctic Oscillation (AO), 56, 57
aromatics, 173, 174
atmospheric science, 193
atomic energy, 18, 97, 277, 278, 372
Atomic Energy Act (AEA) of 1946, 18, 348
Atomic Energy Act of 1954, 18, 372
Atomic Energy Commission (AEC), 18-19
automotive engineering, 147, 211, 221, 349
Avogadro, Amedeo, 13

Babbage, Charles, 80
batteries, 21-26
battery capacity, 41, 43
Bernoulli, Johann, 233
bifuel, 148
binary plants, 178
biochemistry, 1, 81
biodegradable fuel, 26, 31
biodiesel, 26-29
bioelectronics, 349
bioenergetics, 1, 329
bioenergy, 1, 10, 11, 45, 344, 346
bioengineering, 43
bioethanol, 31, 291, 317, 318, 319, 366, 367, 368
bioethanol fuel, 318, 319, 366, 367
biofuels, 8, 11, 29, 31, 32, 43
biofuels industry, 321
biogas, 7, 289
biological conversion, 31
biomass, 10, 69, 284, 285, 289, 290
biomass energy, 29-33
biomass feedstock, 31
biomass resources, 29, 30, 31, 285
biotechnology, 29, 289, 317, 319, 366
bituminous coal, 62, 153
boiler trades, 33
boilers, 33-37
Braungart, Michael, 243
breeder reactors, 37-38
Brownian movement, 97, 98
building envelope (building enclosure), 38-40
building-integrated photovoltaics (BIPV), 276
Bush, George W., 77, 117
buttress dams, 86

caloric theory, 51, 229, 330
calorimeter, 73, 326, 327
calorimetry, 327, 349, 350
Canaries Current, 255
carbon dioxide (CO2), 3, 60, 70, 74, 110, 114, 123, 195, 249
carbon monoxide, 26, 27, 66, 67, 126, 174, 223, 261, 265, 281, 291, 292
Carnot limit, 140
Carrier, Willis Haviland, 368-370

431

Carter, Jimmy, 322
catalysis, 72
catalyst, 3, 27, 72, 73, 162, 163, 174, 265, 266
cathode rays, 225, 332, 333, 334
cellular respiration, 1, 2, 3
cellulose, 43, 44, 117, 366
cellulosic ethanol, 43-45
centrifugal force, 149
centripetal force, 149
chemical energy, 2, 3, 133, 134, 135, 162, 163, 164, 169, 243, 285, 305, 312, 313, 327
chemical engineering, 7, 26, 132, 135, 261, 280, 326, 366
chemistry, 13, 27, 135, 174, 186
Chernobyl, 45-49
China Syndrome, 49-51
China Syndrome, The (movie), 49
chlorofluorocarbons (CFCs), 198
civil engineering, 175
classical mechanics, 103, 111, 359, 378
clathrate, 171, 244
Clausius, Rudolf Julius Emanuel, 51-53
Clean Air Act of 1990, 26, 27
Clean Power Plan, 126
climate change, 11, 15, 32, 51, 58, 59, 67
climate neutral, 58
climate neutrality, 58-60
climatology, 53, 54, 58, 201
Clinton, Bill, 188, 244
coal oil, 153
coercion, 74, 77
coffer dam, 85
cogeneration, 69-72
cold fusion, 72-74
combined heat and power (CHP), 69, 70, 125, 141, 142, 163, 353
combustion, 10, 11, 12, 17, 22, 24, 26, 28, 29, 30, 33, 34, 35, 36, 61, 63, 65, 67
combustion engineering, 69
combustion science, 93, 269
communication campaigns, 74, 76
communication technology, 159
communications, 74-79
Compressed Natural Gas (CNG), 148, 172
compression waves, 359
computer technology, 79, 206
concentrated solar power (CSP), 290
concrete dams, 86
construction trades, 38
Coriolis effect, 168, 255
corn ethanol, 81-84

Corporate Average Fuel Economy standards (CAFE standards), 116
COVID-19 pandemic, 67
crisis communication, 78
crude oil, 7, 31, 153, 154, 155, 173, 246, 249, 261, 262, 263, 264, 265, 266, 267, 269
cybersecurity, 102, 344, 345
cytoplasm, 1, 3

d'Alembert, Jean Le Rond, 13
Davy, Humphry, 13, 146, 332
daylighting, 89-91
Deepwater Horizon oil spill, 127, 266
deforestation, 8, 114, 196, 197, 302
dehumidification, 4, 38
demand-side management (DSM), 91-93
deoxyribonucleic acid (DNA), 1, 199
Department of Housing and Urban Development, 354
desalination, 38, 309, 311
Design with Nature, 187
diamagnetism, 145, 146
Diderot, Denis, 13
dielene, 368, 370
diesel fuel, 11, 26, 27, 28, 31, 95, 154, 155, 173, 174
Diesel, Rudolf Christian Karl, 93-95
diffusion, 76, 146, 203, 228, 318
diodes, 113, 208-211
direct current (DC), 100, 108, 209, 273, 305, 308, 324
distillation, 31, 76, 82, 93, 155, 173, 174, 175, 228, 265, 309, 317, 318, 368
Doppler effect, 161
Dowling, John, 50
Dynamical Theory of the Electromagnetic Field, A, 146

earthfill/earthen dams, 86, 87
ebb generation, 338
ecology, 138, 336
economics, 118, 120, 180, 322
Edison, Thomas Alva, 324, 365
Einstein, Albert, 97-100
Eisenhower, Dwight, 18
EKOenergy, 193
Electric and Hybrid Vehicle Research, Development, and Demonstration Act, 214
electric capacity, 41, 42, 183, 325
electric charges, 103, 106, 165, 167, 333, 350
electric grids, 100-103
Electric Reliability Council of Texas (ERCOT), 107
electric vehicle (EV) battery, 23
electrical trades, 41, 106, 273

electrical energy, 4, 21, 30, 41, 106, 107, 134, 168, 170, 178, 292, 293
electrical engineering, 7, 12, 21, 41, 79, 100, 106, 113, 120, 132, 140, 145, 149, 208, 211, 229, 250, 255, 256, 258, 269, 273, 284, 294, 300, 312, 323, 334, 336, 349, 353, 357, 363, 370
electrical meteorology, 350
electrical trades, 41, 106, 273
electricity transmission, 100, 323
electrochemical cells, 21, 162
electrochemical energy, 21
electrochemical engineering, 161
electrochemical industry, 363, 365
electrochemistry, 21, 145, 349, 350
electrodynamics, 13, 14, 205, 206, 358
electrolysis, 11, 53, 73, 145
electrolyte, 162
electromagnetic (EM) waves, 361
electromagnetic force, 165, 166, 167, 169, 205
electromagnetic radiation, 133, 165, 201, 206, 207, 225, 257, 292, 303, 322
electromagnetic waves, 194, 201, 206, 207, 305, 359, 360, 361
electromagnetism, 12, 13, 14, 99, 103, 111, 145, 146, 160, 165, 169, 205, 332, 333, 335, 336, 349, 350, 359, 360
electromechanical engineering, 49
electronic bonds, 164
electronics, 12, 103, 111, 208, 332, 360
electronics engineering, 147, 206, 323
electronics technology, 79
Elementary Principles in Statistical Mechanics, 186
embodied energy, 108-110
energy capacity, 41, 181, 287, 288
energy conservation, 113-116
energy efficiency, 5, 69, 74, 76, 80, 81, 115, 117, 118, 120
energy efficiency ratio (EER), 4, 6
Energy Independence and Security Act (EISA) of 2007, 116-118
energy intensity, 118-120
Energy Organization Act of 1977, 348
energy payback, 120-122
energy policy, 122-130
Energy Policy Act of 1992, 26, 27, 28
Energy Policy Act of 2005, 83, 117, 191, 251
energy policy studies, 91
energy poverty, 130-132
Energy Reorganization Act of 1974, 19, 348
Energy Research and Development Administration, 19, 348
energy resources, 29, 37, 65, 68, 70, 77, 79, 102, 110

energy security, 8, 116, 117, 123, 211-214
Energy Star program, 188, 191, 192, 345
energy storage, 132-135
energy waste, 74
enthalpy, 135-137
entropy, 137-138
environmental engineering, 21, 45, 85, 175, 243, 289
environmental management, 130, 190, 345
Environmental Protection Agency (EPA), 27, 76, 126, 188, 191, 213, 346
environmental science, 100, 113, 193, 244, 267, 344, 370
environmental studies, 58, 187, 243
environmentalism, 243
ethanol (ethyl alcohol), 317
European Renaissance, 114
European seasonal energy efficiency ratio (ESEER), 6
European Union (EU), 24, 123, 317, 366
exergic energy, 139
exergy, 138-140
external combustion engine (EC engine), 140-143
Exxon Valdez, 266

Faraday, Michael, 145-147
Federal Energy Administration, 348
Federal Energy Management Program (FEMP), 76
Federal Energy Regulatory Commission (FERC), 247
Federal Power Commission, 348
feedstock, 26, 29
fermentation, 3, 11, 29, 44, 82
fermentation chemistry, 43
fermentation engineering, 317
fermentation science, 81
fermentation technology, 366
ferrel cells, 57
fertilizer, 11, 32, 153, 154, 155, 197, 198, 228, 261, 286, 291, 318, 321, 339, 368
fiscal policy, 122
Fischoff, Baruch, 75
Fleishmann, Martin, 73
flex-fuel technology, 149
flexible-fuel (flex-fuel), 147
fluid dynamics, 355
fluid energy transmission, 235-237
fluidized bed combustion (FBC), 36
flywheel energy, 149-151
flywheel energy storage (FES), 149, 150, 151
Ford, Henry, 219, 262, 317
forestry, 10, 29, 44, 366
fossil fuel plants, 253
fossil fuel combustion, 114, 194

fossil fuels, 7, 8, 10, 11, 22, 29, 37, 58, 61, 151-156
fossilization, 152
Fowler, Ralph, 328
Franklin, Benjamin, 156-159
French Revolution, 13, 218, 299
fresnel lens, 89, 91
fresnel reflectors, 301, 302
fuel cells, 12, 21, 128, 134, 161-164
fuel poverty, 130, 132
fuel tanks, 174
fulcrum, 239, 313
furnace oils (fuel oils), 173, 175
fusion energy, 72, 279, 344, 347
Fusion Energy Sciences (FES), 344, 347
fusion power, 347
Future Energy (program), 193

Gaia hypothesis, 187
gas turbines, 71, 223, 296, 312
gasification, 10, 30, 31, 68, 289
gasohol, 82, 319, 321
gasoline, 172-175
geology, 7, 60, 85, 151, 171, 175, 224, 226, 244, 261, 267, 280
geothermal energy, 10, 175-185
Geothermal Energy Association, 181
geothermal engineering, 182
geothermal fluids, 180
geothermal greenhouses, 179
geothermal plants, 288
geothermal power, 10, 167, 169, 177, 181, 293
geothermal resources, 177, 178, 179, 180, 181, 182, 183, 184, 293
"Geothermal Sustainability," 183
geothermal wells, 177, 179
geyser, 175, 177, 178, 181, 288
Gibbs free energy, 135, 136, 185, 204, 205, 327
Gibbs, Josiah Willard, 185-187
global conveyor belt, 53, 57, 201, 203
Global Renewable Energy Fund, 125
global warming, 56, 58, 59, 67, 75, 114, 129, 147, 156, 187, 195, 196, 197, 199, 213, 249, 370
global warming potential (GWP), 195
glycolysis, 3
governmental affairs, 122
Grande Encyclopédie, 13
"Graphical Methods in the Thermodynamics of Fluids," 186
gravity dams, 86
Great London Smog, 67

green buildings, 187-190
green certificates, 190
green energy certification, 190-193
Green Grid, 80
greenhouse effect, 33, 55, 58, 67, 74, 193, 198, 199, 290
Greenhouse Gas Protocol, 199
greenhouse gases (GHGs), 193-199
gross domestic product (GDP), 123
gyre, 255

Hanford Nuclear Reservation, 46
Hannover Principles, 243, 244
harmonics, 159
heat engines, 139, 283, 284, 292
heat transfer, 201-204
helical gears, 239, 240
Helmholtz, Hermann von, 204-206
herringbone gears, 240
hertz (Hz), 160, 206, 208
Hertz, Heinrich, 206-208
High Voltage Direct Current (HVDC), 100-101
horse-related technology, 17
humidification, 4, 5, 6, 38, 40, 311
HVAC Trades, 4, 135, 182, 309, 368
hydraulic engineering, 85, 132, 235, 284, 292
hydraulic fracturing, 154, 214, 264
hydraulic system, 277, 279
hydraulic transmission, 236
hydrocarbon fuels, 22
hydrocarbons, 26, 27, 61, 68, 74, 153, 155, 156, 162, 169, 174, 198, 222, 236, 246, 261, 262, 265, 268, 281
hydrodynamic transmission, 236
hydroelectric power, 10, 42, 85, 167, 287, 292, 336, 338
hydroelectricity, 10, 87, 88, 279, 289, 292
hydrofluorocarbons (HFCs), 198
hydrogen ions, 3
hydrologic cycle, 45, 167, 228, 243, 305
hydropower, 10, 132, 287, 292
hydropower plants, 87, 88, 287, 292
hydropower production, 87, 88
hydrothermal energy, 175-182

Imperial System, 341, 342
Industrial Revolution, 215-221
industrialization, 194, 215, 363, 364
information technology (IT), 80
infrared radiation, 55, 169, 193, 194
Institute of Electrical and Electronics Engineers (IEEE), 208
insulator, 103

intergovernmental affairs, 344
Intergovernmental Panel on Climate Change (IPCC), 195
internal combustion engine, 221-224
International Air Transport Association (IATA), 27
International Atomic Energy Agency (IAEA), 48, 254
International Civil Aviation Organization (ICAO), 127
international climate policy, 128
International Convention for the Prevention of Pollution from Ships (MARPOL), 127
International District Energy Association, 354
International Energy Agency (IEA), 8, 125, 130, 213, 245, 266, 277
International Energy Association, 67
International Institute for Geothermal Research, 183
International Maritime Organization (IMO), 127
International Residential Code, 191
International Standards Organization (ISO), 60
International System of Units (SI), 103, 104, 111, 208, 356, 358, 379
International Thermonuclear Experimental Reactor (ITER), 347
ionosphere, 54, 304, 305
isopropyl alcohol, 282
isotopes, radioactive, 224-226
isotopes, stable, 226-228

Jacobson, Mark, 8
jet fuel, 154, 172, 173, 174, 265
Joule, James Prescott, 229-231

kerosene, 94, 130, 149, 155, 171, 172, 173, 174, 175, 220, 261, 262, 265, 286
kinematics, 378, 379
kinetic energy, 233-234
kinetic theory, 51, 53, 330
Krebs, Hans, 3
Kyoto Protocol, 68, 72, 199

landfilling, 21
Lapp, Ralph, 49
Leadership in Energy and Environmental Design (LEED), 189
LED lightbulbs, 113, 208-211
Leibniz, Gottfried, 233
Light-duty vehicles (LDVs), 163, 211
liquefaction terminals, 247
liquefied natural gas (LNG), 156, 171, 245
liquefied petroleum gas (LPG), 282
Lovelock, James, 81, 187

magma, 175, 182
magnetic resonance imaging (MRI), 346
Manhattan Project, 18, 49, 99, 347
marine construction, 336
marine engineering, 255, 256, 258
materials science, 108, 161, 208, 215, 273, 300, 303, 330, 346
Maxwell, James Clerk, 14, 146, 165, 206, 207, 283, 284, 336
McDonough, William, 188, 243
mechanical energy, 69, 88, 150, 166, 186, 230, 239-241
mechanical engineering, 33, 38, 60, 93, 113, 135, 138, 140, 149, 211, 215, 221, 235, 239, 258, 269, 284, 294, 312, 313, 315, 355, 368, 370
Mechanical Theory of Heat, The, 52, 53
Mercury-Containing and Rechargeable Battery Management Act, 24
meteorology, 53, 350
methane (CH4), 60, 196
Methane Pioneer, 171
metric system, 341, 344, 358
metrology, 341
microbiology, 43, 81
microgrids, 100, 102, 103
mine engineering, 60, 151
mitochondria, 1
MKS system, 341
momentum, 235
Montreal Protocol, 198, 199
Moore's law, 79, 80
Muller, Richard, 49

National Aeronautics and Space Administration (NASA), 162
National Alcohol Program (PRO-ALCOOL), 318, 320
National Electrical Manufacturers Association, 25
National Energy Policy, 128, 347
National Environmental Policy Act (NEPA), 18
National Renewable Energy Laboratory (NREL), 211
National Research Council, 354
National Science Foundation, 74
national security, 123, 344, 346, 348
natural energy flows, 243-244
natural gas, 244-250
Natural Gas Act (1938), 248
Natural Gas Policy Act (1978), 248
natural gas resources, 249
natural resources, 25, 45, 58, 76, 139, 187, 189, 215, 218
Natural Resources Defense Council, 76
natural vegetation, 29, 30

nature conservancy, 188
Naturphilosophie, 205
neap tide, 168, 337
negentropy, 137, 138, 328, 329
Nernst, Walther, 329
New York Mercantile Exchange (NYMEX), 248
Newton, Isaac, 98, 165, 379
nitrous oxide (N2O), 60, 197
noise pollution, 180, 288
nonrenewable energy, 63, 77, 114, 116, 152, 289
North Atlantic Oscillation (NAO), 56
nuclear age, 49
nuclear chemistry, 49, 72
nuclear energy, 7, 19, 50, 75, 97, 116, 117, 123, 127, 167, 168, 180, 279, 303, 312, 334, 347, 372
Nuclear Energy Advisory Committee (NEAC), 347
Nuclear Energy Research Advisory Committee (NERAC), 347
nuclear engineering, 37, 224, 250
nuclear fission, 18, 37, 45, 49, 97, 224, 225, 250, 251, 279, 344
nuclear force, 164, 165, 166, 167, 169
nuclear fuels, 169
nuclear fusion, 73, 279, 303, 305, 344, 347
nuclear medicine, 224
nuclear physics, 97, 344, 346
nuclear power plants, 250-254
nuclear reactors, 18, 37, 251, 253, 254
Nuclear Regulatory Commission (NRC), 19, 251
nuclear research, 18, 347
nuclear science, 226, 250
nuclear technology, 45
nuclear waste management, 253
nuclear weapon, 37, 50, 344, 345, 348

Obama, Barack, 126, 244
ocean currents, 53, 55, 57, 58, 201, 255, 256, 305
ocean energy, 10
ocean thermal energy conversion (OTEC), 167, 168, 256, 257
ocean wave energy, 258-260
oceanography, 53, 201, 255, 256, 336
Office of Energy Efficiency and Renewable Energy (EERE), 345
Office of High Energy Physics (HEP), 345
Office of Nuclear Energy (NE), 347
Ohm, Georg, 358
Ohm's law, 105, 358
oil drilling, 263
oil reserves, 128, 154

oil shale, 7, 264, 267-269
On the Equilibrium of Heterogeneous Substances, 186
organic chemistry, 27, 172
Organisation for Economic Co-operation and Development (OECD), 42, 44, 196
Organization of Petroleum Exporting Countries (OPEC), 67, 147, 261
Otto, Nikolaus August, 269-271
Otto-cycle engine, 269, 270, 271
Overseas Private Investment Corporation, 125
oxaloacetate, 3
oxidation, 3, 61, 145, 221, 268, 282
oxidizer, 221, 222
oxygenates, 173, 174

Pascal's Law, 235, 236
petroleum, 7, 153, 173
petroleum engineering, 151, 171, 172, 244, 261, 267
petroleum fuels, 172-175
pharmacokinetics, 226
photoelectric effect, 97, 98, 113, 206, 207, 273, 306
photosynthesis, 32, 43, 194, 195, 243
photovoltaic (PV) cells, 286, 293, 301, 302, 305, 306
physical chemistry, 51, 72, 161, 277, 326, 328, 353
physical science, 186, 309, 344
physical waves, 161, 361
physics, 7, 12, 37, 51, 145, 149, 156, 159, 164, 185, 201, 204, 206, 226, 229, 239, 277, 299, 328, 330, 332, 334, 341, 360, 375
physiology, 204, 205
pipefitting, 33
Pipeline and Hazardous Materials Safety Administration (PHMSA), 247
Plutarch, 155
plutonium, 37, 38, 99, 253
policy studies, 18, 58, 91, 116, 122, 130, 190, 344
Pons, Stanley, 73
potential energy, 277-280
Power Demonstration Reactor Program, 18
power generation engineering, 69, 363
power grids, 107
pressurized water reactors (PWRs), 252
Pró-Álcool (National Alcohol) program, 318
process engineering, 118, 280, 289
prokaryotes, 1
propane, 280-282
public health, 18, 48, 75, 123, 125, 127, 129
public policy, 322, 355
pyrolysis, 21, 23, 29, 30, 31, 265, 266, 289, 291, 292

quantum (pl. quanta), 97
Quantum Electrodynamics, 97
quantum theory, 97, 98, 334

radio waves, 201, 205, 207, 305, 362
radioactive decay, 165, 167, 177, 183, 227, 293
radioactive waste, 18, 19, 37, 48, 253, 254, 345
radioactivity, 18, 48, 167, 225, 226, 227, 253, 263, 343, 347
radioimmunoassay (RIA), 226
radiological hazards, 18
radiopharmaceuticals, 224, 226
Rankine cycle, 136, 140, 353, 355
Rankine, William John Macquorn, 283-284
rarefactions, 359, 360
refinery operations, 172, 174, 261, 319
refraction, 207, 359, 360
regasification terminals, 247
reheat turbines, 296
renewable energy, 284-288
renewable energy certification, 190
renewable energy resources, 289-294
renewable fuel standard (RFS), 117
renewable identification number (RIN), 190, 191
renewable portfolio standard (RPS), 117
renewable sources, 8, 11, 22, 60, 117, 199, 290, 322
resource management, 21, 91, 100, 122
Rippl method, 87
Rippl, Wenzel, 87
rockfill dams, 86
rotary engine, 221, 222, 294, 297, 314
rotational energy, 134, 149, 150, 295, 372
rotational power, 294-297
Rousseau, Jean-Jacques, 13
Rural Electrification Act of 1936, 372

Sadi Carnot, Nicolas Leonard, 299-300
scalar property, 376
scalar quantity, 233, 234, 376, 377
Second Industrial Revolution, 215, 219, 220
seismology, 182, 246
Sierra Club, 188
siltation, 85, 87
smart grid, 79, 100, 102
Snelling, Walter O., 282
social sciences, 74, 79, 122, 130, 138
solar air heating and cooling, 307, 311
solar cells, 8, 102, 121, 273, 277, 286, 293, 306
solar collector, 244, 286, 292, 303, 305, 306, 310
solar concentrators, 300-303

solar cooling, 311
solar desalination, 311
solar drying, 311
solar electric generating systems (SEGS), 303
Solar Energy Research Institute (SERI), 322
solar energy, 303-309
solar engineering, 300
solar homes, 311
solar photovoltaics (PVs), 114
solar power engineering, 303
solar power, 292
solar radiation, 9, 39, 54, 55, 56, 133, 142, 164, 168, 169, 194, 243, 274, 292, 293, 303, 305, 306, 308, 309, 310
solar thermal power, 134, 167, 292, 303, 305, 309
solar thermal system, 309-312
solar towers, 301
solar water heaters, 9, 311
sound waves, 154, 161, 359, 361
Special Theory of Relativity, 99
spring tide, 168, 337
stationary engineering, 313
steam boiler, 312, 313
steam engine, 313-315
steam power engineering, 140, 312
steam turbine, 312-313
Stirling, Robert, 315-317
Stirling-cycle engine (Stirling engine), 316
Strategic Environmental Assessment (SEA), 129
subtransmission, 107, 108
sugar beet cultivation, 318
sugarcane crops, 319
sugar-ethanol, 321
Sun Day, 322
swamp gas, 196, 245
synthesis gas, 265

tar sands, 267-269
Tesla, Nikola, 323-326
theory of electrolysis, 53
thermal efficiency, 142, 315, 316
thermal energy, 10, 53, 57, 58, 69, 81, 105, 167, 168, 175, 176, 181, 224, 257
thermal equilibrium, 203, 328, 329
thermal solar modules, 307
thermal solar radiation, 194
thermochemistry, 326-328
thermodynamic cycle, 69, 140, 316
thermodynamics, 328-330
thermohaline circulation, 53, 58, 201, 203, 255
Thompson, Benjamin (Count Rumford), 330-332

Thomson, Joseph John, 332-334
Thomson, William (Lord Kelvin), 334-336
tidal barrage schemes, 337, 338, 339
tidal currents, 256, 337, 339
tidal energy, 133, 139, 167, 243, 337
tidal power, 10, 169, 289, 292
tidal power generation, 336-339
tidal stream generators, 339
transportation engineering, 116
transverse wave, 359, 360, 361
Treatise on Natural Philosophy (1867), 336
Treaty of the Meter (*Convention du Mètre*), 341
Truman, Harry S., 18
turbines, 30, 42, 66, 71, 88, 121, 134, 169, 170, 221, 236, 256, 260, 287, 289, 292, 296, 305, 308, 312, 313, 325, 337, 338, 339, 353, 355, 366, 374, 375
two-stroke engine, 221, 270, 271

U.S. Army Corps of Engineers, 347
U.S. Bureau of Mines, 282
U.S. Department of Energy (DOE), 344-348
U.S. Energy Information Administration (EIA), 8, 213
U.S. energy law, 116
U.S. Green Building Council (USGBC), 188
U.S. grid system, 101
U.S. National Energy Technology Laboratory, 102
U.S. nuclear power, 348
U.S. steel industry, 308
U.S. Supreme Court, 126, 325
ultraviolet radiation, 54, 55, 198, 305
United Nations Conference on Climate change, 129
United Nations Convention on the Law of the Sea (UNCLOS), 128
United Nations Framework Convention on Climate Change (UNFCCC), 60, 128
United Nations Millennium Declaration of 2000, 131

vector property, 376
vector quantity, 233
ventilation, 6, 40, 64, 136, 187, 189, 369
vitalism, 204
volcanology, 175
Volta, Alessandro, 349-351

Wankel engine, 222, 296, 297
Wankel, Felix, 222, 297
waste heat, 30, 36, 69, 169, 220, 224, 353-355
water vapor, 4
waterwheel, 7, 279, 287, 289, 294
Watt, James, 355-357
wattage, 77, 357, 358
watt (W), 357-359
wave power, 10, 80, 259, 260
wavelength, 360-363
Westinghouse Air Brake Company (WABCO), 364
Westinghouse, George, 363-366
wheat cultivation, 367, 368
wheat ethanol, 366-368
white certificates, 190
wind energy, 8, 286, 370-375
wind power, 8, 9, 22, 60, 67, 81, 102, 113, 116, 125, 188, 259, 287, 290, 292, 293, 372, 373, 374, 375
wind turbine, 8, 121, 134, 287, 293, 296, 339, 371, 372, 373, 374, 375
work-energy principle, 377
work-energy theorem, 378-380
World Bank funding, 125
World Business Council for Sustainable Development, 199
World Coal Association, 65, 69
World Coal Institute, 64
World Energy Council, 10, 64, 196
World Energy Outlook, 42, 44, 266
World Health Organization (WHO), 131, 254
World Resources Institute, 199
World Solar Decade, 288
World War I, 215, 221, 364
World War II, 11, 66, 177, 246, 256, 282, 318, 347, 354, 370
worm gears, 240

X-ray photons, 167

zeolite, 309
Zero Emission Vehicle (ZEV) Mandate, 214
zeroth law of thermodynamics, 328
zoology, 14